In Case of Accident[1]

*In case of accident notify the laboratory instructor **immediately**.*

Fire

Burning Clothing. Prevent the person from running and fanning the flames. Rolling the person on the floor will help extinguish the flames and prevent inhalation of the flames. If a safety shower is nearby, hold the person under the shower until flames are extinguished and chemicals washed away. Do not use a fire blanket if a shower is nearby. The blanket does not cool and smouldering continues. Remove contaminated clothing. Wrap the person in a blanket to avoid shock. Get prompt medical attention.

Do not, under any circumstances, use a carbon tetrachloride (toxic) fire extinguisher and be very careful using a CO_2 extinguisher (the person may smother).

Burning Reagents. Extinguish all nearby burners and remove combustible material and solvents. Small fires in flasks and beakers can be extinguished by covering the container with an asbestos-wire gauze square, a big beaker, or a watch glass. Use a dry chemical or carbon dioxide fire extinguisher directed at the base of the flames. **Do not use water.**

Burns, either Thermal or Chemical. Flush the burned area with cold water for at least 15 min. Resume if pain returns. Wash off chemicals with a mild detergent and water. Current practice recommends that no neutralizing chemicals, unguents, creams, lotions, or salves be applied. If chemicals are spilled on a person over a large area, quickly remove the contaminated clothing while under the safety shower. Seconds count and time should not be wasted because of modesty. Get prompt medical attention.

Chemicals in the Eye. Flush the eye with copious amounts of water for 15 min using an eye-wash fountain or bottle, or by placing the injured person face up on the floor and pouring water in the open eye. Hold the eye open to wash behind the eyelids. After 15 min of washing obtain prompt medical attention, regardless of the severity of the injury.

CUTS: Minor cuts. This type of cut is most common in the organic laboratory and usually arises from broken glass. Wash the cut, remove any pieces of glass, and apply pressure to stop the bleeding. Get medical attention.

Major cuts. If blood is spurting, place a pad directly on the wound, apply firm pressure, wrap the injured to avoid shock, and get **immediate** medical attention. Never use a tourniquet.

POISONS: Call 800 information (1-800-555-1212) for the telephone number of the nearest Poison Control Center, which is usually also an 800 number.

---

[1]Adapted from *Safety in Academic Chemistry Laboratories*, prepared by the American Chemical Society Committee on Chemical Safety, 1990.

Organic Experiments

Organic
Experiments

Seventh Edition

Louis F. Fieser
Late Professor Emeritus
Harvard University

Kenneth L. Williamson
Mount Holyoke College

D. C. HEATH AND COMPANY
Lexington, Massachusetts   Toronto

Address editorial correspondence to:
D. C. Heath
125 Spring Street
Lexington, MA 02173

Cover: Charlotte Raymond/Science Source/Photo Researchers, Inc.

Published simultaneously in Canada.

Printed in the United States of America.

International Standard Book Number: 0-669-24344-2

Library of Congress Catalog Number: 91-71287

6789-DOC-00 99 98 97

Preface

Organic Experiments, Seventh Edition, presents to the beginning student a series of clear and concise experiments that encourage accurate observation and the development of deductive reasoning.

New to this edition is the section at the end of every experiment entitled "Cleaning Up," which has been written with the intent of focusing students' attention not just on the desired product from a reaction, but also on all of the other substances produced in a typical organic reaction.

Throughout the text most, but not all, of the 60 and 90 MHz nmr spectra have been replaced with 250 MHz proton spectra. A section on 2D nmr is included in the chapter on nmr spectroscopy, and procedures are given for the use of chiral nmr shift reagents to determine the optical purity of the product from the chiral enzymatic reduction of a ketone.

A number of small changes have been made so that *Organic Experiments* reflects the very latest and best of organic chemistry. Nomenclature revision is a continuous process. Gradually names such as isoamyl alcohol are being replaced by their IUPAC equivalents, but slavish adherence to those rules is not followed; phenol is still phenol (and not benzenol).

As a coauthor of *Prudent Practices for the Disposal of Chemicals from Laboratories,* National Academy of Sciences, Washington, D.C., 1983, I have continued to follow closely the rapidly evolving regulatory climate and changes in laboratory safety rules and regulations. The safety information in this text is as current as it can be, but this is a rapidly changing area of chemistry; local rules and regulations must be known and adhered to.

Range of Experiments

Organic Experiments includes a wide range of experiments and experimental procedures:

Pelletized Norit is introduced (Chapter 3). It is easily handled and the progress of decolorization is easily followed.

Several polymers can be synthesized (Chapter 67), enzymes are used to carry out chiral reduction of a ketone (Chapter 64), and a unique synthesis of ferrocene is presented (Chapter 29).

Many of the experiments are classics introduced by Louis Fieser, for example, the isolation of cholesterol from human gallstones, the use of very high boiling solvents to speed syntheses of such compounds as tetraphenylcyclopentadiene and *p*-terphenyl, and the Martius Yellow experiment (Chapter 61).

The Whole Experiment—Disposing of Hazardous Waste

A unique feature of this text is an attempt to focus the student's attention on all materials produced in the experiment, not just the desired "product." One can no longer flush down the drain or place in the waste basket the unwanted materials at the end of an experiment. Now these materials must be disposed of in an environmentally sound manner; often they must be sent to a secure landfill at great expense.

In this text, students are given specific instructions regarding "Cleaning Up" all of the by-products from each experiment. They learn how to convert potentially hazardous waste to material that in many cases can be flushed down the drain. They learn how to reduce the volume of toxic materials in order to cut drastically the costs of disposal. Students must now come to grips with the same problems that confront industry and government.

Cleaning Up   The law states that an unwanted by-product from a reaction is not a waste until the chemist declares it a waste. And once declared a waste, the material cannot be treated further except by a licensed waste treatment facility. So, for example, if all of the dilute dichromate waste from a laboratory were collected together, it could *not* be reduced chemically or reduced in volume before being carefully packed and trucked away by a hazardous waste disposal company. But if each student, *as a part of the experiment*, reduces the Cr^{6+} to Cr^{3+} and precipitates it as the hydroxide, the total volume of hazardous waste becomes extremely small and thus can be cheaply disposed of. Throughout this text, procedures based on the best, current practice, are given for the conversion of hazardous by-products to less hazardous ones.

My own experience in dealing with hazardous wastes dates back to 1982, when I sat on the National Academy of Sciences Committee on Hazardous Substances in the Laboratory. This group met every three months over a two-year period and summarized its findings in *Prudent Practices for the Disposal of Chemicals from Laboratories*. This book is considered an authoritative reference on the disposal of laboratory waste, and many of its procedures are included in this text.

In addition, the whole experimental approach and each "Cleaning Up" section has been reviewed by Blaine C. McKusick, retired assistant director of the Haskell Laboratory, the safety laboratory of E. I. du Pont de Nemours & Company. Dr. McKusick is also a coauthor of *Prudent Practices*.

Acknowledgments

I would like to acknowledge the help of many classes of Chemistry 302 students at Mount Holyoke in developing and refining the experiments in this text.

For her typing skills, preparation of the index, and organization of the *Instructor's Guide*, and for her marvelous tolerance of the writing process, I thank my wife Louise. I am appreciative of the efforts of the entire staff of D.C. Heath.

I am indebted to Blaine C. McKusick, retired assistant director of the Haskell Laboratory of DuPont, who reviewed the overall approach taken toward hazardous laboratory waste and the "Cleaning Up" procedures found at the end of each experiment.

Kenneth L. Williamson

Organic Experiments and Waste Disposal

An unusual feature of this book is the advice at the end of each experiment on how to dispose of its chemical waste. Waste disposal thus becomes part of the experiment, which is not considered finished until the waste products are appropriately taken care of. This is a valuable addition to the book for several reasons.

Although chemical waste from laboratories is less than 0.1% of that generated in the United States, its disposal is nevertheless subject to many of the same federal, state, and local regulations as chemical waste from industry. Accordingly, there are both strong ethical and legal reasons for proper disposal of laboratory wastes. These reasons are backed up by a financial concern, as the cost of waste disposal can become a significant part of the cost of operating a laboratory.

There is yet another reason to include instructions for waste disposal in a teaching laboratory. Students will some day be among those conducting and regulating waste disposal operations and voting on appropriations for them. Learning the principles and methods of sound waste disposal early in their careers will benefit them and society later.

The basics of waste disposal are easy to grasp. Some innocuous water-soluble wastes are flushed down the drain with a large proportion of water. Common inorganic acids and bases are neutralized and flushed down the drain. Containers are provided for several classes of solvents, for example, combustible solvents and halogenated solvents. (The containers are subsequently removed for suitable disposal by licensed waste handlers.) Some toxic substances can be oxidized or reduced to innocuous substances that can then be flushed down the drain; for example, hydrazines, mercaptans, and inorganic cyanides can be thus oxidized by sodium hypochlorite solution, widely available as "household bleach." Dilute solutions of highly toxic cations are expensive to dispose of because of their bulk; precipitation of the cation by a suitable reagent followed by its separation greatly reduces its bulk and cost. These and many other procedures can be found throughout this book.

Chemists often provide great detail in their directions for preparing chemicals so that the synthesis can be repeated, but they seldom say much about how to dispose of the hazardous by-products. Yet the proper disposal of a chemical's by-products is as important as its proper preparation. Dr. Williamson sets a good example by providing explicit directions for such disposal.

Blaine C. McKusick

Contents

1

Introduction

Prelab Exercise: Study the glassware diagrams and be prepared to identify the fractionating column, Claisen distilling head, ordinary distilling head, vacuum adapter, simple bent adapter, calcium chloride tube, Hirsch funnel, and Büchner funnel.

Synthesis and structure determination

Synthesis and structure determination are two major concerns of the organic chemist, and both are dealt with in this book. The rational synthesis of an organic compound, whether it involves the transformation of one functional group into another or a carbon-carbon bond forming reaction, starts with a *reaction*.

Organic reactions usually take place in the liquid phase and are *homogeneous*, in that the reactants are all in one phase. The reactants can be solids and/or liquids dissolved in an appropriate solvent to mediate the reaction. Some reactions are *heterogeneous*—that is, one of the reactants is in the solid phase—and thus require stirring or shaking to bring the reactants in contact with one another. A few heterogeneous reactions involve the reaction of a gas, such as oxygen, carbon dioxide, or hydrogen, with material in solution. Examples of all of these will be found among the experiments in this book.

In an *exothermic* organic reaction, simply mixing the reactants will produce the products; the reaction evolves heat. If it is highly exothermic, one reactant is added slowly to the other and heat is removed by external cooling. Most organic reactions are, however, mildly *endothermic,* which means the reaction mixture must be heated to increase the rate of the

Effect of temperature

reaction. A very useful rule of thumb is that *the rate of an organic reaction doubles with a 10°C rise of temperature*. The late Louis Fieser, an outstanding organic chemist and professor at Harvard University, introduced the idea of changing the traditional solvents of many reactions to high-boiling solvents in order to reduce reaction times. Throughout this book we will use solvents such as triethylene glycol, with a boiling point (bp) of 290°C, to replace ethanol (bp 78°C) and triethylene glycol dimethyl ether (bp 222°C) to replace dimethoxyethane (bp 85°C). The use of these high-boiling solvents can greatly increase the rates of many reactions.

Running an organic reaction is usually the easiest part of a synthesis. The challenge comes in isolating and purifying the product from the reaction because organic reactions seldom give quantitative yields of one pure substance.

"Working up the reaction"

In some cases the solvent and concentrations of reactants are chosen so that, after the reaction mixture has been cooled, the product will *crystallize*. It is then collected by *filtration* and the crystals are washed with an appropriate solvent. If sufficiently pure at that point, the product is dried and collected; otherwise, it is purified by the process of recrystallization or, less commonly, by *sublimation*.

If the product of reaction does not crystallize from the reaction mixture, it is often isolated by the process of *extraction*. This involves adding a solvent to the reaction mixture that will dissolve the product and will be immiscible with the solvent used in the reaction. Shaking the mixture will cause the product to dissolve in the extracting solvent, after which the two layers of liquid are separated and the product isolated from the extraction solvent.

If the product is a liquid, it is isolated by *distillation*, usually after extraction. Occasionally the product can be isolated by the process of *steam distillation* from the reaction mixture.

The apparatus used for these operations is relatively simple. Reactions are carried out in a *round-bottomed flask* (Fig. 1). The reactants are dissolved in an appropriate *solvent* and, depending on the reaction, heated or cooled as the reaction proceeds.

The flask is heated with a *heating mantle* or an electric *flask heater* or a sand bath on a hot plate. The flask is connected via a *standard-taper ground glass joint* to a water-cooled *reflux condenser*. The high heat capacity of water makes it possible to remove the large amount of heat put into the larger volume of refluxing vapor (Fig. 1). If the product of a reaction crystallizes from the reaction mixture on cooling, it is isolated by *filtration*.

Crystals are grown in *Erlenmeyer flasks* (Fig. 2) and removed, if the quantity is small, by filtration on the *Hirsch funnel* (Fig. 3). A modern

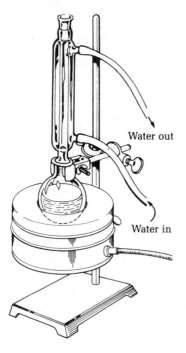

FIG. 1 Reflux apparatus for larger reactions. Liquid boils in flask and condenses on cold inner surface of water-cooled condenser.

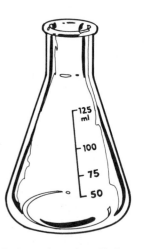

FIG. 2 Erlenmeyer flask with approximate volume graduations.

Hirsch funnel fits into the *filter flask* with no adapter, and is equipped with a *polyethylene frit* for removal of the crystals (Fig. 4). For larger quantities of material, porcelain or plastic *Büchner funnels* are used with pieces of filter paper that fit the bottom of the funnel. A *filter adapter* is used to form a vacuum tight seal between the flask and the funnel (Fig. 5).

Occasionally a solid can be purified by the process of *sublimation*. The solid is heated, usually under vacuum, and the vapor of the solid condenses

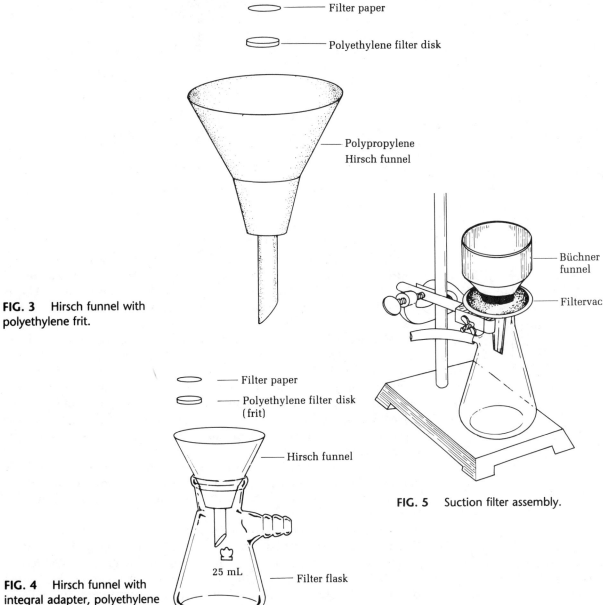

Filter paper

Polyethylene filter disk

Polypropylene
Hirsch funnel

FIG. 3   Hirsch funnel with polyethylene frit.

Filter paper

Polyethylene filter disk
(frit)

Hirsch funnel

25 mL

Filter flask

FIG. 4   Hirsch funnel with integral adapter, polyethylene frit, and 25~mL filter flask.

Büchner funnel

Filtervac

FIG. 5   Suction filter assembly.

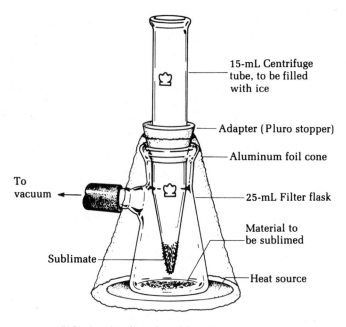

15-mL Centrifuge
tube, to be filled
with ice

Adapter (Pluro stopper)

Aluminum foil cone

To
vacuum

25-mL Filter flask

Material to
be sublimed

Sublimate

Heat source

FIG. 6   Small-scale sublimation apparatus.

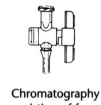

FIG. 7 Chromatography
column consisting of funnel,
tube, base fitted with
polyethylene frit, and Leur valve.

on a cold surface to form crystals in an apparatus constructed from a *centrifuge tube* fitted with a rubber adapter (a Pluro stopper) and pushed into a *filter flask* (Fig. 6). Caffeine can be purified in this manner. This is primarily a small-scale technique, although sublimers holding several grams of solid are available.

Mixtures of solids and occasionally of liquids can be separated and purified by *column chromatography*. A small *chromatography column* is shown in Fig. 7. Larger columns are often made from burettes.

Sometimes the product of a reaction will not crystallize. It may be a liquid, it may be a mixture of compounds, or it may be too soluble in the solvent being used. In this case an immiscible solvent is added, the two layers are shaken to effect *extraction*, and after the layers separate one layer is removed.

A *separatory funnel* is used to effect this process (Fig. 8). The mixture can be shaken in the funnel and then the lower layer removed through the stopcock after the stopper is removed. These funnels are available in sizes from 10 to 5000 mL.

Some of the compounds to be synthesized in these experiments are liquids. On a very small scale, the best way to separate and purify a mixture of liquids is by *gas chromatography*, but this technique is limited to less than 100 mg of material on the usual gas chromatograph. For larger quantities of material *distillation* is used. For this purpose distilling flasks are used. These flasks have a large surface area to allow sufficient heat input to cause the liquid to vaporize rapidly so that it can be distilled and then condensed for collection in a *receiver*. The apparatus consists of a *distilling flask*, a *distillation head*, a *thermometer adapter*, a *thermometer*, a water-cooled *condenser*, and a *distilling adapter* (Fig. 9). *Fractional distillation* is carried out using a small packed *fractionating column* [Fig. 10(f)]. The usual scale for distillation is about 25 mL. The individual components for distillation are shown in Fig. 10.

Some liquids with a relatively high vapor pressure can be isolated and purified by *steam distillation*, a process in which the organic compound codistills with water at a temperature below the boiling point of water. The apparatus for this process are shown in the chapter on steam distillation.

Other apparatus commonly used in the organic laboratory are shown in Fig. 11.

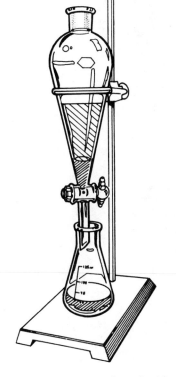

FIG. 8   Separatory funnel with Teflon stopcock.

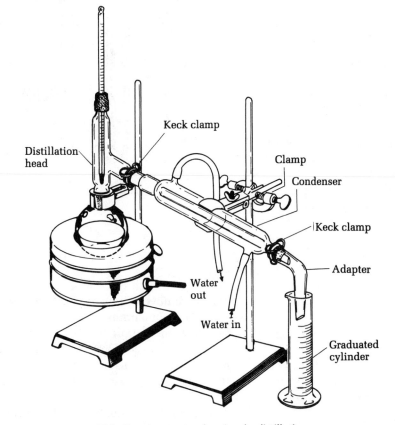

FIG. 9   Apparatus for simple distillation.

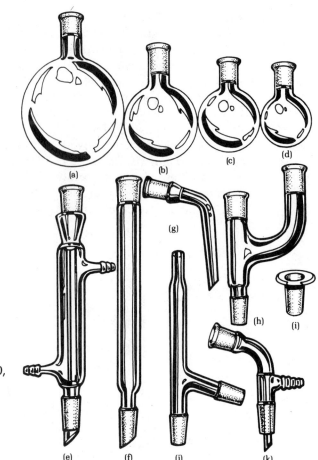

FIG. 10 Standard taper ground glass apparatus (14/20 or 19/22). Round-bottomed flasks of (a) 250, (b) 100, (c) 50, and (d) 25 mL capacity; (e) condenser; (f) distilling column; (g) simple bent adapter; (h) Claisen distilling head; (i) stopper; (j) distilling head; and (k) vacuum adapter.

Check In

Your first duty will be to check in to your assigned desk. The identity of much of the apparatus should already be apparent from the above outline of the experimental processes used in the organic laboratory.

CAUTION: Notify your instructor immediately if you break a thermometer. Mercury is very toxic.

Check to see that your thermometer reads about 22–25°C (20°C = 68°F), normal room temperature. Examine the mercury column to see if the thread is broken—i.e., that the mercury column is continuous from the bulb up. Replace any flasks that have star-shaped cracks. Remember that apparatus with graduations and porcelain apparatus are expensive. Erlenmeyer flasks, beakers, and test tubes are, by comparison, fairly cheap.

Washing and Drying Laboratory Equipment

Clean apparatus immediately

Considerable time can be saved by cleaning each piece of equipment soon after use, for you will know at that point what contaminant is present and be able to select the proper method for removal. A residue is easier to remove before it has dried and hardened. A small amount of organic residue usually can be dissolved with a few milliliters of an appropriate organic solvent. Acetone (bp 56.1°C) has great solvent power and is often effective, but it is extremely flammable and somewhat expensive. Because it is miscible with water and vaporizes readily, it is easy to remove from the vessel. Cleaning after an operation often can be carried out while another experiment is in process.

A *polyethylene bottle* [Fig. 11(l)] is a convenient wash bottle for acetone. The name, symbol, or formula of a solvent can be written on a bottle with a magic marker or wax pencil. For crystallizations, extractions, and quick cleaning of apparatus, it is convenient to have a bottle for each frequently used solvent—95% ethanol, ligroin, dichloromethane, and diethyl ether. A pinhole opposite the spout, which is covered with the finger in use, will prevent the spout from dribbling the solvent.

Pasteur pipettes are very useful for transferring small quantities of liquid, adding reagents dropwise, and carrying out crystallizations. Surprisingly, the acetone used to wash out a dirty Pasteur pipette usually costs more than the pipette itself. Discard used Pasteur pipettes in the special container for waste glass.

Sometimes a flask will not be clean after a washing with detergent and acetone. At that point try an abrasive household cleaner.

To dry a piece of apparatus rapidly, rinse with a few millimeters of acetone and invert over a beaker to drain. **Do not use compressed air**, which contains droplets of oil, water, and particles of rust. Instead draw a slow stream of air through the apparatus using the suction of your water aspirator.

Wash acetone goes in the organic solvents waste container

Insertion of a glass tube into a rubber connector or adapter or hose is easy if the glass is lubricated with a very small drop of glycerol. Grasp the tube very close to the end to be inserted; if it is grasped at a distance, especially at the bend, the pressure applied for insertion may break the tube and result in a serious cut.

If a glass tube or thermometer should become stuck to a rubber connector, it can be removed by painting on glycerol and forcing the pointed tip of an 18-cm spatula between the rubber and glass. Another method is to select a cork borer that fits snugly over the glass tube, moisten it with glycerol, and slowly work it through the connector. When the stuck object is valuable, such as a thermometer, the best policy is to cut the rubber with a sharp knife.

Heat Source

A 10°C rise in temperature will approximately double the rate of an organic reaction. The processes of distillation, sublimation, and crystallization all

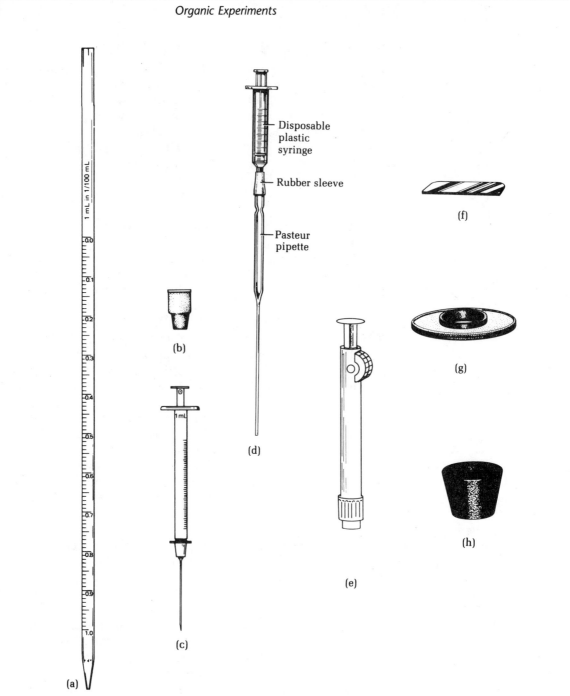

FIG. 11 Miscellaneous apparatus. (a) 1~mL graduated pipette, calibrated in 1/100ths of a mL; (b) septum; (c) 1.0~mL syringe; (d) calibrated Pasteur pipette; (e) pipette pump; (f) glass scorer; (g) Filtervac; (h) set of neoprene adapters; (i) Hirsch funnel with perforated plate in place; (j) thermometer adapter; (k) powder funnel;

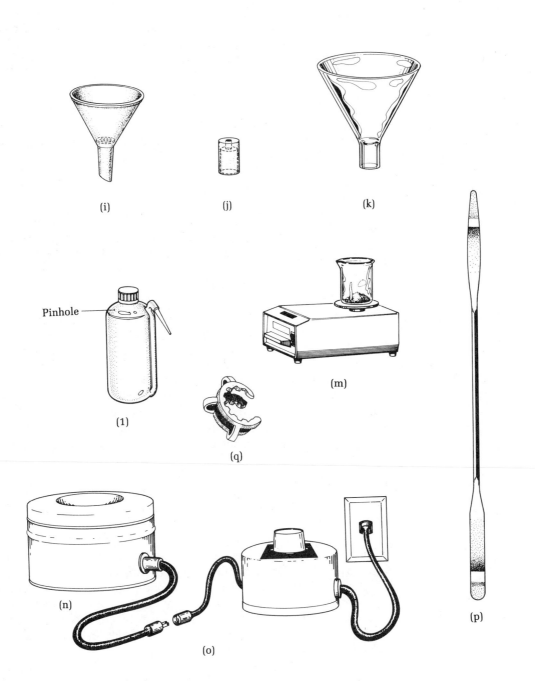

Pinhole

(i)　　　(j)　　　(k)

(l)

(m)

(q)

(n)

(o)

(p)

(l) polyethylene wash bottle; (m) single-pan electronic balance with automatic zeroing and digital readout, 100 g ± 0.001 g capacity; (n) electric flask heater; (o) solid-state control for electric flask heater; (p) stainless steel spatula; (q) Keck clamp.

require heat, which is most conveniently applied from an electrically heated flask heater [see Fig. 11(n)]. If at all possible, for safety reasons, use an electric flask heater or hot plate instead of a Bunsen burner. When the solvent boils below 90°C the most common method for heating flasks is the *steam bath*.

Transfer of a Solid

A convenient funnel

A powder funnel is useful for adding solids to a flask [Fig. 11(k)]. A funnel can also be fashioned from a sheet of weighing paper.

Weighing and Measuring

The single-pan electronic balance [Fig. 11(m)] capable of weighing to either ± 0.01 or ± 0.001 g and having a capacity of 100–250 g is very useful. Weighing is a pleasure with these balances. Although the top-loading digital balances are the easiest to use, a triple beam balance will work just as well.

A container such as a beaker or flask is placed on the pan. At the touch of a bar the digital readout registers zero and the desired quantity of reagent (solid or liquid) can be added as the weight is measured periodically to the nearest milligram or centigram.

It is often convenient to weigh reagents on glossy weighing paper and then transfer the chemical to the reaction container. The success of an experiment often depends on using just the right amount of starting materials and reagents. Inexperienced workers might think that if one milliliter of a reagent will do the job, then two milliliters will do the job twice as well. Such assumptions are usually erroneous.

Liquids can be measured by either volume or weight according to the relationship

$$\text{Volume (mL)} = \frac{\text{Weight (g)}}{\text{Density (g/mL)}}$$

Never pipette by mouth

Modern Erlenmeyer flasks and beakers have approximate volume calibrations fused into the glass, but these are *very* approximate. Somewhat more accurate volumetric measurements are made in the 10-mL graduated cylinders. For volumes less than about 4 mL, use a graduated pipette. **Never** apply suction to a pipette by mouth. The pipette can be fitted with a small rubber bulb. A Pasteur pipette can be converted into a calibrated pipette with the addition of a plastic syringe body [see Fig. 11(d)] or you can calibrate it at 0.5, 1.0, and 1.5 ml and put three file scratches on the tube; this eliminates the need to use a syringe with this Pasteur pipette in the future. Also see the Pasteur pipette calibration marks in the back of this book. You should find among your equipment a 1-mL pipette, calibrated in hundredths of a milliliter [Fig. 11(a)]. Determine whether it is designed to *deliver* 1 mL or to *contain* 1 mL between the top and bottom calibration marks. For our purposes the latter is the better pipette.

Because the viscosity, surface tension, and wetting characteristics of organic liquids are different from those of water, the so-called automatic pipette (designed for aqueous solutions) gives poor accuracy in measuring organic liquids. *Syringes* and *pipette pumps* [Fig. 11(c), (e)], on the other hand, are quite useful and frequent use will be made of them. Several reactions that require especially dry or oxygen-free atmosphere will be run in sealed systems. Reagents can be added to the system via syringe through a rubber *septum* [Fig. 11(b)].

Tares

The tare of a container is its weight when empty. Throughout this laboratory course it will be necessary to know the tares of containers so that the weights of the compounds within can be calculated. If identifying marks can be placed on the containers (e.g., with a diamond stylus) you may want to record tares for frequently used containers in your laboratory notebook.

Tare = weight of empty container

To be strictly correct we should use the word *mass* instead of *weight* because gravitational acceleration is not constant at all places on earth. But electronic balances record weights, unlike two-pan or triple-beam balances, which record masses.

The Laboratory Notebook

A complete, accurate record is an essential part of laboratory work. Failure to keep such a record means laboratory labor lost. An adequate record includes the procedure (what was done), observations (what happened), and conclusions (what the results mean).

Use a lined, paperbound, $8\frac{1}{2} \times 11$ in. notebook and record all data in ink. Allow space at the front for a table of contents, number the pages throughout, and date each page as you use it. Reserve the left-hand page for calculations and numerical data, and use the right-hand page for notes. Never record **anything** on scraps of paper to be recorded later in the notebook. Do not erase, remove, or obliterate notes; simply draw a single line through incorrect entries.

Never record anything on scraps of paper

The notebook should contain a statement or title for each experiment followed by *balanced equations* for all principal and side reactions, and, where relevant, mechanisms of the reactions. Consult your textbook for supplementary information on the class of compounds or type of reaction involved. Give a *reference to the procedure* used; do not copy verbatim the procedure in the laboratory manual.

Before coming to the lab to do preparative experiments, prepare a *table of reagents* (in your notebook) to be used and the *products* expected, with their *physical properties*. (An illustrative table appears with the first preparative equipment, the preparation of 1-bromobutane.) From your table, use the molar ratios of reactants and determine the *limiting reagent* and calculate the *theoretical yield* (in grams) of the desired product (see Chapter 15). Enter all data in your notebook (left-hand page).

Include an outline of the procedure and method of purification of the product in a *flow sheet,* which lists all possible products, by-products, unused reagents, solvents, etc., that appear in the crude reaction mixture. On the flow sheet diagram indicate how each of these is removed—for example, by extraction, various washing procedures, distillation, or crystallization. With this information entered in the notebook before coming to the laboratory, you will be ready to carry out the experiments with the utmost efficiency. Plan your time before the laboratory period. Often two or three experiments can be run simultaneously.

The laboratory notebook
What you did.
How you did it.
What you observed.
Your conclusions.

When working in the laboratory, record everything you do and everything you observe **as it happens.** The recorded observations constitute the most important part of the laboratory record, as they form the basis for the conclusions you will draw at the end of each experiment. Record the physical properties of the product, the yield in grams, and the percentage yield. Analyze your results. When things do not turn out as expected, explain why. When your record of an experiment is complete, another chemist should be able to understand your account and determine what you did, how you did it, and what conclusions you reached. In other words, from the information in your notebook a chemist should be able to repeat your work.

2

Laboratory Safety and Waste Disposal

Prelab Exercise: Read this chapter carefully. Locate the emergency eye-wash station, safety shower, and fire extinguisher in your laboratory. Check your safety glasses or goggles for size and transparency. Learn which reactions must be carried out in the hood. Learn to use your laboratory fire extinguisher; learn how to summon help and how to put out a clothing fire. Learn first aid procedures for acid and alkali spills on the skin. Learn how to tell if your laboratory hood is working properly. Learn which operations under reduced pressure require special precautions. Check to see that compressed gas cylinders in your lab are firmly fastened to benches or walls. Learn the procedures for properly disposing of solid and liquid waste in your laboratory.

The organic chemistry laboratory is an excellent place to learn and practice safety. Commonsense procedures practiced here also apply to other laboratories as well as the shop, kitchen, and studio.

General laboratory safety information particularly applicable to this organic chemistry laboratory course is presented in this chapter. It is not comprehensive. Throughout this text you will find specific cautions and safety information presented as margin notes printed in red. For a relatively brief and more thorough discussion of all of the topics in this chapter you should read the first 35 pages of *Safety in Academic Chemistry Laboratories,* American Chemical Society, Washington, D.C., 1990.

Important General Rules

Know the safety rules of your particular laboratory. Know the locations of emergency eye washes and safety showers. Never eat, drink, or smoke in the laboratory. Don't work alone. Perform no unauthorized experiments and don't distract your fellow workers; horseplay has no place in the laboratory.

Eye protection
Don't wear contact lenses

Eye protection is extremely important. Safety glasses of some type must be worn at all times. Contact lenses should not be worn because reagents can get under a lens and cause damage to the eye before the lens can be removed. It is very difficult to remove a contact lens from the eye after a chemical splash has occurred.

Ordinary prescription eyeglasses don't offer adequate protection. Laboratory safety glasses should be of plastic or tempered glass. If you do not have such glasses, wear goggles that afford protection from splashes and objects coming from the side as well as the front. If plastic safety glasses are permitted in your laboratory, they should have side shields (see Fig. 1).

Dress sensibly

Dress sensibly in the laboratory. Wear shoes, not sandals or cloth-top sneakers. Confine long hair and loose clothes. Don't wear shorts. Don't use

FIG. 1 Safety goggles and safety glasses.

mouth suction to fill a pipette, and wash your hands before leaving the laboratory. Don't use a solvent to remove chemicals from skin. This will only hasten the absorption of the chemical through the skin.

Working with Flammable Substances

Relative flammability of organic solvents

Flammable substances are the most common hazard of the organic laboratory; two factors can make this laboratory much safer than its predecessor: making the scale of the experiments as small as possible and not using burners. Diethyl ether (bp 35°C), the most flammable substance you will usually work with in this course, has an ignition temperature of 160°C, which means that a hot plate at that temperature will cause it to burn. For comparison, *n*-hexane (bp 69°C), a constituent of gasoline, has an ignition temperature of 225°C. The flash points of these organic liquids—that is, the temperatures at which they will catch fire if exposed to a flame or spark— are below −20°C. These are very flammable liquids; however, if you are careful, they are not difficult to work with. Except for water, almost all of the liquids you will use in the laboratory will be flammable.

FIG. 2 Solvent safety can.

Bulk solvents should be stored in and dispensed from *safety cans* (see Fig. 2). These and other liquids will burn in the presence of the proper amount of their flammable vapors, oxygen, and a source of ignition (most commonly a flame or spark). It is usually difficult to remove oxygen from a fire, although it is possible to put out a fire in a beaker or a flask by simply covering the vessel with a flat object, thus cutting off the supply of air. Your lab notebook might do in an emergency. The best solution is to pay close attention to sources of ignition—open flame, sparks, and hot surfaces. Remember the vapors of flammable liquids are **always** heavier than air and thus will travel along bench tops and down drain troughs and will remain in sinks. For this reason all flames within the vicinity of a flammable liquid must be extinguished. Adequate ventilation is one of the best ways to prevent flammable vapors from accumulating. Work in an exhaust hood when manipulating large quantities of flammable liquids.

Flammable vapors travel along bench tops

Should a person's clothing catch fire and a safety shower is close at hand, shove the person under it. Otherwise, shove the person down and roll him or her over to extinguish the flames. It is extremely important to prevent the victim from running or standing because the greatest harm comes from breathing the hot vapors that rise past the mouth. The safety shower might then be used to extinguish glowing cloth that is no longer aflame. A so-called fire blanket should not be used—it tends to funnel flames past the victim's mouth, and clothing continues to char beneath it. However, it is useful for retaining warmth to ward off shock after the flames are out.

FIG. 3   Carbon dioxide fire extinguisher.

An organic chemistry laboratory should be equipped with a carbon dioxide or dry chemical (monoammonium phosphate) *fire extinguisher* (see Fig. 3). To use this type of extinguisher, lift it from its support, pull the ring to break the seal, raise the horn, aim it at the base of the fire, and squeeze the handle. Do not hold onto the horn because it will become extremely cold. Do not replace the extinguisher; report the incident so the extinguisher can be refilled.

When disposing of certain chemicals, be alert for the possibility of *spontaneous combustion*. This may occur in oily rags; organic materials exposed to strong oxidizing agents such as nitric acid, permanganate ion, and peroxides; alkali metals such as sodium; or very finely divided metals such as zinc dust and platinum catalysts. Fires sometimes start when these chemicals are left in contact with filter paper.

Working with Hazardous Chemicals

If you do not know the properties of a chemical you will be working with, it is wise to regard the chemical as hazardous. The *flammability* of organic substances poses the most serious hazard in the organic laboratory. There is the possibility that storage containers in the laboratory may contribute to a fire. Large quantities of organic solvents should not be stored in glass bottles. Use safety cans. Do not store chemicals on the floor.

Flammable vapors plus air in a confined space are explosive

A flammable liquid can often be vaporized to form, with air, a mixture that is *explosive* in a confined space. The beginning chemist is sometimes surprised to learn that diethyl ether is more likely to cause a laboratory fire or explosion than a worker's accidental anesthesia. The chances of being confined in a laboratory with a high enough concentration of ether to cause loss of consciousness are extremely small. A spark in such a room would probably eradicate the building.

The probability of forming an explosive mixture of volatile organic liquids with air is much greater than that of producing an explosive solid or liquid. The chief functional groups that render compounds explosive are the *peroxide, acetylide, azide, diazonium, nitroso, nitro,* and *ozonide* groups (see Fig. 4). Not all members of these groups are equally sensitive to shock

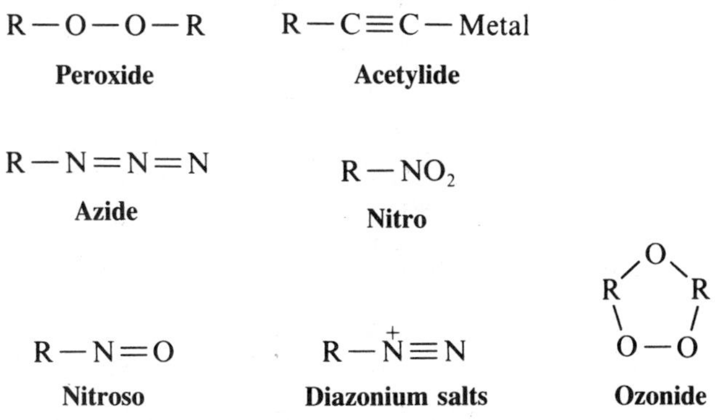

$$R-O-O-R \qquad R-C\equiv C-\text{Metal}$$

Peroxide   **Acetylide**

$$R-N=N=N \qquad R-NO_2$$

Azide   **Nitro**

FIG. 4   Functional groups that can be explosive in some compounds.

$$R-N=O \qquad R-\overset{+}{N}\equiv N \qquad \begin{array}{c} R \diagdown^{O}\diagup R \\ O-O \end{array}$$

Nitroso   **Diazonium salts**   **Ozonide**

or heat. You would find it difficult to detonate trinitrotoluene (TNT) in the laboratory, but nitroglycerine is treacherously explosive. Peroxides present special problems that are dealt with below.

Safety glasses must *be worn at all times*

You will need to contend with the corrosiveness of many of the reagents you will handle. The danger here is principally to the eyes. Proper eye protection is *mandatory* and even small-scale experiments can be hazardous to the eyes. It takes only a single drop of a corrosive reagent to do lasting damage. Handling concentrated acids and alkalis, dehydrating agents, and oxidizing agents calls for commonsense care to avoid spills and splashes and to avoid breathing the often corrosive vapors.

Certain organic chemicals present problems with acute toxicity from short-duration exposure and chronic toxicity from long-term or repeated exposure. Exposure can come about through ingestion, contact with the skin, or, most commonly, inhalation. Currently, great attention is being focused on chemicals that are teratogens (chemicals that often have no effect on a pregnant woman but cause abnormalities in a fetus), mutagens (chemicals causing changes in the structure of the DNA, which can lead to mutations in offspring), and carcinogens (cancer-causing chemicals).

Peroxides

Ethers form explosive peroxides

Certain functional groups can make an organic molecule become sensitive to heat and shock, such that it will explode. Chemists work with these functional groups only when there are no good alternatives. One of these functional groups, the *peroxide* group, is particularly insidious because it can form spontaneously when oxygen and light are present (see Fig. 5). Ethers, especially *cyclic ethers* and those made from primary or secondary alcohols (such as tetrahydrofuran, diethyl ether, and diisopropyl ether),

 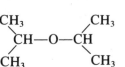

Tetrahydrofuran   **Diisopropyl ether**   **Dioxane**   **Benzylic Compounds**

 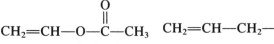

Cyclohexene   **Vinyl acetate**   **Allylic Compounds**   **Aldehydes**

$$\overset{O}{\overset{\|}{RCR}}$$

Ketones

FIG. 5   Some compounds that form peroxides.

form peroxides. Other compounds that form peroxides are *aldehydes, alkenes* that have allylic hydrogen atoms (such as cyclohexene), compounds having benzylic hydrogens on a tertiary carbon atom (such as isopropyl benzene), and vinyl compounds (such as vinyl acetate). Peroxides are low-power explosives but are extremely sensitive to shock, sparks, light, heat, friction, and impact. The biggest danger from peroxide impurities comes when the peroxide-forming compound is distilled. The peroxide has a higher boiling point than the parent compound and remains in the distilling flask as a residue that can become overheated and explode. This is one reason why it is very poor practice to distill anything to dryness.

Don't distill to dryness

Detection of peroxides	Removal of peroxides
To a solution of 0.01 g of sodium iodide in 0.1 mL of glacial acid, add 0.1 mL of the liquid suspected of containing a peroxide. If the mixture turns brown, a high concentration of peroxide is present; if it turns yellow, a low concentration of peroxide is present.	Pouring the solvent through a column of activated alumina will simultaneously remove peroxides and dry the solvent. Do not allow the column to dry out while in use. When the alumina column is no longer effective, wash the column with 5% aqueous ferrous sulfate and discard it as nonhazardous waste.

Problems with peroxide formation are especially critical for ethers. Ethers form peroxides readily and, because they are frequently used as solvents, they are often used in quantity and then removed to leave reaction products. Cans of diethyl ether should be dated when opened and if not used within one month should be treated for peroxides or disposed of.

You may have occasion to use *30% hydrogen peroxide*. This material causes severe burns if it contacts the skin, and it decomposes violently if contaminated with metals or their salts. Be particularly careful not to contaminate the reagent bottle.

Working with Corrosive Substances

Handle strong acids, alkalis, dehydrating agents, and oxidizing agents carefully so as to avoid contact with the skin and eyes and to avoid breathing the corrosive vapors that attack the respiratory tract. All strong concentrated acids attack the skin and eyes. *Concentrated sulfuric acid* is both a dehydrating agent and a strong acid and will cause very severe burns. *Nitric acid* and *chromic acid* (used in cleaning solutions) also cause bad burns. *Hydrofluoric acid* is especially harmful, causing deep, painful, and slow-healing wounds. It should be used only after thorough instruction.

Sodium, potassium, and *ammonium hydroxides* are common bases you will encounter. The first two are extremely damaging to the eye, and ammonium hydroxide is a severe bronchial irritant. Like sulfuric acid,

Add H_2SO_4, P_2O_5, CaO, and NaOH to water, not the reverse

sodium hydroxide, *phosphorous pentoxide*, and *calcium oxide* are powerful dehydrating agents. Their great affinity for water will cause burns to the skin. Because they release a great deal of heat when they react with water, to avoid spattering they should always be added to water rather than water being added to them. That is, the heavier substance should always be added to the lighter one so that rapid mixing results as a consequence of the law of gravity.

You will receive special instruction when it comes time to handle *metallic sodium, lithium aluminum hydride,* and *sodium hydride,* substances that can react explosively with water.

$HClO_4$ Perchloric acid

Among the strong oxidizing agents, *perchloric acid* is probably the most hazardous. It can form heavy metal and organic *perchlorates* that are *explosive,* and it can react explosively if it comes in contact with organic compounds.

Should one of these substances get on the skin or in the eyes, wash the affected area with very large quantities of water, using the safety shower and/or eye-wash fountain (Fig. 6). Do not attempt to neutralize the reagent chemically. Remove contaminated clothing so that thorough washing can take place. Take care to wash the reagent from under the fingernails.

Wipe up spilled hydroxide pellets rapidly

Take care not to let the reagents, such as sulfuric acid, run down the outside of a bottle or flask and come in contact with the fingers. Wipe up spills immediately with a very damp sponge, especially in the area around the balances. Pellets of sodium and potassium hydroxide are very hygroscopic and will dissolve in the water they pick up from the air; therefore, they should be wiped up very quickly. When working with larger quantities of these corrosive chemicals, wear protective gloves; with still larger quantities, use a face mask, gloves, and a Neoprene apron. The corrosive vapors can be avoided by carrying out work in a good exhaust hood.

Working with Toxic Substances

Many chemicals have very specific toxic effects. They interfere with the body's metabolism in a known way. For example, the cyanide ion combines irreversibly with hemoglobin to form cyanomethemoglobin, which can no longer carry oxygen. Aniline acts in the same way. Carbon tetrachloride and some other halogenated compounds cause liver and kidney failure. Carcinogenic and mutagenic substances deserve special attention because of their long-term insidious effects. The ability of certain carcinogens to cause cancer is very great; for example, special precautions are needed in handling aflatoxin B_1. In other cases, such as with dioxane, the hazard is so low that no special precautions are needed beyond reasonable normal care in the laboratory.

Women of child-bearing age should be careful when handling any substance of unknown properties. Certain substances are highly suspect teratogens and will cause abnormalities in an embryo or fetus. Among these are benzene, toluene, xylene, aniline, nitrobenzene, phenol, formaldehyde, dimethylformamide (DMF), dimethyl sulfoxide (DMSO), polychlorinated biphenyls (PCBs), estradiol, hydrogen sulfide, carbon disulfide, carbon monoxide, nitrites, nitrous oxide, organolead and mercury compounds, and the notorious sedative thalidomide. Some of these substances will be used in subsequent experiments. Use care. Of course, the leading known cause of embryotoxic effects is ethyl alcohol in the form of maternal alcoholism. The amount of ethanol vapor inhaled in the laboratory or absorbed through the skin is so small it is unlikely to have these morbid effects.

It is impossible to avoid handling every known or suspected toxic substance, so it is wise to know what measures should be taken. Because the eating of food or the consumption of beverages in the laboratory is strictly forbidden and because one should never taste material in the laboratory, the possibility of poisoning by mouth is remote. Be more careful than your predecessors—the hallucinogenic properties of LSD and **all** artificial sweeteners were discovered by accident. The two most important measures to be taken then are avoiding skin contact by wearing the proper type of protective gloves and avoiding inhalation by working in a good exhaust hood. A very thorough treatment of ventilation in the organic laboratory is found in *Microscale Organic Laboratory,* by D. W. Mayo, R. M. Pike, and S. S. Butcher.

Many of the chemicals used in this course will be unfamiliar to you. Their properties can be looked up in reference books, a very useful one being the *Aldrich Catalog Handbook of Fine Chemicals*. It is interesting to note that 1,4-dichlorobenzene is listed as a "toxic irritant" and naphthalene is listed as an "irritant." Both are used as moth balls. Camphor, used in vaporizers, is classified as a "flammable solid irritant." Salicylic acid, which we will use to synthesize aspirin (Chapter 26) is listed as "moisture-sensitive toxic." Aspirin (acetylsalicyclic acid) is classified as an "irritant." Caffeine, which we will isolate from tea or cola syrup (Chapter 8), is classified as "toxic." Substances not so familiar to you—1-naphthol and benzoic acid—are classified respectively as "toxic irritants" and "irritant." To put things in some perspective, nicotine is classified as "highly toxic."

Consult Armour, Browne, and Weir, *Hazardous Chemicals, Information and Disposal Guide,* for information on truly hazardous chemicals.

Because you have not had previous experience working with organic chemicals, most of the experiments you will carry out in this course will not involve the use of known carcinogens, although you will work routinely with flammable, corrosive, and toxic substances. A few experiments involve the use of substances that are suspected of being carcinogenic, such as hydrazine. If you pay proper attention to the rules of safety, you should find working with these substances no more hazardous than working with ammonia or nitric acid. The single, short-duration exposure you might receive from a suspected carcinogen, should an accident occur, would probably have no long-term consequences. The reason for taking the precautions noted in each experiment is to learn, from the beginning, good safety habits.

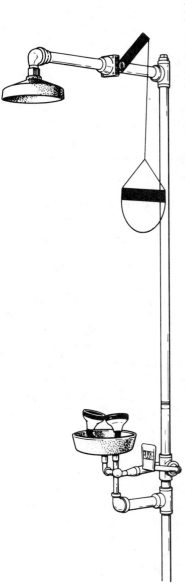

FIG. 6   Emergency shower and eye-wash station.

Using the Laboratory Hood

Modern practice dictates that in laboratories where workers spend most of their time working with chemicals, there should be one exhaust hood for every two people. This precaution is often not possible in the beginning organic chemistry laboratory, however. In this course you will find that for some experiments the hood must be used and for others it is advisable; in these instances it may be necessary to schedule experimental work around access to the hoods.

Keep the hood sash closed

The hood offers a number of advantages for work with toxic and flammable substances. Not only does it draw off the toxic and flammable fumes, it also affords an excellent physical barrier on all four sides of a reacting system when the sash is pulled down. And should a chemical spill occur, it is nicely contained within the hood.

It is your responsibility each time you use a hood to see that it is working properly. You should find some type of indicating device that will give you this information on the hood itself. A simple propeller on a cork works well (Fig. 7). The hood is a back-up device. Don't use it alone to dispose of chemicals by evaporation; use an aspirator tube or carry out a distillation. Toxic and flammable fumes should be trapped or condensed in some way and disposed of in the prescribed manner. Except when you are actually carrying out manipulations on the experimental apparatus, the sash should be pulled down. The water, gas, and electrical controls should be on the outside of the hood so it is not necessary to open the hood to adjust them. The ability of the hood to remove vapors is greatly enhanced if the apparatus is kept as close to the back of the hood as possible. Everything should be at least 15 cm back from the hood sash. Chemicals should not be stored permanently in the hood but should be removed to ventilated storage areas. If the hood is cluttered with chemicals, you will not have good, smooth air flow or adequate room for experiments.

Working at Reduced Pressure

Implosion

Whenever a vessel or system is evacuated, an implosion could result from atmospheric pressure on the empty vessel. It makes little difference whether the vacuum is perfect or just 10 mm Hg; the pressure difference is almost the same (760 mm Hg versus 750 mm Hg). An implosion may occur

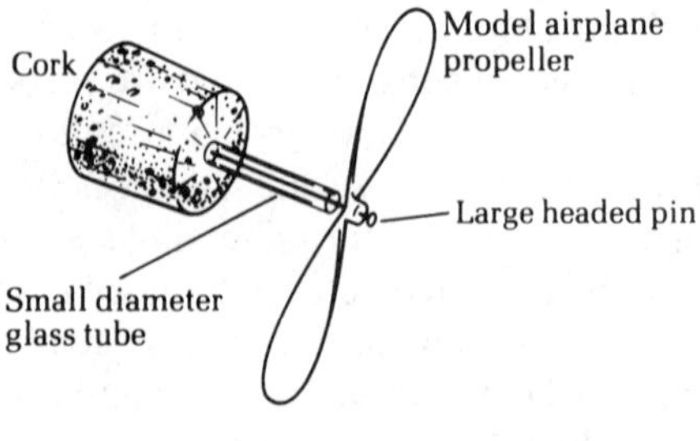

FIG. 7 Air flow indicator for hoods. The indicator should be permanently mounted in the hood and should be spinning whenever the hood is in operation.

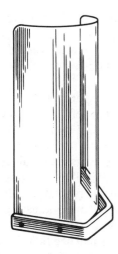

FIG. 8   Safety shield.

Always clamp gas cylinders

if there is a star crack in the flask, or if the flask is scratched or etched. Only with heavy-walled flasks specifically designed for vacuum filtration is the use of a safety shield (Fig. 8) ordinarily unnecessary.

Dewar flasks (thermos bottles) are often found in the laboratory without shielding. They should be wrapped with friction tape or covered with plastic net to prevent the glass from flying about in case of an implosion (Fig. 9). Similarly, vacuum desiccators should be wrapped with tape before being evacuated.

Working with Compressed Gas Cylinders

Many reactions are carried out under an inert atmosphere so that the reactants and/or products will not react with oxygen or moisture in the air. Nitrogen and argon are the inert gases most frequently used. Oxygen is widely used both as a reactant and to provide a hot flame for glassblowing and welding. It is used in the oxidative coupling of alkynes (Chapter 50). Helium is the carrier gas used in gas chromatography. Some other gases commonly used in the laboratory are ammonia, often used as a solvent; chlorine, used for chlorination reactions; acetylene, used in combination with oxygen for welding; and hydrogen used for high- and low-pressure hydrogenation reactions.

The following rule applies to all compressed gases: Compressed gas cylinders should be firmly secured at all times. For temporary use, a clamp that attaches to the laboratory bench top and has a belt for the cylinder will suffice (Fig. 10). Eyebolts and chains should be used to secure cylinders in permanent installations.

FIG. 9   Dewar flask with safety net in place.

FIG. 10   Gas cylinder clamp.

A variety of outlet threads are used on gas cylinders to prevent incompatible gases from becoming mixed because of an interchange of connections. Both right- and left-handed external and internal threads are used. Left-handed nuts are notched to differentiate them from right-handed nuts. Right-handed threads are used on nonfuel and oxidizing gases, and left-handed threads are used on fuel gases, such as hydrogen.

Cylinders come equipped with caps that should be left in place during storage and transportation. These caps can be removed by hand. Under these caps is a hand wheel valve. It can be opened by turning the wheel counterclockwise; however, because most compressed gases in full cylinders are under very high pressure (commonly up to 3000 lb/in²), a pressure regulator must be attached to the cylinder. This pressure regulator is almost always of the diaphragm type and has two gauges, one indicating the pressure in the cylinder, the other the outlet pressure (Fig. 11). On the outlet, low-pressure side of the regulator is located a small needle valve and then the outlet connector. After connecting the regulator to the cylinder, unscrew the diaphragm valve (turn it counterclockwise) before opening the hand wheel valve on the top of the cylinder. This valve should be opened only as far as necessary. For most gas flow rates in the laboratory, this will be a very small amount. The gas flow and/or pressure is increased by turning the two-flanged diaphragm valve **clockwise**. When the apparatus is not being used, turn off the hand wheel valve (clockwise) on the top of the cylinder. Before removing the regulator from the cylinder, reduce the flow or pressure to zero. Cylinders should never be emptied to zero pressure and left with the valve open because the residual contents will become contam-

Clockwise *movement of diaphragm valve handle* increases *pressure*

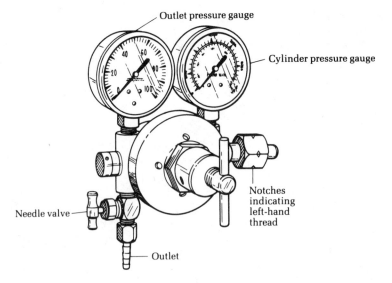

Outlet pressure gauge

Cylinder pressure gauge

Notches indicating left-hand thread

Needle valve

Outlet

FIG. 11   Gas pressure regulator. Turn *clockwise* to increase outlet pressure.

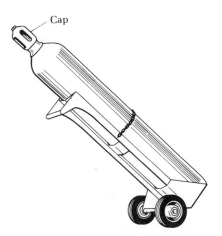

FIG. 12   Gas cylinder cart.

inated with air. Empty cylinders should be labeled "empty," capped, and returned to the storage area, separated from full cylinders. Gas cylinders should never be dragged or rolled from place to place but should be fastened into and moved in a cart designed for the purpose (Fig. 12).

Waste Disposal—Cleaning Up

Spilled solids should simply be swept up and placed in the appropriate solid waste container. This should be done promptly because many solids are hygroscopic and become difficult if not impossible to sweep up in a short time. This is particularly true of sodium hydroxide and potassium hydroxide.

Spilled acids should be neutralized. Use sodium carbonate or, for larger spills, cement or limestone. For bases use sodium bisulfate. If the spilled material is very volatile, clear the area and let it evaporate, provided there is no chance of igniting flammable vapors. Other liquids can be taken up into such absorbents as vermiculite, diatomaceous earth, dry sand, or paper towels. Be particularly careful in wiping up spills with paper towels. If a strong oxidizer is present, the towels can later ignite. Bits of sodium metal will also cause paper towels to ignite. Sodium metal is best destroyed with *n*-butyl alcohol. Unless you are sure the spilled liquid is not toxic, wear gloves when using paper towels or a sponge to remove the liquid.

Clean up spills rapidly

Mercury requires special measures—see instructor

Cleaning Up   In the not-too-distant past it was common practice to wash all unwanted liquids from the organic laboratory down the drain and to place all solid waste in the trash basket. Never a wise practice, for environmental reasons this is no longer allowed by law.

Organic reactions usually employ a solvent and often involve the use of a strong acid, a strong base, an oxidant, a reductant, or a catalyst. None of

Waste containers:
Nonhazardous solid waste
Organic solvents
Halogenated organic
solvents
Hazardous waste (various
types)

these should be washed down the drain or placed in the waste basket. We will place the material we finally classify as waste in containers labeled for nonhazardous solid waste, organic solvents, halogenated organic solvents, and hazardous wastes of various types.

Nonhazardous waste encompasses such solids as paper, corks, sand, alumina, and sodium sulfate. These ultimately will end up in a sanitary landfill (the local dump). Any chemicals that are leached by rainwater from this landfill must not be harmful to the environment. In the *organic solvents* container are placed the solvents that are used for recrystallization and for running reactions, cleaning apparatus, etc. These solvents can contain dissolved, solid, nonhazardous organic solids. This solution will go to an incinerator where it will be burned. If the solvent is a halogenated one (e.g., dichloromethane) or contains halogenated material, it must go in the *halogenated organic solvents* container. Ultimately this will go to a special incinerator equipped with a scrubber to remove HCl from the combustion gasses. The final container is for various *hazardous wastes*. Since hazardous wastes are often incompatible (oxidants with reductants, cyanides with acids, etc.), there may be several different containers for these in the laboratory, e.g., for phosphorus compounds, heavy metal hydroxides, mercury salts, etc.

Some hazardous wastes are concentrated nitric acid, platinum catalyst, sodium hydrosulfite (a reducing agent), and Cr^{6+} (an oxidizing agent). To dispose of small quantities of a hazardous waste, e.g., concentrated sulfuric acid, the material must be carefully packed in bottles and placed in a 55-gal drum called a lab pack, to which is added an inert packing material.

Waste disposal is very
expensive

The lab pack is carefully documented and then hauled off by a bonded, licensed, and heavily regulated waste disposal company to a site where such waste is disposed of. Formerly, many hazardous wastes were disposed of by burial in a "secure landfill." The kinds of hazardous waste that can be thus disposed of have become extremely limited in recent years, and much of the waste undergoes various kinds of treatment at the disposal site (e.g., neutralization, incineration, reduction) to put it in a form that can be safely buried in a secure landfill or flushed to a sewer. There are relatively few places for approved disposal of hazardous waste. For example, there are none in New England, so most hazardous waste from this area is trucked to South Carolina! The charge to small generators of waste is usually based on the volume of waste. So, 1000 mL of a 2% cyanide solution would cost much more to dispose of than 20 g of solid cyanide, even though the total amount of this poisonous substance is the same. It now costs much more to dispose of most hazardous chemicals than it does to purchase them new.

The Law: A waste is not a
waste until the laboratory
worker declares it a waste.

The law states that a material is not a waste until the laboratory worker declares it a waste. So for pedagogical and practical reasons, we would like you to regard the chemical treatment of the by-products of each reaction in this text as a part of the experiment.

The area of waste disposal is changing rapidly. Many different laws apply—local, state, and federal. What may be permissible to wash down the drain or evaporate in the hood in one jurisdiction may be illegal in another, so before carrying out this part of the experiment check with your college or university waste disposal officer.

Cleaning Up: reducing the volume of hazardous waste or converting hazardous waste to less hazardous or nonhazardous waste

In the section entitled "Cleaning Up" at the end of each experiment the goal is to reduce the volume of hazardous waste, to convert hazardous waste to less hazardous waste, or to convert it to nonhazardous waste. The simplest example is concentrated sulfuric acid. As a by-product from a reaction, it is obviously hazardous. But after careful dilution with water and neutralization with sodium carbonate, the sulfuric acid becomes a dilute solution of sodium sulfate, which in almost every locale can be flushed down the drain with a large excess of water. Anything flushed down the drain must be accompanied by a large excess of water. Similarly, concentrated base can be neutralized, oxidants such as Cr^{6+} can be reduced, and reductants such as hydrosulfite can be oxidized (by hypochlorite—household bleach). Dilute solutions of heavy metal ions can be precipitated as their insoluble sulfides or hydroxides. The precipitate may still be a hazardous waste, but it will have a much smaller volume.

Disposing of solids wet with organic solvents: alumina and anhydrous sodium sulfate

One type of hazardous waste is unique: a harmless solid that is damp with an organic solvent. Alumina from a chromatography column or sodium sulfate used to dry an ether solution are examples. Being solids they obviously can't go in the organic solvents container, and being flammable they can't go in the nonhazardous waste container. A solution to this problem is to spread the solid out in the hood to let the solvent evaporate. You can then place the solid in the nonhazardous waste container. The saving in waste disposal costs by this operation is enormous.

Our goal in "Cleaning Up" is to make you more aware of *all* aspects of an experiment. Waste disposal is now an extremely important aspect. Check to be sure the procedure you use is legal in your location. Three sources of information have been used as the basis of the procedures at the end of each experiment: the *Aldrich Catalog Handbook of Fine Chemicals,* which gives brief disposal procedures for every chemical in their catalog; *Prudent Practices for the Disposal of Chemicals from Laboratories,* National Academy Press, Washington, D.C., 1983; and *Hazardous Chemicals, Information and Disposal Guide,* M. A. Armour, L. M. Browne, and G. L. Weir, Department of Chemistry, University of Alberta, Edmonton, Alberta, Canada T6G 2G2, 3rd ed., 1988. The last title should be on the bookshelf of every laboratory. This 300-page book gives detailed information about 287 hazardous substances, including physical properties, hazardous reactions, physiological properties and health hazards, spillage disposal, and waste disposal. Many of the treatment procedures in "Cleaning Up" are adaptations of these procedures. *Destruction of Hazardous Chemicals in the Laboratory,* G. Lunn and E. B. Sansone, Wiley, New York, N.Y., 1990, complements this book.

Questions

1. Write a balanced equation for the reaction between iodide ion, a peroxide, and hydrogen ion. What causes the orange or brown color?

2. Why does the horn of the carbon dioxide fire extinguisher become cold when the extinguisher is used?

3. Why is water not used to put out most fires in the organic laboratory?

3

Crystallization

Prelab Exercise: Write an expanded outline for the seven-step process of crystallization

Crystallization: the most important purification method, especially for small-scale experiments

Crystallization is the most important method for the purification of solid organic compounds. A crystalline organic substance is made up of a three-dimensional array of molecules held together primarily by van der Waals forces. These intramolecular attractions are fairly weak; most organic solids melt in the range of room temperature to 250°C.

Crystals can be grown from the molten state just as water is frozen into ice, but it is not easy to remove impurities from crystals made in this way. Thus most purifications in the laboratory involve dissolving the material to be purified in the appropriate hot solvent. As the solvent cools, the solution becomes saturated with respect to the substance, which then crystallizes. As the perfectly regular array of a crystal is formed, foreign molecules are excluded and thus the crystal is one pure substance. Soluble impurities stay in solution because they are not concentrated enough to saturate the solution. The crystals are collected by *filtration,* the surface of the crystals is washed with cold solvent to remove the adhering impurities, and then the crystals are dried. This process is carried out on an enormous scale in the commercial purification of sugar.

In the organic laboratory, crystallization is usually the most rapid and convenient method for purifying the products of a reaction. Initially you will be told which solvent to use to crystallize a given substance and how much of it to use; later on you will judge how much solvent is needed; and finally the choice of both the solvent and its volume will be left to you. It takes both experience and knowledge to pick the correct solvent for a given purification.

The Seven Steps of Crystallization

The process of crystallization can be broken into seven discrete steps: choosing the solvent, dissolving the *solute, decolorizing* the solution, removing suspended solids, crystallizing the solute, collecting and washing the crystals, and drying the product. The process involves dissolving the impure substance in an appropriate hot solvent, removing some impurities by decolorizing and/or filtering the hot solution, allowing the substance to crystallize as the temperature of the solution falls, removing the crystallization solvent, and drying the resulting purified crystals.

How crystallization starts

Crystallization is initiated at a point of nucleation—a seed crystal, a speck of dust, or a scratch on the wall of the test tube if the solution is supersaturated with respect to the substance being crystallized (the solute). Supersaturation will occur if a hot, saturated solution cools and crystals do not form. Large crystals, which are easy to isolate, are formed by nucleation and then slow cooling of the hot solution.

1. Choosing the Solvent and Solvent Pairs

Similia similibus solvuntur

In choosing the solvent the chemist is guided by the dictum "like dissolves like." Even the nonchemist knows that oil and water do not mix and that sugar and salt dissolve in water but not in oil. Hydrocarbon solvents such as hexane will dissolve hydrocarbons and other nonpolar compounds, and hydroxylic solvents such as water and ethanol will dissolve polar compounds. Often it is difficult to decide, simply by looking at the structure of a molecule, just how polar or nonpolar it is and therefore which solvent would be best. Therefore, the solvent is often chosen by experimentation.

The ideal solvent

The best crystallization solvent (and none is ideal) will dissolve the solute when the solution is hot but not when the solution is cold; it will either not dissolve the impurities at all or it will dissolve them very well (so they won't crystallize out along with the solute); it will not react with the solute; and it will be nonflammable, nontoxic, inexpensive, and very volatile (so it can be removed from the crystals).

Some common solvents and their properties are presented in Table 1 in order of decreasing polarity of the solvent. Solvents adjacent to each other in the list will dissolve in each other, i.e., they are miscible with each other, and each solvent will, in general, dissolve substances that are similar to it in chemical structure. These solvents are used both for crystallization and as solvents in which reactions are carried out.

Miscible: capable of being mixed

Procedure

Picking a Solvent.   To pick a solvent for crystallization, put a few crystals of the impure solute in a small test tube or centrifuge tube and add a very small drop of the solvent. Allow it to flow down the side of the tube and onto the crystals. If the crystals dissolve instantly at room temperature, that solvent cannot be used for crystallization because too much of the solute will remain in solution at low temperatures. If the crystals do not dissolve at room temperature, warm the tube on the hot sand bath and observe the crystals. If they do not go into solution, add a drop more solvent. If the crystals go into solution at the boiling point of the solvent and then crystallize when the tube is cooled, you have found a good crystallization solvent. If not, remove the solvent by evaporation and try another solvent. In this trial-and-error process it is easiest to try low-boiling solvents first, because they can be removed most easily. Occasionally no single satisfactory solvent can be found, so mixed solvents, or *solvent pairs,* are used.

TABLE 1   Crystallization Solvents

Solvent	Boiling Point (°C)	Remarks
Water (H_2O)	100	The solvent of choice because it is cheap, nonflammable, nontoxic, and will dissolve a large variety of polar organic molecules. Its high boiling point and high heat of vaporization make it somewhat difficult to remove from the crystals.
Acetic acid (CH_3COOH)	118	Will react with alcohols and amines. Difficult to remove. Not a common solvent for recrystallizations, although used as a solvent when carrying out oxidation reactions.
Dimethyl sulfoxide (DMSO) Methyl sulfoxide (CH_3SOCH_3)	189	Also not a commonly used solvent for crystallization, but used for reactions.
Methanol (CH_3OH)	64	A very good solvent, used often for crystallization. Will dissolve molecules of higher polarity than will the other alcohols.
95% Ethanol (CH_3CH_2OH)	78	One of the most commonly used crystallization solvents. Its high boiling point makes it a better solvent for the less polar molecules than methanol. Evaporates readily from the crystals. Esters may undergo interchange of alcohol groups on recrystallization.
Acetone (CH_3COCH_3)	56	An excellent solvent, but its low boiling point means there is not much difference in solubility of a compound at its boiling point and at room temperature.
2-Butanone, Methyl ethyl ketone, MEK ($CH_3COCH_2CH_3$)	80	An excellent solvent with many of the most desirable properties of a good crystallization solvent.
Ethyl acetate ($CH_3COOC_2H_5$)	78	Another excellent solvent that has about the right combination of moderately high boiling point and yet the volatility needed to remove it from crystals.
Dichloromethane, methylene chloride (CH_2Cl_2)	40	Although a common extraction solvent, dichloromethane boils too low to make it a good crystallization solvent. It is useful in a solvent pair with ligroin.
Diethyl ether, ether ($CH_3CH_2OCH_2CH_3$)	35	Its boiling point is too low for crystallization, although it is an extremely good solvent and fairly inert. Used in solvent pair with ligroin.
Methyl t-butyl ether ($CH_3OC(CH_3)_3$)	52	A new solvent that is very inexpensive because of its large-scale industrial use as an antiknock agent in gasoline. Does not easily form peroxides.

Note: The solvents in this table are listed in decreasing order of polarity. Adjacent solvents in the list will in general be miscible with each other.

TABLE 1   *continued*

Solvent	Boiling Point (°C)	Remarks
Dioxane, $C_4H_8O_2$	101	A very good solvent, not too difficult to remove from crystals; a mild carcinogen, forms peroxides.
Toluene ($C_6H_5CH_3$)	111	An excellent solvent that has replaced the formerly widely used benzene (a weak carcinogen) for crystallization of aryl compounds. Because of its boiling point it is not easily removed from crystals.
Pentane (C_5H_{12})	36	A widely used solvent for nonpolar substances. Not often used alone for crystallization, but good in combination with a number of other solvents as part of a solvent pair.
Hexane (C_6H_{14})	69	Frequently used to crystallize nonpolar substances. It is inert and has the correct balance between boiling point and volatility. Often used as part of a solvent pair.
Cyclohexane (C_6H_{12})	81	Similar in all respects to hexane.
Petroleum ether	30–60	A mixture of hydrocarbons of which pentane is a chief component. Used interchangeably with pentane because it is cheap. Unlike diethyl ether, it is not an ether in the modern chemical sense.
Ligroin	60–90	A mixture of hydrocarbons with the properties of hexane and cyclohexane. A very commonly used crystallization solvent.

Solvent Pairs.   To use a mixed solvent dissolve the crystals in the better solvent and add the poorer solvent to the hot solution until it becomes cloudy and the solution is saturated with the solute. The two solvents must, of course, be miscible with each other. Some useful solvent pairs are given in Table 2.

TABLE 2

Solvent Pairs	
acetic acid–water	ethyl acetate–cyclohexane
ethanol–water	acetone–ligroin
acetone–water	ethyl acetate–ligroin
dioxane–water	diethyl ether–ligroin
acetone–ethanol	dichloromethane–ligroin
ethanol–diethyl ether	toluene–ligroin

FIG. 1 Swirling of a solution to mix contents and help dissolve material to be crystallized.

Activated charcoal = decolorizing carbon = Norit

2. Dissolving the Solute

Procedure

Place the substance to be crystallized in an Erlenmeyer flask (never use a beaker), add enough solvent to cover the crystals, and then heat the flask on a steam bath (if the solvent boils below 90°C) or a hot plate until the solvent boils. Stir the mixture or, better, swirl it (Fig. 1) to promote dissolution. Add solvent gradually, keeping it at the boil, until all of the solute dissolves. Addition of a boiling stick or a boiling chip to the solution once most of the solid is gone will promote even boiling. It is not difficult to superheat the solution, i.e., heat it above the boiling point with no boiling taking place. Once the solution does boil it does so with explosive violence. Never add a boiling chip or boiling stick to a hot solution. A glass rod with a flattened end can sometimes be of use in crushing large particles of solute to speed up the dissolving process. Be sure no flames are nearby when working with flammable solvents.

Be careful not to add too much solvent. Note how rapidly most of the material dissolves and then stop adding solvent when you suspect that almost all of the desired material has dissolved. It is best to err on the side of too little solvent rather than too much. Undissolved material noted at this point could be an insoluble impurity that never will dissolve. Allow the solvent to boil, and if no further material dissolves, proceed to Step 4 to remove suspended solids from the solution by filtration, or if the solution is colored, go to Step 3 to carry out the decolorization process. If the solution is clear, proceed to Step 5, Crystallizing the Solute.

3. Decolorizing the Solution

The vast majority of pure organic chemicals are colorless or a light shade of yellow. Occasionally a chemical reaction will produce high molecular weight by-products that are highly colored. The impurities can be adsorbed onto the surface of activated charcoal by simply boiling the solution with charcoal. Activated charcoal has an extremely large surface area per gram (several hundred square meters) and can bind a large number of molecules to this surface. On a commercial scale the impurities in brown sugar are adsorbed onto charcoal in the process of refining sugar.

In the past, laboratory manuals have advocated the use of finely powdered activated charcoal for removal of colored impurities. This has two drawbacks. Because the charcoal is so finely divided it can only be separated from the solution by filtration through paper, and even then some of the finer particles pass through the filter paper. And the presence of the charcoal completely obscures the color of the solution, so that adding the correct amount of charcoal is mostly a matter of luck. If too little charcoal is added, the solution will still be colored after filtration, making repetition necessary; if too much is added, it will absorb some of the product in addition to the impurities. We have found that charcoal extruded as short

Pelletized Norit

cylindrical pieces measuring about 0.8 × 3 mm made by the Norit Company solves both of these problems. It works just as well as the finely divided powder and it does not obscure the color of the solution. It can be added in small portions until the solution is decolorized and the size of the pieces makes it easy to remove from the solution.[1]

Add a small amount (0.1% of the solute weight is sufficient) of pelletized Norit to the colored solution and then boil the solution for a few minutes. Be careful not to add the charcoal pieces to a superheated solution; the charcoal functions like hundreds of boiling chips and will cause the solution to boil over. Remove the Norit by filtration as described in Step 4.

4. Filtering Suspended Solids

The filtration of a hot, saturated solution to remove solid impurities or charcoal can be done in a number of ways. Processes include gravity filtration, pressure filtration, decantation, or removal of the solvent using a Pasteur pipette. Vacuum filtration is not used because the hot solvent will cool during the process and the product will crystallize in the filter.

Procedure

Decant: to pour off. A fast, easy separation procedure.

(A) Decantation.   It is often possible to pour off (decant) the hot solution leaving the insoluble material behind. This is especially easy if the solid is granular like sodium sulfate. The solid remaining in the flask and the inside of the flask should be rinsed with a few milliliters of the solvent in order to recover as much of the product as possible.

(B) Gravity Filtration.   The most common method for the removal of insoluble solid material is gravity filtration through a fluted filter paper (Fig. 2). This is the method of choice for the removal of finely divided charcoal, dust, lint, etc. The following equipment will be needed for this process: three Erlenmeyer flasks on a steam bath or hot plate—one to contain the solution to be filtered, one to contain a few milliliters of solvent and a stemless funnel, and the third to contain several milliliters of the crystallizing solvent to be used for rinsing purposes—a fluted piece of filter paper, a towel for holding the hot flask and drying out the stemless funnel, and boiling chips for all solutions.

A piece of filter paper is fluted as shown in Fig. 3 and is then placed in a stemless funnel. Appropriate sizes of Erlenmeyer flasks, stemless funnels, and filter paper are shown in Fig. 4. The funnel is stemless so that the saturated solution being filtered will not have a chance to cool and clog the

FIG. 2   Gravity filtration of hot solution through fluted filter paper.

---

1. Available from Aldrich Chemical Co., 940 West St. Paul Ave., Milwaukee, WI 53233 Catalog number 32942-8 as Norit RO 0.8. This form of Norit is an extrudate 0.8 mm dia. It has a surface area of 1000 m^2/g, a total pore volume of 1.1 mL/g.

stem with crystals. The filter paper should fit entirely inside the rim of the funnel; it is fluted to allow rapid filtration. Test to see that the funnel is stable in the neck of the Erlenmeyer flask. If it is not, support it with a ring attached to a ring stand. A few milliliters of solvent and a boiling chip should be placed in the flask into which the solution is to be filtered. This solvent is brought to a boil on the steam bath or hot plate along with the solution to be filtered.

The solution to be filtered should be saturated with the solute at the boiling point. Note the volume and then add 10% more solvent. The resulting slightly dilute solution is not as likely to crystallize out in the funnel in the process of filtration. Bring the solution to be filtered to a boil, grasp the flask in a towel, and pour the solution into the filter paper in the stemless funnel (Fig. 2).

The funnel should be warm in order to prevent crystallization from occurring in the funnel. This can be accomplished in two ways: (1) invert the funnel over a steam bath for a few seconds, then pick up the funnel with a towel, wipe it perfectly dry, place it on top of the Erlenmeyer flask, and add the fluted filter paper, or (2) place the stemless funnel in the neck of the Erlenmeyer flask and allow the solvent to reflux into the funnel, thereby warming it.

Pour the solution to be filtered at a steady rate into the fluted filter paper. Check to see whether crystallization is occurring in the filter. If it does, add boiling solvent (from the third Erlenmeyer flask heated on the steam bath or hot plate) until the crystals dissolve, dilute the solution being filtered, and carry on. Rinse the flask that contained the solution to be filtered with a few milliliters of boiling solvent and rinse the fluted filter paper with this same solvent.

Since the filtrate has been diluted in order to prevent it from crystallizing during the filtration process the excess solvent must now be removed

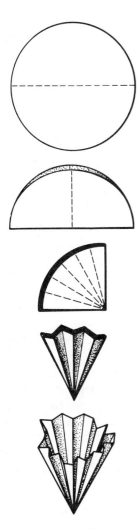

FIG. 3   Fluting a filter paper.

FIG. 4   Assemblies for gravity filtration. Stemless funnels have diameters of 2.5, 4.2, 5.0, and 6.0 cm.

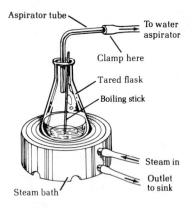

Aspirator tube

To water aspirator

Clamp here

Tared flask

Boiling stick

Steam in

Outlet to sink

Steam bath

FIG. 5 Aspirator tube in use. A boiling stick may be necessary to promote even boiling.

Slow cooling is important

by boiling the solution. The process can be speeded up somewhat by blowing a slow current of air into the flask in the hood or using an aspirator tube to pull vapors into the aspirator (Fig. 5).

5. Crystallizing the Solute

On both a macroscale and a microscale the crystallization process should normally start from a solution that is saturated with the solute at the boiling point. If it has been necessary to remove impurities or charcoal by filtration the solution has been diluted. To concentrate the solution simply boil off the solvent under an aspirator tube as shown in Fig. 5 or blow off solvent using a gentle stream of air or, better, nitrogen in the hood. Be sure to have a boiling chip or a boiling stick in the solution during this process and then do not forget to remove it before initiating crystallization.

Once it has been ascertained that the hot solution is saturated with the compound just below the boiling point of the solvent, it is allowed to cool slowly to room temperature. Crystallization should begin immediately. If it does not, add a seed crystal or scratch the inside of the tube with a glass rod at the liquid-air interface. Crystallization must start on some nucleation center. A minute crystal of the desired compound saved from the crude material will suffice. If a seed crystal is not available, crystallization can be started on the rough surface of a fresh scratch on the inside of the container.

Once it is ascertained that crystallization has started, the solution must be cooled slowly without disturbing the container in order that large crystals can form. The Erlenmeyer flask is set atop a cork ring or other insulator and allowed to cool spontaneously to room temperature. If the flask is moved during crystallization, many nuclei will form and the crystals will be small and will have a large surface area. They will not be so easy to filter and wash clean of mother liquor. Once crystallization ceases at room temperature, the flask should be placed in ice to cool further. Take care to clamp the flask in the ice bath so it does not tip over.

6. Collecting and Washing the Crystals

Once crystallization is complete, the crystals must be separated from the ice-cold *mother liquor* (the *filtrate*), washed with ice-cold solvent, and dried.

(A) Filtration on the Büchner Funnel. The Büchner funnel is used for large-scale filtrations. If the quantity of material is small (<2 g), the Hirsch funnel can be used. Properly matched Büchner funnels, filter paper, and flasks are shown in Fig. 6. The Hirsch funnel shown in the figure is an old-style porcelain type.

Place a piece of filter paper in the bottom of the Büchner funnel. Wet it with solvent, and be sure it lies flat so crystals cannot escape around the

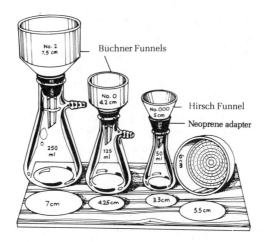

FIG. 6 Matching filter assemblies. The 6.0-cm polypropylene Büchner funnel (*right*) resists breakage and can be disassembled for cleaning.

Clamp the filter flask

edge and under the filter paper. Then with the vacuum off, pour the cold slurry of crystals into the center of the filter paper. Apply the vacuum as soon as the liquid disappears from the crystals then break the vacuum to the flask by disconnecting the hose. Rinse the Erlenmeyer flask with cold solvent. Add this to the crystals, and reapply the vacuum just until the liquid disappears from the crystals. Repeat this process as many times as necessary, and then leave the vacuum on to dry the crystals.

(B) Filtration Using the Hirsch Funnel. When the quantity of material to be collected is small, then the material is collected on the Hirsch funnel.

The Williamson/Kontes Hirsch funnel is unique. It is made of polypropylene and has an integral molded stopper that fits the 25-mL filter flask. It comes fitted with a 20-micron polyethylene fritted disk, which is not meant to be disposable (Fig. 7). While products can be collected directly on this

FIG. 7 Polypropylene Hirsch funnel, 25-mL filter flask, and polyethylene filter disk.

disk it is good practice to place an 11- or 12-mm dia. piece of #1 filter paper on the disk. In this way the frit will not become clogged with insoluble impurities. A piece of filter paper is mandatory on porcelain Hirsch funnels.

Clamp the clean, dry 25-mL filter flask in an ice bath to prevent it from falling over and place the Hirsch funnel with filter paper in the flask. Wet the filter paper with the solvent used in the crystallization, turn on the water aspirator (see below), and ascertain that the filter paper is pulled down onto the frit. Pour and scrape the crystals and mother liquor onto the Hirsch funnel, and as soon as the liquid is gone from the crystals, break the vacuum at the filter flask by removing the rubber hose. Cool the filter flask in ice. The filtrate can be used to rinse out the container that contained the crystals. Again break the vacuum as soon as all of the liquid has disappeared from the crystals; this prevents impurities from drying on the crystals. The reason for cooling the filter flask is to keep the mother liquor cold so that it will not dissolve the crystals on the Hirsch funnel. With a very few drops of ice-cold solvent, rinse the crystallization flask. That container should still be ice cold. Place the ice-cold solvent on the crystals and then reapply the vacuum. As soon as the liquid is pulled from the crystals break the vacuum. Repeat this washing process as many times as necessary to remove colored material or other impurities from the crystals. In some cases only one very small wash will be needed. After the crystals have been washed with ice-cold solvent, the vacuum can be left on to dry the crystals.

Break the vacuum, add a very small quantity of ice-cold wash solvent, reapply vacuum

The Water Aspirator and Trap.   The most common way to produce a vacuum in the organic laboratory for filtration purposes is by employing a *water aspirator*. Air is efficiently entrained in the water rushing through the aspirator so that it will produce a vacuum roughly equal to the vapor pressure of the water going through it (17 torr at 20°C, 5 torr at 4°C). A check valve is built into the aspirator, but even so when the water is turned off it may back into the evacuated system. For this reason a *trap* is always installed in the line (Fig. 8). *The water passing through the aspirator should always be turned on full force*. The system can be opened to the atmosphere by removing the hose from the small filter flask or by opening the screw clamp on the trap. Open the system, then turn off the water to avoid having water sucked back into the filter trap. Thin rubber tubing on the top of the trap will collapse and bend over when a good vacuum is established. You will, in time, learn to hear the difference in the sound of an aspirator when it is pulling a vacuum and when it is working on an open system.

Use a trap

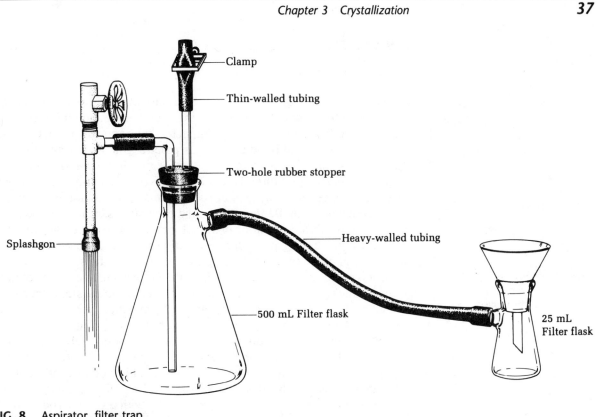

Clamp

Thin-walled tubing

Two-hole rubber stopper

Heavy-walled tubing

Splashgon

500 mL Filter flask

25 mL
Filter flask

FIG. 8   Aspirator, filter trap, and Hirsch funnel. Clamp small filter flask to prevent its turning over.

Collecting a Second Crop of Crystals.   Regardless of the method used to collect the crystals on either a macroscale or a microscale, the filtrate and washings can be combined and evaporated to the point of saturation to obtain a second crop of crystals—hence the necessity for having a clean receptacle for the filtrate. This second crop will increase the overall yield, but the crystals will not usually be as pure as the first crop.

7. Drying the Product

Once the crystals have been washed on the Hirsch funnel or the Büchner funnel, press them down with a clean cork or other flat object and allow air to pass through them until they are substantially dry. Final drying can be done under reduced pressure (Fig. 9). The crystals can then be turned out of the funnel and squeezed between sheets of filter paper to remove the last bit of solvent before final drying on a watch glass.

FIG. 9   Drying a solid by reduced air pressure.

Experiments

1. Solubility Tests

Test Compounds:

Resorcinol

Anthracene

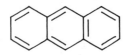

Benzoic acid

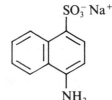

**4-Amino-1-
naphthalenesulfonic acid,
sodium salt**

To test the solubility of a solid, transfer an amount roughly estimated to be about 10 mg (the amount that forms a symmetrical mound on the end of a stainless steel spatula) into a 10×75-mm test tube and add about 0.25 mL of solvent from a calibrated dropper or pipette. Stir with a fire-polished stirring rod (4-mm), break up any lumps, and determine if the solid is readily soluble at room temperature. If the substance is readily soluble in methanol, ethanol, acetone, or acetic acid at room temperature, add a few drops of water from a wash bottle to see if a solid precipitates. If it does, heat the mixture, adjust the composition of the solvent pair to produce a hot solution saturated at the boiling point, let the solution stand undisturbed, and note the character of the crystals that form. If the substance fails to dissolve in a given solvent at room temperature, heat the suspension and see if solution occurs. If the solvent is flammable, heat the test tube on the steam bath or in a small beaker of water kept warm on the steam bath or a hot plate. If the solid completely dissolves, it can be declared readily soluble in the hot solvent; if some but not all dissolves, it is said to be moderately soluble, and further small amounts of solvent should then be added until solution is complete. When a substance has been dissolved in hot solvent, cool the solution by holding the flask under the tap and, if necessary, induce crystallization by rubbing the walls of the tube with a stirring rod to make sure that the concentration permits crystallization. Then reheat to dissolve the solid, let the solution stand undisturbed, and inspect the character of the ultimate crystals.

Make solubility tests on the test compounds shown to the left in each of the solvents listed. Note the degree of solubility in the solvents, cold and hot, and suggest suitable solvents, solvent-pairs, or other expedients for crystallization of each substance. Record the crystal form, at least to the extent of distinguishing between needles (pointed crystals), plates (flat and thin), and prisms. How do your observations conform to the generalization that like dissolves like?

Solvents:
Water—hydroxylic, ionic
Toluene—an aromatic hydrocarbon
Ligroin—a mixture of aliphatic hydrocarbons

Cleaning Up[2]   Place organic solvents and solutions of the compounds in the organic solvents container. Dilute the aqueous solutions with water and flush down the drain.

---

2. For this and all other ''Cleaning Up'' sections, see Chapter 2 for a complete discussion of waste disposal procedures.

Handwritten margin notes:
Volume of water
and p dis?

2. Crystallization of Pure Phthalic Acid and Naphthalene

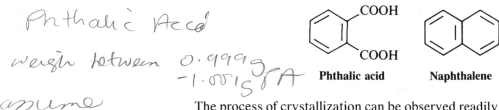

Phthalic acid   **Naphthalene**

Handwritten margin notes (left):
Phthalic Acid
weigh between 0.999 g
—1.00'1 g PA

assume

1.00 g PA × (100 ml H₂O)

18 g PA = 5.6 ml H₂O

5.6 ml × 1.2 = 6.7 ml H₂O

% recovery

= mass recovered × 100%
* 1.00 g PA*

The process of crystallization can be observed readily using phthalic acid. In the reference book *The Handbook of Chemistry and Physics,* in the table "Physical Constants of Organic Compounds," the entry for phthalic acid gives the following solubility data (in grams of solute per 100 mL of solvent). The superscripts refer to temperature in °C:

Water	Alcohol	Ether, etc.
0.54^{14}	11.71^{18}	0.69^{15} eth., i. chl.
18^{99}		

The large difference in solubility in water as a function of temperature suggests this as the solvent of choice. The solubility in alcohol is high at room temperature. Ether is difficult to use because it is so volatile; the compound is insoluble is chloroform (i. chl.).

Procedure

Crystallize 1.0 g of phthalic acid from the minimum volume of water, using the above data to calculate the required volume. Add the solid to the smallest practical Erlenmeyer flask and then, using a Pasteur pipette, add water dropwise from a full 10-mL graduated cylinder. A boiling stick (a stick of wood) facilitates even boiling and will prevent bumping. After a portion of the water has been added, gently heat the solution to boiling on a hot plate or over a Bunsen burner. As soon as boiling begins, continue to add water dropwise until all the solid just dissolves. Place the flask on a cork ring or other insulator and allow it to cool undisturbed to room temperature, during which time the crystallization process can be observed. Slow cooling favors large crystals. Then cool the flask in an ice bath, decant (pour off) the mother liquor (the liquid remaining with the crystals), and remove the last traces of liquid with a Pasteur pipette. Scrape the crystals onto a filter paper using a stainless steel spatula, squeeze the crystals between sheets of filter paper to remove traces of moisture, and allow the crystals to dry. Compare the calculated volume of water with the volume of water actually used to dissolve the acid. Calculate the percent recovery of dry, recrystallized phthalic acid.

Cleaning Up   Dilute the filtrate with water and flush the solution down the drain. Phthalic acid is not considered toxic to the environment.

3. Decolorization of Brown Sugar (Sucrose, $C_{12}H_{22}O_{11}$)

Raw sugar is refined commercially with the aid of decolorizing charcoal. The clarified solution is seeded generously with small sugar crystals, and excess water removed under vacuum to facilitate crystallization. The pure white crystalline product is collected by centrifugation. Brown sugar is partially refined sugar and can be decolorized easily using charcoal.

Dissolve 15 g of dark brown sugar in 30 mL of water in a 50 mL Erlenmeyer flask by heating and stirring. Pour half of the solution into another 50-mL flask. Heat one of the solutions nearly to the boiling point, allow it to cool slightly, and add to it 250 mg (0.25 g) of decolorizing charcoal (Norit pellets). Bring the solution back to near the boiling point for two minutes, then filter the hot solution into an Erlenmeyer flask through a fluted filter paper held in a previously heated funnel. Treat the other half of the sugar solution in exactly the same way but use only 50 mg of decolorizing charcoal. In collaboration with a fellow student try heating the solutions for only 15 s after addition of the charcoal. Compare your results.

Cleaning Up   Decant (pour off) the aqueous layer. Place the Norit in the nonhazardous solid waste container. The sugar solution can be flushed down the drain.

4. Recrystallization of Naphthalene from a Mixed Solvent

Do not try to grasp Erlenmeyer flasks with a test tube holder

Add 2.0 g of impure naphthalene[3] to a 50-mL Erlenmeyer flask along with 3 mL of methanol and a boiling stick to promote even boiling. Heat the mixture to boiling over a steam bath or hot plate and then add methanol dropwise until the naphthalene just dissolves when the solvent is boiling. The total volume of methanol should be 4 mL. Remove the flask from the heat and cool it rapidly in an ice bath. Note that the contents of the flask set to a solid mass, which would be impossible to handle. Add enough methanol to bring the total volume to 25 mL, heat the solution to the boiling point, remove the flask from the heat, allow it to cool slightly, and add 30 mg of decolorizing charcoal pellets to remove the colored impurity in the solution. Heat the solution to the boiling point for two minutes; if the color is not gone add more Norit and boil again, then filter through a fluted filter paper in a previously warmed stemless funnel into a 50-mL Erlenmeyer flask.

Support the funnel in a ring stand

Sometimes filtration is slow because the funnel fits so snugly into the mouth of the flask that a back pressure develops. If you note that raising the funnel increases the flow of filtrate, fold a small strip of paper two or three times and insert it between the funnel and flask. Wash the used flask with 2 mL of hot methanol and use this liquid to wash the filter paper, transferring the solvent with a Pasteur pipette in a succession of drops around the upper rim

---

3. A mixture of 100 g of naphthalene, 0.3 g of a dye such as congo red, and perhaps sand, magnesium sulfate, dust, etc.

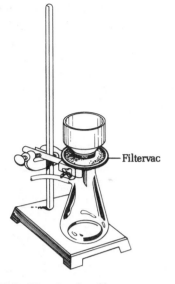
—Filtervac

FIG. 10   Suction filter assembly clamped to provide firm support. The funnel must be pressed down on the Filtervac to establish reduced pressure in the flask.

of the filter paper. When the filtration is complete, the volume of methanol should be 15 mL. If it is not, evaporate excess methanol.

Because the filtrate is far from being saturated with naphthalene at this point it will not yield crystals on cooling; however, the solubility of naphthalene in methanol can be greatly reduced by addition of water. Heat the solution to the boiling point and add water dropwise from a 10-mL graduated cylinder, using a Pasteur pipette (or use a precalibrated pipette). After each addition of water the solution will turn cloudy for an instant. Swirl the contents of the flask and heat to redissolve any precipitated naphthalene. After the addition of 3.5 mL of water the solution will be almost saturated with naphthalene at the boiling point of the solvent. Remove the flask from the heat and place it on a cork ring or other insulating surface to cool, without being disturbed, to room temperature.

Immerse the flask in an ice bath along with another flask containing methanol and water in the ratio of 15:3.5. This cold solvent will be used for washing the crystals. The cold crystallization mixture is collected by vacuum filtration on a small Büchner funnel (50-mm) (Fig. 10). The water flowing through the aspirator should always be turned on full force. In collecting the product by suction filtration use a spatula to dislodge crystals and ease them out of the flask. If crystals still remain in the flask, some filtrate can be poured back into the crystallization flask as a rinse for washing as often as desired, since it is saturated with solute. To free the crystals from contaminating mother liquor, break the suction and pour a few milliliters of the fresh cold solvent mixture into the Büchner funnel and immediately reapply suction. Repeat this process until the crystals and the filtrate are free of color. Press the crystals with a clean cork to eliminate excess solvent, pull air through the filter cake for a few minutes, and then put the large flat platelike crystals out on a filter paper to dry. The yield of pure white crystalline naphthalene should be about 1.6 g. The mother liquor contains about 0.25 g and about 0.15 g is retained in the charcoal and on the filter paper.

Cleaning Up   Place the Norit in the nonhazardous solid waste container. The methanol filtrate and washings are placed in the organic solvents container.

5. Purification of an Unknown

Bear in mind the seven-step crystallization procedure:

1. Choose the solvent.
2. Dissolve the solute.
3. Decolorize the solution (if necessary).
4. Filter suspended solids (if necessary).
5. Crystallize the solute.
6. Collect and wash the crystals.
7. Dry the product.

You are to purify 2.0 g of an unknown provided by the instructor. Conduct tests for solubility and ability to crystallize in several organic solvents, solvent pairs, and water. Conserve your unknown by using very small quantities for solubility tests. If only a drop or two of solvent is used the solvent can be evaporated by heating the test tube on the steam bath or sand bath and the residue can be used for another test. Submit as much pure product as possible with evidence of its purity (i.e., the melting point). From the posted list identify the unknown.

Cleaning Up   Place decolorizing charcoal, if used, and filter paper in the nonhazardous solid waste container. Put organic solvents in the organic solvents container, and flush aqueous solutions down the drain.

Crystallization Problems and Their Solutions

Induction of Crystallization

Occasionally a sample will not crystallize from solution on cooling, even though the solution is saturated with the solute at elevated temperature. The easiest method for inducing crystallization is to add to the supersaturated solution a seed crystal that has been saved from the crude

Seeding

material (if it was crystalline before recrystallization was attempted). In a probably apocryphal tale, the great sugar chemist Emil Fischer merely had to wave his beard over a recalcitrant solution and the appropriate seed crystals would drop out, causing crystallization to occur. In the absence of seed crystals, crystallization can often be induced by scratching the inside of the flask with a stirring rod at the air-liquid interface. One theory holds

Scratching

that part of the freshly scratched glass surface has angles and planes corresponding to the crystal structure, and crystals start growing on these spots. Often crystallization is very slow to begin and placing the sample in a refrigerator overnight will bring success. Other expedients are to change the solvent (usually to a poorer one) and to place the sample in an open container where slow evaporation and dust from the air may help induce crystallization.

Oils and "Oiling Out"

Some saturated solutions, especially those containing water, when they cool deposit not crystals but small droplets referred to as oils. Should these droplets subsequently crystallize and be collected they will be found to be

Crystallize at a lower temperature

rather impure. Should the temperature of the saturated solution be above the melting point of the solute when it starts to come out of solution the solute will, of necessity, be deposited as an oil. Similarly, the melting point of the desired compound may be depressed to a point such that a low-melting eutectic mixture of the solute and the solvent comes out of solution. The simplest remedy for this latter problem is to lower the temperature at which the solution becomes saturated with the solute by simply adding more solvent. In extreme cases it may be necessary to lower this temperature well below room temperature by cooling the solution with dry ice.

Crystallization Summary

1. **Choosing the solvent.** "Like dissolves like." Some common solvents are water, methanol, ethanol, ligroin, and toluene. When you use a solvent pair, dissolve the solute in the better solvent and add the poorer solvent to the hot solution until saturation occurs. Some common solvent pairs are ethanol–water, diethyl ether–ligroin, and toluene–ligroin.

2. **Dissolving the solute.** To the crushed or ground solute in an Erlenmeyer flask or reaction tube add solvent; heat the mixture to boiling. Add more solvent as necessary to obtain a hot, saturated solution.

3. **Decolorizing the solution.** If it is necessary to remove colored impurities, cool the solution to near room temperature and add more solvent to prevent crystallization from occurring. Add decolorizing charcoal in the form of pelletized Norit to the cooled solution, and then heat it to boiling for a few minutes, taking care to swirl the solution to prevent bumping. Remove the Norit by filtration, then concentrate the filtrate.

4. **Filtering suspended solids.** If it is necessary to remove suspended solids, dilute the hot solution slightly to prevent crystallization from occurring during filtration. Filter the hot solution. Add solvent if crystallization begins in the funnel. Concentrate the filtrate to obtain a saturated solution.

5. **Crystallizing the solute.** Let the hot saturated solution cool spontaneously to room temperature. Do not disturb the solution. Then cool it in ice. If crystallization does not occur, scratch the inside of the container or add seed crystals.

6. **Collecting and washing the crystals.** Collect the crystals using the Pasteur pipette method or by vacuum filtration on a Hirsch funnel or a Büchner funnel. If the latter technique is employed, wet the filter paper with solvent, apply vacuum, break vacuum, add crystals and liquid, apply vacuum until solvent just disappears, break vacuum, add cold wash solvent, apply vacuum, and repeat until crystals are clean and filtrate comes through clear.

7. **Drying the product.** Press the product on the filter to remove solvent. Then remove it from the filter, squeeze it between sheets of filter paper to remove more solvent, and spread it on a watch glass to dry.

Questions

1. A sample of naphthalene, which should be pure white, was found to have a greyish color after the usual purification procedure. The melting point was correct and the melting point range small. Explain the grey color.

2. How many milliliters of boiling water are required to dissolve 25 g of phthalic acid? If the solution were cooled to 14°C, how many grams of phthalic acid would crystallize out?

3. What is the reason for using activated carbon during a crystallization?

4. If a little activated charcoal does a good job removing impurities in a crystallization, why not use a lot?

5. Under what circumstances is it wise to use a mixture of solvents to carry out a crystallization?

6. Why is gravity filtration and not suction filtration used to remove suspended impurities and charcoal from a hot solution?

7. Why is a fluted filter paper used in gravity filtration?

8. Why are stemless funnels used instead of long-stem funnels to filter hot solutions through fluted filter paper?

9. Why is the final product from the crystallization process isolated by vacuum filtration and not by gravity filtration?

Melting Points and Boiling Points

Prelab Exercise: Predict what the melting points of the three urea-cinnamic acid mixtures will be.

Part 1. Melting Points

The melting point of a pure solid organic compound is one of its characteristic physical properties, along with molecular weight, boiling point, refractive index, and density. A pure solid will melt reproducibly over a narrow range of temperatures, typically less than 1°C. The process of determining this melting "point" is done on a truly micro scale using less than 1 mg of material; the apparatus is very simple, consisting of a thermometer, a capillary tube to hold the sample, and a heating bath.

Melting points—a micro technique

Melting points are determined for three reasons. If the compound is a known one the melting point will help to characterize the sample in hand. If the compound is new then the melting point is recorded in order to allow future characterization by others. And finally the range of the melting point is indicative of the purity of the compound; an impure compound will melt over a wide range of temperatures. Recrystallization of the compound will purify it and the melting point range will decrease. In addition, the entire range will be displaced upward. For example, an impure sample might melt from 120–124°C and after recrystallization melt at 125–125.5°C. A solid is considered pure if the melting point does not rise after recrystallization.

Characterization

An indication of purity

A crystal is an orderly arrangement of molecules in a solid. As heat is added to the solid, the molecules will vibrate and perhaps rotate but still remain a solid. At a characteristic temperature it will suddenly acquire the necessary energy to overcome the forces that attract one molecule to another and it will undergo translational motion—in other words, it will become a liquid.

The forces by which one molecule is attracted to another include ionic attraction, van der Waals forces, hydrogen bonds, and dipole–dipole attraction. Most, but by no means all, organic molecules are covalent in nature and melt at temperatures below 300°C. Typical inorganic compounds are ionic and have much higher melting points, e.g., sodium chloride melts at 800°C. Ionic organic molecules often decompose before melting, as do compounds having strong hydrogen bonds such as sucrose.

Melting point generalizations

Other factors being equal, larger molecules melt at higher temperatures than smaller ones. Among structural isomers the more symmetrical will have the higher melting point. Among optical isomers the R and S enantiomers will have the same melting points; but the racemate, the

A phase diagram

mixture of equal parts of R and S, will usually have a different melting point. Molecules that can form hydrogen bonds will usually have higher melting points than their counterparts of similar molecular weight.

    The melting point behavior of impure compounds is best understood by consideration of a simple binary mixture of compounds X and Y (Fig. 1). This melting point-composition diagram shows the melting point behavior as a function of composition. The melting point of a pure compound is the temperature at which the vapor pressures of the solid and liquid are equal. But in dealing with a mixture the situation is different. Consider the case of a mixture of 75% X and 25% Y. At a temperature below ET, the eutectic temperature, the mixture is solid Y and solid X. At the eutectic temperature the solid begins to melt. The melt is a solution of Y dissolved in liquid X. The vapor pressure of the solution of X and Y together is less than that of pure X

Melting point depression

at the melting point; therefore, the temperature at which X will melt is lower when mixed with Y. This is an application of Raoult's Law (Chapter 5). As the temperature is raised, more and more of solid X melts until it is all gone at point M (temperature m). The melting point range is thus from ET to m. In practice it is very difficult to detect point ET when a melting point is determined in a capillary because it represents the point at which an infinitesimal amount of the liquid solution has started to melt.

    In this hypothetical example the liquid solution becomes saturated with Y at point EP. This is the point at which X and Y and their liquid solutions are in equilibrium. A mixture of X and Y containing 60% X will

The eutectic point

appear to have a sharp melting point at temperature ET. This point, EP, is the eutectic point.

    In general the melting point range of a mixture of compounds is broad and the breadth of the range is an indication of purity. The chances of

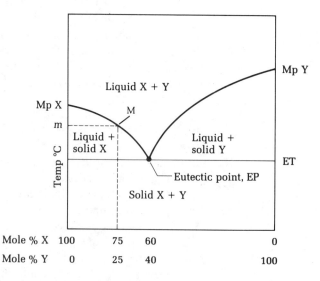

FIG. 1 Melting point-composition diagram for mixtures of the solids X and Y.

accidentally coming upon the eutectic composition are small. Recrystallization will enrich the predominant compound while excluding the impurity and therefore the melting point range will decrease.

It should be apparent that the impurity must be soluble in the compound, so an insoluble impurity such as sand or charcoal will not depress the melting point. The impurity does not need to be a solid. It can be a liquid such as water (if it is soluble) or an organic solvent, such as the one used to recrystallize the compound; hence the necessity for drying the compound before determining the melting point.

Mixed melting points

Advantage is taken of the depression of melting points of mixtures to prove whether two compounds having the same melting points are identical. If X and Y are identical, then a mixture of the two will have the same melting point; but if X and Y are not identical, then a small amount of X in Y or of Y in X will cause the melting point to be lowered.

Apparatus

The apparatus needed for determining an accurate melting point need not be elaborate; the same results are obtained on the simplest as on the most complex devices.

Thomas-Hoover Uni-Melt

The Thomas-Hoover Uni-Melt apparatus (Fig. 2) will accommodate seven capillaries in a small, magnified, lighted beaker of high-boiling silicone oil that is stirred and heated electrically. The heating rate is controlled with a variable transformer that is part of the apparatus. The rising mercury column of the thermometer can be observed with an optional traveling periscope device so the eye need not move away from the capillary. For industrial analytical and control work there is even an apparatus (Mettler) that automatically determines the melting point and displays the result in digital form.

Mel-Temp

The Mel-Temp apparatus (Fig. 3) consists of an electrically heated aluminum block that accommodates three capillaries. The sample is illuminated through the lower port and observed with a 6-power lens through the upper port. The heating rate can be controlled, and with a special thermometer the apparatus can be used up to 500°C, far above the useful limit of silicone oil (about 350°C). It is also available with a digital thermometer.

Test tube

Using a 4-in. test tube half filled with silicone oil and a stirrer made from heavy wire, melting points can be determined on an electric heater filled with sand (Fig. 4, p. 49). Once the sand is warmed up it has a very large temperature gradient from top to bottom, so the temperature and heating rate can be adjusted according to the depth of the test tube in the sand.

A simple device is illustrated in Fig. 5, p. 49. The thermometer is fitted through a cork, a section of which is cut away for pressure release and so that the scale is visible. A single-edge razor blade is convenient for cutting, and the cut can be smoothed or deepened with a triangular file. The curvature of the walls of the flask causes convection currents in the heating liquid to rise evenly along the walls and then descend and converge at the

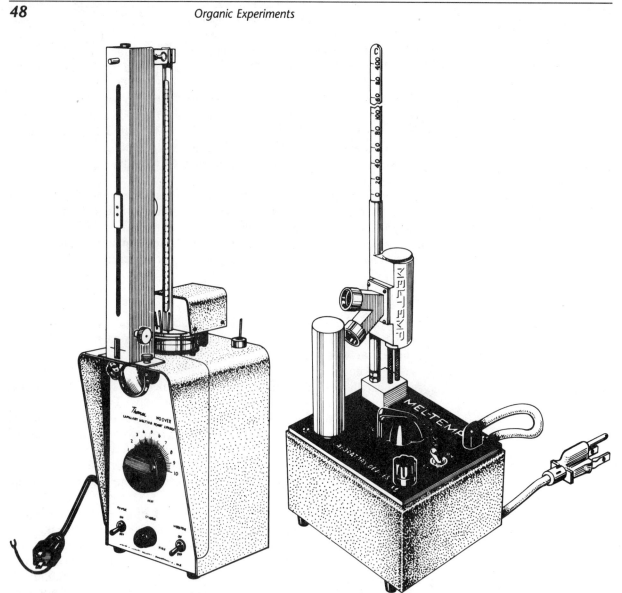

FIG. 2 Thomas-Hoover Uni-Melt melting point apparatus.

FIG. 3 Mel-Temp melting point apparatus.

Handle flames with great care in the organic laboratory

center; hence, the thermometer must be centered in the flask. The long neck prevents spilling and fuming and minimizes error due to stem exposure. The bulb of the flask (dry!) is three-quarters filled with silicone oil. The flask is mounted as shown in Fig. 5, with the bulb close to the chimney of a microburner. Careful control of heat input required in taking a melting point is accomplished both by regulating the gas supply and by raising or lowering the heating bath. This same apparatus can be heated more safely with the electrically heated sand bath.

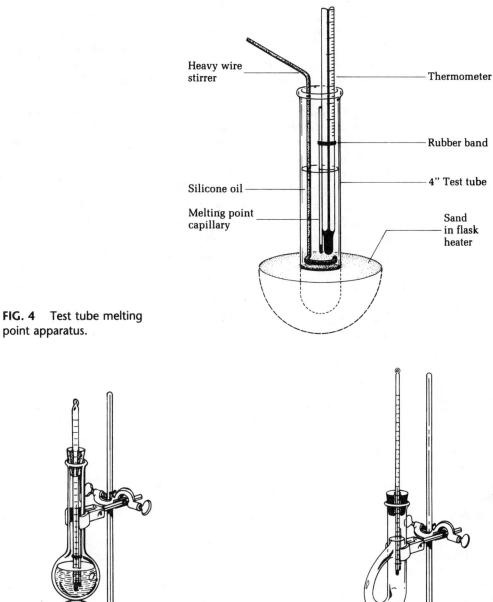

Heavy wire stirrer

Thermometer

Rubber band

4" Test tube

Silicone oil

Melting point capillary

Sand in flask heater

FIG. 4 Test tube melting point apparatus.

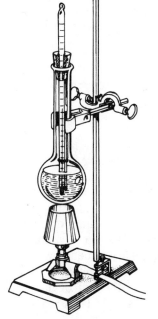

FIG. 5 Apparatus for melting point determination.

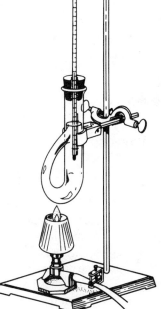

FIG. 6 Thiele apparatus.

Thiele tube

Beaker

The Thiele apparatus (Fig. 6) achieves stirring and uniform heat distribution by convection. It is filled to the base of the neck with silicone oil (the oil expands on heating) and outfitted with the same type of slotted cork described above. The tube is heated at the base of the bend. The bulb of the thermometer should be halfway down the tube, as shown in Fig. 6, to assure uniform heating.

Melting points are also easily determined in a beaker, as seen in Fig. 7. The glass rod used for stirring has a circular base and a handle, as seen in the figure. Alternatively, the beaker can be heated on a hot plate and stirred magnetically using a Teflon-covered stirring bar.

Do not discard the oil used in the apparatus because it will be necessary to determine a number of melting points in future experiments.

Capillaries can be obtained commercially or can be made by drawing out 12-mm soft-glass tubing. The tubing is rotated in the hottest part of the Bunsen burner flame until it is very soft and begins to sag. It should not be drawn out during heating, but is removed from the flame and after a moment's hesitation drawn steadily and not too rapidly to arm's length. With some practice it is possible to produce 10–15 good tubes in a single drawing. The long capillary tube can be cut into 100-mm lengths with a glass scorer. Each tube is sealed by rotating the end in the edge of a small flame, as seen in Fig. 8.

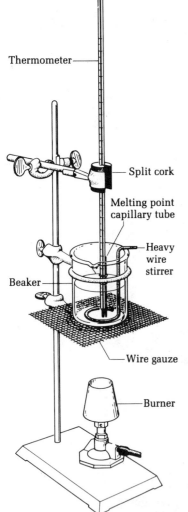

Thermometer

Split cork

Melting point
capillary tube

Heavy
wire
stirrer

Beaker

Wire gauze

Burner

FIG. 7 Beaker melting point apparatus.

FIG. 8 Sealing a melting point capillary tube.

Filling Melting Point Capillaries

The dry sample is ground to a fine powder on a watch glass or a piece of glassine paper on a hard surface using the flat portion of a spatula. It is formed into a small pile and the melting point capillary forced down into the pile. The sample is shaken into the closed end of the capillary by rapping sharply on a hard surface or by dropping it down a 2-ft length of glass tubing onto a hard surface. The height of the sample should be no more than 2–3 mm.

Sealed Capillaries

Samples that sublime

Some samples sublime (go from the solid directly to the vapor phase without appearing to melt) or undergo rapid air oxidation and decompose at the melting point. These samples should be sealed under vacuum. This can be accomplished by forcing a capillary through a hole previously made in a rubber septum and evacuating the capillary using the water aspirator or a mechanical vacuum pump (Fig. 9). Using the flame from a small micro-burner the tube is gently heated about 15 mm above the tightly packed sample. This will cause any material in this region to sublime away. It is then heated more strongly in the same place to collapse the tube, taking care that the tube is straight when it cools. It is also possible to seal the end of a Pasteur pipette, add the sample, pack it down, and seal off a sample under vacuum in the same way.

Handle flame with great care in the organic laboratory

Determining the Melting Point

The accuracy of the melting point depends on the accuracy of the thermometer, so the first exercise in this experiment will be to calibrate the thermometer. Melting points of pure, known compounds will be determined and deviations recorded so that a correction can be applied to future melting points. Be forewarned, however, that the thermometers are usually fairly accurate.

The most critical factor in determining an accurate melting point is the rate of heating. At the melting point the temperature rise should not be greater than 1°C per minute. This may seem extraordinarily slow, but it is necessary in order that heat from the bath be transferred equally to the sample and to the glass and mercury of the thermometer.

From experience you know the rate at which ice melts. Consider doing a melting point experiment on an ice cube. Because water melts at 0°C, you would need to have a melting point bath a few degrees below zero. To observe the true melting point of the ice cube you would need to raise the temperature extraordinarily slowly. The ice cube would appear to begin to melt at 0°C and, if you waited for temperature equilibrium to be established, it would all be melted at 0.5°C. If you were impatient and raised the temperature too rapidly, the ice might appear to melt over the range 0 to 20°C. Similarly, melting points determined in capillaries will not be accurate if the rate of heating is too fast.

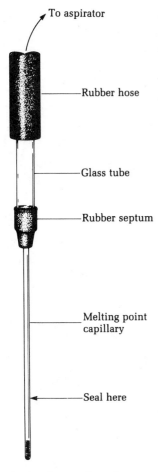

To aspirator

Rubber hose

Glass tube

Rubber septum

Melting point capillary

Seal here

FIG. 9   Evacuation of a melting point capillary prior to sealing.

> **The rate of heating is the most important factor in obtaining accurate melting points. Heat no faster than 1°C per minute.**

Experiments

1. Calibration of the Thermometer

Determine the melting point of standard substances (Table 1) over the temperature range of interest. The difference between the values found and those expected constitutes the correction that must be applied to future temperature readings. If the thermometer has been calibrated previously, then determine one or more melting points of known substances to familiarize yourself with the technique. If the determinations do not agree within 1°C, then repeat the process.

TABLE 1 Melting Point Standards

Compound	Structure	Melting Point (°C)
Naphthalene		80–82
Urea	$\overset{\displaystyle O}{\overset{\|}{H_2NCNH_2}}$	132.5–133
Sulfanilamide		164–165
4-Toluic Acid		180–182
Anthracene		214–217
Caffeine (evacuated capillary)		234–236.5

2. Melting Points of Pure Urea and Cinnamic Acid

$$\text{CH}=\text{CH}-\text{COOH}$$

Cinnamic acid

Using a metal spatula crush the sample to a fine powder on a hard surface such as a watch glass. Push a melting point capillary into the powder and force the powder down in the capillary by tapping the capillary or by dropping it through a long glass tube held vertically and resting on a hard surface. The column of solid should be no more than 2–3 mm in height and should be tightly packed.

Except for the Thomas-Hoover and Mel-Temp apparatus, the capillary is held to the thermometer with a rubber band made by cutting a slice off the end of a piece of $\frac{3}{16}$-in. rubber tubing. This rubber band must be above the level of the oil bath; otherwise, it will break in the hot oil. Insertion of a fresh tube under the rubber band is facilitated by leaving the used tube in place. The sample should be close to and on a level with the center of the thermometer bulb.

Heat rapidly to within 20°C of the melting point

If the approximate melting temperature is known, the bath can be heated rapidly until the temperature is about 20°C below this point, but the heating during the last 15–20°C should slow down considerably so that the rate of heating at the melting point is no more than 1°C per minute while the sample is melting. As the melting point is approached the sample may shrink because of crystal structure changes. However, the melting process begins when the first drops of liquid are seen in the capillary and ends when the last trace of solid disappears. For a pure compound this whole process may occur over a range of only 0.5°C; hence the necessity of having the temperature rise slowly during the determination.

Two or three melting points at once

If determinations are to be done on two or three samples that differ in melting point by as much as 10°C, two or three capillaries can be secured to the thermometer together and the melting points observed in succession without removal of the thermometer from the bath. As a precaution against interchange of tubes while they are being attached, use some system of identification, such as one, two, and three dots made with a marking pencil.

Determine the melting point of either urea (mp 132.5–133°C) or cinnamic acid (mp 132.5–133°C). Repeat the determination, and if the two determinations do not check within 1°C, do a third one.

3. Melting Points of Urea-Cinnamic Acid Mixtures

Make mixtures of urea and cinnamic acid in the approximate proportions 1:4, 1:1, and 4:1 by putting side by side the correct number of equal-sized small piles of the two substances and then mixing them. Grind the mixture thoroughly for at least a minute on a watch glass using a metal spatula. Note the ranges of melting of the three mixtures and use the temperatures of complete liquefaction to construct a rough diagram of mp versus composition.

4. Unknowns

Determine the melting point of one or more of the following unknowns to be selected by the instructor (Table 2) and on the basis of the melting point identify the substance. Prepare two capillaries of each unknown. Run a very fast determination on the first sample to ascertain the approximate melting point and then cool the melting point bath to just below the melting point and make a slow, careful determination using the other capillary.

TABLE 2   Melting Point Unknowns

Compound	Melting Point (°C)
Benzophenone	49–51
Maleic anhydride	54–56
Naphthalene	80–82
Acetanilide	113.5–114
Benzoic acid	121.5–122
Urea	132.5–133
4-Nitrotoluene	148–149
Salicylic acid	158.5–159
Sulfanilamide	165–166
Succinic acid	184.5–185
3,5-Dinitrobenzoic acid	205–207
p-Terphenyl	210–211

Part 2. Boiling Points

The boiling point of a pure organic liquid is one of its characteristic physical properties, just like its density, molecular weight, and refractive index, and the melting point of a solid. The boiling point is used to characterize a new organic liquid and knowledge of the boiling point helps to compare one organic liquid with another, as in the process of identifying an unknown organic substance.

Comparison of boiling points with melting points is instructive. The process of determining the boiling point is more complex than that for the melting point: it requires more material and because it is affected less by impurities, it is not as good an indication of purity. Boiling points can be determined on a few microliters of a liquid, but on a small scale it is difficult to determine the boiling point *range*. This requires enough material to distill—about 1 to 2 mL. Like the melting point, the boiling point of a liquid is affected by the forces that attract one molecule to another—ionic attraction, van der Waals forces, dipole–dipole interactions, and hydrogen bonding.

Structure and Boiling Point

In a homologous series of molecules the boiling point increases in a perfectly regular manner. The normal saturated hydrocarbons have boiling points ranging from $-162°C$ for methane to $330°C$ for n-$C_{19}H_{40}$. It is convenient to remember that n-heptane with a molecular weight of 100 has a boiling point near 100°C (98.4°C). A spherical molecule such as 2,2-dimethylpropane has a lower boiling point than n-pentane because it cannot have as many points of attraction to adjacent molecules. For molecules of the same molecular weight, those with dipoles, such as carbonyl groups, will have higher boiling points than those without, and molecules that can form hydrogen bonds will boil even higher. The boiling point of such molecules depends on the number of hydrogen bonds that can be formed, so that an alcohol with one hydroxyl group will boil lower than one with two if they both have the same molecular weight. A number of other generalizations can be made about boiling point behavior as a function of structure; you will learn of these throughout your study of organic chemistry.

Boiling Point as a Function of Pressure

Since the boiling point of a pure liquid is defined as the temperature at which the vapor pressure of the liquid exactly equals the pressure exerted upon it, the boiling point will be a function of atmospheric pressure. At an altitude of 14,000 ft the boiling point of water is 81°C. At pressures near that of the atmosphere at sea level (760 mm), the boiling point of most liquids decreases about 0.5°C for each 10-mm decrease in atmospheric pressure. This generalization does not hold for greatly reduced pressures because the boiling point decreases as a nonlinear function of pressure (see Fig. 1 in Chapter 5). Under these conditions a nomograph relating observed boiling point, boiling point at 760 mm, and pressure in mm should be consulted (see Fig. 10 in Chapter 7). This nomograph is not highly accurate; the change in boiling point as a function of pressure also depends on the type of compound (polar, nonpolar, hydrogen bonding, etc.). Consult the *Handbook of Chemistry and Physics* for the correction of boiling points to standard pressure.

The Laboratory Thermometer

Most laboratory thermometers have a mark around the stem that is three inches (76 mm) from the bottom of the bulb. This is the immersion line; the thermometer will record accurate temperatures if immersed to this line. If the thermometer happens to be of the total immersion type then the readings will need to be corrected when taking melting and boiling points because only a few inches of the thermometer are being heated. Again the procedure for making this "emergent stem correction for liquid-in-glass thermometers" can be found in the *Handbook of Chemistry and Physics*. Should you break a mercury thermometer, immediately inform your instructor, who will use special apparatus to clean up the mercury. Mercury vapor is very toxic.

Prevention of Superheating—Boiling Sticks and Boiling Stones

A very clean liquid in a very clean vessel will superheat and not boil when subjected to a temperature above its boiling point. This means that a thermometer placed in the liquid will register a temperature higher than the boiling point of the liquid. If boiling does occur under these conditions, it occurs with explosive violence. To avoid this problem boiling stones are always added to liquids before heating them to boiling—whether to determine a boiling point or to carry out a reaction or distillation. The stones provide the nuclei on which the bubble of vapor indicative of a boiling liquid can form. Some boiling stones, also called boiling chips, are porous unglazed porcelain. This material is filled with air in numerous fine capillaries. Upon heating, this air expands to form the fine bubbles on which even boiling can take place. Once the liquid cools it will fill these capillaries and the boiling chip becomes ineffective, so another must be added each time the liquid is heated to boiling. Sticks of wood—so-called applicator sticks about 1.5 mm in diameter—also promote even boiling and, unlike stones, are easy to remove from the solution.

Apparatus and Technique

When enough material is available the best method for determining the boiling point of a liquid is to distill it (Chapter 5). Distillation allows the boiling range to be determined and thus gives an indication of purity.

For smaller quantities of material the apparatus is very similar to that used for determining melting points. A very simple method is to place 0.2 mL of the liquid along with a boiling stone in a 10 × 75 mm test tube, clamp a thermometer so the bulb is just above the level of the liquid, and then heat the liquid with a sand bath or oil bath. It is very important that no part of the thermometer touch the test tube. Heating is regulated so that the boiling liquid refluxes (condenses and drips down) about one or two inches up the thermometer, but does not boil out the top of the tube. Droplets of liquid must drip from the thermometer bulb in order to heat the mercury thoroughly. The boiling point is the highest temperature recorded by the thermometer and maintained over about a 1-min time interval.

For smaller quantities the tube is attached to the side of the thermometer (Fig. 10) and heated with a liquid bath. The tube, which can be from

FIG. 10   Smaller-scale boiling point.

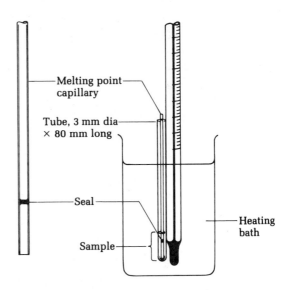

Smaller-scale boiling point apparatus

tubing 3–5 mm in diameter, contains a small inverted capillary. This is made by cutting a 6-mm piece from the sealed end of a melting point capillary, inverting it, and sealing it again to the capillary. A cm/mm ruler is printed on the inside cover of this book.

When the sample is heated in this device the air in the inverted capillary will expand and an occasional bubble will escape. At the boiling point a continuous and rapid stream of bubbles will emerge from the inverted capillary. At this point the heating is stopped and the bath allowed to cool. A time will come when bubbling ceases and the liquid just begins to rise in the inverted capillary. The temperature at which this happens is recorded. The liquid is allowed to partially fill the small capillary and the heat is applied carefully until the first bubble comes from the capillary, and that temperature is recorded. The two temperatures approximate the boiling point range for the liquid. The explanation: As the liquid was being heated the air expanded in the inverted capillary and was replaced by vapor of the liquid. The liquid was actually slightly superheated when rapid bubbles emerged from the capillary, but on cooling the point was reached at which the pressure on the inside of the capillary matched the outside (atmospheric) pressure. This is, by definition, the boiling point.

Microscale boiling point apparatus

It is possible, with care, to make this same boiling point device on an even smaller scale. A Pasteur pipette is heated in a microburner flame until the glass at one point is very soft. The pipette is lifted from the flame and after a moment's hesitation drawn out steadily and not too rapidly for a foot or so to produce a very fine capillary. One end of this capillary is sealed; a piece about 6 mm long is broken off and the sealed end attached to a 3-in.

piece of capillary. The sample is added to an ordinary melting point capillary and shaken or centrifuged to the bottom, and the fine capillary just made is added to the melting point capillary. It will be very similar in appearance to Fig. 10. The boiling point is determined as described above, using a Thomas-Hoover or similar liquid bath melting point apparatus.

Questions

1. What effect would poor circulation of the melting point bath liquid have on the observed melting point?

2. What is the effect of an insoluble impurity, such as sodium sulfate, on the observed melting point of a compound?

3. Three test tubes, labeled A, B, and C, contain substances with approximately the same melting points. How could you prove the test tubes contain three different chemical compounds?

4. One of the most common causes of inaccurate melting points is too rapid heating of the melting point bath. Under these circumstances how will the observed melting point compare with the true melting point?

5. Strictly speaking, why is it incorrect to speak of a melting *point*?

6. What effect would the incomplete drying of a sample (e.g., the incomplete removal of a recrystallization solvent) have on the melting point?

7. Which would be expected to have the higher boiling point, *t*-butyl alcohol (2-methyl-2-propanol) or *n*-butyl alcohol (1-butanol)?

Reference

A. Weissberger and B. W. Rossiter (eds). *Physical Methods of Chemistry,* Vol. 1, Part V, Wiley-Interscience, New York, 1971.

5

Distillation

Prelab Exercise: Predict what a plot of temperature vs. volume of distillate will look like for the simple distillation and the fractional distillation of (a) a cyclohexane-toluene mixture, (b) an ethanol-water mixture.

The origins of distillation are lost in antiquity as man in his thirst for more potent beverages found that dilute solutions of alcohol from fermentation could be separated into alcohol-rich and water-rich portions by heating the solution to boiling and condensing the vapors above the boiling liquid—the process of distillation. Since ethyl alcohol, ethanol, boils at 78°C and water boils at 100°C, one might naively assume that heating a 50:50 mixture of ethanol and water to 78°C would cause the ethanol molecules to leave the solution as a vapor that could be condensed to give pure ethanol. Such is not the case. A mixture of 50:50 ethanol:water boils near 87°C and the vapor above it is not 100% ethanol.

   Consider a simpler mixture, cyclohexane and toluene. The vapor pressures as a function of temperature are plotted in Fig. 1. When the vapor pressure of the liquid equals the applied pressure the liquid boils, so this diagram shows that at 760 mm pressure, standard atmospheric pressure, these pure liquids boil at 78 and 111°C, respectively. If one of these pure liquids were to be distilled, it would be found that the boiling point of the

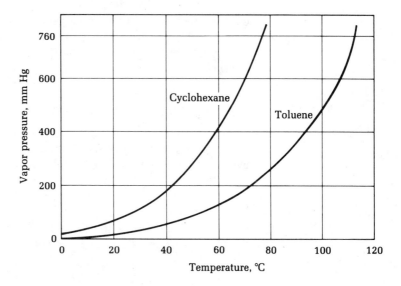

FIG. 1 Vapor pressure vs. temperature for cyclohexane and toluene.

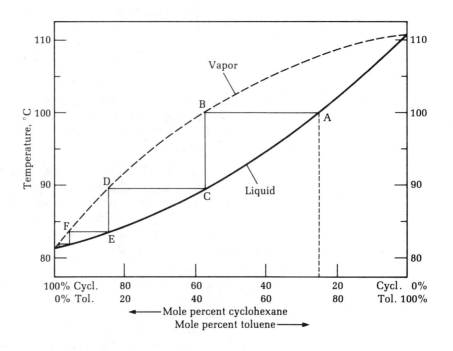

FIG. 2 Boiling point-composition curves for cyclohexane-toluene mixtures.

liquid would equal the temperature of the vapor and that the temperature of the vapor would remain constant throughout the distillation.

A mixture of two liquids will usually have a bp that is between the bps of the pure liquids

Figure 2 is a boiling point–composition diagram for the cyclohexane-toluene system. If a mixture of 75 mole percent toluene and 25 mole percent cyclohexane is heated, we find from Fig. 2 that it boils at 100°C, or point A. Above a binary mixture of cyclohexane and toluene the vapor pressure has contributions from each component. Raoult's law states that the vapor pressure of the cyclohexane is equal to the product of the vapor pressure of pure cyclohexane and the mole fraction of cyclohexane in the liquid mixture:

Raoult's Law

$$P_c = P_c^\circ N_c$$

where P_c is the partial pressure of cyclohexane, P_c° is the vapor pressure of pure cyclohexane at the given temperature, and N_c is the mole fraction of cyclohexane in the mixture. Similarly for toluene:

$$P_t = P_t^\circ N_t$$

and the total vapor pressure above the solution, P_{Tot}, is given by the sum of the partial pressures due to cyclohexane and toluene:

$$P_{Tot} = P_c + P_t$$

Dalton's law states that the mole fraction of cyclohexane in the vapor at a given temperature is equal to the partial pressure of the cyclohexane at that temperature divided by the total pressure:

$$X_c = P_c/\text{total vapor pressure}$$

At 100°C cyclohexane has a partial pressure of 433 mm and toluene a partial pressure of 327 mm; the sum of the partial pressures is 760 mm and so the liquid boils. If some of the liquid in equilibrium with this boiling mixture were condensed and analyzed, it would be found to be 433/760 or 57 mole percent cyclohexane (point B, Fig. 2). This is the best separation that can be achieved on simple distillation of this mixture. As the simple distillation proceeds, the boiling point of the mixture moves toward 110°C along the line from A, and the vapor composition becomes richer in toluene as it moves from B to 110°C. In order to obtain pure cyclohexane, it would be necessary to condense the liquid at B and redistill it. When this is done it is found that the liquid boils at 90°C (point C) and the vapor equilibrium with this liquid is about 85 mole percent cyclohexane (point D). So to separate a mixture of cyclohexane and toluene, a series of fractions would be collected and each of these partially redistilled. If this fractional distillation were done enough times the two components could be separated.

This series of redistillations can be done "automatically" in a fractionating column. Perhaps the easiest to understand is the bubble cap column used to fractionally distill crude oil. These columns dominate the skyline of oil refineries, some being 150 ft high and capable of distilling 200,000 barrels of crude oil per day. The crude oil enters the column as a hot vapor (Fig. 3). Some of this vapor with high boiling components condenses on one of the plates. The more volatile substances travel through the bubble cap to the next higher plate where some of the less volatile components condense. As high boiling liquid material accumulates on a plate it descends through the overflow pipe to the next lower plate and vapor rises through the bubble cap to the next higher plate. The temperature of the vapor that is rising through a cap is above the boiling point of the liquid on that plate. As bubbling takes place heat is exchanged and the less volatile components on that plate vaporize and go on to the next plate. The composition of the liquid on a plate is the same as that of the vapor coming from the plate below. So on each plate a simple distillation takes place. At equilibrium, vapor containing low boiling material is ascending and high boiling liquid is descending through the column.

Figure 2 shows that the condensations and redistillations in a bubble cap column consisting of three plates correspond to moving on the boiling point–composition diagram from point A to point E.

In the laboratory the successive condensations and distillations that occur in the bubble cap column take place in a distilling column. The column is packed with some material on which heat exchange between ascending vapor and descending liquid can take place. A large surface area

Fractional distillation

Overflow pipe

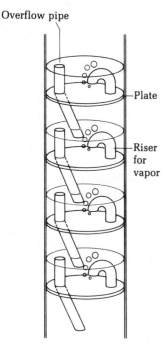

Plate

Riser for vapor

FIG. 3   Bubble plate distilling column.

Heat exchange between ascending vapor and descending liquid

Column packing

Height equivalent to a theoretical plate (HETP)

Equilibration is slow

Good fractional distillation takes a long time

for this packing is desirable, but the packing cannot be so dense that pressure changes take place within the column causing nonequilibrium conditions. Also, if the column packing has a very large surface area, it will absorb (hold up) much of the material being distilled. A number of different packings for distilling columns have been tried—glass beads, glass helices, carborundum chips, etc. We find one of the best packings is a copper or steel sponge (Chore Boy). It is easy to put into the column, does not come out of the column as beads do, has a large surface area, good heat transfer characteristics, and low holdup. It can be used in both the microscale and macroscale apparatus.

The ability of different column packings to separate two materials of differing boiling points is evaluated by calculating the number of theoretical plates, each theoretical plate corresponding to one distillation and condensation as discussed above. Other things being equal, the number of theoretical plates is proportional to the height of the column, so various packings are evaluated according to the height equivalent to a theoretical plate (HETP); the smaller the HETP, the more plates the column will have and the more efficient it will be. The calculation is made by analyzing the proportion of lower to higher boiling material at the top of the column and in the distillation pot.[1]

Although not obvious, the most important variable contributing to a good fractional distillation is the rate at which the distillation is carried out. A series of simple distillations take place within a fractionating column and it is important that complete equilibrium be attained between the ascending vapors and the descending liquid. This process is not instantaneous. It should be an adiabatic process; that is, heat should be transferred from the ascending vapor to the descending liquid with no net loss or gain of heat. In larger, more complex distilling columns, a means is provided for adjusting the ratio between the amount of material that boils up and condenses (refluxes) and is returned to the column (thus allowing equilibrium to take place) and the amount that is removed as distillate. A reflux ratio of 30:1 or 50:1 would not be uncommon for a 40-plate column; distillation would take several hours.

Carrying out a fractional distillation on the truly micro scale (<1 mg) is impossible, and even impossible on a small scale (10–400 mg). In the present microscale experiment the distillation will be carried out on a 4-mL scale. As seen in later chapters various types of chromatography are employed for the separation of micro and semimicro quantities of material while distillation is the best method for separating more than a few grams of material.

---

1. See *Technique of Organic Chemistry*, Vol. IV, "Distillation," A. Weissberger (ed.), Wiley-Interscience, New York, 1951.

Azeotropes

Not all liquids form ideal solutions and conform to Raoult's law. Ethanol and water are such liquids. Because of molecular interaction, a mixture of 95.5% (by weight) of ethanol and 4.5% of water boils *below* (78.15°C) the boiling point of pure ethanol (78.3°C). Thus, no matter how efficient the distilling apparatus, 100% ethanol cannot be obtained by distillation of a mixture of, say, 75% water and 25% ethanol. A mixture of liquids of a certain definite composition that distills at a constant temperature without change in composition is called an azeotrope; 95% ethanol is such an azeotrope. The boiling point–composition curve for the ethanol-water mixture is seen in Fig. 4. To prepare 100% ethanol the water can be removed chemically (reaction with calcium oxide) or by removal of the water as an azeotrope (with still another liquid). An azeotropic mixture of 32.4% ethanol and 67.6% benzene (bp 80.1°C) boils at 68.2°C. A ternary azeotrope (bp 64.9°C) contains 74.1% benzene, 18.5% ethanol, and 7.4% water. Absolute alcohol (100% ethanol) is made by addition of benzene to 95% alcohol and removal of the water in the volatile benzene-water-alcohol azeotrope.

The ethanol/water azeotrope

The ethanol and water form a minimum boiling azeotrope. Other substances, such as formic acid (bp 100.7°C) and water (bp 100°C), form

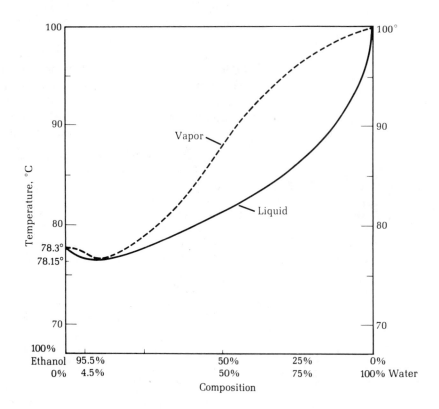

FIG. 4   Boiling point-composition curves for ethanol-water mixtures.

maximum boiling azeotropes. For these two compounds the azeotrope boils at 107.3°C.

A pure liquid has a constant boiling point. A change in boiling point during distillation is an indication of impurity. The converse proposition, however, is not always true, and constancy of a boiling point does not necessarily mean that the liquid consists of only one compound. For instance, two miscible liquids of similar chemical structure that boil at the same temperature individually will have nearly the same boiling point as a mixture. And, as noted previously, azeotropes have constant boiling points that can be either above or below the boiling points of the individual components.

When a solution of sugar in water is distilled, the boiling point recorded on a thermometer located in the vapor phase is 100°C (at 760 torr) throughout the distillation, whereas the temperature of the boiling sugar solution itself is initially somewhat above 100°C and continues to rise as the concentration of sugar in the remaining solution increases. The vapor pressure of the solution is dependent upon the number of water molecules present in a given volume; and hence with increasing concentration of nonvolatile sugar molecules and decreasing concentration of water, the vapor pressure at a given temperature decreases and a higher temperature is required for boiling. However, sugar molecules do not leave the solution, and the drop clinging to the thermometer is pure water in equilibrium with pure water vapor.

When a distillation is carried out in a system open to the air and the boiling point is thus dependent on existing air pressure, the prevailing barometric pressure should be noted and allowance made for appreciable deviations from the accepted boiling point temperature (see Table 1). Distillation can also be done at the lower pressures that can be achieved by an oil pump or an aspirator with substantial reduction of boiling point.

A constant bp on distillation does not guarantee that the distillate is one pure compound

Distilling a mixture of sugar and water

Bp changes with pressure

TABLE 1   Variation in Boiling Point with Pressure

Pressure (mm)	Water (°C)	Benzene (°C)
780	100.7	81.2
770	100.4	80.8
760	100.0	80.1
750	99.6	79.9
740	99.2	79.5
584*	92.8	71.2

*Instituto de Quimica, Mexico City, altitude 7700 ft (2310 m).

Experiments

1. Calibration of Thermometer

If you have not previously carried out a calibration, test the 0°C point of your thermometer with a well-stirred mixture of crushed ice and distilled water. To check the 100°C point, put 2 mL of water in a test tube with a boiling chip to prevent bumping and boil the water gently over a hot sand bath with the thermometer in the vapor from the boiling water. Take care to see that the thermometer does not touch the side of the test tube. Then immerse the bulb of the thermometer in the liquid and see if you can observe superheating. Check the atmospheric pressure to determine the true boiling point of the water.

2. Simple Distillation

Apparatus

CAUTION: Cyclohexane and toluene are flammable; make sure distilling apparatus is airtight.

In any distillation the flask should not be more than two-thirds full at the start. Great care should be taken not to distill to dryness because, in some cases, high-boiling explosive peroxides can become concentrated.

Assemble the apparatus for simple distillations shown in Fig. 5, starting with the support ring, followed by the electric flask heater [Fig. 11(n) in Chapter 1] and then the flask. One or two boiling stones are put in the flask to promote even boiling. Each ground joint is greased by putting three or four stripes of grease lengthwise around the male joint and pressing the joint firmly into the other without twisting. The air is thus eliminated and the joint will appear almost transparent. (Do not use excess grease as it will contaminate the product.) Water enters the condenser at the tublature nearest the receiver. Because of the large heat capacity of water only a very small stream (3 mm dia.) is needed; too much water pressure will cause the tubing to pop off. A heavy rubber band or a Keck clamp can be used to hold the condenser to the distillation head. Note that the bulb of the thermometer is below the opening into the side arm of the distillation head.

Cyclohexane
bp 81.4°C,
den 0.78, MW 84.16

Toluene
bp 110.8°C
den 0.87, MW 92.13

Do not add a boiling chip to a hot liquid. It may boil over.

(A) Simple Distillation of a Cyclohexane–Toluene Mixture.   Place a mixture of 30 mL of cyclohexane and 30 mL of toluene and a boiling chip in a dry 100-mL round-bottomed flask and assemble the apparatus for simple distillation. After making sure all connections are tight, heat the flask strongly until boiling starts. Then adjust the heat until the distillate drops at a regular rate of about one drop per second. Record both the temperature and the volume of distillate at regular intervals. After 50 mL of distillate are collected, discontinue the distillation. Record the barometric pressure, make any thermometer correction necessary, and plot boiling point versus volume of distillate. Save the distillate for fractional distillation.

Dispose of cyclohexane and toluene in the container provided. Do not pour them down the drain.

Cleaning Up   The pot residue should be placed in the organic solvents container. The distillate can also be placed there or it can be recycled.

(B) Simple Distillation of an Ethanol–Water Mixture.   In a 500-mL round-bottomed flask place 200 mL of a 20% aqueous solution of ethanol. Follow the procedure (above) for the distillation of a cyclohexane–toluene mixture. Discontinue the distillation after 50 mL of distillate have been

FIG. 5　Apparatus for simple distillation.

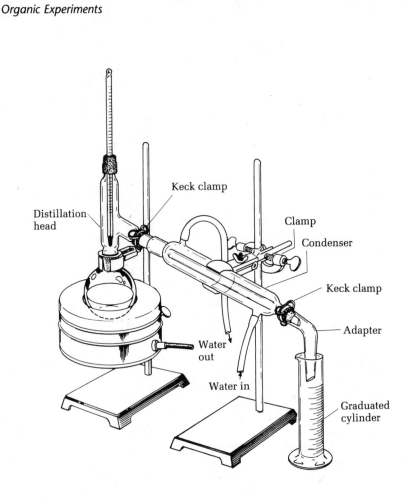

Distillation head

Keck clamp

Clamp

Condenser

Keck clamp

Adapter

Water out

Water in

Graduated cylinder

A 10–12% solution of ethanol in water is produced by fermentation. See Chapter 18.

collected. In the hood place three drops of distillate on a Pyrex watch glass and try to ignite it with the blue cone of a microburner flame. Does it burn? Is any unburned residue observed?

Cleaning Up　The pot residue and distillate can be flushed down the drain.

3. Fractional Distillation

Apparatus

Assemble the apparatus shown in Figs. 6 and 7. The fractionating column is packed with one-fourth to one-third of a metal sponge. The column should be perfectly vertical and it should be insulated with glass wool covered with aluminum foil with the shiny side in. However, in order to observe what is taking place within the column, insulation is omitted for this experiment.

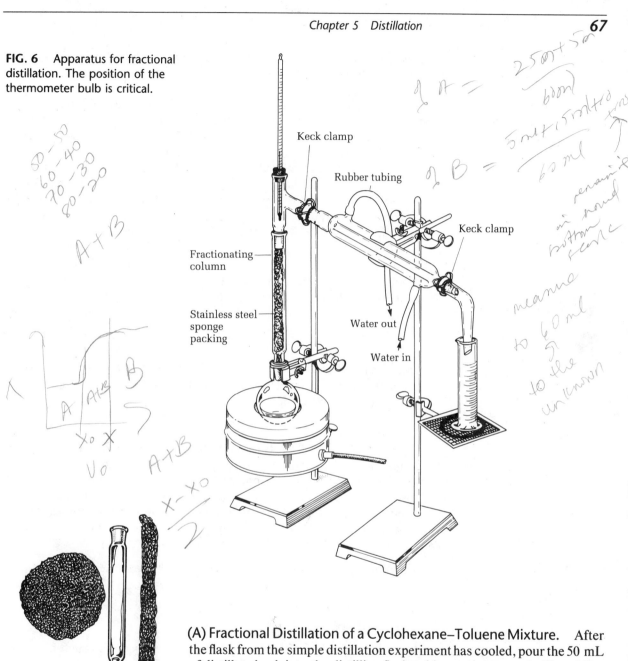

FIG. 6 Apparatus for fractional distillation. The position of the thermometer bulb is critical.

Keck clamp

Rubber tubing

Keck clamp

Fractionating column

Stainless steel sponge packing

Water out

Water in

FIG. 7 Fractionating column and packing. Use $\frac{1}{3}$ of sponge (Chore Boy).

(A) Fractional Distillation of a Cyclohexane–Toluene Mixture. After the flask from the simple distillation experiment has cooled, pour the 50 mL of distillate back into the distilling flask, add one or two new boiling chips, and assemble the apparatus for fractional distillation. The stillhead delivers into a short condenser fitted with a bent adapter leading into a 10-mL graduated cylinder. Gradually turn up the heat to the electric flask heater until the mixture of cyclohexane and toluene just begins to boil. As soon as boiling starts, turn down the power. Heat slowly at first. A ring of condensate will rise slowly through the column; if you cannot at first see this

*Best fractionation by slow
and steady heating*

ring, locate it by cautiously touching the column with the fingers. The rise
should be very gradual, in order that the column can acquire a uniform
temperature gradient. Do not apply more heat until you are sure that the
ring of condensate has stopped rising; then increase the heat gradually. In a
properly conducted operation, the vapor-condensate mixture reaches the
top of the column only after several minutes. Once distillation has com-
menced, it should continue steadily without any drop in temperature at a
rate not greater than 1 mL in 1.5–2 min. Observe the flow and keep it steady
by slight increases in heat as required. Protect the column from drafts.
Record the temperature as each milliliter of distillate collects, and make
more frequent readings when the temperature starts to rise abruptly. Each
time the graduated cylinder fills, quickly empty it into a series of labeled
25-mL Erlenmeyer flasks. Stop the distillation when a second constant
temperature is reached. Plot a distillation curve and record what you
observed inside the column in the course of the fractionation. Combine the
fractions which you think are pure and turn in the product in a bottle labeled
with your name, desk number, the name of the product, the bp range, and
the weight.

Cleaning Up   The pot residue should be placed in the organic solvents
container. The cyclohexane and toluene fractions can be placed there also
or they can be recycled.

(B) Fractional Distillation of an Ethanol–Water Mixture.   Place the
50 mL of distillate from the simple distillation experiment in a 100-mL
round-bottomed flask, add one or two boiling chips, and assemble the
apparatus for fractional distillation. Follow the procedure (above) for the
fractional distillation of a cyclohexane–toluene mixture. Repeat the ignition test. Is any difference noted?

Cleaning Up   The pot residue and distillate can be flushed down the drain.

4. Unknowns

You will be supplied with an unknown, prepared by the instructor, that is a
mixture of two solvents from those listed in Table 2. The solvents in the
mixture will be mutually soluble and differ in boiling point by more than
20°C. Fractionate the unknown and identify the components from the
boiling points. Prepare a distillation curve.

Cleaning Up   Organic material goes in the organic solvents container.
Water and aqueous solutions can be flushed down the drain.

TABLE 2   Some Properties of Common Solvents

Solvent	Boiling Point (°C)
Acetone	56.5
Methanol	64.7
Hexane	68.8
1-Butanol	117.2
2-Methyl-2-propanol	82.2
Water	100.0
Toluene*	110.6

*Methanol and toluene form an azeotrope, bp 63.8°C (69% methanol).

[handwritten margin note: unknown will not have bpt of 10°C]

Questions

1. In the simple distillation experiment (2), can you account for the boiling point of your product in terms of the known boiling points of the pure components of your mixture?

2. From the plot of boiling point versus volume of distillate in the simple distillation experiment, what can you conclude about the purity of your product?

3. From the boiling point versus volume of distillate plot in the fractional distillation of the cyclohexane–toluene mixture (3), what conclusion can you draw about the homogeneity of the distillate?

4. From the boiling point versus volume of distillate in the fractional distillation of the ethanol–water mixture (3), what conclusion can you draw about the homogeneity of the distillate? Does it have a constant boiling point? Is it a pure substance because it has a constant boiling point?

5. What is the effect on the boiling point of a solution (e.g., water) produced by a soluble nonvolatile substance (e.g., sodium chloride)? What is the effect of an insoluble substance such as sand or charcoal? What is the temperature of the vapor above these two boiling solutions?

6. In the distillation of a pure substance (e.g., water), why doesn't all the water vaporize at once when the boiling point is reached?

7. In fractional distillation, liquid can be seen running from the bottom of the distillation column back into the distilling flask. What effect does this returning condensate have on the fractional distillation?

8. Why is it dangerous to attempt to carry out a distillation in a completely closed apparatus (one with no vent to the atmosphere)?

9. Why is better separation of two liquids achieved by slow rather than fast distillation?

10. Explain why a packed fractionating column is more efficient than an unpacked one.

11. In the distillation of the cyclohexane–toluene mixture, the first few drops of distillate may be cloudy. Explain.

6

Steam Distillation

Prelab Exercise: If a mixture of toluene and water is distilled at 97°, what weight of water would be necessary to carry over 1 g of toluene?

Each liquid exerts its own vapor pressure

The bp of immiscible liquids is below the bp of either pure liquid

Used for isolation of perfume and flavor oils

When a mixture of cyclohexane and toluene are distilled (Chapter 5) the boiling point of these two miscible liquids is *between* the boiling points of each of the pure components. By contrast, if a mixture of benzene and water (immiscible liquids) is distilled, the boiling point of the mixture will be found *below* the boiling point of each pure component. Since the two liquids are essentially insoluble in each other, the benzene molecules in a droplet of benzene are not diluted by water molecules from nearby water droplets, and hence the vapor pressure exerted by the benzene is the same as that of benzene alone at the existing temperature. The same is true of the water present. Because they are immiscible, the two liquids independently exert pressures against the common external pressure, and when the sum of the two partial pressures equals the external pressure boiling occurs. Benzene has a vapor pressure of 760 torr at 80.1°, and if it is mixed with water the combined vapor pressure must equal 760 torr at some temperature below 80.1°. This temperature, the boiling point of the mixture, can be calculated from known values of the vapor pressures of the separate liquids at that temperature. Vapor pressures found for water and benzene in the range 50–80° are plotted in Fig. 1. The dotted line cuts the two curves at points where the sum of the vapor pressures is 760 torr; hence this temperature is the boiling point of the mixture (69.3°).

Practical use can sometimes be made of the fact that many water-insoluble liquids and solids behave as benzene does when mixed with water, volatilizing at temperatures below their boiling points. Thus, naphthalene, a solid, boils at 218° but distills with water at a temperature below 100°. Since naphthalene is not very volatile, considerable water is required to entrain it, and the conventional way of conducting the distillation is to pass steam into a boiling flask containing naphthalene and water. The process is called *steam distillation*. With more volatile compounds, or with a small amount of material, the substance can be heated with water in a simple distillation flask and the steam generate *in situ*.

Some high-boiling substances decompose before the boiling point is reached and, if impure, cannot be purified by ordinary distillation. However, they can be freed from contaminating substances by steam distillation at a lower temperature, at which they are stable. Steam distillation also offers the advantage of selectivity, since some water-insoluble substances are volatile with steam and others are not, and some volatilize so very slowly that sharp separation is possible. The technique is useful in process-

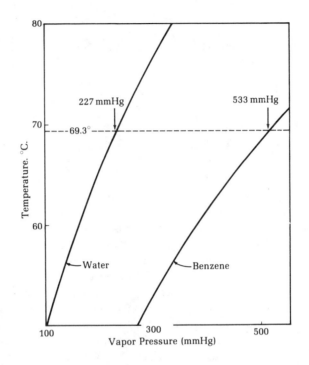

FIG. 1 Vapor pressure vs. temperature curves for water and benzene.

ing natural oils and resins, which can be separated into steam-volatile and nonsteam-volatile fractions. It is useful for recovery of a nonsteam-volatile solid from its solution in a high boiling solvent such as nitrobenzene, bp 210°; all traces of the solvent can be eliminated and the temperature can be kept low.

The boiling point remains constant during a steam distillation so long as adequate amounts of both water and the organic component are present to saturate the vapor space. Determination of the boiling point and correction for any deviation from normal atmospheric pressure permits calculation of the amount of water required for distillation of a given amount of organic substance. According to Dalton's law the molecular proportion of the two

Dalton's law components in the distillate is equal to the ratio of their vapor pressures (p) in the boiling mixture; the more volatile component contributes the greater number of molecules to the vapor phase. Thus,

$$\frac{\text{Moles of water}}{\text{Moles of substance}} = \frac{p_{\text{water}}}{p_{\text{substance}}}$$

The vapor pressure of water (p_{water}) at the boiling temperature in question can be found by interpolation of the data of Table 1, and that of the organic substance is, of course, equal to $760 - p_{\text{water}}$. Hence, the weight of water required per gram of substance is given by the expression

$$\frac{\text{Wt. of water per}}{\text{g of substance}} = \frac{18 \times p_{\text{water}}}{\text{MW of substance} \times (760 - p_{\text{water}})}$$

TABLE 1   Vapor Pressure of Water in mm of Mercury

$t°C$	p	$t°C$	p	$t°C$	p	$t°C$	p
60	149.3	70	233.7	80	355.1	90	525.8
61	156.4	71	243.9	81	369.7	91	546.0
62	163.8	72	254.6	82	384.9	92	567.0
63	171.4	73	265.7	83	400.6	93	588.6
64	179.3	74	277.2	84	416.8	94	610.9
65	187.5	75	289.1	85	433.6	95	633.9
66	196.1	76	301.4	86	450.9	96	657.6
67	205.0	77	314.1	87	468.7	97	682.1
68	214.2	78	327.3	88	487.1	98	707.3
69	223.7	79	341.0	89	506.1	99	733.2

From the data given in Fig. 1 for benzene-water, the fact that the mixture boils at 69.3°C, and the molecular weight 78.11 for benzene, the water required for steam distillation of 1 g of benzene is only $227 \times \frac{18}{533} \times \frac{1}{78} = 0.10$ g. Nitrobenzene (bp 210°, MW 123.11) steam distills at 99° and requires 4.0 g of water per gram. The low molecular weight of water makes water a favorable liquid for two-phase distillation of organic compounds.

Steam distillation will be employed in a number of experiments in this text. On a small scale steam is generated in the flask that contains the substance to be steam distilled by simply boiling a mixture of water and the immiscible substance.

Isolation of Citral

Terpenes

Citral is an example of a very large group of *natural products* called *terpenes*. They are responsible for the characteristic odors of plants such as eucalyptus, pine, mint, peppermint, and lemon. The odors of camphor, menthol, lavender, rose, and hundreds of other fragrances are due to terpenes, which have ten carbon atoms with double bonds, and aldehyde, ketone, or alcohol functional groups. (See Fig. 2.)

In nature these terpenes all arise from a common precursor, iso-pentenyl pyrophosphate. At one time they were thought to come from the simple diene, isoprene (2-methyl-1,3-butadiene), because the skeletons of terpenes can be dissected into isoprene units, having five carbon atoms

The isoprene rule

arranged as in 2-methylbutane. These isoprene units are almost always arranged in a "head-to-tail fashion."

In the present experiment citral is isolated by steam distillation of lemongrass oil, which is used to make lemongrass tea, a popular drink in Mexico. The distillate contains 90% citral and 10% neral, the isomer about the 2,3-bond. Citral is used in a commercial synthesis of vitamin A.

Lemongrass oil contains a number of substances; simple or fractional distillation would not be a practical method for obtaining pure citral. And because it boils at 229°C, it has a tendency to polymerize, oxidize, and

FIG. 2 Structures of some terpenes.

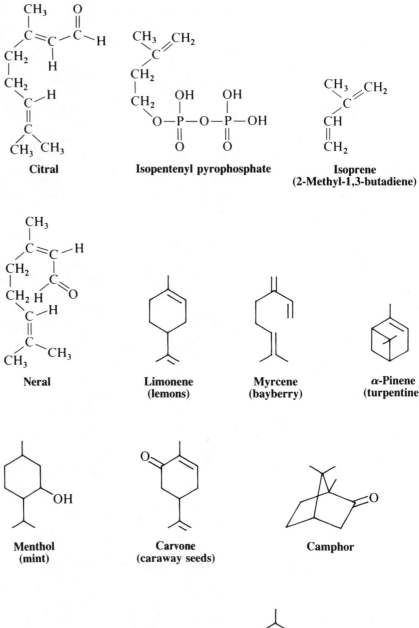

Citral

Isopentenyl pyrophosphate

Isoprene
(2-Methyl-1,3-butadiene)

Neral

Limonene
(lemons)

Myrcene
(bayberry)

α-Pinene
(turpentine)

Menthol
(mint)

Carvone
(caraway seeds)

Camphor

1,8-Cineole
(eucalyptus)

Pulegone
(pennyroyal oil)

decompose during distillation. For example, heating with potassium bisulfate, an acidic compound, converts citral to 1-methyl-4-isopropylbenzene (*p*-cymene).

Steam distillation is thus a very gentle method for isolating citral. The distillation takes place below the boiling point of water. The distillate consists of a mixture of citral (and some neral) and water. It is isolated by shaking the mixture with ether. The citral dissolves in the ether, which is immiscible with water, and the two layers are separated. The ether is dried (it dissolves some water) and evaporated to leave citral.

Store the citral in the smallest possible container in the dark for later characterization. The homogeneity of the substance can be investigated using thin-layer chromatography (Chapter 9), gas chromatography (Chapter 12), or high-performance liquid chromatography (Chapter 16). Chemically the molecule can be characterized by reaction with bromine and also permanganate, which shows it contains double bonds, and by the Tollens test, which confirms the presence of an aldehyde (Chapter 30). An infrared spectrum (Chapter 19) would confirm the presence of the aldehyde; an ultraviolet spectrum (Chapter 21) would indicate it is a conjugated aldehyde; and an nmr spectrum (Chapter 20) would clearly show the aldehyde proton, the three methyl groups, and the two olefinic protons. Finally the molecule can be reacted with another substance to form a crystalline derivative for further identification (Chapter 70). You will probably not be required to carry out all of these analyses; however, these are the major tools available to the organic chemist for ascertaining purity and determining the structure of an organic molecule.

Steam distillation: the isolation of heat-sensitive compounds

Citral and neral are geometric isomers

Experiments

1. Recovery of a Dissolved Substance

Steam Distillation Apparatus

In the assembly shown in Fig. 3 steam is passed into a 250-mL, round-bottomed flask through a section of 6-mm glass tubing fitted into a stillhead with a piece of 5-mm rubber tubing connected to a trap, which in turn is connected to the steam line. The trap serves two purposes: it allows water, which is in the steam line, to be removed before it reaches the round-bottomed flask, and adjustment of the clamp on the hose at the bottom of the trap allows precise control of the steam flow. The stopper in the trap should be wired on, as shown, as a precaution. A bent adapter attached to a *long* condenser delivers the condensate into a 250-mL Erlenmeyer flask. Measure 50 mL of a 0.2% solution of anthracene in toluene into the 250-mL round-bottomed flask and add 100 mL of water. For an initial distillation to determine the boiling point and composition of the toluene–water azeotrope, fit the stillhead with a thermometer instead of the steam-inlet tube (see Fig. 5 in Chapter 5). Heat the mixture with a hot plate and sand bath or

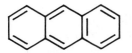

Anthracene
mp 216°C

Toluene
bp 111°C
den 0.866

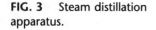

FIG. 3 Steam distillation apparatus.

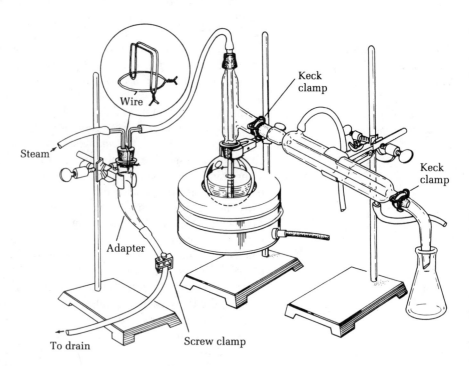

electric flask heater, distill about 50 mL of the azeotrope, and record a value for the boiling point. After removing the heat, pour the distillate into a graduate and measure the volumes of toluene and water. Calculate the weight of water per gram of toluene and compare the result with the theoretical value calculated from the vapor pressure of water at the observed boiling point (see Table 1 on page 73).

CAUTION: Live steam causes severe burns. If apparatus leaks steam, turn off steam valve immediately.

Replace the thermometer with the steam-inlet tube. To start the steam distillation, heat the flask containing the mixture with a Thermowell to prevent water from condensing in the flask to the point where water and product splash over into the receiver. Then turn on the steam valve, making sure the screw clamp on the bottom of the trap is open. Slowly close the clamp, and allow steam to pass into the flask. Try to adjust the rate of steam addition and the rate of heating so the water level in the flask remains constant. Unlike ordinary distillations, steam distillations are usually run as fast as possible, with proper care to avoid having material splash into the receiver and to avoid having steam escape uncondensed.

Continue distillation by passing in steam until the distillate is clear and then until fluorescence appearing in the stillhead indicates that a trace of anthracene is beginning to distill. Stop the steam distillation by first opening the clamp at the bottom of the trap and then turning off the steam valve. Grasp the round-bottomed flask with a towel when disconnecting it and, using the clamp to support it, cool it under the tap. The bulk of the anthracene can be dislodged from the flask walls and collected on a small suction filter. To recover any remaining anthracene, add a little acetone to the flask, warm on the steam bath to dissolve the material, add water to precipitate it, and then collect the precipitate on the same suction filter. About 80% of the hydrocarbon in the original toluene solution should be recoverable. When dry, crystallize the material from about 1 mL of toluene and observe that the crystals are more intensely fluorescent than the solution or the amorphous solid. The characteristic fluorescence is quenched by mere traces of impurities.

Cleaning Up   Place the toluene in the organic solvents container. The aqueous layer can be flushed down the drain.

2. Isolation of Citral

Citral, a fragrant terpene aldehyde made up of two isoprene units, is the main component of the steam-volatile fraction of lemon grass oil and is used in a commercial synthesis of vitamin A.

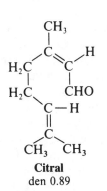

Citral
den 0.89

Using a graduate, measure out 10 mL of lemon grass oil (*not* lemon oil) into the 250-mL boiling flask. Rinse the remaining contents of the graduate into the flask with a little ether. Add 100 mL of water, make connections as in Fig. 3, heat the flask with a small flame, and pass in steam. Distill as rapidly as the cooling facilities allow and continue until droplets of oil no longer appear at the tip of the condenser (about 250 mL of distillate).

Pour 50 mL of ether into a 125-mL separatory funnel, cool the distillate if necessary, and pour a part of it into the funnel. Shake, let the layers separate, discard the lower layer, add another portion of distillate, and repeat. When the last portion of distillate has been added, rinse the flask with a little ether to recover adhering citral. Use the techniques described in Chapter 8 for drying, filtering, and evaporating the ether. Take the tare of a 1-g tincture bottle, transfer the citral to it with a Pasteur pipette, and determine the weight and the yield from the lemon grass oil. Label the bottle and store (in the dark) for later testing for the presence of functional groups.

Cleaning Up   The aqueous layer can be flushed down the drain. Any ether goes in the organic solvents container. Allow ether to evaporate from the sodium sulfate in the hood, then place it in the nonhazardous solid waste container.

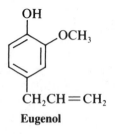

Eugenol

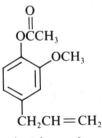

Acetyleugenol

3. Isolation of Eugenol from Cloves

In the fifteenth century Theophrastus Bombastus von Hohenheim, otherwise known as Paracelsus, urged all chemists to make extractives for medicinal purposes rather than try to transmute base metals into gold. He believed every extractive had a quintessence that was the most effective part in effecting cures. There was then an upsurge in the isolation of essential oils from plant materials, some of which are still used for medicinal purposes. Among these are camphor, quinine, oil of cloves, cedarwood, turpentine, cinnamon, gum benzoin, and myrrh. Oil of cloves, which consists almost entirely of eugenol and its acetate, is a food flavoring agent as well as a dental anesthetic. The Food and Drug Administration has declared clove oil to be the most effective nonprescription remedy for toothache.

The biggest market for essential oils is for perfumes, and, as might be expected, prices for these oils reflect their rarity. Recently, worldwide production of orange oil was 1500 tons and it sold for $0.75 per lb, while 400 tons of clove oil sold for $14.00 per lb, and 10 tons of jasmine oil sold for $2000 per lb. These three oils represent the most common extractive processes: orange oil is obtained by expression (squeezing) of the peel in presses; clove oil is obtained by steam distillation, as will be performed in this experiment; and jasmine oil is obtained by extraction of the flower petals using ethanol.

Cloves are the flower buds from a tropical tree that can grow to 50 feet in height. The buds are hand-picked in the Moluccas (the Spice Islands) of Indonesia, the East Indies, and the islands of the Indian Ocean, where a tree can yield up to 75 lb of the sun-dried buds we know as cloves.

Cloves contain between 14% and 20% by weight of essential oil of which about half can be isolated. The principal constituent of clove oil is eugenol and its acetyl derivative. Eugenol boils at 255°C, but being insoluble in water, it will form an azeotrope with water and steam distill at a temperature slightly below the boiling point of water. Being a phenol, eugenol will dissolve in aqueous alkali to form the phenolate anion. This forms the basis for its separation from its acetyl derivative. The relative amounts of eugenol and acetyleugenol and the effectiveness of the alkali extraction in separating them can be analyzed by thin-layer chromatography.

Procedure

Place 25 g of whole cloves in a 250-mL round-bottomed flask, add 100 mL of water, and set up an apparatus for steam distillation (Fig. 3) or simple distillation (see Fig. 5 in Chapter 5). In the latter procedure steam is generated *in situ*. Heat the flask strongly until boiling starts, then reduce the flame just enough to prevent foam from being carried over into the receiver. Instead of using a graduated cylinder as a receiver, use an Erlenmeyer flask and transfer the distillate periodically to the graduated cylinder; then, if any material does foam over, the entire distillate will not be contaminated. Collect 60 mL of distillate, remove the flame, and add 60 mL of water to the flask. Resume the distillation and collect an additional 60 mL of distillate.

The eugenol is very sparingly soluble in water and is easily extracted from the distillate with dichloromethane. Place the 120 mL of distillate in a 250-mL separatory funnel and extract with three 15-mL portions of dichloromethane. In this extraction very gentle shaking will fail to remove all of the product, and long and vigorous shaking will produce an emulsion of the organic layer and water. The separatory funnel will appear to have three layers in it. It is better to err on the side of vigorous shaking and draw off the clear lower layer to the emulsion line for the first two extractions. For the third extraction shake the mixture less vigorously and allow a longer period of time for the two layers to separate.

Combine the dichloromethane extracts (the aqueous layer can be discarded) and add just enough anhydrous sodium sulfate so that the drying agent no longer clumps together, but appears to be a dry powder as it settles in the solution. This may require as little as 2 g of drying agent. Swirl the flask for a minute or two to complete the drying process and then decant the solvents into an Erlenmeyer flask. It is quite easy to decant the dichloromethane from the drying agent, and therefore it will not be necessary to set up a filtration apparatus to make this separation.

Measure the volume of the dichloromethane in a dry graduated cylinder, place exactly one-fifth of the solution in a tared Erlenmeyer flask, add a wood boiling stick (easier to remove than a boiling chip), and evaporate the solution on a steam bath in a hood. The residue of crude clove oil will be used in the thin-layer chromatography analysis. From its weight, the total yield of crude clove oil can be calculated.

See Chapter 8 for the theory of acid/base extractions

To separate eugenol from acetyleugenol, extract the remaining four-fifths of the dichloromethane solution (about 30 mL) with 5% aqueous sodium hydroxide solution. Carry out this extraction three times, using 10-mL portions of sodium hydroxide each time. Dry the dichloromethane layer over anhydrous sodium sulfate, decant the solution into a tared Erlenmeyer flask, and evaporate the solvent. The residue should consist of acetyleugenol and other steam-volatile neutral compounds from cloves.

Acidify the combined aqueous extracts to pH 1 with concentrated hydrochloric acid (use Congo red or litmus paper), and then extract the liberated eugenol with three 8-mL portions of dichloromethane. Dry the combined extracts, decant into a tared Erlenmeyer flask, and evaporate the solution on a steam bath. Calculate the percent yields of eugenol and acetyleugenol on the basis of the weight of cloves used.

Analyze your products by infrared spectroscopy and thin-layer chromatography. Obtain an infrared spectrum of either eugenol or acetyleugenol using the thin film method. Compare your spectrum with the spectrum of a neighbor who has examined the other compound.

Use plastic sheets precoated with silica gel for thin-layer chromatography. One piece the size of a microscope slide should suffice. Spot crude clove oil, eugenol, and acetyleugenol (1% solutions) side by side about 5 mm from one end of the slide. The spots should be very small. Immerse the end of the slide in a dichloromethane-hexane mixture (1:2 or 2:1) about 3 mm deep in a covered beaker. After running the chromatogram observe the spots under an ultraviolet lamp or by developing in an iodine chamber. Calculate the yields of eugenol and acetyleugenol based on the dry weight of cloves.

Cleaning Up   Combine all aqueous layers, neutralize with sodium carbonate, dilute with water, and flush down the drain. Any solutions containing dichloromethane should be placed in the halogenated organic solvents container. Allow the solvent to evaporate from the sodium sulfate in the hood and then place the drying agent in the nonhazardous solid waste container.

Questions

1. Assign the peaks in the ^{1}H nmr spectrum of eugenol to specific protons in the molecule. See Fig. 4. The OH peak is at 5.1 ppm.

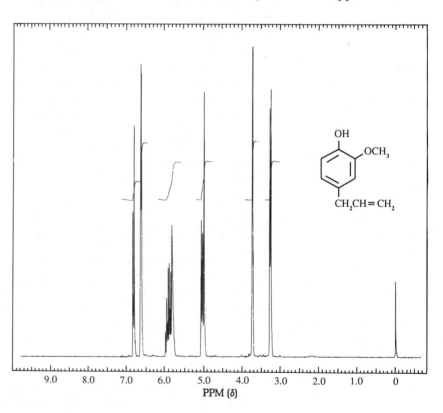

FIG. 4 ^{1}H nmr spectrum of eugenol (250 MHz).

2. A mixture of ethyl iodide (C_2H_5I, bp 72.3°C) and water boils at 63.7°C. What weight of ethyl iodide would be carried over by 1 g of steam during steam distillation?

3. Iodobenzene (C_6H_5I, bp 188°C) steam distills at a temperature of 98.2°C. How many molecules of water are required to carry over one molecule of iodobenzene? How many grams per gram of iodobenzene?

4. The condensate from a steam distillation contains 8 g of an unknown compound and 18 g of water. The mixture steam distilled at 98°C. What is the molecular weight of the unknown?

5. Bearing in mind that a sealed container filled with steam will develop a vacuum if the container is cooled, explain what will happen if a steam distillation is stopped by turning off the steam valve before opening the screw clamp on the adapter trap.

7

Vacuum Distillation and Sublimation

Prelab Exercise: Compare the operation of a barometer, used to measure atmospheric pressure, and a mercury manometer, used to measure the pressure in an evacuated system.

Part 1. Vacuum Distillation

Many substances cannot be distilled satisfactorily in the ordinary way, either because they boil at such high temperatures that decomposition occurs or because they are sensitive to oxidation. In such cases purification can be accomplished by distillation at diminished pressure. A few of the many pieces of apparatus for vacuum distillation are described on the following pages. Round-bottomed Pyrex ware and thick-walled suction flasks are not liable to collapse, but even so safety glasses should be worn at all times when carrying out this type of distillation.

Macroscale Vacuum Distillation Assemblies

Carry out vacuum distillation behind a safety shield. Check apparatus for scratches and cracks

A typical vacuum distillation apparatus is illustrated in Fig. 1. It is constructed of a round-bottomed flask (often called the "pot") containing the material to be distilled, a Claisen distilling head fitted with a hair-fine capillary mounted through a rubber tubing sleeve, and a thermometer with the bulb extending below the side arm opening. The condenser fits into a vacuum adapter that is connected to the receiver and, via heavy-walled rubber tubing, to a mercury manometer and thence to the trap and water aspirator.

Prevention of bumping

Liquids usually bump vigorously when boiled at reduced pressure and most boiling stones lose their activity in an evacuated system; it is therefore essential to make a special provision for controlling the bumping. This is done by allowing a fine stream of air bubbles to be drawn into the boiling liquid through a glass tube drawn to a hair-fine capillary. The capillary should be so fine that even under vacuum only a few bubbles of air are drawn in each second; smooth boiling will be promoted and the pressure will remain low. The capillary should extend to the very bottom of the flask and it should be slender and flexible so that it will whip back and forth in the boiling liquid. Another method used to prevent bumping, when small quantities of material are being distilled, is to introduce sufficient glass wool into the flask to fill a part of the space above the liquid.

The capillary is made in three operations. First, a 6-in. length of 6-mm glass tubing is rotated and heated in a small area over a *very hot* flame to

FIG. 1 Vacuum distillation apparatus.

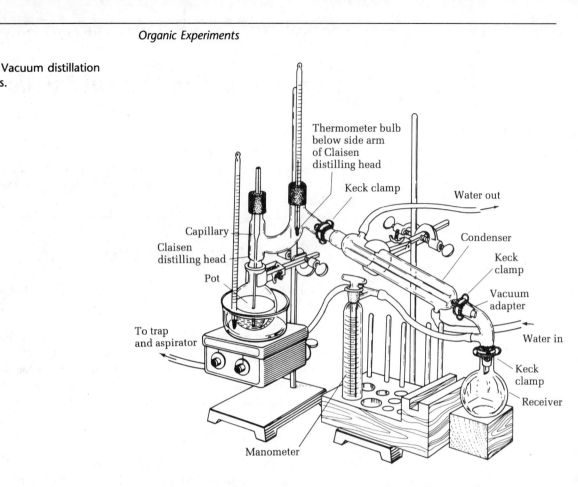

Thermometer bulb below side arm of Claisen distilling head

Keck clamp

Water out

Capillary

Claisen distilling head

Pot

Condenser

Keck clamp

Vacuum adapter

To trap and aspirator

Water in

Keck clamp

Receiver

Manometer

Making a hair-fine capillary

Handle flames in the organic laboratory with great care

Heating baths

collapse the glass and thicken the side walls as seen in Fig. 2(a). The tube is removed from the flame, allowed to cool slightly and then drawn into a thick-walled, coarse diameter capillary [Fig. 2(b)]. This coarse capillary is heated at point X over the wing top of a Bunsen burner turned 90°C. When the glass is very soft, but not soft enough to collapse the tube entirely, the tubing is lifted from the flame and without hesitation drawn smoothly and rapidly into a hair-fine capillary by stretching the hands as far as they will reach (about 2 m) [Fig. 2(c)]. The two capillaries so produced can be snapped off to the desired length. To ascertain that there is indeed a hole in the capillary, place the end beneath a low-viscosity liquid such as acetone or ether and blow in the large end. A stream of very small bubbles should be seen. If the stream of bubbles is not extremely fine make a new capillary. The flow of air through the capillary *cannot* be controlled by attaching a rubber tube and clamp to the top of the capillary tube. Should the right-hand capillary of Fig. 2(c) break when in use, it can be fused to a scrap of glass (for use as a handle) and heated again at point Y [Fig. 2(c)]. In this way the capillary can be redrawn many times.

The pot is heated with a heating bath rather than a Thermowell to promote even boiling and make possible accurate determination of the boiling point. The bath is filled with a suitable heat transfer liquid (water,

FIG. 2 Capillary for vacuum distillation.

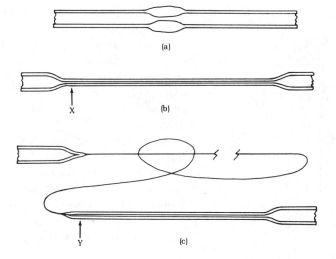

cottonseed oil, silicone oil, or molten metal) and heated to a temperature about 20°C higher than that at which the substance in the flask distills. The bath temperature is kept constant throughout the distillation. The surface of the liquid in the flask should be below that of the heating medium, for this condition lessens the tendency to bump. Heating of the flask is begun only after the system has been evacuated to the desired pressure; otherwise the liquid might boil too suddenly on reduction of the pressure.

Changing fractions

To change fractions the following must be done in sequence: Remove the source of heat, release the vacuum, change the receiver, restore the vacuum to the former pressure, resume heating.

The pressure of the system is measured with a closed-end mercury manometer. The manometer (Fig. 3) is connected to the system by turning the stopcock until the V-groove in the stopcock is aligned with the side arm. To avoid contamination of the manometer it should be connected to the system only when a reading is being made. The pressure, in mm Hg, is given by the height, in mm, of the central mercury column above the reservoir of mercury and represents the difference in pressure between the nearly perfect vacuum in the center tube (closed at the top, open at the bottom) and the large volume of the manometer, which is at the pressure of the system.

Mercury manometer

A better vacuum distillation apparatus is shown in Fig. 4. The distillation neck of the Claisen adapter is longer than other adapters and has a series of indentations made from four directions, so that the points nearly meet in the center. These indentations increase the surface area over which rising vapor can come to equilibrium with descending liquid and it then serves as a fractionating column (a Vigreux column). A column packed with a metal sponge has a great tendency to become filled with liquid (flood) at reduced pressure. The apparatus illustrated in Fig. 4 also has a fraction collector, which allows the removal of a fraction without disturbing the vacuum in the system. While the receiver is being changed the distillate

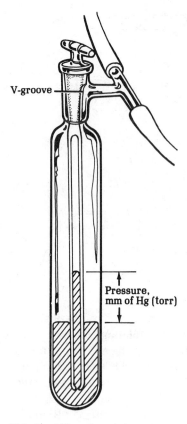

collects in the small reservoir A. The clean receiver is evacuated by another aspirator at tube B before being connected again to the system.

If only a few milliliters of a liquid are to be distilled, the apparatus shown in Fig. 5 has the advantage of low hold-up—that is, not much liquid is lost wetting the surface area of the apparatus. The fraction collector illustrated is known as a "cow." Rotation of the cow about the standard taper joint will allow four fractions to be collected without interrupting the vacuum.

A distillation head of the type shown in Fig. 6 allows fractions to be removed without disturbing the vacuum, and it also allows control of the reflux ratio (Chapter 5) by manipulation of the condenser and stopcock A. These can be adjusted to remove all material that condenses or only a small fraction, with the bulk of the liquid being returned to the distilling column to establish equilibrium between descending liquid and ascending vapor. In this way liquids with small boiling-point differences can be separated.

Microscale Vacuum Distillation Assemblies

On a truly micro scale (<10 mg) simple distillation is not practical because of mechanical losses. On a small scale (10–500 mg), distillation is still not a good method for isolating material, again because of mechanical losses; but bulb-to-bulb distillations can serve to rid a sample of all low-boiling material and nonvolatile impurities. In research, preparative-scale gas chromatography (Chapters 12, 13) and high-performance liquid chromatography

FIG. 3 Closed-end mercury manometer.

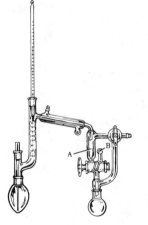

FIG. 4 Vacuum distillation apparatus with Vigreux column and fraction collector.

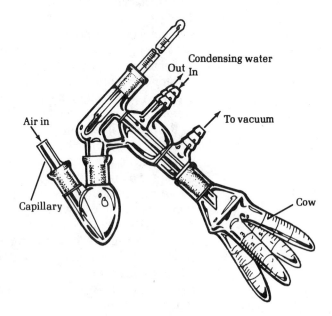

FIG. 5 Semi-micro vacuum distillation apparatus.

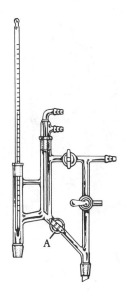

FIG. 6   Vacuum distillation head.

(HPLC) (Chapter 16) have supplanted vacuum distillation of very small quantities of material.

Fractional distillation of small quantities of liquids is even more difficult. The necessity for maintaining an equilibrium between ascending vapors and descending liquid means there will inevitably be large mechanical losses. The best apparatus for the fractional distillation of about 0.5 to 20 mL under reduced pressure is the so-called micro spinning band column, which has a hold-up of about 0.1 mL. Again, preparative-scale gas chromatography and HPLC are the best means for separating mixtures of liquids.

The Water Aspirator in Vacuum Distillation

A water aspirator in good order gives a vacuum nearly corresponding to the vapor pressure of the water flowing through it. Polypropylene aspirators give good service and are not subject to corrosion as are the brass ones. If a manometer is not available, and the assembly is free of leaks and the trap and lines clean and dry, an approximate estimate of the pressure can be made by measuring the water temperature and reading the pressure from Table 1.

TABLE 1   Vapor Pressure of Water at Different Temperatures

t (°C)	p (mm Hg)	t (°C)	p (mm Hg)	t (°C)	p (mm Hg)	t (°C)	p (mm Hg)
0	4.58	20	17.41	24	22.18	28	28.10
5	6.53	21	18.50	25	23.54	29	29.78
10	9.18	22	19.66	26	24.99	30	31.55
15	12.73	23	20.88	27	26.50	35	41.85

The Rotary Oil Pump

To obtain pressures below 10 mm Hg a mechanical vacuum pump of the type illustrated in Fig. 7(a), is used. A pump of this type in good condition can give pressures as low as 0.1 mm Hg. These low pressures are measured with a tilting type McLeod gauge [Fig. 7(b)]. When a reading is being made the gauge is tilted to the vertical position shown and the pressure is read as the difference between the heights of the two columns of mercury. Between readings the gauge is rotated clockwise 90°.

Dewar should be wrapped with tape or otherwise protected from implosion

Never use a mechanical vacuum pump before placing a mixture of dry ice and isopropyl alcohol in a Dewar flask [Fig. 7(c)] around the trap and never pump corrosive vapors (e.g., HCl gas) into the pump. Should this happen change the pump oil immediately. With care, it will give many years of good service. The dry ice trap condenses organic vapors and water vapor, both of which would otherwise contaminate the vacuum pump oil and exert enough vapor pressure to destroy a good vacuum.

FIG. 7   (a) Rotary oil pump,
(b) Tilting McLeod gauge,
(c) Vacuum system.

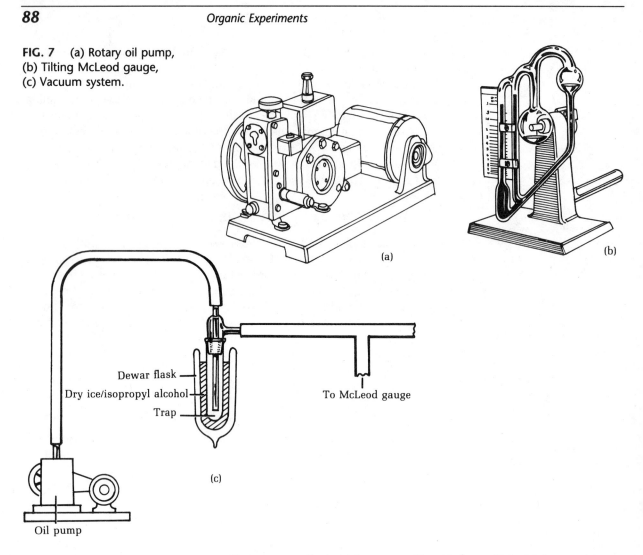

Dewar flask

Dry ice/isopropyl alcohol

Trap

To McLeod gauge

Oil pump

(c)

For an exceedingly high vacuum (5×10^{-8} mm Hg) a high-speed three-stage mercury diffusion pump is used (Fig. 8).

Relationship Between Boiling Point and Pressure

It is not possible to calculate the boiling point of a substance at some reduced pressure from a knowledge of the boiling temperature at 760 mm Hg, for the relationship between boiling point and pressure varies from compound to compound and is somewhat unpredictable. It is true, however, that boiling point curves for organic substances have much the same general disposition, as illustrated by the two lower curves in Fig. 9. These are similar and do not differ greatly from the curve for water. For substances boiling in the region 150–250°C at 760 mm Hg, the boiling point at 20 mm Hg is 100–120°C lower than at 760 mm Hg. Benzaldehyde, which is very sensitive to air oxidation at the normal boiling point of 178°C, distills at

76°C at 20 mm Hg, and the concentration of oxygen in the rarefied atmosphere is just $\frac{20}{760}$, or 3% of that in an ordinary distillation.

The curves all show a sharp upward inclination in the region of very low pressure. The lowering of the boiling point attending a reduction in pressure is much more pronounced at low than at high pressures. A drop in the atmospheric pressure of 10 mm Hg lowers the normal boiling point of an

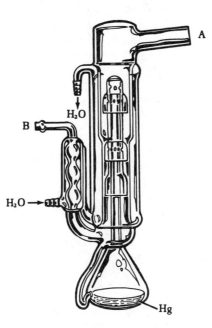

FIG. 8   High-speed three-stage mercury diffusion pump capable of producing a vacuum of 5 × 10^{-8} mm Hg. Mercury is boiled in the flask; the vapor rises in the center tube, is deflected downward in the inverted cups, and entrains gas molecules, which diffuse in from the space to be evacuated, A. The mercury condenses to a liquid and is returned to the flask; the gas molecules are removed at B by an ordinary rotary vacuum pump.

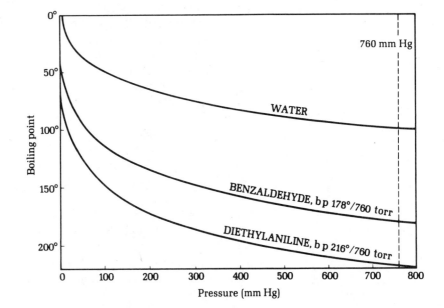

FIG. 9   Boiling point curves.

ordinary liquid by less than a degree, but a reduction of pressure from 20 to 10 mm Hg causes a drop of about 15°C in the boiling point. The effect at pressures below 1 mm is still more striking, and with development of practical forms of the highly efficient oil vapor or mercury vapor diffusion pump, distillation at a pressure of a few thousandths or ten thousandths of a millimeter has become a standard operation in many research laboratories. High-vacuum distillation, that is, at a pressure below 1 mm Hg, affords a useful means of purifying extremely sensitive or very slightly volatile substances. Table 2 indicates the order of magnitude of the reduction in boiling point attainable by operating in different ways and illustrates the importance of keeping vacuum pumps in good repair.

TABLE 2　Distillation of a (Hypothetical) Substance at Various Pressures

Method		Pressure (mm Hg)	bp (°C)
Ordinary distillation		760	250
Aspirator {	summer	25	144
	winter	15	134
	poor condition	10	124
Rotary oil pump {	good condition	3	99
	excellent condition	1	89
Mercury vapor pump		0.01	30

The boiling point of a substance at various pressures can be estimated from a pressure-temperature nomograph such as the one shown in Fig. 10. If the boiling point of a substance at 760 mm Hg is known, e.g., 300°C (Column B), and the new pressure measured, e.g., 10 mm Hg (Column C), then a straight line connecting these values on Columns B and C when extended intersects Column A to give the observed bp of 160°C. Conversely, a substance observed to boil at 50°C (Column A) at 1.0 mm Hg (Column C) will boil at approximately 212°C at atmospheric pressure (Column B).

Vacuum distillation is not confined to the purification of substances that are liquid at ordinary temperatures but often can be used to advantage for solid substances. The operation is conducted for a different purpose and by a different technique. A solid is seldom distilled to cause a separation of constituents of different degrees of volatility but rather to purify the solid. It is often possible in one vacuum distillation to remove foreign coloring matter and tar without appreciable loss of product, whereas several wasteful crystallizations might be required to attain the same purity. It is often good practice to distill a crude product and then to crystallize it. Time is saved in the latter operation because the hot solution usually requires neither filtration nor clarification. The solid must be dry and a test should be

FIG. 10 Pressure-temperature nomograph.

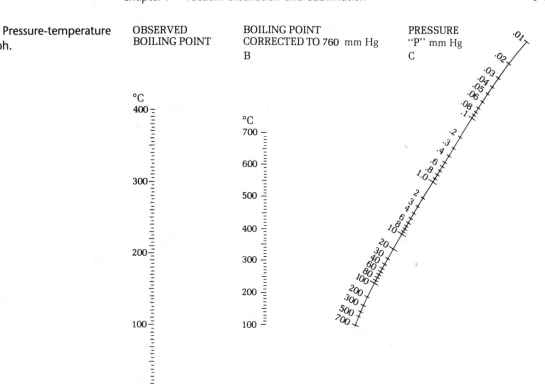

made to determine if it will distill without decomposition at the pressure of the pump available. That a compound lacks the required stability at high temperatures is sometimes indicated by the structure, but a high melting point should not be taken as an indication that distillation will fail. Substances melting as high as 300°C have been distilled with success at the pressure of an ordinary rotary vacuum pump.

It is not necessary to observe the boiling point in distillations of this kind because the purity and identity of the distillate can be checked by melting point determinations. The omission of the customary thermometer simplifies the technique. A simple and useful assembly is shown in Fig. 11. A rather stout capillary tube carrying an adjustable vent at the top is fitted into the neck of the two-bulb flask by means of a rubber stopper and the suction pump is connected through a trap at the other bulb. It is not necessary to insert in the mouth a rubber stopper of just the right size; a somewhat larger stopper may be put on backwards as shown, as it will be held in place by atmospheric pressure. The same scheme can be used for the other stopper. Water cooling is unnecessary. If some cooling of the receiving bulb is required it is best to use an air blast. Because the connection between the distilling and the receiving flask is of glass, any material that solidifies and tends to plug the side arm can be melted with a free flame. A

FIG. 11 Two-bulb flask for the distillation of solids.

heating bath should not be used; it is best to heat the flask with a rather large flame. Hold the burner in the hand and play the flame in a rotary motion around the side walls of the flask. This allows less bumping than when the flask is heated from the bottom. If there is much frothing at the start of the heating, direct the flame upon the upper walls and the neck of the flask. If the liquid froths over into the receiving bulb, the flask is tilted to such a position that this bulb can drain through the connecting tube back into the distillation bulb when suitably warmed.

At the end of the distillation the vacuum is broken by carefully opening the pinchcock, and the contents of the receiving bulb melted and poured out. This method of emptying the bulb is sometimes inadvisable because the hot, molten material may be susceptible to air oxidation. In such a case, the material is allowed to solidify and cool completely before the vacuum is broken. The solid is then chipped out with a clean knife or with a strong nickel spatula and the last traces recovered with the solvent to be used in the crystallization.

Molecular Distillation

At 1×10^{-3} mm Hg the mean free path of nitrogen molecules is 56 mm at 25°C. This means that an average N_2 molecule can travel 56 mm before bumping into another N_2 molecule. In specially designed apparatus it is possible to distill almost any volatile molecule by operating at a very low pressure and having the condensing surface close to the material being distilled. The thermally energized molecule escapes from the liquid, moves about 10 mm without encountering any other molecules, and condenses on a cold surface. A simple apparatus for molecular distillation is illustrated in Fig. 12. The bottom of the apparatus is placed on a hot surface and the distillate moves a very short distance before condensing. If it is not too viscous it will drip from the point on the glass condensing surface and run down to the receiver.

The efficiency of molecular distillation apparatus is less than one theoretical plate (see Chapter 5). It is used to remove very volatile sub-

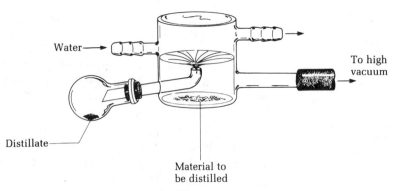

FIG. 12   Molecular distillation apparatus.

stances such as solvents and to separate volatile, high boiling substances at low temperatures from nonvolatile impurities.

In more sophisticated apparatus of quite different design the liquid to be distilled falls down a heated surface, which is again close to the condenser. This moving film prevents a buildup of nonvolatile material on the surface of the material to be distilled, which would cause the distillation to cease. Vitamin A is distilled commercially in this manner.

Part 2. Sublimation

Sublimation is the process whereby a solid evaporates from a warm surface and condenses on a cold surface, again as a solid (Figs. 13, 15). This technique is particularly useful for the small-scale purification of solids because there is so little loss of material in transfer. If the substance has the correct properties, sublimation is preferred over crystallization when the amount of material to be purified weighs less than 100 mg.

As demonstrated in the first experiment, sublimation can occur readily at atmospheric pressure. For substances with lower vapor pressures vacuum sublimation is used. At very low pressure the sublimation becomes very similar to molecular distillation, where the molecule leaves the warm solid and passes unobstructed to a cold condensing surface and condenses in the form of a solid.

Since sublimation occurs from the surface of the warm solid, impurities can accumulate and slow down or even stop the sublimation, in

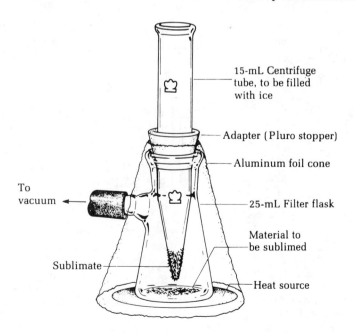

FIG. 13 Small-scale sublimation apparatus.

15-mL Centrifuge tube, to be filled with ice

Adapter (Pluro stopper)

Aluminum foil cone

To vacuum

25-mL Filter flask

Material to be sublimed

Sublimate

Heat source

which case it is necessary to open the apparatus, grind the impure solid to a fine powder, and restart the sublimation.

Sublimation is a technique that is much easier to carry out on a small scale than on a large one. It would be unusual for a research chemist to sublime more than about 10 g of material, because the sublimate tends to fall back to the bottom of the apparatus. In an apparatus such as that illustrated in Fig. 14(a), the solid to be sublimed is placed in the lower flask and connected via a lubricant-free rubber O-ring to the condenser, which in turn is connected to a vacuum pump. The lower flask is immersed in an oil bath at the appropriate temperature and the product sublimed and condensed onto the cool walls of the condenser. The parts of the apparatus are gently separated and the condenser inverted; the vacuum connection serves as a convenient funnel for product removal. For large-scale work the sublimator of Fig. 14(b) is used. The inner well is filled with a coolant (ice or dry ice). The sublimate clings to this cool surface, from which it can be removed by scraping and dissolving in an appropriate solvent.

Lyophilization

Lyophilization, also called freeze-drying, is the process of subliming a solvent, usually water, with the object of recovering the solid that remains after the solvent is removed. This technique is extensively used to recover heat- and oxygen-sensitive substances of natural origin, such as proteins, vitamins, nucleic acids, and other biochemicals from dilute aqueous solution. The aqueous solution of the substance to be lyophilized is usually

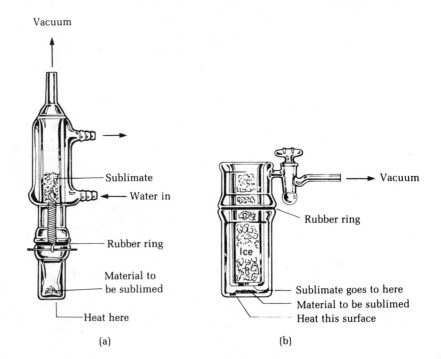

FIG. 14   (a) Mallory sublimator, (b) large vacuum sublimator.

(a)   (b)

frozen to prevent bumping and then subjected to a vacuum of about 1×10^{-3} mm Hg. The water sublimes and condenses as ice on the surface of a large and very cold condenser. The sample remains frozen during this entire process, without any external cooling being supplied, because of the very high heat of vaporization of water; thus the temperature of the sample never exceeds 0°C. Freeze-drying on a large scale is employed to make "instant" coffee, tea, soup, rice, and all sorts of dehydrated foods for backpackers.

The Kugelrohr

The Kugelrohr (Ger., bulb-tube) is a widely used piece of research apparatus that consists of a series of bulbs that can be rotated and heated under vacuum. Bulb-to-bulb distillation frees the desired compound of very low-boiling and very high-boiling or nonvolatile impurities. The crude mixture is placed in bulb A (Fig. 15) in the heated glass chamber B. At C is a shutter mechanism that holds in the heat but allows the bulbs to be moved out one-by-one as distillation proceeds. The lowest-boiling material collects in bulb D; then bulb C is moved out of the heated chamber, the temperature increased, and the next fraction collected in bulb C. Finally this process is repeated for bulb B. The bulbs rotate under vacuum using a mechanism such as the one used on rotary evaporators. The same apparatus is used for bulb-to-bulb fractional sublimation.

FIG. 15 Kugelrohr bulb-to-bulb distillation apparatus.

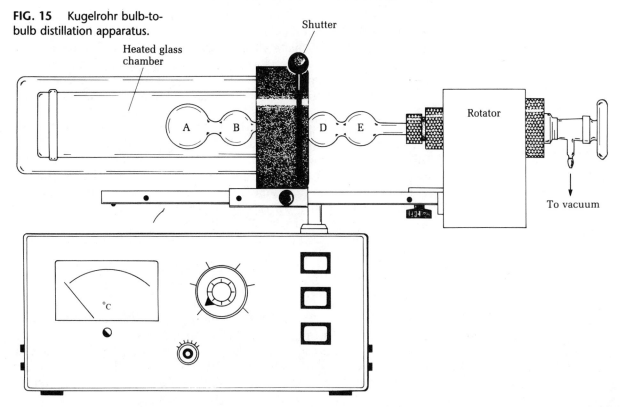

Experiment_____

Sublimation of an Unknown Substance

Apparatus and Technique

The apparatus consists of a 15-mL centrifuge tube thrust through an adapter (use a drop of glycerol to lubricate the adapter) fitted in a 25-mL filter flask.

The sample is either ground to a fine powder and uniformly distributed on the bottom of the flask or is introduced into the flask as a solution and the solvent evaporated to deposit the substance on the bottom of the filter flask. If the compound sublimes easily, care must be exercised using the latter technique to ensure that the sample does not evaporate as the last of the solvent is being removed.

The 15-mL centrifuge tube and adapter are placed in the flask so that the tip of the centrifuge tube is about 3 to 8 mm above the bottom of the flask (Fig. 13). The flask is clamped with a three-prong clamp. Many substances sublime at atmospheric pressure (see following experiment). For vacuum sublimation, the sidearm of the flask leads to an aspirator or vacuum pump for reduced pressure sublimation. The vacuum source is turned on and the centrifuge tube filled with ice and water. The ice is not added before applying the vacuum so atmospheric moisture will not condense on the tube.

The filter flask is warmed cautiously on a hot sand bath until the product just begins to sublime. The heat is maintained at that temperature until sublimation is complete. Because some product will collect on the cool upper parts of the flask, it should be fitted with a loose cone of aluminum foil to direct heat there and cause the material to collect on the surface of the centrifuge tube. Alternatively, tilt the flask and rotate it in the hot sand or use a heat gun to warm the flask.

Once sublimation is judged complete the ice water is removed from the centrifuge tube with a pipette and replaced with water at room temperature. This will prevent moisture from condensing on the product once the vacuum is turned off and the tube removed from the flask. The product is scraped from the centrifuge tube with a metal spatula onto a piece of glazed paper. It is much easier to scrape the product from a centrifuge tube than from a round-bottomed test tube. The last traces can be removed by washing off the tube with a few drops of an appropriate solvent, if that is deemed necessary.

Procedure

A very simple sublimation apparatus: an inverted stemless funnel on a watch glass

Into the bottom of a 25-mL filter flask place 50 mg of an impure unknown taken from the list in Table 3. These substances can be sublimed at atmospheric pressure although some will sublime more rapidly at reduced pressure. Close the flask with a rubber pipette bulb, and then place ice water in the centrifuge tube. Cautiously warm the flask until sublimation starts

TABLE 3   Sublimation Unknowns

Substance	Mp (°C)	Substance	Mp (°C)
1,4-Dichlorobenzene	55	Benzoic acid	122
Naphthalene	82	Salicyclic acid	159
1-Naphthol	96	Camphor	177
Acetanilide	114	Caffeine	235

Roll the filter flask in the sand bath or heat it with a heat gun to drive material from the upper walls of the flask

and then maintain that temperature throughout the sublimation. Fit the flask with a loose cone of aluminum foil to direct heat up the sides of the flask, causing the product to collect on the tube. Once sublimation is complete remove the ice water from the centrifuge tube and replace it with water at room temperature. Collect the product, determine its weight and the percent recovery, and from the melting point identify the unknown. Hand in the product in a neatly labeled vial.

Cleaning Up   Wash material from the apparatus with a minimum quantity of acetone which, except for the chloro compound, can be placed in the organic solvents container. The 1,4-dichlorobenzene solution must be placed in the halogenated organic waste container.

8

Extraction: Isolation of Caffeine from Tea and Cola Syrup

Extraction is one of humankind's oldest chemical operations. The preparation of a cup of coffee or tea involves the extraction of flavor and odor components from dried vegetable matter with hot water. Aqueous extracts of bay leaves, stick cinnamon, peppercorns, and cloves are used as food flavorings, along with alcoholic extracts of vanilla and almond. For the last century and a half, organic chemists have been extracting, isolating, purifying, and then characterizing the myriad compounds produced by plants that have been used for centuries as drugs and perfumes—substances such as quinine from cinchona bark, morphine from the opium poppy, cocaine from coca leaves, and menthol from peppermint oil. In research a Soxhlet extractor is often used (Fig. 1). The organic chemist commonly employs, in addition to solid/liquid extraction, two other types of extraction: liquid/liquid extraction and acid/base extraction.

Liquid/Liquid Extraction

After a chemical reaction has been carried out the organic product is often separated from inorganic substances by liquid/liquid extraction. For example, in the synthesis of 1-bromobutane (Chapter 15)

$$2 \; CH_3CH_2CH_2CH_2OH + 2 \; NaBr + H_2SO_4 \rightarrow 2 \; CH_3CH_2CH_2CH_2Br$$
$$+ \; 2 \; H_2O + Na_2SO_4$$

1-butanol is heated with an aqueous solution of sodium bromide and sulfuric acid to produce the product and sodium sulfate. The 1-bromobutane is isolated from the reaction mixture by extraction with ether, a solvent in which 1-bromobutane is soluble and in which water and sodium sulfate are insoluble. The extraction is accomplished by simply adding ether to the aqueous mixture and shaking it. The ether is less dense than water and floats on top; it is removed and evaporated to leave the bromo compound free of inorganic substances.

Properties of Extraction Solvents

The solvent used for extraction should have many properties of a satisfactory recrystallization solvent. It should readily dissolve the substance to be extracted; it should have a low boiling point so that it can readily be

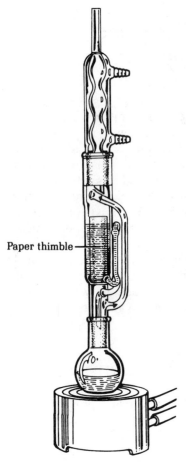

FIG. 1   Soxhlet extractor. For exhaustive extraction of solid mixtures, and even of dried leaves or seeds, the solid is packed into a filter paper thimble. Solvent vapor rises in the large diameter tube on the right, and condensed solvent drops onto the solid contained in the filter paper thimble, leaches out soluble material and, after initiating an automatic siphon, carries it to the boiling flask where nonvolatile extracted material accumulates. The same solvent is repeatedly vaporized; substances of very slight solubility can be extracted by prolonged operation.

Paper thimble

removed; it should not react with the solute or the other solvent; it should not be flammable or toxic; and it should be relatively inexpensive. In addition, it should not be miscible with water (the usual second phase). No solvent meets all these criteria, but several come close. Diethyl ether, usually referred to simply as ether, is probably the most common solvent used for extraction, but it is extremely flammable.

Ether has high solvent power for hydrocarbons and for oxygen-containing compounds and is so highly volatile (bp 34.6°) that it is easily removed from an extract at a temperature so low that even highly sensitive compounds are not likely to decompose.

Ether is useful for isolation of natural products that occur in animal and plant tissues having high water content. Although often preferred for research work because of these properties, ether is avoided in industrial processes because of the fire hazard, high solubility in water, losses in solvent recovery incident to volatility, and the oxidation on long exposure to air to a peroxide, which in a dry state may explode. Alternative water-immiscible solvents sometimes preferred, even though they do not match all the favorable properties of ether, are petroleum ether, ligroin, benzene, carbon tetrachloride, chloroform, dichloromethane, 1,2-dichloroethane, and 1-butanol. The chlorinated hydrocarbon solvents are heavier than water and hence, after equilibration of the aqueous and nonaqueous phases, the heavier lower layer is drawn off and the upper aqueous layer is extracted further and discarded. Chlorinated hydrocarbon solvents have the advantage of freedom from fire hazard, but their higher cost militates against their general use.

Distribution Coefficient

The extraction of a compound such as 1-butanol, which is slightly soluble in water as well as very soluble in ether, is an equilibrium process governed by the solubilities of the alcohol in the two solvents. The ratio of the solubilities is known as the distribution coefficient, also called the partition coefficient, k, and is an equilibrium constant with a certain value for a given substance, pair of solvents, and temperature.

To a good approximation the *concentration* of the solute in each solvent can be correlated with the *solubility* of the solute in the pure solvent, a figure that can be found readily in tables of solubility in reference books.

$$k = \frac{\text{Concentration of C in ether}}{\text{Concentration of C in water}} \cong \frac{\text{Solubility of C in ether (g/100 mL)}}{\text{Solubility of C in water (g/100 mL)}}$$

Consider a compound, A, which dissolves in ether to the extent of 12 g/100 mL and dissolves in water to the extent of 6 g/100 mL.

$$k = \frac{12 \text{ g/100 mL ether}}{6 \text{ g/100 mL water}} = 2$$

If a solution of 6 g of A in 100 mL of water is shaken with 100 mL of ether then

$$k = \frac{(x \text{ grams of A}/100 \text{ mL ether})}{(6 - x \text{ grams of A}/100 \text{ mL water})} = 2$$

from which

$$x = 4.0 \text{ g of A in the ether layer}$$

$$6 - x = 2.0 \text{ g left in the water layer}$$

It is, however, more efficient to extract the 100 mL of aqueous solution twice with 50-mL portions of ether rather than once with a 100-mL portion.

$$k = \frac{(x \text{ g of A}/50 \text{ mL})}{(6 - x \text{ g of A}/100 \text{ mL})} = 2$$

from which

$$x = 3.0 \text{ g in ether layer}$$

$$6 - x = 3.0 \text{ g in water layer}$$

Several small extractions are better than one large one

If this 3.0 g/100 mL of water is extracted once more with 50 mL of ether we can calculate that 1.5 g will be in the ether layer, leaving 1.5 g in the water layer. So two extractions with 50-mL portions of ether will extract 3.0 g + 1.5 g = 4.5 g of A, while one extraction with a 100-mL portion of ether removes only 4.0 g of A. Three extractions with $33\frac{1}{3}$-mL portions of ether would extract 4.7 g. Obviously there is a point at which the increased amount of A extracted does not repay the effort of multiple extractions, but keep in mind that several small-scale extractions are more effective than one large-scale extraction.

Acid/Base Extraction

The third type of extraction, acid/base extraction, involves carrying out simple acid/base reactions in order to separate strong organic acids, weak organic acids, neutral organic compounds, and basic organic substances. The chemistry involved is given in the equations that follow, using benzoic acid, phenol, naphthalene, and aniline as examples of the four types of compounds.

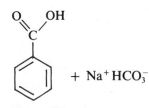
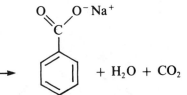

Benzoic acid
$pK_a = 4.17$
Covalent, sol. in org. solvents

Sodium benzoate
Ionic, sol. in water

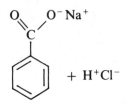
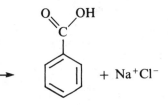

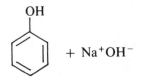
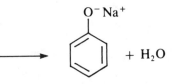

Phenol
$pK_a = 10$
Covalent, sol. in org. solvents

Sodium phenoxide
Ionic, sol. in water

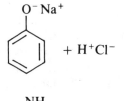

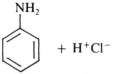

Aniline
$pK_b = 9.30$
Covalent, sol. in org. solvents

Anilinium chloride
Ionic, sol. in water

Here is the strategy: the four organic compounds are dissolved in ether. The ether solution is shaken with a saturated aqueous solution of sodium bicarbonate, a weak base. This will react only with the strong acid, benzoic acid, to form the ionic salt, sodium benzoate, which dissolves in the aqueous layer and is removed. The ether solution now contains just phenol, naphthalene, and aniline. A 10% aqueous solution of sodium hydroxide is added and the mixture shaken. The hydroxide, a strong base, will react only with the phenol, a weak acid, to form sodium phenoxide, an ionic compound that dissolves in the aqueous layer and is removed. The ether now contains only naphthalene and aniline. Shaking it with dilute hydrochloric acid removes the aniline, a base, as the ionic anilinium chloride. The aqueous layer is removed. Evaporation of the ether now leaves naphthalene, the neutral compound. The other three compounds are recovered by adding acid to the sodium benzoate and sodium phenolate and base to the anilinium chloride to regenerate the covalent compounds benzoic acid, phenol, and aniline. These operations are conveniently represented in a flow diagram (Fig. 2).

The ability to separate strong from weak acids depends on the acidity constants of the acids and the basicity constants of the bases as follows. In the first equation consider the ionization of benzoic acid, which has an equilibrium constant, K_a, of 6.8×10^{-5}. The conversion of benzoic acid to the benzoate anion in Eq. 4 is governed by the equilibrium constant, K (Eq. 5), obtained by combining the third and fourth equations.

$$C_6H_5COOH + H_2O \rightleftharpoons C_6H_5COO^- + H_3O^+ \tag{1}$$

$$K_a = \frac{[C_6H_5COO^-][H_3O^+]}{[C_6H_5COOH]} = 6.8 \times 10^{-5}, \; pK_a = 4.17 \tag{2}$$

$$K_w = [H_3O^+][OH^-] = 10^{-14} \tag{3}$$

$$C_6H_5COOH + OH^- \rightleftharpoons C_6H_5COO^- + H_2O \tag{4}$$

$$K = \frac{[C_6H_5COO^-]}{[C_6H_5COOH][OH^-]} = \frac{K_a}{K_w} = \frac{6.8 \times 10^{-5}}{10^{-14}} = 3.2 \times 10^8 \tag{5}$$

If 99% of the benzoic acid is converted to $C_6H_5COO^-$

$$\frac{[C_6H_5COO^-]}{[C_6H_5COOH]} = \frac{99}{1} \tag{6}$$

then from Eq. 5 the hydroxide ion concentration would need to be 3.2×10^{-7} M. Because saturated $NaHCO_3$ has $[OH^-] = 3 \times 10^{-4}$ M, the hydroxide ion concentration is high enough to convert benzoic acid completely to sodium benzoate.

FIG. 2   Flow diagram for the separation of a strong acid, a weak acid, a neutral compound, and a base–benzoic acid, phenol, naphthalene, and aniline.

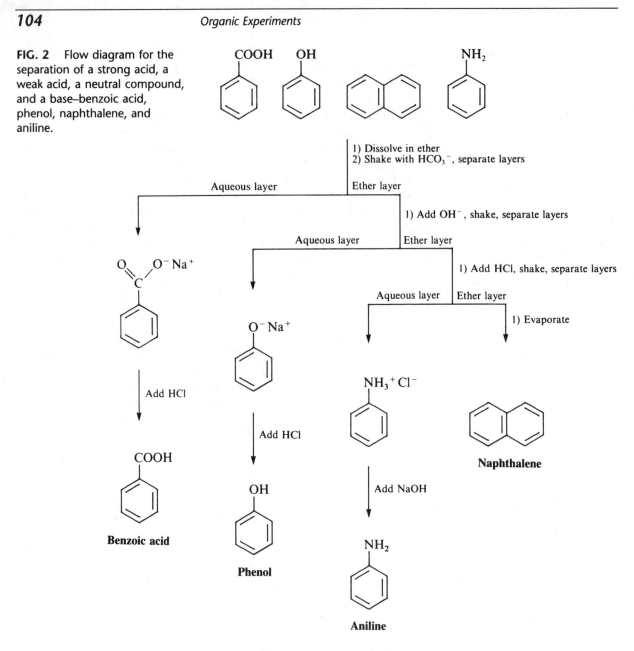

For phenol with a K_a of 10^{-10} the minimum hydroxide ion concentration that will produce the phenoxide anion in 99% conversion is 10^{-2} M. The concentration of hydroxide in 10% sodium hydroxide solution is 10^{-1} M and so phenol in strong base is entirely converted to the water-soluble salt.

Liquid/liquid extraction and acid/base extraction are employed in the majority of organic reactions because it is unusual to have the product crystallize from the reaction mixture or to be able to distill the reaction product directly from the reaction mixture.

Practical Considerations

Emulsions

Imagine trying to extract a soap solution, e.g., a nonfoaming dishwasher detergent, into an organic solvent. A few shakes with an organic solvent and you would have an absolutely intractable emulsion. An emulsion is a suspension of one liquid as droplets in another. Detergents stabilize emulsions, and so any time a detergent-like molecule happens to be in the material being extracted there is the danger of forming emulsions. Substances of this type are commonly found in nature, so one must be particularly wary of emulsion formation when making organic extracts of aqueous plant material, such as caffeine from tea. Emulsions, once formed, can be quite stable. You would be quite surprised to open your refrigerator one morning and see a layer of clarified butter floating on the top of a perfectly clear aqueous solution that had once been milk, but that is the classic example of an emulsion.

Shake gently to avoid emulsions

Prevention is the best cure for emulsions. This means shaking the solution to be extracted *very gently* until you see that the two layers will separate readily. If a bit of emulsion forms it may break simply on standing for a sufficient length of time. Making the aqueous layer highly ionic will help. Add as much sodium chloride as will dissolve and shake the mixture gently. Vacuum filtration sometimes works and, when the organic layer is the lower layer, filtration through silicone-impregnated filter paper is an aid. Centrifugation works very well for breaking emulsions. This is easy on a small scale, but often the equipment is not available for large-scale centrifugation of organic liquids.

Pressure Buildup

The heat of the hand will cause pressure buildup in an extraction mixture that contains a very volatile solvent such as ether or dichloromethane. The extraction container, be it a test tube or a separatory funnel, must be opened carefully to vent this pressure.

Sodium bicarbonate solution is often used to neutralize acids when carrying out acid/base extractions. The result is the formation of carbon dioxide, which can cause foaming and high pressure buildup. Whenever bicarbonate is used add it very gradually with thorough mixing and frequent venting of the extraction device. If a large amount of acid is to be neutralized with bicarbonate the process should be carried out in a beaker.

Removal of Water

The organic solvents used for extraction dissolve not only the compound being extracted but also water. Evaporation of the solvent then leaves the desired compound contaminated with water. At room temperature water dissolves 7.5% of ether by weight and ether dissolves 1.5% of water. But ether is virtually insoluble in water saturated with sodium chloride (36.7 g/ 100 mL). If ether that contains dissolved water is shaken with a saturated aqueous solution of sodium chloride, water will be transferred from the ether to the aqueous layer. So, strange as it may seem, ethereal extracts routinely are dried by shaking them with saturated sodium chloride solution.

Saturated aqueous sodium chloride solution removes water from ether

Solvents such as dichloromethane do not dissolve nearly as much water and so are dried over a chemical drying agent. There are many choices of chemical drying agents for this purpose and the choice of which one to use is governed by four factors: the possibility of reaction with the substance being extracted, the speed with which it removes water from the solvent, the efficiency of the process, and the ease of recovery from the drying agent.

Some very good but specialized and reactive drying agents are potassium hydroxide, anhydrous potassium carbonate, sodium metal, calcium hydride, lithium aluminum hydride, and phosphorus pentoxide. Substances that are essentially neutral and unreactive and are widely used as drying agents include anhydrous calcium sulfate (Drierite), magnesium sulfate, molecular sieves, calcium chloride, and sodium sulfate.

Drierite, $CaSO_4$

Drierite, a specially prepared form of calcium sulfate, is a fast and effective drying agent. However, it is difficult to ascertain whether enough has been used. An indicating type of Drierite is impregnated with cobalt chloride, which turns from blue to red when it is saturated with water. This works well when gases are being dried, but it should not be used for liquid extractions because the cobalt chloride dissolves in many protonic solvents.

Magnesium sulfate, $MgSO_4$

Magnesium sulfate is also a fast and fairly effective drying agent, but it is so finely powdered that it always requires careful filtration for removal.

Molecular sieves, zeolites

Molecular sieves are sodium alumino-silicates (zeolites) that have well-defined pore sizes. The 4A size adsorbs water to the exclusion of almost all organic substances and is a fast and effective drying agent, but like Drierite it is impossible to ascertain by appearance whether enough has been used. Molecular sieves in the form of 1/16-in. pellets are often used to dry solvents by simply adding them to the container.

Calcium chloride, $CaCl_2$

Calcium chloride is a very fast and effective drying agent, but may react with alcohols, phenols, amides, and carbonyl-containing compounds. Advantage is sometimes taken of this property to remove not only water from a solvent but, for example, a contaminating alcohol.

Sodium sulfate, Na_2SO_4*. The drying agent of choice for small-scale experiments.*

Sodium sulfate is the drying agent used almost exclusively in this book. It has a very high capacity for water, but is slow and not highly efficient in the removal of water. It has two advantages: first, it is granular, and the solvent being dried can be poured off (decanted) from the drying agent, or a Pasteur pipette can be used to draw off the drying agent without any filtration; and second, it has a tendency to clump together on the bottom of the container when excess water is present, but will fall freely through the solution when enough has been added. This latter property makes it easy to ascertain the correct quantity of drying agent to use. Complete drying of the solution is achieved by subsequent use of Drierite.

Experiments

Apparatus and Operations

In macroscale experiments a frequently used method of working up a reaction mixture is to dilute the mixture with water and extract with ether in a separatory funnel (Fig. 3). When the stoppered funnel is shaken to distribute the components between the immiscible solvents ether and water, pressure always develops through volatilization of ether from the heat of the hands, and the liberation of a gas (CO_2) can increase the pressure. Consequently, the funnel is grasped so that the stopper is held in place by one hand and the stopcock by the other, as illustrated. After a brief shake or two the funnel is held in the inverted position shown and the stopcock opened cautiously (with the funnel stem pointed away from nearby people) to release pressure. The mixture can then be shaken more vigorously and pressure released as necessary. When equilibration is judged to be complete, the slight, constant terminal pressure due to ether is released, the stopper is rinsed with a few drops of ether delivered by a Pasteur pipette, and the layers are allowed to separate. The organic reaction product is distributed wholly or largely into the upper ether layer, whereas inorganic salts, acids, and bases pass into the water layer, which can be drawn off and discarded. If the reaction was conducted in alcohol or some other water-soluble solvent the bulk of the solvent is removed in the water layer and the rest can be eliminated in two or three washings with 1–2 volumes of water conducted with the techniques used in the first equilibration. The separatory funnel should be supported in a ring stand as shown in Figure 4.

If acetic acid were used as the reaction solvent it would also be distributed largely into the aqueous phase, but if the reaction product is a neutral substance the residual acetic acid in the ether can be removed by one washing with excess 5% sodium bicarbonate solution. If the reaction product is a higher molecular weight acid, for example benzoic acid (C_6H_5COOH), it will stay in the ether layer, while acetic acid is being removed by repeated washing with water; the benzoic acid can then be separated from neutral by-products by extraction with sodium bicarbonate or sodium hydroxide solution and acidification of the extract. Acids of high molecular weight are extracted only slowly by sodium bicarbonate, so sodium carbonate is used in its place; however, carbonate is more prone than bicarbonate to produce emulsions. Sometimes an emulsion in the lower layer can be settled by twirling the separatory funnel by its stem. An emulsion in the upper layer can be broken by grasping the funnel by the neck and swirling it. Because the tendency to emulsify increases with

Extinguish all flames when working with ether!

Ether vapors are heavier than air and can travel along bench tops, run down drain troughs, and collect in sinks. Be extremely careful to avoid flames when working with ether.

Foaming occurs when bicarbonate is added to acid; use care.

FIG. 3 Correct positions for holding a separatory funnel when shaking.

removal of electrolytes and solvents, a little sodium chloride or hydrochloric acid solution is added with each portion of wash water. If the layers are largely clear but an emulsion persists at the interface, the clear part of the water layer can be drawn off and the emulsion run into a second funnel and shaken with fresh ether.

Before adding a liquid to the separatory funnel, check the stopcock. If it is glass see that it is properly greased, bearing in mind that too much grease will clog the hole in the stopcock and also contaminate the extract. If the stopcock is Teflon see that it is adjusted to a tight fit in the bore. Store the separatory funnel with the Teflon stopcock loosened to prevent sticking. Since Teflon has a much larger temperature coefficient of expansion than glass, a stuck stopcock can be loosened by cooling the stopcock in ice or dry ice. Do not store liquids in the separatory funnel; they often leak or cause the stopper or stopcock to freeze. In order to have sufficient room for mixing the layers, fill the separatory funnel no more than three-fourths full. Withdraw the lower layer from the separatory funnel through the stopcock and pour the upper layer out through the neck.

All too often the inexperienced chemist discards the wrong layer when using a separatory funnel. Through incomplete neutralization a desired component may still remain in the aqueous layer or the densities of the layers may change. Cautious workers save all layers until the desired product has been isolated. The organic layer is not always the top layer. If in doubt, test the layers by adding a few drops of each to water in a test tube.

1. Separation of Acidic and Neutral Substances

A mixture of equal proportions of benzoic acid, 2-naphthol, and hydroquinone dimethyl ether (1,4-dimethoxybenzene) is to be separated by extraction from ether. Note the detailed directions for extraction carefully. Prepare a flow sheet (see Fig. 2) for this sequence of operations. In the next experiment you will work out your own extraction procedure.

Dissolve 3 g of the mixture in 30 mL of ether and transfer the mixture to a 125-mL separatory funnel using a little ether to complete the transfer. Add 10 mL of water and note which layer is organic and which is aqueous. Add 10 mL of a 10% aqueous solution of sodium bicarbonate to the funnel. Stopper the funnel and cautiously mix the contents. Vent the liberated carbon dioxide and then shake the mixture thoroughly with frequent venting of the funnel. Allow the layers to separate completely and then draw off the lower layer into a 50-mL Erlenmeyer flask (labeled Flask 1). What does this layer contain?

Add 10 mL of 5% aqueous sodium hydroxide to the separatory funnel, shake the mixture thoroughly, allow the layers to separate, and draw off the lower layer into a 25-mL Erlenmeyer flask (labeled Flask 2). Add an additional 5 mL of water to the separatory funnel, shake the mixture as before, and add this to Flask 2. What does Flask 2 contain?

Add 15 mL of a saturated aqueous solution of sodium chloride to the separatory funnel, shake the mixture thoroughly, allow the layers to

---

Handwritten annotations (left margin):

1.00 g of each   (Total solid 3.00 g)

neutral   OCH₃ ... OCH₃   para   + 1,4-dimethoxybenzene

acidic   benzoic acid   COOH   pKₐ = 5

2-naphthol

125 erlenmeyer flask

dissolve in ether

$pH = -log[H^+]$
$pK_a = acidity\ constant$
$pK_b = basicity\ constant$

separatory funnel

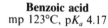
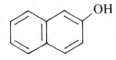

Benzoic acid
mp 123°C, pK_a 4.17

2-Naphthol
mp 123°C, pK_a 9.5

= pKₐ ≈ 10

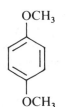

1,4-Dimethoxybenzene
(Hydroquinone dimethyl ether)
mp 57°C

Extinguish all flames when
working with ether! The best
method for removing ether is
by simple distillation.
Dispose of waste ether in
container provided.

separate, and draw off the lower layer, which can be discarded. What is the purpose of adding saturated sodium chloride solution? Carefully pour the ether layer into a 50-mL Erlenmeyer flask (labeled Flask 3) from the top of the separatory funnel, taking great care not to allow any water droplets to be transferred. Add about 4 g of anhydrous sodium sulfate to the ether extract and set it aside.

Acidify the contents of Flask 2 by dropwise addition of concentrated hydrochloric acid while testing with litmus paper. Cool the flask in an ice bath.

Cautiously add concentrated hydrochloric acid dropwise to Flask 1 until the contents are acidic to litmus and then cool the flask in ice.

Decant (pour off) the ether from Flask 3 into a tared (previously weighed) flask taking care to leave all of the drying agent behind. Wash the drying agent with additional ether to ensure complete transfer of the product. If decantation is difficult then remove the drying agent by gravity filtration (see Fig. 2 in Chapter 3). Put a boiling stick in the flask and evaporate the ether in the hood. An aspirator tube can be used for this purpose (see Fig. 5). Determine the weight of the crude *p*-dimethoxybenzene and then recrystallize it from methanol. See Chapter 3 for detailed instructions on how to carry out crystallization.

Isolate the 2-naphthol from Flask 2 employing vacuum filtration on a Hirsch funnel (see Fig. 2 in Chapter 3) and wash it on the filter with a small quantity of ice water. Determine the weight of the crude product and then recrystallize it from boiling water. Similarly isolate, weigh, and recrystallize from boiling water the benzoic acid in Flask 1. The solubility of benzoic acid in water is 1.9 g/L at 0°C and 68 g/L at 95°C.

Dry the purified products, determine their melting points and weights, and calculate the percent recovery of each substance, bearing in mind that the original mixture contained 1 g of each compound. Hand in the three products in neatly labeled vials.

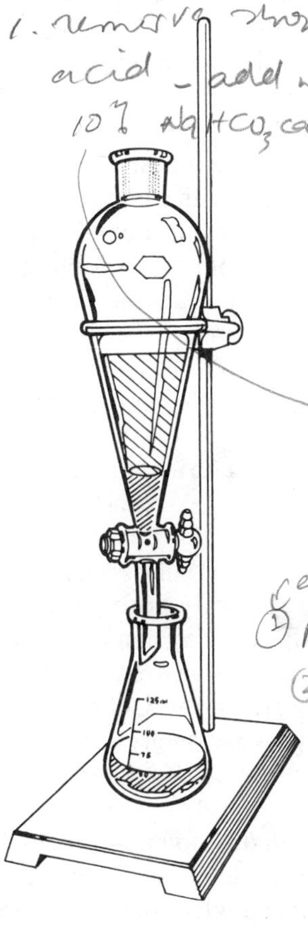

FIG. 4   Separatory funnel with Teflon stopcock.

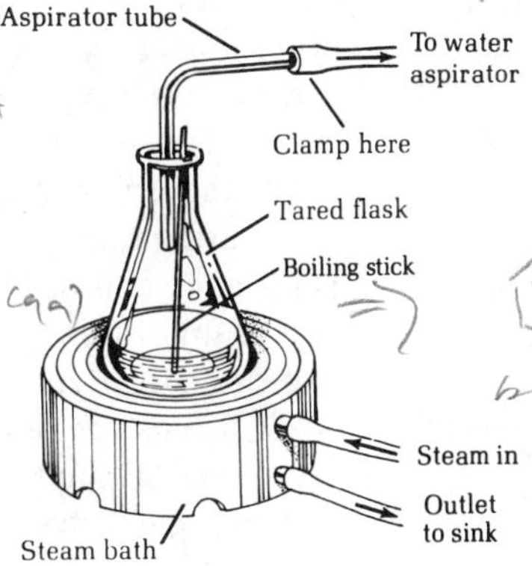

FIG. 5   Aspirator tube in use.

Aspirator tube
To water aspirator
Clamp here
Tared flask
Boiling stick
Steam in
Outlet to sink
Steam bath

Cleaning Up Combine all aqueous layers, washes, and filtrates. Dilute with water, neutralize using either sodium carbonate or dilute hydrochloric acid. Methanol filtrate and any ether go in the organic solvents container. Allow ether to evaporate from the sodium sulfate in the hood. It can be placed in the nonhazardous solid waste container.

2. Separation of a Neutral and Basic Substance

Naphthalene
mp 80°C

A mixture of equal parts of a neutral substance, naphthalene, and a basic substance, 4-chloroaniline, are to be separated by extraction from an ether solution. The base will dissolve in hydrochloric acid while the neutral naphthalene will remain in the ether solution. 4-Chloroaniline is insoluble in cold water but will dissolve to some extent in hot water and is soluble in ethanol. Naphthalene can be purified as described in Chapter 3.

Handle aromatic amines with care. Most are toxic and some are carcinogenic. Avoid breathing the dust and vapor from the solid and keep the compounds off the skin, best done by wearing gloves.

Plan a procedure for separating 2.0 g of the mixture into its components and have the plan checked by the instructor before proceeding. A flow sheet is a convenient way to present the plan. Using solubility tests, select the correct solvent or mixture of solvents to crystallize 4-chloroaniline. Determine the weights and melting points of the isolated and purified products and calculate the percent recovery of each. Turn in the products in neatly labeled vials.

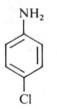

4-Chloroaniline
mp 68–71°C, pK_b 10.0

Cleaning Up Combine all aqueous filtrates and solutions, neutralize them, and flush the resulting solution down the drain with a large excess of water. Used ether should be placed in the organic solvents container, and the sodium sulfate, once the solvent has evaporated from it, can be placed in the nonhazardous solid waste container. Any 4-chloroaniline should be placed in the chlorinated organic compounds container.

3. Extraction of Caffeine from Tea

Tea and coffee have been popular beverages for centuries, primarily because they contain the stimulant caffeine. It stimulates respiration, the heart, and the central nervous system, and is a diuretic (promotes urination). It can cause nervousness and insomnia and, like many drugs, can be addictive, making it difficult to reduce the daily dose. A regular coffee drinker who consumes as few as four cups per day can experience headache, insomnia, and even nausea upon withdrawal from the drug.

Caffeine may be the most widely abused drug in the United States. During the course of a day an average person may unwittingly consume up to a gram of this substance. The caffeine content of some common foods and drugs is given in Table 1.

Caffeine belongs to a large class of compounds known as alkaloids. These are of plant origin, contain basic nitrogen, often have a bitter taste and complex structure, and usually have physiological activity. Their

TABLE 1   Caffeine Content of Common Foods and Drugs

Coffee	80 to 125 mg per cup
Coffee, decaffeinated	2 to 4 mg per cup
Tea	30 to 75 mg per cup
Cocoa	5 to 40 mg per cup
Milk chocolate	6 mg per oz
Baking chocolate	35 mg per oz
Coca-Cola	46 mg per 12 oz
Anacin, Bromo-Seltzer, Midol	32 mg per tablet
Excedrin, extra strength	65 mg per tablet
Dexatrim, Dietac, Vivarin	200 mg per tablet
Dristan	16 mg per tablet
No-Doz	100 mg per tablet

names usually end in "ine"; many are quite familiar by name if not chemical structure—nicotine, cocaine, morphine, strychnine.

Tea leaves contain tannins, which are acidic, as well as a number of colored compounds and a small amount of undecomposed chlorophyll (soluble in dichloromethane). In order to ensure that the acidic substances remain water soluble and that the caffeine will be present as the free base, sodium carbonate is added to the extraction medium.

The solubility of caffeine in water is 2.2 mg/mL at 25°C, 180 mg/mL at 80°C, and 670 mg/mL at 100°C. It is quite soluble in dichloromethane, the solvent used in this experiment to extract the caffeine from water.

Caffeine can be extracted easily from tea bags. The procedure one would use to make a cup of tea—simply "steeping" the tea with very hot water for about 7 min—extracts most of the caffeine. There is no advantage to boiling the tea leaves with water for 20 min. Since caffeine is a white, slightly bitter, odorless, crystalline solid, it is obvious that water extracts more than just caffeine. When the brown aqueous solution is subsequently extracted with dichloromethane, primarily caffeine dissolves in the organic solvent. Evaporation of the solvent leaves crude caffeine, which on sublimation yields a relatively pure product. When the concentrated tea solution is extracted with dichloromethane, emulsions can form very easily. There are substances in tea that cause small droplets of the organic layer to remain suspended in the aqueous layer. This emulsion formation results from vigorous shaking. To avoid this problem, it might seem that one would boil the tea leaves with dichloromethane first and then extract the caffeine from the dichloromethane solution with water. In fact this does not work. Boiling 25 g of tea leaves with 50 mL of dichloromethane gives only 0.05 g of residue after evaporation of the solvent. Subsequent extractions give less material. Hot water causes the tea leaves to swell and is obviously a much more efficient extraction solvent. An attempt to sublime caffeine directly from tea leaves was also unsuccessful.

Procedure

To an Erlenmeyer flask containing 25 g of tea leaves (or 10 tea bags) add 225 mL of vigorously boiling water. Allow the mixture to stand for 7 min and then decant into another Erlenmeyer flask. To the hot tea leaves add another 50 mL of hot water and then immediately decant and combine with the first extract. Very little, if any, additional caffeine is extracted by boiling the tea leaves for 20 min. Decantation works nearly as well as vacuum filtration and is much faster.

Rock the separatory funnel very gently to avoid emulsions

Cool the aqueous solution to near room temperature and extract it twice with 30 mL-portions of dichloromethane. Take great care not to shake the separatory funnel so vigorously as to cause emulsion formation, bearing in mind that if it is not shaken vigorously enough the caffeine will not be extracted into the organic layer. Use a gentle rocking motion of the separatory funnel. Drain off the dichloromethane layer on the first extraction; include the emulsion layer on the second extraction. Dry the combined dichloromethane solutions and any emulsion layer with anhydrous sodium sulfate. Add sufficient drying agent until it no longer clumps together on the bottom of the flask. Carefully decant or filter the dichloromethane solution into a tared (previously weighed) Erlenmeyer or distilling flask. Silicone-impregnated filter paper passes dichloromethane and retains water. Wash the drying agent with a further portion of solvent, and evaporate or distill the solvent. A wooden stick is better than a boiling chip to promote smooth boiling because it is easily removed once the solvent is gone. The residue of greenish-white crystalline caffeine should weigh about 0.25 g.

Dispose of used dichloromethane in the container provided

Cleaning Up   The filtrate can be diluted with water and washed down the drain. Any dichloromethane collected goes into the halogenated organic waste container. After the solvent is allowed to evaporate from the sodium sulfate in the hood, it can be placed in the nonhazardous solid waste container, otherwise it goes in the hazardous waste container. The tea leaves go in the nonhazardous solid waste container.

4. Extraction of Caffeine from Cola Syrup

Coca-Cola was originally flavored with extracts from the leaves of the coca plant and the kola nut. Coca is grown in northern South America; the Indians of Peru and Bolivia have for centuries chewed the leaves to relieve the pangs of hunger and high mountain cold. The cocaine from the leaves causes local anesthesia of the stomach. It has limited use as a local anesthetic for surgery on the eye, nose, and throat. Unfortunately it is now a widely abused illicit drug. Kola nuts contain about 3% caffeine as well as a number of other alkaloids. The kola tree is in the same family as the cacao tree from which cocoa and chocolate are obtained. Modern cola drinks do not contain cocaine; however, Coca-Cola contains 43 mg of caffeine per

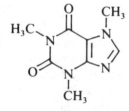

Caffeine
mp 238°C

12-oz serving. The acidic taste of many soft drinks comes from citric, tartaric, phosphoric, and benzoic acids.

Automatic soft drink dispensing machines mix a syrup with carbonated water. In the following experiment caffeine is extracted from concentrated cola syrup.

Procedure

Add 10 mL of concentrated ammonium hydroxide to a mixture of 50 mL of commercial cola syrup and 50 mL of water. Place the mixture in a separatory funnel, add 50 mL of dichloromethane, and swirl the mixture and tip the funnel back and forth for at least 5 min. Do not shake the solutions together as in a normal extraction because an emulsion will form and the layers will not separate. An emulsion is made up of droplets of one phase suspended in the other (milk is an emulsion). Separate the layers. Repeat the extraction with a second 50-mL portion of dichloromethane. From your knowledge of the density of dichloromethane and water you should be able to predict which is the top layer and which is the bottom layer. If in doubt, add a few drops of each layer to water. The aqueous layer will be soluble and the organic layer will not. Combine the dichloromethane extracts and any emulsion that has formed in a 125 mL-Erlenmeyer flask, and add anhydrous powdered sodium sulfate to remove water from the solution. Add the drying agent until it no longer clumps together at the bottom of the flask but swirls freely in solution. Swirl the flask with the drying agent from time to time over a 10-min period. Carefully decant (pour off) the dichloromethane or remove it by filtration through a fluted filter paper, add about 5 mL more solvent to the drying agent to wash it, and decant this also. Combine the dried dichloromethane solutions in a tared flask and remove the dichloromethane by distillation or evaporation on the steam bath. Remember to add a wooden boiling stick to the solution to promote even boiling. Determine the weight of the crude product.

Chlorinated solvents are toxic, insoluble in water, expensive, and should never be poured down the drain.

Crystallization of Caffeine

To recrystallize the caffeine dissolve it in 5 mL of hot acetone, transfer it with a Pasteur pipette to a small Erlenmeyer flask, and, while it is hot, add ligroin to the solution until a faint cloudiness appears. Set the flask aside and allow it to cool slowly to room temperature. This mixed solvent method of recrystallization depends on the fact that caffeine is much more soluble in acetone than ligroin, so a combination of the two solvents can be found where the solution is saturated in caffeine (the cloud point). Cool the solution containing the crystals and remove them by vacuum filtration, employing the Hirsch funnel or a very small Büchner funnel. Use a few drops of ligroin to transfer the crystals and wash the crystals. If you wish to obtain a second crop of crystals, collect the filtrate in a test tube, concentrate it to the cloud point using the aspirator tube (Fig. 5 in Chapter 3), and repeat the crystallization process.

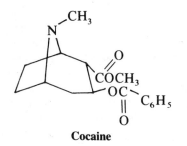

Cocaine

Cleaning Up Combine all aqueous filtrates and solutions, neutralize them, and flush the resulting solution down the drain. Used dichloromethane should be placed in the halogenated waste container, and the sodium sulfate, once the solvent has evaporated from it, can be placed in the nonhazardous solid waste container. The ligroin-acetone filtrates should be placed in the organic solvents container.

Sublimation of Caffeine

Sublimation is a fast and easy way to purify caffeine. Using the apparatus depicted in Fig. 13 in Chapter 7, sublime the crude caffeine at atmospheric pressure following the procedure in Part 2 of Chapter 7.

5. Caffeine Salicylate

Preparation of a derivative of caffeine

One way to confirm the identity of an organic compound is to prepare a derivative of it. Caffeine melts and sublimes at 238°C. It is an organic base and can therefore accept a proton from an acid to form a salt. The salt formed when caffeine combines with hydrochloric acid, like many amine salts, does not have a sharp melting point; it simply decomposes when heated. But the salt formed from salicyclic acid, even though ionic, has a sharp melting point and can thus be used to help characterize caffeine.

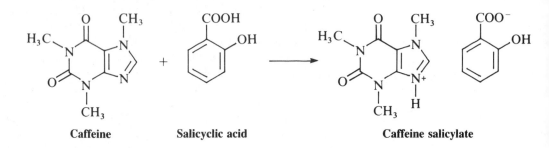

Caffeine   **Salicyclic acid**   **Caffeine salicylate**

Procedure

To 50 mg of sublimed caffeine in a tared test tube add 38 mg of salicyclic acid and 2.5 mL of dichloromethane. Heat the mixture to boiling and add petroleum ether (a poor solvent for the product) dropwise until the mixture just turns cloudy, indicating the solution is saturated. If too much petroleum ether is added then clarify it by adding a very small quantity of dichloromethane. Insulate the tube in order to allow it to cool slowly to room temperature, and then cool it in ice. The needle-like crystals are isolated by removing the solvent while the reaction tube is in the ice bath. Evaporate the last traces of solvent under vacuum and determine the weight of the derivative and its melting point. Caffeine salicylate is reported to melt at 137°C.

Crystallization from mixed solvents

Cleaning Up   Place the filtrate in the halogenated organic solvents container.

Questions_____

1. Suppose a reaction mixture, when diluted with water, afforded 300 mL of an aqueous solution of 30 g of the reaction product malononitrile, $CH_2(CN)_2$, which is to be isolated by extraction with ether. The solubility of malononitrile in ether at room temperature is 20.0 g per 100 mL, and in water is 13.3 g per 100 mL. What weight of malononitrile would be recovered by extraction with (a) three 100-mL portions of ether; (b) one 300-mL portion of ether? *Suggestion:* For each extraction let x equal the weight extracted into the ether layer. In case (a) the concentration in the ether layer is $x/100$, and in the water layer is $(30 - x)/300$; the ratio of these quantities is equal to $k = 20/13.3$.

2. Why is it necessary to remove the stopper from a separatory funnel when liquid is being drained from it through the stopcock?

3. The pK_a of p-nitrophenol is 7.15. Would you expect this to dissolve in sodium bicarbonate solution? The pK_a of 2,5-dinitrophenol is 5.15. Will it dissolve in bicarbonate solution?

4. The distribution coefficient, $k =$ (conc. in ligroin/conc. in water), between ligroin and water for solute A is 7.5. What weight of A would be removed from a solution of 10 g of A in 100 mL of water by a single extraction with 100 mL of ligroin? What weight of A would be removed by four successive extractions with 25-mL portions of ligroin? How much ligroin would be required to remove 98.5% of A in a single extraction?

5. In Experiment 1 how many moles of benzoic acid are present? How many moles of sodium bicarbonate are contained in 1 mL of a 10% aqueous solution? (A 10% solution has 1 g of solute in 9 mL of solvent.) Is the amount of sodium bicarbonate sufficient to react with all of the benzoic acid?

6. To isolate the benzoic acid from the bicarbonate solution, it is acidified with concentrated hydrochloric acid in Experiment 1. What volume of acid is needed to neutralize the bicarbonate? The concentration of hydrochloric acid is expressed in various ways on the inside back cover of this laboratory manual.

7. How many moles of 2-naphthol are in the mixture to be separated in Experiment 1? How many moles of sodium hydroxide are contained in 1 mL of 5% sodium hydroxide solution? (Assume the density of the solution is 1.0.) What volume of concentrated hydrochloric acid is needed to neutralize this amount of sodium hydroxide solution?

FIG. 6 ^{1}H nmr spectrum of caffeine (250 MHz).

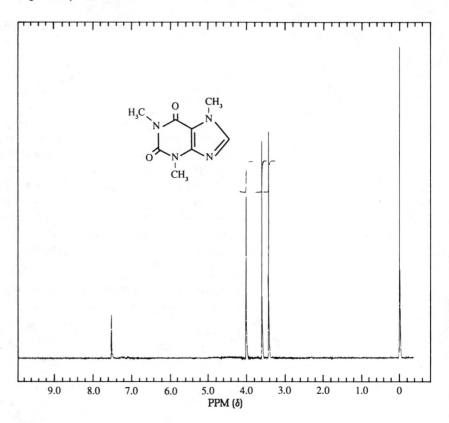

9

Thin-Layer Chromatography: Analysis of Analgesics and Isolation of Lycopene from Tomato Paste

Prelab Exercise: Compare thin-layer chromatography (TLC) with column chromatography with regard to (1) quantity of material that can be separated, (2) the speed, (3) the solvent systems, and (4) the ability to separate compounds.

TLC requires micrograms of material

Thin-layer chromatography (TLC) is a sensitive, fast, simple, and inexpensive analytical technique that will be used repeatedly in carrying out organic experiments. It is a micro technique; as little as 10^{-9} g of material can be detected, although the usual sample size is from 1 to 100×10^{-6} g.

TLC involves spotting the sample to be analyzed near one end of a sheet of glass or plastic that is coated with a thin layer of an adsorbent. The sheet, which can be the size of a microscope slide, is placed on end in a covered jar containing a shallow layer of solvent. As the solvent rises by capillary action up through the adsorbent, differential partitioning occurs between the components of the mixture dissolved in the solvent and the stationary adsorbent phase. The more strongly a given component of the mixture is adsorbed onto the stationary phase, the less time it will spend in the mobile phase and the more slowly it will migrate up the TLC plate.

Uses of Thin-layer Chromatography

1. **To determine the number of components in a mixture.** TLC affords a quick and easy method for analyzing such things as a crude reaction mixture, an extract from some plant substance, or a painkiller. Knowing the number and relative amounts of the components aids in planning further analytical and separation steps.
2. **To determine the identity of two substances.** If two substances spotted on the same TLC plate give spots in identical locations, they *may* be identical. If the spot positions are not the same the substances cannot be the same. It is possible for two closely related compounds that are not identical to have the same positions on a TLC plate.
3. **To monitor the progress of a reaction.** By sampling a reaction from time to time it is possible to watch the reactants disappear and the products appear using TLC. Thus, the optimum time to halt the reaction can be determined, and the effect of changing such variables as temperature, concentrations, and solvents can be followed without the necessity of isolating the product.
4. **To determine the effectiveness of a purification.** The effectiveness of

distillation, crystallization, extraction, and other separation and purification methods can be monitored using TLC, with the caveat that a single spot does not *guarantee* a single substance.

5. **To determine the appropriate conditions for a column chromatographic separation.** Thin-layer chromatography is generally unsatisfactory for purifying and isolating macroscopic quantities of material; however, the adsorbents most commonly used for TLC—silica gel and alumina—are used for column chromatography, discussed in the next chapter. Column chromatography is used to separate and purify up to about a gram of a solid mixture. The correct adsorbent and solvent used to carry out the chromatography can be determined rapidly by TLC.

6. **To monitor column chromatography.** As column chromatography is carried out the solvent is collected in a number of small flasks. Unless the desired compound is colored the various fractions must be analyzed in some way to determine which ones have the desired components of the mixture. TLC is a fast and effective method for doing this.

Adsorbents and Solvents

The two most common coatings for thin-layer chromatography plates are alumina, Al_2O_3, and silica gel, SiO_2. These are the same adsorbents most commonly used in column chromatography (Chapter 10) for the purification of macroscopic quantities of material. Of the two, alumina, when anhydrous, is the more active, i.e., it will adsorb substances more strongly. It is thus the adsorbent of choice when the separation involves relatively nonpolar substrates such as hydrocarbons, alkyl halides, ethers, aldehydes, and ketones. To separate the more polar substrates such as alcohols, carboxylic acids, and amines, the less active adsorbent, silica gel, is used. In an extreme situation very polar substances on alumina do not migrate very far from the starting point (give low *Rf* values) and nonpolar compounds travel with the solvent front (give high *Rf* values) if chromatographed on silica gel. These extremes of behavior are markedly affected, however, by the solvents used to carry out the chromatography. A polar solvent will carry along with it polar substrates, and nonpolar solvents will do the same with nonpolar compounds—another example of the generalization "like dissolves like."

Avoid the use of benzene, carbon tetrachloride, and chloroform. Benzene is a carcinogen, the others are suspect carcinogens.

In Table 1 are listed common solvents used in chromatography, both thin-layer and column. Only the environmentally safe solvents are listed; the polarities of such solvents as benzene, carbon tetrachloride, or chloroform can be matched by other less toxic solvents. In general these solvents are characterized by having low boiling points and low viscosities that allow them to migrate rapidly. They are listed in order of increasing polarity. A solvent more polar than methanol is seldom needed. Often just two solvents are used in varying proportions; the polarity of the mixture is a weighted average of the two. Ligroin–ether mixtures are often employed in this way.

TABLE 1   Chromatography Solvents

Solvent	Bp (°C)
Petroleum ether (pentanes)	35–60
Ligroin (hexanes)	60–80
Diethyl ether	35
Dichloromethane	40
Ethyl acetate	77
Acetone	56
2-Propanol	82
Ethanol	78
Methanol	65
Water	100
Acetic acid	118

The order in which solutes migrate on thin-layer chromatography is the same as the order of solvent polarity. The largest *Rf* values are shown by the least polar solutes. In Table 2 the solutes are arranged in order of increasing polarity.

TABLE 2   Order of Solute Migration on Chromatography

Solute	Solute
Fastest	
Alkanes	Ketones
Alkyl halides	Aldehydes
Alkenes	Amines
Dienes	Alcohols
Aromatic hydrocarbons	Phenols
Aromatic halides	Carboxylic acids
Ethers	Sulfonic acids
Esters	*Slowest*

Apparatus and Procedure

A very dilute solution (1%) of the test substance is employed for TLC.

A 1% solution of the substance to be examined is spotted onto the plate about 1 cm from the bottom end, and the plate is inserted into a beaker or a 4-oz wide-mouth bottle containing 4 mL of an organic solvent. The bottle is lined with filter paper wet with solvent to saturate the atmosphere within the container. The top is put in place and the time noted. The solvent travels rapidly up in the thin layer by capillary action, and if the substance is a pure colored compound, one soon sees a spot traveling either along with the solvent front or, more usually, at some distance behind the solvent front.

Colored compounds

And colorless ones

TLC tests for
1. Completeness of reaction
2. Purity of product
3. Side reactions

One can remove the slide, quickly mark the front before the solvent evaporates, and calculate the *Rf* value. The *Rf* value is the ratio of the distance the spot travels from the point of origin to the distance the solvent travels (Fig. 1).

If two colored compounds are present and an appropriate solvent is selected, two spots will appear.

Fortunately the method is not limited to colored substances. Any organic compound capable of being eluted from alumina will form a spot, which soon becomes visible when the solvent is let evaporate and the plate let stand in a stoppered 4-oz bottle containing a few crystals of iodine. Iodine vapor is adsorbed by the organic compound to form a brown spot. A spot should be outlined at once with a pencil because it will soon disappear as the iodine sublimes away; brief return to the iodine chamber will regenerate the spot. The order of elution and the elution power for solvents are the same as for column chromatography.

The use of commercial TLC sheets such as Eastman silica gel with fluorescent indicator (No. 13181) is strongly recommended. These poly(ethylene terephthalate) sheets are coated with silica gel using polyacrylic acid as a binder. A fluorescent indicator has been added to the silica gel so that when the sheet is observed under 254-nm ultraviolet light, spots that either quench or enhance fluorescence can be seen. Iodine can also be used to visualize spots. The coating on these sheets is only 100 microns thick, so very small spots must be applied. Unlike student-prepared plates, these coated sheets (cut to 1 × 3-in. size with scissors) give very consistent results. A light pencil mark 1 cm from the end will guide spotting. A supply of these little precut sheets makes it a simple matter to examine most of the reactions in this book for completeness of reaction, purity of product, and side reactions.

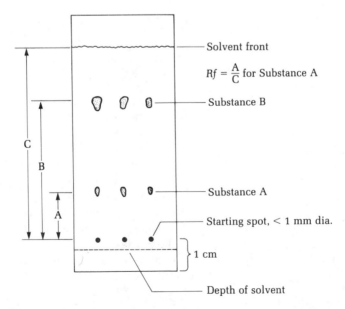

$$Rf = \frac{A}{C} \text{ for Substance A}$$

FIG. 1   Thin-layer chromatography plate.

Spotting Test Solutions

This is done with micropipettes made by drawing open-end mp capillaries in a microburner flame (Fig. 2). The bore should be of such a size that, when the pipette is dipped deep into ligroin, the liquid flows in to form a tiny thread which, when the pipette is withdrawn, does not flow out to form a drop. To spot a test solution, let a 2–3 cm column of solution flow into the pipette, hold this vertically over a coated plate, aim it at a point on the right side of the plate and about 1 cm from the bottom, and lower the pipette until the tip just touches the adsorbent and liquid flows onto the plate; withdraw when the spot is about 1 mm in diameter. Make a second 1-mm spot on the left side of the plate, let it dry, and make two more applications of the same size (1-mm) at the same place. Determine whether the large or the small spot gives the better results.

FIG. 2   Micropipette.

Making TLC Plates

The adsorbent recommended for making TLC plates is a preparation of finely divided silica gel with gypsum binder and fluorescent indicator (Aldrich 28855-1) or alumina containing plaster of paris as binder (Fluka[1]). A slurry of this material in water can be applied to microscope slides by simple techniques of dipping (A) or coating (B); for a small class, or for occasional preparation of a few slides by an individual, the more economical coating method is recommended.

(A) By Dipping.   Place 15 g of silica gel or alumina formulated for TLC in a 40 × 80-mm weighing bottle[2] and stir with a glass rod or with a magnetic stirrer while gradually pouring in 75 mL of distilled water. Stir until lumps are eliminated and a completely homogeneous slurry results. Grasp a pair of clean slides[3] at one end with the thumb and forefinger; dip them in the slurry of adsorbent (Fig. 3); withdraw with an even, unhurried stroke; and touch a corner of the slides to the mouth of the weighing bottle to allow excess fluid to return to the container. Then dry the working surface by mounting the slides above a 70-watt hot plate, using as support two pairs of 1 × 7-cm strips of blotting paper (or a pair of applicator sticks). Drying takes about a minute and a half and is evident by inspection. Remove the plate with a forceps as soon as it is dry, for cooling takes longer. If you dip

FIG. 3   Preparation of TLC plates by dipping. The two microscope slides are grasped at one end, dipped into the silica gel slurry, removed and separated, and dried and activated by heating.

1. Aluminumoxid Fluka für Dünnschicht Chromatographie, Typ D5, Fluka AG, Buchs, S. G. Switzerland; Fluka Alumina is distributed by Tridorn Chemical, Inc., 255 Oser Av., Hauppauge, N.Y. 11787.

2. Kimble Exax, 15146.

3. The slides should be washed with water and a detergent, rinsed with tap water and then with distilled water, and placed on a clean towel to dry. Dry slides should be touched only on the sides, for a touch on the working surface may spoil a chromatogram.

and dry 8–10 slides in succession, finished plates will be ready for use when you are through. Alternatively, dry the slides in an oven at 110°C. This heating activates the alumina. Coated slides dried at room temperature do not yield spots as intense as those that have been heat-dried.

Keep the storage bottle or adsorbent closed when not in use. Note that this is not a stable emulsion, but that the solid settles rapidly on standing. Stir thoroughly with a rod before each reuse.

Plates ready for use, as well as clean dry slides, are conveniently stored in a microscope slide box (25 slides).

(B) By Coating. Place 2 g of silica gel or alumina and 10 mL of distilled water in a 25-mL Erlenmeyer flask, stopper the flask, and shake to produce an even slurry. Keep the flask stoppered when not in use. Place a clean, dry slide on a block of wood or a box, with the slide projecting about 1 cm on the left-hand side (if you are right-handed) so that it can be grasped easily on the two sides. Swirl the flask to mix the slurry and draw a portion into a medicine dropper. Hold the dropper vertically and, starting at the right end of the slide, apply emulsion until the entire upper face is covered; make further applications to repair pin holes and eliminate bubbles. Grasp the left end of the slide with a forceps and even the emulsion layer by tilting the slide to the left to cause a flow, and then to the right; tilt again to the front and to the rear. Dry the slide on a hot plate as you did when preparing slides by procedure A. The adsorbent should be about 0.25 mm thick.

Visualization of the Chromatogram

Iodine chamber

If the substance being chromatographed is colored then it is possible to detect the components visually. Colorless substances can be detected by placing the dry TLC plate in a jar containing a few crystals of iodine. The iodine vapor will be preferentially adsorbed by the substances on the plate and they will appear as brown spots on a lighter-colored background. The plate is removed from the jar and the outline of the spots traced lightly in pencil because the iodine will soon evaporate.

Don't look into lamp

Plates that have been impregnated with a fluorescent indicator will show dark spots for the compounds under an ultraviolet light due to quenching of the fluorescence by the substance on the plate. Again trace the spots lightly in pencil while the plate is under the uv light. Don't look directly into the light; it will damage the eyes.

A large number of specialized spray reagents have been developed that give specific colors for certain types of compounds, and there is a large amount of literature on the solvents and adsorbents to use for the separation of given types of material.[4]

---

4. Egon Stahl, "Thin-Layer Chromatography," Springer Verlag, New York, 1969.

Experiments

1. Analgesics

Analgesics are substances that relieve pain. The most common of these is aspirin, a component of more than 100 nonprescription drugs. In Chapter 26, the history and background of this most popular drug is discussed. In the present experiment analgesic tablets will be analyzed by thin-layer chromatography to determine which analgesics they contain and whether they contain caffeine, which is often added to counteract the sedative effects of the analgesic.

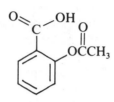

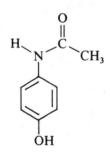

Aspirin
Acetylsalicyclic acid

Acetaminophen
4-Acetamidophenol

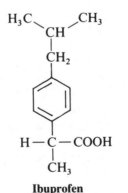

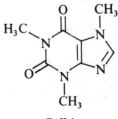

Ibuprofen
2-(4-Isobutylphenyl)propionic acid

Caffeine

In addition to aspirin and caffeine the most common components of analgesics are, at present, acetaminophen and ibuprofen (Notrin). Phenacetin, the P of the APC tablet and a former component of Empirin, has been removed from the market because of deleterious side effects. In addition to one or more of these substances, each tablet contains a binder, often starch or silica gel. And to counteract the acidic properties of aspirin an inorganic buffering agent is added to some analgesics. Inspection of labels will reveal that most cold remedies and decongestants contain both aspirin and caffeine in addition to the primary ingredient.

The usual strategy for identifying an unknown by TLC is to run chromatograms of known substances (the standards) and the unknown at the same time. If the unknown has one or more spots that correspond to spots with the same *Rf*'s as the standards, then those substances are probably present.

Proprietary drugs that contain one or more of the common analgesics and sometimes caffeine are sold under the names of Bayer Aspirin, Anacin, Datril, Advil, Excedrin, Extra Strength Excedrin, Tylenol, and Vanquish.

Procedure

Following the procedure outlined above, draw a light pencil line about 1 cm from the end of a chromatographic plate and on this line spot aspirin, acetaminophen, ibuprofen, and caffeine, which are available as reference standards. Make each spot as small as possible, preferably less than 0.5 mm in diameter. Examine the plate under the uv light to see that enough of the compound has been applied; if not, add more. On a separate plate run the unknown and one or more of the standards.

The unknown sample is prepared by crushing a part of a tablet, adding this powder to a test tube or small vial along with an appropriate amount of ethanol, and then mixing the suspension. Not all of the tablet will dissolve, but enough will go into solution to spot the plate. The binder—starch or silica—will not dissolve. Try to prepare a 1% solution of the unknown.

Use as the solvent for the chromatogram a mixture of 95% ethyl acetate and 5% acetic acid (Fig. 4 or 5). After the solvent has risen to near the top of the plate, mark the solvent front with a pencil, remove the plate from the developing chamber, and allow the solvent to dry. Examine the plate under uv light to see the components as dark spots against a bright green-blue background. Outline the spots with a pencil. The spots can also be visualized by putting the plate in an iodine chamber made by placing a few crystals of iodine in the bottom of a capped 4-oz jar. Calculate the *Rf* values for the spots and identify the components in the unknown.

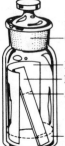

4-oz Wide-mouth bottle (glass stoppered or screw capped)

Paper liner
Microscope slide
Solvent front

4 mL of solvent

FIG. 4   A method of developing thin-layer chromatographic plates.

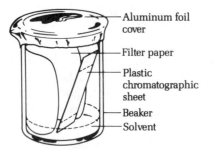

Aluminum foil cover

Filter paper

Plastic chromatographic sheet

Beaker
Solvent

FIG. 5   An alternate method for developing TLC plates.

Cleaning Up Solvents should be placed in the organic solvents container, and dry, used chromatographic plates can be discarded in the non-hazardous solid waste container.

2. Plant Pigments

The botanist Michael Tswett discovered the technique of chromatography and applied it, as the name implies, to colored plant pigments. The leaves of plants contain, in addition to chlorophyll-a and -b, other pigments that are revealed in the fall when the leaf dies and the chlorophyll rapidly decomposes. Among the most abundant of the other pigments are the carotenoids, which include the carotenes and their oxygenated homologs, the xanthophylls. The bright orange β-carotene is the most important of these because it is transformed in the liver to Vitamin A, which is required for night vision.

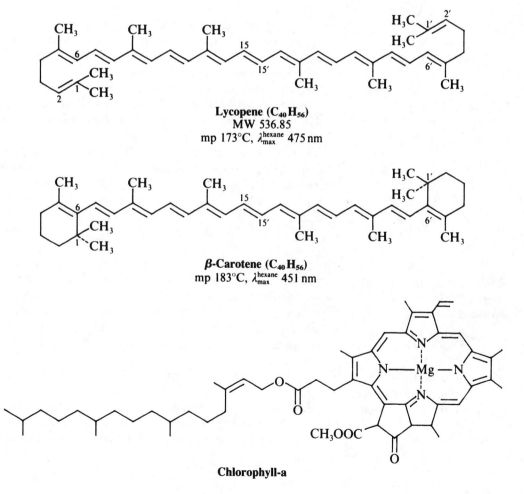

Lycopene (C$_{40}$H$_{56}$)
MW 536.85
mp 173°C, λ_{max}^{hexane} 475 nm

β-Carotene (C$_{40}$H$_{56}$)
mp 183°C, λ_{max}^{hexane} 451 nm

Chlorophyll-a

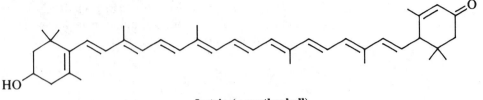

Lutein (a xanthophyll)
3′-Dehydrolutein (a xanthophyll)

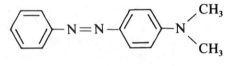

Butter Yellow
(carcinogenic)

Cows eat fresh, green grass that contains carotene, but they do not metabolize the carotene entirely, and so it ends up in their milk. Butter made from this milk is therefore yellow. In the winter the silage cows eat does not contain carotene because it readily undergoes air oxidation, and the butter made at that time is white. For some time an azo dye called Butter Yellow was added to winter butter to give it the accustomed color, but the dye was found to be a carcinogen. Now winter butter is colored with synthetic carotene, as is all margarine.

Lycopene, the red pigment of the tomato, is a C_{40}-carotenoid made up of eight isoprene units. β-Carotene, the yellow pigment of the carrot, is an isomer of lycopene in which the double bonds at C_1—C_2 and C_1'—C_2' are replaced by bonds extending from C_1 to C_6 and from C_1' to C_6' to form rings. The chromophore in each case is a system of eleven all-*trans* conjugated double bonds; the closing of the two rings renders β-carotene less highly pigmented than lycopene.

Lycopene and β-carotene from tomato paste and strained carrots

Fresh tomato fruit is about 96% water, and R. Willstätter and H. R. Escher isolated from this source 20 mg of lycopene per kg of fruit. They then found a more convenient source in commercial tomato paste, from which seeds and skin have been eliminated and the water content reduced by evaporation in vacuum to a content of 26% solids, and isolated 150 mg of lycopene per kg of paste. The expected yield in the present experiment is 0.075 mg.

As an interesting variation, try extraction of lycopene from commercial catsup

A jar of strained carrots sold as baby food serves as a convenient source of β-carotene. The German investigators isolated 1 g of β-carotene per kg of "dried" shredded carrots of unstated water content.

The following procedure calls for dehydration of tomato or carrot paste with ethanol and extraction with dichloromethane, an efficient solvent for lipids.

Procedure

In a small mortar grind 2 g of green or brightly colored fall leaves (don't use ivy or waxy leaves) with 10 mL of ethanol, pour off the ethanol, which serves to break up and dehydrate the plant cells, and grind the leaves successively with three 1-mL portions of dichloromethane that are decanted or withdrawn with a Pasteur pipette and placed in a test tube. The pigments of interest are extracted by the dichloromethane. Alternatively, place 0.5 g of carrot paste (baby food) or tomato paste in a test tube, stir and shake the paste with 3 mL of ethanol until the paste has a somewhat dry or fluffy appearance, remove the ethanol, and extract the dehydrated paste with three 1-mL portions of dichloromethane. Stir and shake the plant material with the solvent in order to extract as much of the pigments as possible.

These pigments are very sensitive to light-catalyzed photochemical air oxidation. Work quickly, keep containers stoppered where possible, and protect solutions from undue exposure to light.

Fill the tube containing the dichloromethane extract from leaves or vegetable paste with a saturated sodium chloride solution and shake the mixture. Remove the aqueous layer and to the dichloromethane solution add anhydrous sodium sulfate until the drying agent no longer clumps together. Shake the mixture with the drying agent for about 5 min and then withdraw the solvent with a Pasteur pipette and place it in a test tube. Add to the solvent a few pieces of Drierite to complete the drying process. Gently stir the mixture for about 5 min, transfer the solvent to a test tube, wash off the drying agent with more solvent, and then evaporate the combined dichloromethane solutions under a stream of nitrogen while warming the tube in the hand or in a beaker of warm water. Carry out this evaporation in the hood.

*These hydrocarbons air-oxidize with **great** rapidity*

Immediately cork the tube filled with nitrogen and then add a drop or two of dichloromethane to dissolve the pigments for TLC analysis. Carry out the analysis without delay by spotting the mixture on a TLC plate about 1 cm from the bottom and 8 mm from the edge. Make one spot concentrated by repeatedly touching the plate, but ensure that the spot is as small as possible—less than 1.0 mm in diameter. The other spot can be of lower concentration. Develop the plate with 70:30 hexane:acetone. With other plates try cyclohexane and toluene as eluents and also hexane/ethanol mixtures of various compositions. The container in which the chromatography is carried out should be lined with filter paper that is wet with the solvent so the atmosphere in the container will be saturated with solvent vapor. On completion of elution, mark the solvent front with a pencil and outline the colored spots. Examine the plate under the uv light. Are any new spots seen? Report colors and *Rf* values for all of your spots, and identify each as lycopene, carotene, chlorophyll, or xanthophyll.

Cleaning Up   The ethanol used for dehydration of the plant material can be flushed down the drain along with the saturated sodium chloride solution. Recovered and unused dichloromethane should be placed in the halogenated organic waste container. The solvents used for TLC should be placed in the organic solvents container. The drying agents, once free of solvents, can be placed in the nonhazardous solid waste container along with the used plant material and TLC plates.

3. Colorless Compounds

You are now to apply the thin-layer technique to a group of colorless compounds. The spots may be visualized under an ultraviolet light if the plates have been coated with a fluorescent indicator, or chromatograms may be developed in a 4-oz bottle containing crystals of iodine. During development, spots appear rapidly, but remember that they also disappear rapidly. Therefore, outline each spot with a pencil immediately on withdrawal of the plate from the iodine chamber. Solvents suggested are as follows:

Cyclohexane	Toluene (3 mL)-Dichloromethane (1 mL)
Toluene	9:1 Toluene-Methanol (use 4 mL)

Make your own selections

Compounds for trial are to be selected from the following list (all 1% solutions in toluene):

1. Anthracene*
2. Cholesterol
3. 2,7-Dimethyl-3,5-octadiyne-2,7-diol
4. Diphenylacetylene
5. *trans,trans*-1,4-Diphenyl-1,3-butadiene*
6. *p*-Di-*t*-butylbenzene
7. 1,4-Di-*t*-butyl-2,5-dimethoxybenzene
8. *trans*-Stilbene
9. 1,2,3,4-Tetraphenylnaphthalene*
10. Tetraphenylthiophene
11. *p*-Terphenyl*
12. Triphenylmethanol
13. Triptycene
    *(Fluorescent under uv light.)

Except for tetraphenylthiophene, the structures for all of these molecules will be found in this book

It is up to you to make selections and to plan your own experiments. Do as many as time permits. One plan would be to select a pair of compounds estimated to be separable and which have *Rf* values determinable with the same solvent. One can assume that a hydroxyl compound will travel less rapidly with a hydrocarbon solvent than a hydroxyl-free compound, and so

Preliminary trials on used plates

you will know what to expect if the solvent contains a hydroxylic component. An aliphatic solvent should carry along an aromatic compound with aliphatic substituents better than one without such groups. However, instead of relying on assumptions, you can do brief preliminary experiments on used plates on which previous spots are visible or outlined. If you spot a pair of compounds on such a plate and let the solvent rise about 3 cm from the starting line before development, you may be able to tell if a certain solvent is appropriate for a given sample. If so, run a complete chromatogram on the two compounds on a fresh plate. If separation of the two seems feasible, put two spots of one compound on a plate, let the solvent evaporate, and put spots of the second compound over the first ones. Run a chromatogram and see if you can detect two spots in either lane (with colorless compounds, it is advisable not to attempt a three-lane chromatogram until you have acquired considerable practice and skill).

Cleaning Up   Solvents should be placed in the organic solvents container, and dry, used chromatographic plates can be discarded in the nonhazardous solid waste container.

Discussion

If you have investigated hydroxylated compounds, you doubtless have found that it is reasonably easy to separate a hydroxylated from a nonhydroxylated compound, or a diol from a mono-ol. How, by a simple reaction followed by a thin-layer chromatogram, could you separate cholesterol from triphenylmethanol? Heating a sample of each with acetic anhydride-pyridine for 5 min on the steam bath, followed by chromatography, should do it. A first trial of a new reaction leaves questions about what has happened and how much, if any, starting material is present. A comparative chromatogram of reaction mixture with starting material may tell the story. How crude is a crude reaction product? How many components are present? The thin-layer technique may give the answers to these questions and suggest how best to process the product. A preparative column chromatogram may afford a large number of fractions of eluent (say 1 to 30). Some fractions probably contain nothing and should be discarded, while others should be combined for evaporation and workup. How can you identify the good and the useless fractions? Take a few used plates and put numbered circles on clean places of each; spot samples of each of the fractions; and, without any chromatography, develop the plates with iodine. Negative fractions for discard will be obvious and the pattern alone of positive fractions may allow you to infer which fractions can be combined. Thin-layer chromatograms of the first and last fractions of each suspected group would then show whether or not your inferences are correct.

Fluorescence

Do not look into a uv lamp

Four of the compounds listed in Section 3 are fluorescent under ultraviolet light, and such compounds give colorless spots that can be picked up on a chromatogram by fluorescence (after removal from the uv-absorbing glass bottle). If a uv-light source is available, spot the four compounds on a used plate and observe the fluorescence. Take this opportunity to examine a white shirt or handkerchief under uv light to see if it contains a brightener, that is, a fluorescent white dye or optical bleach. These substances are added to counteract the yellow color that repeated washing gives to cloth. Brighteners of the type of Calcofluor White MR, a sulfonated *trans*-stilbene derivative, are commonly used in detergent formulations for cotton; the substituted coumarin derivative formulated is typical of brighteners used for nylon, acetate, and wool. Detergents normally contain 0.1–0.2% of optical bleach. The amount of dye on a freshly laundered shirt is approximately 0.01% of the weight of the fabric.

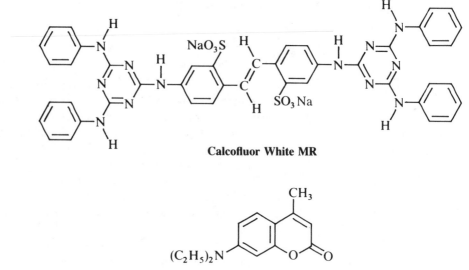

Calcofluor White MR

7-Diethylamino-4-methylcoumarin

Questions

1. Why might it be very difficult to visualize the separation of *cis*- and *trans*-2-butene by thin-layer chromatography?

2. What error is introduced into the determination of an *Rf* value if the top is left off of the developing chamber?

3. What problem will ensue if the level of the developing liquid is higher than the applied spot in a TLC analysis?

4. In what order (from top to bottom) would you expect to find naphthalene, butyric acid, and phenyl acetate on a silica gel TLC plate developed with dichloromethane?

5. In carrying out an analysis of a mixture, what do you expect to see when the TLC plate has been allowed to remain in the developing chamber too long, so that the solvent front has reached the top of the plate?

6. Arrange the following in order of increasing R_f on thin-layer chromatography: acetic acid, acetaldehyde, 2-octanone, decane, and 1-butanol.

7. Why must the spot applied to a TLC plate be above the level of the developing solvent?

8. What will be the result of applying too much compound to a TLC plate?

9. Why is it necessary to run TLC in a closed container and to have the interior vapor saturated with the solvent?

10. What will be the appearance of a TLC plate if a solvent of too low polarity is used for the development? too high polarity?

11. A TLC plate showed two spots of R_f 0.25 and 0.26. The plate was removed from the developing chamber, dried carefully, and returned to the developing chamber. What would you expect to see after the second development was complete?

10

Column Chromatography: Acetyl Ferrocene, Cholesteryl Acetate, and Fluorenone

Column chromatography is one of the most useful methods for the separation and purification of both solids and liquids when carrying out small-scale experiments. It becomes expensive and time-consuming, however, when more than about 10 g of material must be purified.

The application in the present experiment is typical: a reaction is carried out, it does not go to completion, and so column chromatography is used to separate the product from starting material, reagents, and by-products.

The theory of column chromatography is analogous to that of thin-layer chromatography. The most common adsorbents—silica gel and alumina—are the same ones used in TLC. The sample is dissolved in a small quantity of solvent (the eluent) and applied to the top of the column. The eluent, instead of rising by capillary action up a thin layer, flows down through the column filled with the adsorbent. Just as in TLC, there is an equilibrium established between the solute adsorbed on the silica gel or alumina and the eluting solvent flowing down through the column. Under some conditions the solute may be partitioning between an adsorbed solvent and the elution solvent; the partition coefficient, just as in the extraction process, determines the efficiency of separation in chromatography. The partition coefficient is determined by the solubility of the solute in the two phases, as was discussed in the extraction experiment (Chapter 8).

Three mutual interactions must be considered in column chromatography: the polarity of the sample, the polarity of the eluting solvent, and the activity of the adsorbent.

Adsorbent

A large number of adsorbents have been used for column chromatography—cellulose, sugar, starch, inorganic carbonates—but most separations employ alumina (Al_2O_3) and silica gel (SiO_2). Alumina comes in three forms: acidic, neutral, and basic. The neutral form of Brockmann Activity II or III, 150 mesh, is most commonly employed. The surface area of this alumina is about 150 m^2/g. Alumina as purchased will usually be Activity I, meaning it will strongly adsorb solutes. It must be deactivated by adding water, shaking, and allowing the mixture to reach equilibrium over an hour or so. The amount of water needed to achieve certain activities is given in Table 1.

TABLE 1 Alumina Activity

Brockmann Activity	I	II	III	IV	V
Percent by weight of water	0	3	6	10	15

The activity of the alumina on TLC plates is usually about III. Silica gel for column chromatography, 70–230 mesh, has a surface area of about 500 m^2/g and comes in only one activity.

Solvents

The elutropic series for a number of solvents is given in Table 2. The solvents are arranged in increasing polarity, with *n*-pentane the least polar. This is the order of ability of these solvents to dissolve polar organic compounds and to dislodge a polar substance adsorbed onto either silica gel or alumina, with *n*-pentane having the lowest solvent power.

As a practical matter the following sequence of solvents is recommended in an investigation of unknown mixtures: elute first with petroleum ether; then ligroin, followed by ligroin containing 1%, 2%, 5%, 10%, 25%, and 50% ether; pure ether; ether and dichloromethane mixtures, followed by dichloromethane and methanol mixtures. A sudden change in solvent polarity will cause heat evolution as the alumina or silica gel adsorbs the new solvent. This will cause undesirable vapor pockets and cracks in the column.

Petroleum ether: mostly isomeric pentanes. Ligroin: mostly isomeric hexanes.

TABLE 2 Elutropic Series for Solvents

n-Pentane (first)
Petroleum ether
Cyclohexane
Ligroin
Carbon disulfide
Ethyl ether
Dichloromethane
Tetrahydrofuran
Dioxane
Ethyl acetate
2-Propanol
Ethanol
Methanol
Acetic acid (last)

Solutes

The ease with which different classes of compounds elute from a column is indicated in Table 3. The order is similar to that of the eluting solvents—another application of "like dissolves like."

TABLE 3 Elution Order for Solutes

Alkanes (first)
Alkenes
Dienes
Aromatic hydrocarbons
Ethers
Esters
Ketones
Aldehydes
Amines
Alcohols
Phenols
Acids (last)

Sample and Column Size

In general the amount of alumina or silica gel used should weigh at least 30 times as much as the sample, and the column, when packed, should have a height at least 10 times the diameter. The density of silica gel is 0.4 g/mL and the density of alumina is 0.9 g/mL, so the optimum size for any column can be calculated.

Packing the Chromatography Column

Uniform packing of the chromatography column is critical to the success of this technique. The sample is applied as a pure liquid or, if it is a solid, as a very concentrated solution in the solvent that will dissolve it best, regardless of polarity. As elution takes place, this narrow band of sample will separate into several bands corresponding to the number of components in the mixture and their relative polarities and molecular weights. It is essential that the components move through the column as a narrow horizontal band in order to come off the column in the least volume of solvent and not overlap with other components of the mixture. Therefore, the column should be vertical and the packing should be perfectly uniform, without voids caused by air bubbles.

The preferred method for packing silica gel and alumina columns is the slurry method, whereby a slurry of the adsorbent and the first eluting solvent is made and poured into the column. When nothing is known about the mixture being separated, the column is prepared in petroleum ether, the least polar of the eluting solvents.

Procedure

Extinguish all flames; work in laboratory hood

The column can be prepared using a 50-mL burette such as the one shown in Fig. 1, or using the less expensive and equally satisfactory chromatographic tube shown in Fig. 2, in which the flow of solvent is controlled by a screw pinchclamp. Weigh out the required amount of silica gel (12.5 g in the first experiment), close the pinchclamp on the tube, and fill about half full with 90:10 ligroin–ether. With a wooden dowel or glass rod push a small plug of glass wool through the liquid to the bottom of the tube, dust in through a funnel enough sand to form a 1-cm layer over the glass wool, and level the surface by tapping the tube. Unclamp the tube. With the right hand grasp both the top of the tube and the funnel so that the whole assembly can be shaken to dislodge silica gel that may stick to the walls, and with the left

Dry packing

hand pour in the silica gel slowly (Fig. 3) while tapping the column with a rubber stopper fitted on the end of a pencil. If necessary, use a Pasteur pipette full of 90:10 ligroin–ether to wash down any silica gel that adheres to the walls of the column above the liquid. When the silica gel has settled, add a little sand to provide a protective layer at the top. Open the pinchclamp, let the solvent level fall until it is just a little above the upper layer of sand, and then stop the flow.

Slurry packing

Alternatively, the silica gel can be added to the column (half filled with ligroin) by slurrying the silica gel with 90:10 ligroin–ether in a beaker. The

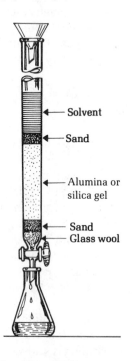

— Solvent

— Sand

— Alumina or silica gel

— Sand
— Glass wool

FIG. 1 Macroscale chromatographic column.

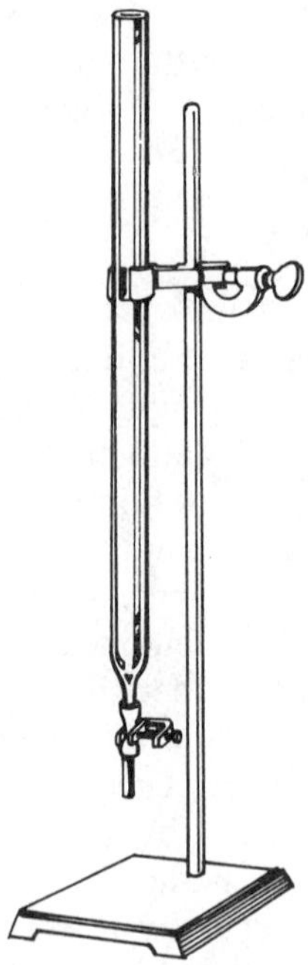

FIG. 2 Chromatographic tube on ring stand.

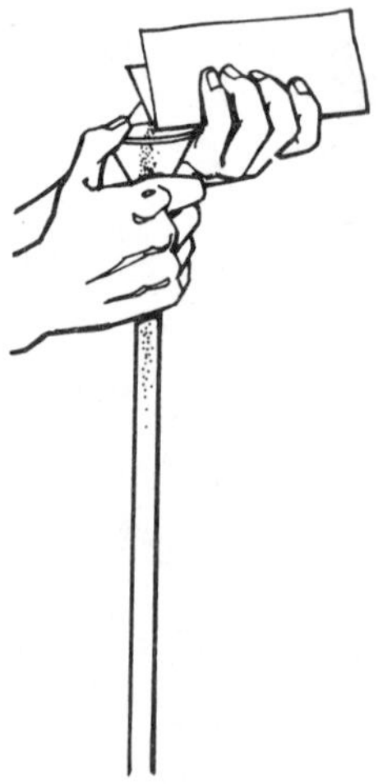

FIG. 3 A useful technique for filling a chromatographic tube with silica gel.

powder is stirred to suspend it in the solvent and immediately poured through a wide-mouth funnel into the chromatographic tube. Rap the column with a rubber stopper to cause the silica gel to settle and to remove bubbles. Add a protective layer of sand to the top. The column is now ready for use.

Prepare several Erlenmeyer flasks as receivers by taring (weighing) each one carefully and marking them with numbers on the etched circle.

After use, the tube is conveniently emptied by pointing the open end into a beaker, opening the pinchclamp, and applying gentle air pressure to the tip. If the plug of glass wool remains in the tube after the alumina leaves, wet it with acetone and reapply air pressure.

Experiments

1. Chromatography of a Mixture of Ferrocene and Acetylferrocene

Both of these compounds are colored (see Chapter 38 for the preparation of acetylferrocene) so it is easy to follow the progress of the chromatographic separation.

Prepare the column exactly as described above using activity III alumina. Then add a solution of 0.4 g of a 50:50 mixture of acetylferrocene (Caution toxic) and ferrocene that has been dissolved in the minimum quantity of dichloromethane, following the above procedure for adding the sample.

Carefully add petroleum ether to the column, open the valve and elute the two compounds. The first to be eluted, ferrocene, will be seen as a yellow band. Collect this in a tared 50-mL flask. Any crystalline material seen at the tip of the valve should be washed into the flask with a drop or two of ether. Without allowing the column to run dry, add a 50:50 mixture of petroleum ether and diethyl ether and elute the acetylferrocene, which will be seen as an orange band. Collect it in a tared 50-mL flask. Spot a thin-layer silica gel chromatography plate with these two solutions. Evaporate the solvents from the two flasks and determine the weights of the residues.

Recrystallize the products from the minimum quantities of hot hexane or ligroin. Isolate the crystals, dry them, and determine their weights and melting points. Calculate the percent recovery of the crude and recrystallized products based on the 0.2 g of each in the original mixture.

The thin-layer chromatography plate is eluted with 30:1 toluene-absolute ethanol. Do you detect any contamination of one compound by the other?

Cleaning Up   Empty the chromatography column onto a piece of aluminum foil in the hood. After the solvent has evaporated place alumina and sand in the nonhazardous waste container. Evaporate the crystallization mother liquors to dryness and place the residue in the hazardous waste container.

Ferrocene
MW 186.04
mp 172–174°C

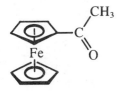

Acetylferrocene
MW 228.08, mp 85–86°C

Caution: Toxic

2. Acetylation of Cholesterol

Cholesterol is a solid alcohol; the average human body contains about 200 g distributed in brain, spinal cord, and nerve tissue and occasionally clogging the arteries and the gall bladder (see Chapter 22 for background and procedure for isolating cholesterol from human gallstones).

In the present experiment cholesterol is dissolved in acetic acid and allowed to react with acetic anhydride to form the ester, cholesteryl acetate. The reaction does not take place rapidly and consequently does not go to completion under the conditions of this experiment. Thus, when the reaction is over, both unreacted cholesterol and the product, cholesteryl acetate, are present. Separating these by fractional crystallization would be extremely difficult; but because they differ in polarity (the hydroxyl group of the cholesterol is the more strongly adsorbed on alumina), they are easily separated by column chromatography. Both molecules are colorless and hence cannot be detected visually. Each fraction should be sampled for thin-layer chromatography. In that way not only the presence but also the purity of each fraction can be assessed. It is also possible to put a drop of each fraction on a watch glass and evaporate it to see if the fraction contains product. Solid will also appear on the tip of the column while a compound is being eluted.

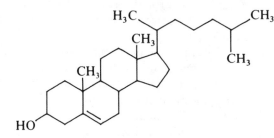

Cholesterol

$$+ \; CH_3\overset{O}{\overset{\|}{C}}O\overset{O}{\overset{\|}{C}}CH_3 \longrightarrow$$

Acetic anhydride

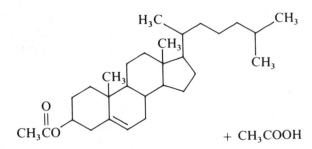

$+ \; CH_3COOH$

Cholesteryl acetate

Procedure

Carry out procedure in laboratory hood

Cover 0.5 g of cholesterol with 5 mL of acetic acid in a small Erlenmeyer flask, swirl, and note that the initially thin slurry soon sets to a stiff paste of the molecular compound $C_{27}H_{45}OH \cdot CH_3CO_2H$. Add 1 mL of acetic anhydride and heat the mixture on the steam bath for any convenient period of time from 15 min to 1 h; record the actual heating period. While the reaction takes place, prepare the chromatographic column. Cool, add 20 mL of water, and extract with two 25-mL portions of ether. Wash the combined ethereal extracts twice with 15-mL portions of water and once with 25 mL of 10% sodium hydroxide, dry by shaking the ether extracts with 25 mL of saturated sodium chloride solution, then dry the ether over anhydrous sodium sulfate for 10 min in an Erlenmeyer flask, filter, and evaporate the ether. Save a few crystals of this material for TLC (thin-layer chromatography) analysis. Dissolve the residue in 3–4 mL of ether, transfer the solution with a capillary dropping tube onto a column of 12.5 g of silica gel, and rinse the flask with another small portion of ether.[1]

In order to apply the ether solution to the top of the sand and avoid having it coat the interior of the column, pipette the solution down a 6-mm dia. glass tube that is resting on the top of the sand. Label a series of 50-mL Erlenmeyer flasks as fractions 1–10. Open the pinchclamp, run the eluant solution into a 50-mL Erlenmeyer flask, and as soon as the solvent in the column has fallen to the level of the upper layer of sand fill the column with a part of a measured 125 mL of 70:30 ligroin–ether. When about 25 mL of eluant has collected in the flask (fraction 1), change to a fresh flask; add a boiling stone to the first flask, and evaporate the solution to dryness on the steam bath under an aspirator tube (Fig. 4). Evacuation using the aspirator helps to remove last traces of ligroin (see Fig. 5). If this fraction 1 is negative (no residue), use the flask for collecting further fractions. Continue adding the ligroin–ether mixture until the 125-mL portion is exhausted and then use 100 mL of a 1:1 ligroin–ether mixture. A convenient bubbler (Fig. 6) made from a 125-mL Erlenmeyer flask, a short piece of 10-mm dia. glass tubing, and a cork will automatically add solvent. A separatory funnel with stopper and partially open stopcock serves the same purpose. Collect and evaporate successive 25-mL fractions of eluant. Save any flask that has any visible solid residue. The ideal method for removal of solvents involves the use of a rotary evaporator (Fig. 7). Analyze the original mixture and each fraction by TLC (see Chapter 9) on silica gel plates (Eastman No. 13181) using 1:1 ether–ligroin to develop the plates and either ultraviolet light or iodine vapor to visualize the spots.

Cholesteryl acetate (mp 115°C) and cholesterol (mp 149°C) should appear, respectively, in early and late fractions with a few empty fractions

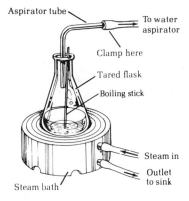

FIG. 4   Aspirator tube in use. A boiling stick may be necessary to promote even boiling.

FIG. 5   Drying a solid by reduced air pressure.

1. Ideally, the material to be adsorbed is dissolved in ligroin, the solvent of least eluant power. The present mixture is not soluble enough in ligroin and so ether is used, but the volume is kept to a minimum.

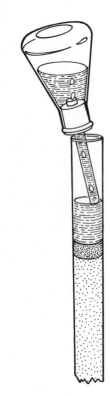

FIG. 6   Bubbler for adding solvent automatically.

FIG. 7   Rotary evaporator. The rate of evaporation with this apparatus is very fast due to the thin film of liquid over the entire inner surface of the rotating flask, which is heated under a vacuum. Foaming and bumping are also greatly reduced.

(no residue) in between. If so, combine consecutive fractions of early and late material and determine the weights and melting points. Calculate the percentage of the acetylated material compared to the total recovered, and compare your result with those of others in your class employing different reaction periods.

Cleaning Up   Acetic acid, aqueous layers, and saturated sodium chloride layers from the extraction, after neutralization, can be flushed down the drain with water. Ether, ligroin, and TLC solvents should be placed in the organic solvents container and the drying agent, once free of solvent, can be placed in the nonhazardous solid waste container. Ligroin and ether from the chromatography go into the organic solvents container. If possible, spread out the silica gel in the hood to dry. It can then be placed in the nonhazardous solid waste container. If it is wet with organic solvents, it is a hazardous solid waste and must be disposed of, at great expense, in a secure landfill.

3. Fluorene and Fluorenone

The 9-position of fluorene is unusually reactive for a hydrocarbon. The protons on this carbon atom are acidic by virtue of being doubly benzylic and consequently this carbon can be oxidized by several reagents, includ-

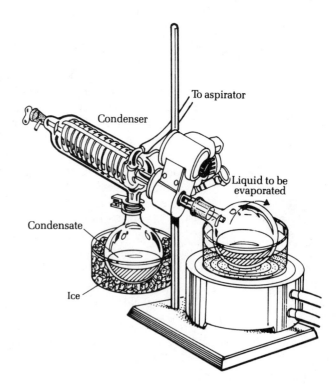

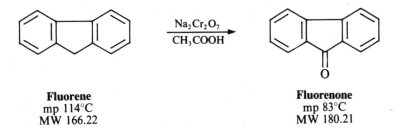

Fluorene
mp 114°C
MW 166.22

Fluorenone
mp 83°C
MW 180.21

ing elemental oxygen. In the present experiment the very powerful and versatile oxidizing agent Cr(VI), in the form of chromium trioxide, is used to carry out this oxidation. Cr(VI) in a variety of other forms is used to carry out about a dozen oxidation reactions in this text. The *dust* of Cr(VI) salts is reported to be a carcinogen, so avoid breathing it.

Procedure

CAUTION: Sodium dichromate is toxic. The dust is corrosive to nasal passages and skin and is a cancer suspect agent. Hot acetic acid is very corrosive to skin; handle with care.

In a 250-mL Erlenmeyer flask dissolve 5.0 g of practical grade fluorene in 25 mL of acetic acid by heating on the steam bath with occasional swirling. In a 125-mL Erlenmeyer flask dissolve 15 g of sodium dichromate dihydrate in 50 mL of acetic acid by swirling and heating on a hot plate. Adjust the temperature of the dichromate solution to 80°, transfer the thermometer and adjust the fluorene-acetic acid solution to 80°, and then, **in the hood,** pour in the dichromate solution. Note the time and the temperature of the solution, and heat on the steam bath for 30 min. Observe the maximum and final temperature, and then cool the solution and add 150 mL of water. Swirl the mixture for a full two minutes to coagulate the product and so promote rapid filtration, and collect the yellow solid in an 8.5-cm Büchner funnel (in case filtration is slow, empty the funnel and flask into a beaker and stir vigorously for a few minutes). Wash the filter cake well with water and then suck the filter cake as dry as possible. Either let the product dry overnight, or dry it quickly as follows: put the moist solid into a 50-mL Erlenmeyer flask, add ether (20 mL) and swirl to dissolve, and add anhydrous sodium sulfate (10 g) to scavenge the water. Decant the ethereal solution through a cone of anhydrous sodium sulfate in a funnel into a 125-mL Erlenmeyer flask, and rinse the flask and funnel with ether. Evaporate on the steam bath under an aspirator, heat until the ether is all removed, and pour the hot oil into a 50-mL beaker to cool and solidify. Scrape out the yellow solid. Yield: 4.0 g.

Handle dichromate and acetic acid in laboratory hood; wear gloves

Cleaning Up   The filtrate probably contains unreacted dichromate. To destroy it add 3 M sulfuric acid until the pH is 1 and then complete the reduction by adding solid sodium thiosulfate until the solution becomes cloudy and blue-colored. Neutralize with sodium carbonate and filter the flocculent precipitate of Cr(OH)$_3$ through Celite in a Büchner funnel. The filtrate can be diluted with water and flushed down the drain while the precipitate and Celite should be placed in the heavy metal hazardous waste container.

4. Separation of Fluorene and Fluorenone

Prepare a column of 12.5 g of alumina, run out excess solvent, and pour onto the column a solution of 0.5 g of fluorene–fluorenone mixture. Elute at first with ligroin and use tared 50-mL flasks as receivers. The yellow color of fluorenone provides one index of the course of the fractionation, and the appearance of solid around the delivery tip provides another. Wash the solid on the tip frequently into the receiver with ether. When you think that one component has been eluted completely, change to another receiver until you judge that the second component is beginning to appear. Then, when you are sure the second component is being eluted, change to a 1:1 ligroin–ether mixture and continue until the column is exhausted. It is possible to collect practically all the two components in the two receiving flasks, with only a negligible intermediate fraction. After evaporation of solvent, evacuate each flask under vacuum (Fig. 5) and determine the weight and melting point of the products. A convenient method for evaporating fractions is to use a rotary evaporator (Fig. 7).

Cleaning Up   All organic material from this experiment can go in the organic solvents container. The alumina absorbent, if free of organic solvents, can go in the nonhazardous solid waste container. If it is wet with organic solvents, it is a hazardous solid waste. Spread it in the hood to dry.

Questions_____

1. Predict the order of elution of a mixture of triphenylmethanol, biphenyl, benzoic acid, and methyl benzoate from an alumina column.

2. What would be the effect of collecting larger fractions when carrying out either of the experiments described?

3. What would have been the result if a large quantity of petroleum ether alone were used as the eluent in either of the experiments described?

4. Once the chromatographic column has been prepared, why is it important to allow the level of the liquid in the column to drop to the level of the alumina before applying the solution of the compound to be separated?

5. A chemist started to carry out column chromatography on a Friday afternoon, got to the point at which the two compounds being separated were about three-fourths of the way down the column, and then returned on Monday to find that the compounds came off the column as a mixture. Speculate on the reason for this. The column had not run dry over the weekend.

11

Alkenes from Alcohols: Cyclohexene from Cyclohexanol

Prelab Exercise: Prepare a detailed flow sheet for the preparation of cyclohexene, indicating at each step which layer contains the desired product.

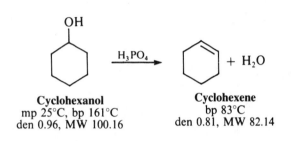

Cyclohexanol
mp 25°C, bp 161°C
den 0.96, MW 100.16

Cyclohexene
bp 83°C
den 0.81, MW 82.14

Dehydration of cyclohexanol to cyclohexene can be accomplished by pyrolysis of the cyclic secondary alcohol with an acid catalyst at a moderate temperature or by distillation over alumina or silica gel. The procedure selected for this experiment involves catalysis by phosphoric acid; sulfuric acid is no more efficient, causes charring, and gives rise to sulfur dioxide. When a mixture of cyclohexanol and phosphoric acid is heated in a flask equipped with a fractionating column, the formation of water is soon evident. On further heating, the water and the cyclohexene formed distill together by the principle of steam distillation, and any high-boiling cyclohexanol that may volatilize is returned to the flask. However, after dehydration is complete and the bulk of the product has distilled, the column remains saturated with water-cyclohexene that merely refluxes and does not distill. Hence, for recovery of otherwise lost reaction product, a chaser solvent is added and distillation is continued. A suitable chaser solvent is the water-immiscible, aromatic solvent toluene, boiling point 110°C; as it steam-distills it carries over the more volatile cyclohexene. When the total water-insoluble layer is separated, dried, and redistilled through the dried column the chaser again drives the cyclohexene from the column; the difference in boiling points is such that a sharp separation is possible. The holdup in the metal sponge-packed column is so great (about 1.0 mL) that if a chaser solvent is not used in the procedure the yield will be much lower.

The mechanism of this reaction involves initial rapid protonation of the hydroxyl group by the phosphoric acid:

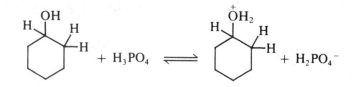

This is followed by loss of water to give the unstable secondary carbonium ion, which quickly loses a proton to water or the conjugate acid to give the alkene:

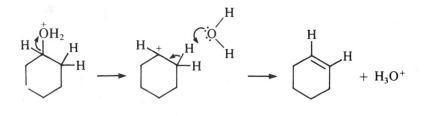

Experiments

1. Preparation of Cyclohexene

Handle phosphoric acid with care as it is corrosive to tissue

Introduce 20.0 g of cyclohexanol (technical grade), 5 mL of 85% phosphoric acid, and a boiling stone into a 100-mL round-bottomed flask and shake to mix the layers. Note the evolution of heat. Use the arrangement for fractional distillation shown in Fig. 6 in Chapter 5 but modified by use of a bent adapter delivering into an ice-cooled test tube in a 125-mL Erlenmeyer receiver, as shown in Fig. 11.1.

Use of a chaser

Note the initial effect of heating the mixture, and then distill until the residue in the flask has a volume of 5–10 mL and very little distillate is being formed; note the temperature range. Then let the assembly cool a little, remove the thermometer briefly, and pour 20 mL of toluene (the chaser solvent) into the top of the column through a long-stemmed funnel. Note the amount of the upper layer in the boiling flask and distill again until the volume of the layer has been reduced by about half. Pour the contents of the test tube into a small separatory funnel and rinse with a little chaser solvent; use this solvent for rinsing in subsequent operations. Wash the mixture with an equal volume of saturated sodium chloride solution, separate the water layer, run the upper layer into a clean flask, and add 5 g of anhydrous sodium sulfate (10 × 75-mm test tube-full) to dry it. Before the final distillation note the barometric pressure, apply any thermometer corrections necessary, and determine the reading expected for a boiling point of

Note: Cyclohexene is
extremely flammable; keep
away from open flame

83°C. Dry the boiling flask, column, and condenser, decant the dried liquid into the flask through a stemless funnel plugged with a bit of cotton, and fractionally distill, with all precautions against evaporation losses. The ring of condensate should rise very slowly as it approaches the top of the fractionating column in order that the thermometer may record the true boiling point soon after distillation starts. Record both the corrected boiling point of the bulk of the cyclohexene fraction and the temperature range,

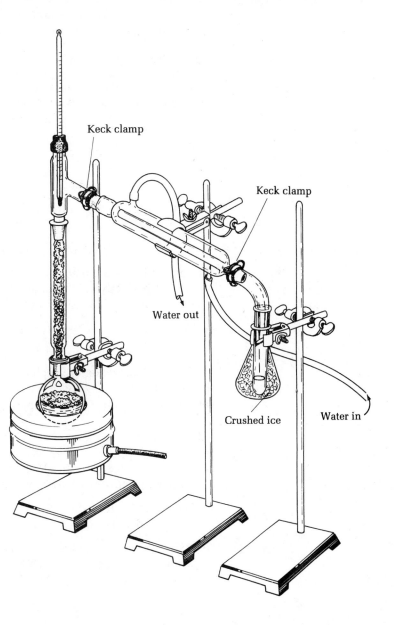

Keck clamp

Keck clamp

Water out

Crushed ice

Water in

FIG. 1 Fractionation into an ice-cooled receiver.

which should not be more than 2°C. If the cyclohexene has been dried thoroughly it will be clear and colorless; if wet it will be cloudy. A typical student yield is 13.2 g. Report your yield in grams and your percent yield.

Cleaning Up　　The aqueous solutions (pot residues and washes) should be diluted with water and neutralized before flushing down the drain with a large excess of water. The ethanol wash and sodium chloride solution can also be flushed down the drain, while the acetone wash and all toluene-containing solutions should be placed in the organic solvents container. Once free of solvent, the sodium sulfate can be placed in the nonhazardous solid waste container. Allow it to dry on a tray in the hood.

Yield Calculations

Rarely do organic reactions give 100% yields of one pure product; an important objective of every experiment is a high yield of the desired product. The present experiment uses 20 g of starting cyclohexanol. This corresponds to 0.2 mole because the molecular weight of cyclohexanol is 100:

$$\frac{20.0 \text{ g}}{100.16 \text{ g/mole}} = 0.20 \text{ mole} = 200 \text{ millimoles}$$

If the reaction gave a 100% yield of cyclohexene then 0.20 mole of the alkene would be produced. The weight of 0.20 mole of cyclohexene is

$$0.20 \text{ mole} \times 82.14 \text{ g/mole} = 16.4 \text{ g}$$

We call the 16.4 g the *theoretical yield;* it could be obtained if the reaction proceeded perfectly. If the actual yield is only 8.2 g, then the reaction would be said to give a 50% yield:

$$\frac{8.2 \text{ g}}{16.4 \text{ g}} \times 100 = 50\%$$

Typical student yields are included throughout this text. These are *not* theoretical yields; they suggest what an average or above-average student can expect to obtain for the experiment.

Questions

1. Assign the peaks in the ^{1}H nmr spectrum of cyclohexene (Fig. 2) to specific groups of protons on the molecule.

2. What product(s) would be obtained by the dehydration of 2-heptanol? Of 2-methyl-1-cyclohexanol?

3. Mixing cyclohexanol with phosphoric acid is an exothermic process while the production of cyclohexene is endothermic. Referring to the two chemical reactions on p. 144, construct an energy diagram show-

ing the course of this reaction. Label the diagram with the starting alcohol, the oxonium ion (the protonated alcohol), the carbocation, and the product.

4. Is it reasonable that the ¹H spectrum of cyclohexene (Fig. 2) should closely resemble the ¹³C spectrum (Fig. 3)?

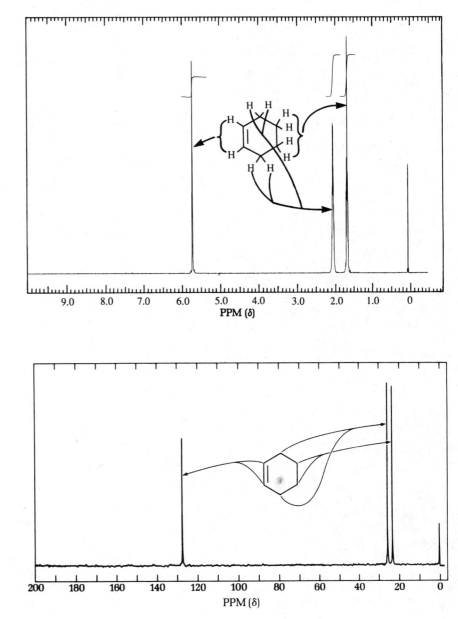

FIG. 2   ¹H nmr spectrum of cyclohexene (250 MHz).

FIG. 3   ¹³C nmr spectrum of cyclohexene.

FIG. 4 ^{13}C nmr spectrum of cyclohexanol.

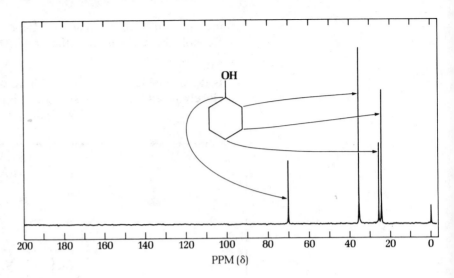

12

Alkenes from Alcohols: Analysis of a Mixture by Gas Chromatography

Prelab Exercise: If the dehydration of 2-methyl-2-butanol occurred on a purely statistical basis, what would be the relative proportions of 2~methyl-1-butene and 2~methyl-2-butene?

Separation of mixtures of volatile samples

Gas chromatography (gc), also called vapor phase chromatography (vpc) and gas-liquid chromatography (glc), is a means of separating volatile mixtures, the components of which may differ in boiling points by only a few tenths of a degree. The gc process is similar to fractional distillation, but instead of a glass column 25 cm long packed with a stainless-steel sponge, the gc column used is a 3–10 m long coiled metal tube (dia. 6 mm), packed with ground firebrick. The firebrick serves as an inert support for a very high-boiling liquid (essentially nonvolatile), such as silicone oil and low-molecular-weight polymers like Carbowax. These are the *liquids* of gas-*liquid* chromatography and are referred to as the stationary phase. The sample (1–25 *micro*liters) is injected through a silicone rubber septum into the column, which is being swept with a current of helium (ca. 200 mL/min). The sample first dissolves in the high-boiling liquid phase and then the more volatile components of the sample evaporate from the liquid and pass into the gas phase. Helium, the carrier gas, carries these components along the column a short distance where they again dissolve in the liquid phase before reevaporation (Fig. 1).

High-boiling liquid phase

The thermal conductivity detector

Eventually the carrier gas, which is a very good thermal conductor, and the sample reach the detector, an electrically heated tungsten wire. As long as pure helium is flowing over the detector the temperature of the wire is relatively low and the wire has a low resistance to the flow of electric current. Organic molecules have lower thermal conductivities than helium. Hence, when a mixture of helium and an organic sample flows over the detector wire it is cooled less efficiently and heats up. When the wire is hot its electrical resistance becomes higher and offers higher resistance to the flow of current. The detector wire actually is one leg of a Wheatstone bridge, connected to a chart recorder that records, as a peak, the amount of current necessary to again balance the bridge. The record produced by the recorder is called a **chromatogram.**

The number and relative amounts of components in a mixture

A gas chromatogram is simply a recording of current versus time (which is equivalent to a certain volume of helium) (Fig. 2). In the illustration the smaller peak from component A has the shorter retention time, T_1, and so is a more volatile substance than component B (if the stationary phase is an inert liquid such as silicone oil). The areas under the two peaks

FIG. 1 Gas chromatography:
Diagrammatic partition between
gas and liquid phases.

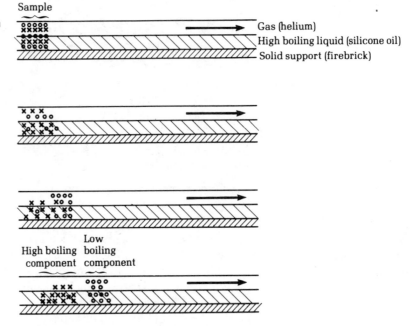

Sample

Gas (helium)
High boiling liquid (silicone oil)
Solid support (firebrick)

High boiling   Low
component   boiling
component

are directly proportional to the molar amounts of A and B in the mixture (provided they are structurally similar). The retention time of a given component is a function of the column temperature, the helium flow rate, and the nature of the stationary phase. Hundreds of stationary phases are available; picking the correct one to carry out a given analysis is somewhat of an art, but widely used stationary phases are silicone oil and silicone rubber, both of which can be used at temperatures up to 300°C and separate mixtures on the basis of boiling point differences of the mixture's components. More specialized stationary phases will, for instance, allow alkanes to pass through readily (short retention time) while holding back (long retention time) alcohols by hydrogen bonding to the liquid phase.

The gas chromatograph

A diagram of a typical gas chromatograph is shown in Fig. 3. The carrier gas, usually helium, enters the chromatograph at ca. 60 lb/sq in. The sample (1 to 25 microliters) is injected through a rubber septum using a small

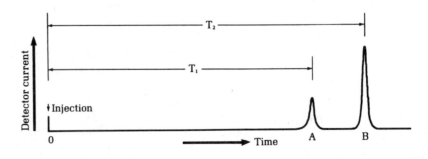

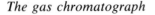

T_2

T_1

Detector current

↑Injection

0   Time   A   B

FIG. 2 Gas chromatogram.

FIG. 3   Diagram of a gas chromatograph.

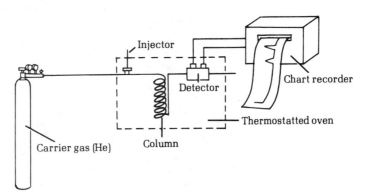

FIG. 4   One of the commercially available chromatographs.

Collecting a sample for an infrared spectrum

hypodermic syringe (Fig. 4). Handle the syringe with care. The syringe needles and plungers are thin and delicate; take care not to bend either one. Be wary of the injection port. It is very hot. The sample immediately passes through the column and then the detector. Injector, column, and detector are all enclosed in a thermostatted oven, which can be maintained at any temperature up to 300°C. In this way samples that would not volatilize enough at room temperature can be analyzed.

Gas chromatography determines the number of components and their relative amounts in a very small sample. The small sample size is an advantage in many cases, but it precludes isolating the separated components. Some specialized chromatographs can separate samples as large as 0.5 mL per injection and automatically collect each fraction in a separate container. At the other extreme gas chromatographs equipped with flame ionization detectors can detect micrograms of sample, such as traces of pesticides in food or of drugs in blood and urine. Clearly a gas chromatogram gives little information about the chemical nature of the sample being detected. However, it is sometimes possible to collect enough sample at the exit port of the chromatograph to obtain an infrared spectrum. As the peak for the compound of interest appears on the chart paper, a 2-mm dia. glass tube, 3 in. long and packed with glass wool, is inserted into the rubber septum at the exit port. The sample, if it is not too volatile, will condense in the cold glass tube. Subsequently, the sample is washed out with a drop or two of solvent and an infrared spectrum obtained.

Dehydration

1. 2-Methyl-1-butene and 2-Methyl-2-butene[1]

The dilute sulfuric acid catalyzed dehydration of 2-methyl-2-butanol (*t*-amyl alcohol) proceeds readily to give a mixture of alkenes that can be analyzed by gas chromatography. The mechanism of this reaction involves

1. C. W. Schimelpfenig, *J. Chem. Ed.*, **39**, 310 (1962).

the intermediate formation of the relatively stable tertiary carbocation followed by loss of a proton either from a primary carbon atom to give the terminal olefin, 2-methyl-1-butene, or from a secondary carbon to give 2-methyl-2-butene.

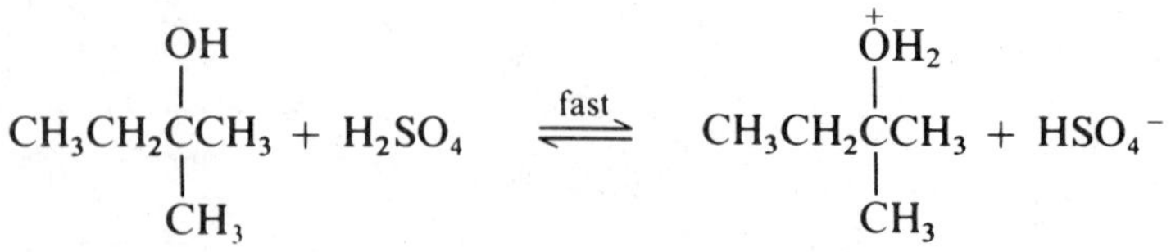

$$CH_3CH_2\overset{\overset{\displaystyle OH}{|}}{\underset{\underset{\displaystyle CH_3}{|}}{C}}CH_3 + H_2SO_4 \underset{\text{fast}}{\rightleftharpoons} CH_3CH_2\overset{\overset{\displaystyle \overset{+}{O}H_2}{|}}{\underset{\underset{\displaystyle CH_3}{|}}{C}}CH_3 + HSO_4{}^-$$

2-Methyl-2-butanol
bp 102°C
den 0.805
MW 88.15

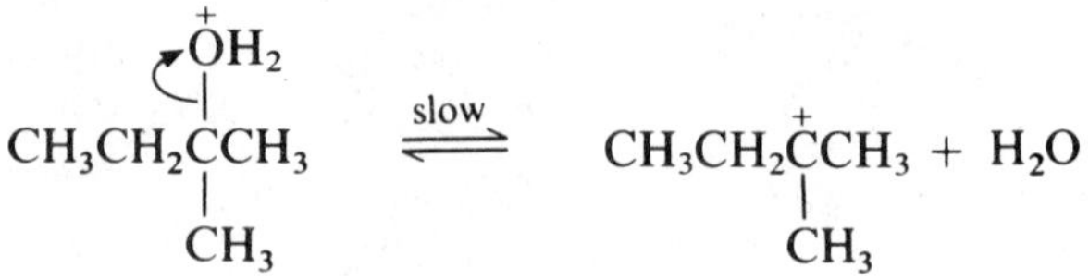

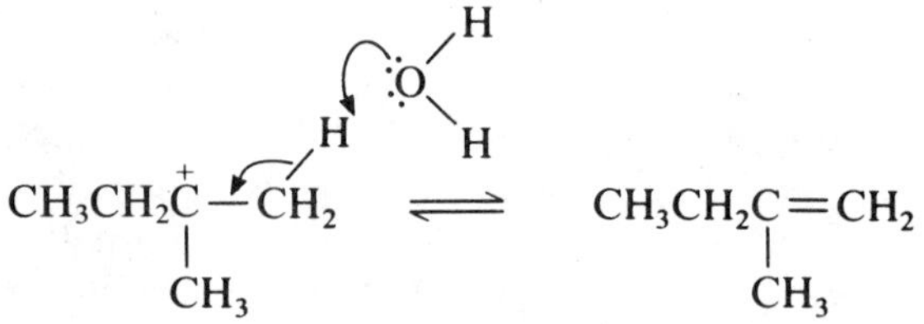

$$CH_3CH_2\overset{\overset{\displaystyle \overset{+}{O}H_2}{|}}{\underset{\underset{\displaystyle CH_3}{|}}{C}}CH_3 \underset{\text{slow}}{\rightleftharpoons} CH_3CH_2\overset{+}{\underset{\underset{\displaystyle CH_3}{|}}{C}}CH_3 + H_2O$$

$$CH_3CH_2\overset{+}{C}-CH_2 \rightleftharpoons CH_3CH_2C=CH_2$$

2-Methyl-1-butene
bp 31.16°, den 0.662
MW 70.14

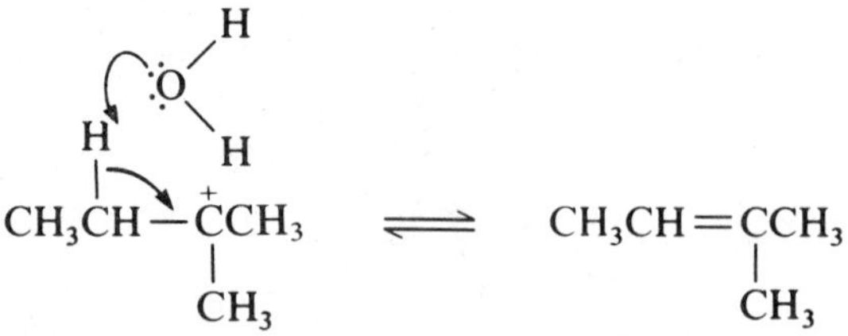

$$CH_3CH-\overset{+}{C}CH_3 \rightleftharpoons CH_3CH=CCH_3$$

2-Methyl-2-butene
bp 38.57°C
den 0.662
MW 70.14

2-Methyl-2-butanol can also be dehydrated in high yield using iodine as a catalyst:

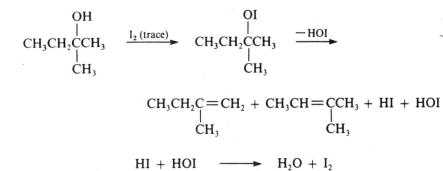

$$HI + HOI \longrightarrow H_2O + I_2$$

Each step of this E_1 elimination reaction is reversible and thus the reaction is driven to completion by removing one of the products, the alkene. In these reactions several alkenes can be produced. The Saytzeff rule states that the more substituted alkene is the more stable and thus the one formed in larger amount. And the *trans* isomer is more stable than the *cis* isomer. With this information it should be possible to deduce which peaks on the gas chromatogram correspond to a given alkene and to predict the ratios of the products.

In analyzing your results from this experiment, consider the fact that the carbocation can lose any of six primary hydrogen atoms but only two secondary hydrogen atoms to give the product olefins.

Dehydration Procedure

Pour 36 mL of water into a 250-mL round-bottomed flask, and cool in an ice-water bath while **slowly** pouring in 18 mL of concentrated sulfuric acid. Cool this "1:2" acid further with swirling while slowly pouring in 36 mL (30 g) of 2-methyl-2-butanol. Shake the mixture thoroughly and then mount the flask for fractional distillation over a flask heater, as in Fig. 6 in Chapter 5, with the arrangement for ice cooling of the distillate seen in Fig. 1 in Chapter 11. Cooling is needed because the olefin is volatile. Use a long condenser and a rapid stream of cooling water. Heat the flask slowly with an electric flask heater until distillation of the hydrocarbon is complete. If the ice-cooled test tube will not hold 30 mL, be prepared to collect half of the distillate in another ice-cooled test tube. Transfer the distillate to a separatory funnel and shake with about 10 mL of 10% sodium hydroxide solution to remove any traces of sulfurous acid. The aqueous solution sinks to the bottom and is drawn off. Dry the hydrocarbon layer by adding sufficient anhydrous sodium sulfate until the drying agent no longer clumps together. After about 5 min remove the drying agent by gravity filtration or careful decantation into a *dry* 50-mL round-bottomed flask, and distill the dried product through a fractionating column (see Fig. 1 in Chapter 11), taking the same precautions as before to avoid evaporation losses. Rinse the fractionating column with acetone and dry it with a stream of air to remove water from the first distillation. Collect in a tared (previously weighed) bottle the portion boiling at 30–43°C. The yield reported in the literature is 84%; the average student yield is about 50%.

Cleaning Up   The pot residue from the reaction is combined with the sodium hydroxide wash and neutralized with sodium carbonate. The pot residue from the distillation of the product is combined with the acetone used to wash out the apparatus and placed in the organic solvents container. If the sodium sulfate is dry, it can be placed in the nonhazardous solid waste container. If it is wet with organic solvents, it must be placed in the hazardous solid waste container for solvent-contaminated drying agent.

Gas Chromatographic Analysis

Be sure the helium cylinder is attached to a bench or wall.
Use care. The injection port is hot.

In your notebook record all information relevant to this analysis: column diameter and length, column packing, carrier gas and its flow rate, filament current, temperature of column, detector and injection port, sample size, attenuation, and chart speed.

Inject a few microliters of product into a gas chromatograph maintained at room temperature and equipped with a 6 mm dia. × 3 m column packed with 10% SE-30 silicone rubber on Chromosorb-W or a similar inert packing. Mark the chart paper at the time of injection. In a few minutes two peaks should appear. After using the chromatograph, turn off the recorder and cap the pen. Make no unauthorized adjustments on the gas chromatograph. From your knowledge of the mechanisms of dehydration of secondary alcohols, which olefin should predominate? Does this agree with the boiling points? (In general, the compound with the shorter retention time has the lower boiling point.) Measure the relative areas under the two peaks. One way to perform this integration is to cut out the peaks with scissors and weigh the two pieces of paper separately on an analytical balance. Although time-consuming, this method gives very precise results. If the peaks are symmetrical, their areas can be approximated by simply multiplying the height of the peak by its width at half-height.

Questions

1. Write the structure of the three olefins produced by the dehydration of 3-methyl-3-pentanol.

2. When 2-methylpropene is bubbled into dilute sulfuric acid at room temperature, it appears to dissolve. What new substance has been formed?

3. A student wished to prepare ethylene gas by dehydration of ethanol at 140°C using sulfuric acid as the dehydrating agent. A low-boiling liquid was obtained instead of ethylene. What was the liquid and how might the reaction conditions be changed to give ethylene?

4. What would be the effect of increasing the carrier gas flow rate on the retention time?

5. What would be the effect of raising the column temperature on the retention time?

6. What would be the effect of raising the temperature or increasing the carrier gas flow rate on the ability to resolve two closely spaced peaks?

7. If you were to try a column one-half the length of the one you actually used, how do you think the retention times of the butenes would be affected? How do you think the separation of the peaks would be affected? How do you think the width of each peak would be affected?

8. From your knowledge of the dehydration of tertiary alcohols, which olefin should predominate in the product of the dehydration of 2-methyl-2-butanol and why?

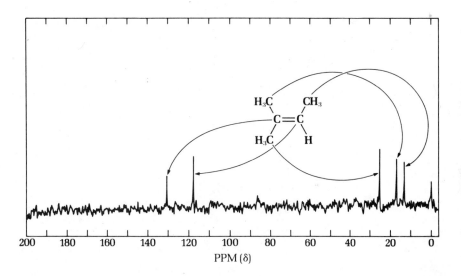

FIG. 5 ^{13}C nmr spectrum of 2-methyl-2-butene (22.6 MHz).

Alkanes and Alkenes: Radical Initiated Chlorination of 1-Chlorobutane

Most of the alkanes from petroleum are used to produce energy by combustion, but a few percent are converted to industrially useful compounds by controlled reaction with oxygen or chlorine. The alkanes are inert to attack by most chemical reagents and will react with oxygen and halogens only under the special conditions of radical initiated reactions.

At room temperature an alkane such as butane will not react with chlorine. In order for a reaction to occur

$$CH_3CH_2CH_2CH_3 + Cl_2 \longrightarrow CH_3CH_2CH_2CH_2Cl + HCl$$

a precisely oriented four-center collision of the Cl_2 molecule with the butane molecule must occur with two bonds broken and two bonds formed simultaneously:

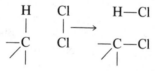

It is unlikely that the necessarily precise orientation of the two reacting molecules will be found. They must, of course, be within bonding distance of each other as well, so steric factors play a part.

If a concerted four-center reaction won't work, then an alternative possibility is a stepwise mechanism:

$$:\ddot{C}l\!:\!\ddot{C}l\!: \quad \xrightarrow{\text{slow}} \quad 2\ :\ddot{C}l\cdot \qquad (1)$$

$$CH_3CH_2CH_2CH_3 \quad \xrightarrow{\text{slow}} \quad CH_3CH_2CH_2CH_2\cdot + H\cdot \qquad (2)$$

$$:\ddot{C}l\cdot + CH_3CH_2CH_2CH_2\cdot \quad \xrightarrow{\text{fast}} \quad CH_3CH_2CH_2CH_2Cl \qquad (3)$$

$$:\ddot{C}l\cdot + H\cdot \quad \xrightarrow{\text{fast}} \quad H:\ddot{C}l\!: \qquad (4)$$

For this series of reactions to occur, the chlorine molecule must dissociate into two chlorine atoms. Because chlorine has a bond energy of

58 kcal mole^{-1}, we would not expect any significant numbers of molecules to dissociate at room temperature. Thermal motion at 25°C can only break bonds having energies less than 30–35 kcal mole^{-1}. Although thermal dissociation would require a high temperature, the dissociation of chlorine into atoms (chlorine radicals) can be caused by violet and ultraviolet light:

$$Cl_2 \xrightarrow{h\nu} 2 \; :\ddot{\underset{..}{C}}l \cdot$$

The photon energy of red light is 48 kcal mole^{-1} while light of 300 nm (ultraviolet) has a photon energy of 96 kcal mole^{-1}.

The chlorine radical can react with butane by abstraction of a hydrogen atom:

$$CH_3CH_2CH_2CH_3 + :\ddot{\underset{..}{C}}l \cdot \rightleftharpoons CH_3CH_2CH_2CH_2 \cdot + HCl$$

a reaction that is very slightly exothermic (and thus written as a reversible reaction). The butyl radical can react with chlorine:

$$CH_3CH_2CH_2CH_2 \cdot + Cl_2 \longrightarrow CH_3CH_2CH_2CH_2Cl + :\ddot{\underset{..}{C}}l \cdot$$

a reaction that evolves 26 kcal mole^{-1} of energy. The net result of these two reactions is a reaction of Cl_2 with butane "catalyzed" by $:\ddot{C}l \cdot$, the chlorine radical. The whole process can be terminated by the reaction of radicals with each other:

$$CH_3CH_2CH_2CH_2 \cdot + :\ddot{\underset{..}{C}}l \cdot \longrightarrow CH_3CH_2CH_2CH_2Cl$$

$$2 \; CH_3CH_2CH_2CH_2 \cdot \longrightarrow CH_3(CH_2)_6CH_3$$

$$2 \; :\ddot{\underset{..}{C}}l \cdot \longrightarrow Cl_2$$

To summarize, the process of light-induced radical chlorination involves three steps: chain initiation in which chlorine radicals are produced, chain propagation that involves no net consumption of chlorine radicals, and chain termination that destroys radicals.

$$Cl_2 \longrightarrow 2 \; :\ddot{\underset{..}{C}}l \cdot \qquad \textit{Chain initiation}$$

$$\left. \begin{array}{l} CH_3CH_2CH_2CH_3 + :\ddot{\underset{..}{C}}l \cdot \rightleftharpoons CH_3CH_2CH_2CH_2 \cdot + HCl \\ CH_3CH_2CH_2CH_2 \cdot + Cl_2 \longrightarrow CH_3CH_2CH_2CH_2Cl + :\ddot{\underset{..}{C}}l \cdot \end{array} \right\} \begin{array}{l} \textit{Chain} \\ \textit{propagation} \end{array}$$

$$\left. \begin{array}{l} CH_3CH_2CH_2CH_2 \cdot + :\ddot{\underset{..}{C}}l \cdot \longrightarrow CH_3CH_2CH_2CH_2Cl \\ 2 \; CH_3CH_2CH_2CH_2 \cdot \qquad \longrightarrow CH_3(CH_2)_6CH_3 \\ 2 \; :\ddot{\underset{..}{C}}l \cdot \longrightarrow Cl_2 \end{array} \right\} \begin{array}{l} \textit{Chain} \\ \textit{termination} \end{array}$$

In the example the product is shown to be 1-chlorobutane, but in fact the reaction produces a mixture of 1-chlorobutane and 2-chlorobutane. If the reaction were to occur purely by chance, we would expect the ratio of products to be 6:4 because there are six primary hydrogens and four secondary hydrogens on butane. But because a secondary C—H bond is weaker than a primary C—H bond (95 vs 98 kcal mole^{-1}), we might expect more 2-chlorobutane than chance would dictate. In the chlorination of 2-methylbutane:

	Found	Statistical Expectation	
		Ratio	Percent
$CH_3-\overset{\overset{\displaystyle CH_3}{\displaystyle |}}{\underset{\underset{\displaystyle H}{\displaystyle |}}{C}}-CH_2CH_3 \xrightarrow[300°]{Cl_2} CH_3-\overset{\overset{\displaystyle CH_3}{\displaystyle |}}{\underset{\underset{\displaystyle H}{\displaystyle |}}{C}}-CHCl-CH_3$	33%	2/12	17
$ClCH_2-\overset{\overset{\displaystyle CH_3}{\displaystyle |}}{\underset{\underset{\displaystyle H}{\displaystyle |}}{C}}-CH_2-CH_3$	30%	6/12	50
$CH_3-\overset{\overset{\displaystyle CH_3}{\displaystyle |}}{\underset{\underset{\displaystyle Cl}{\displaystyle |}}{C}}-CH_2-CH_3$	22%	1/12	8
$CH_3-\overset{\overset{\displaystyle CH_2}{\displaystyle |}}{\underset{\underset{\displaystyle H}{\displaystyle |}}{C}}-CH_2-CH_2Cl$	15%	3/12	25

As can be seen, the relatively weak tertiary C—H bond (92 kcal mole^{-1}) gives rise to 22% of product in contrast to the 8% expected on the basis of a random attack of $:\ddot{C}l\cdot$ on the starting material.

The relative reactivities of the various hydrogens of 2-methylbutane on a per-hydrogen basis (referred to the primary hydrogens of C-4 as 1.0) can be calculated:

$$\frac{\text{C-2 tertiary}}{\text{C-4 primary}} = \frac{33/1}{15/3} = \frac{6.6}{1}$$

$$\frac{\text{C-3 secondary}}{\text{C-4 primary}} = \frac{33/2}{15/3} = \frac{3.3}{1}$$

A reaction of this type is of little use unless you happen to need the four products in the ratios found and can manage to separate them (their boiling points are very similar). But industrially the radical chlorination of methane and ethane are important reactions and the products can be separated easily:

$$CH_4 \xrightarrow[\Delta]{Cl_2} CH_3Cl \ + \ CH_2Cl_2 \ + \ CHCl_3 \ + \ CCl_4$$

	Chloro-methane	Dichloro-methane	Trichloro-methane	Tetrachloro-methane
Common names:	Methyl chloride	Methylene chloride	Chloroform	Carbon Tetrachloride
Boiling points:	−24°C	40°C	62°C	77°C

In the first experiment we will chlorinate 1-chlorobutane because it is easier to handle in the laboratory than gaseous butane and we will use sulfuryl chloride as our source of chlorine radicals because it is easier to handle than gaseous chlorine. Instead of using light to initiate the reaction we will use a chemical initiator, 2,2′-azobis-(2-methylpropionitrile). This azo compound (R—N=N—R) decomposes at moderate temperatures (80–100°C) to give two relatively stable radicals and nitrogen gas:

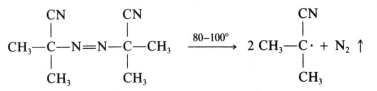

2,2′-Azobis-(2-methylpropionitrile)
MW 164.21, mp 102–103° (dec.)

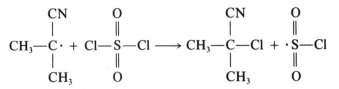

Sulfuryl chloride

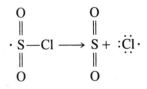

Before conducting the experiment, try to predict the ratio of products

The radical monochlorination of 1-chlorobutane can give four products: 1,1-, 1,2-, 1,3-, and 1,4-dichlorobutane. If the reaction occurred completely at random we would expect products in the ratios of the number of hydrogen atoms on each carbon, i.e., 2:2:2:3, respectively (22%, 22%, 22%, 33%). The object of the present experiment is to carry out the radical chlorination of 1-chlorobutane and then to determine the ratio of products using gas-liquid chromatography.

$$CH_3CH_2CH_2CH_2Cl + SO_2Cl_2 \xrightarrow{R\cdot} CH_3CH_2CH_2CHCl_2 + CH_3CH_2CHClCH_2Cl$$

1-Chlorobutane
MW 92.57
den 0.886
bp 77–78°C

Sulfuryl chloride
MW 134.97
den 1.67
bp 69°C

bp 114°C   bp 124°C

$$+ CH_3CHClCH_2CH_2Cl + CH_2ClCH_2CH_2CH_2Cl$$

bp 134°C   bp 162°C

$$+ SO_2 + HCl$$

Experiments

1. Free-Radical Chlorination of 1-Chlorobutane

Conduct this experiment in a laboratory hood

To a 100-mL round-bottomed flask add 1-chlorobutane (25 mL, 21.6 g, 0.23 mole), sulfuryl chloride (8 mL, 13.5 g, 0.10 mole), 2,2'-azobis-(2-methylpropionitrile) (0.1 g) and a boiling chip. Equip the flask with a condenser and gas trap as seen in Fig. 1. Heat the mixture to gentle reflux on the steam bath for 20 min. Remove the flask from the steam bath, allow it to cool somewhat and *quickly*, to minimize the escape of sulfur dioxide and hydrochloric acid, lift the condenser from the flask and add a second 0.1-g portion of the initiator. Heat the reaction mixture for an additional 10 min, remove the flask and condenser from the steam bath, and cool the flask in a beaker of water. Pour the contents of the flask through a funnel into about 50 mL of water in a small separatory funnel, shake the mixture, and separate the two phases. Wash the organic phase with two 20-mL portions of 5% sodium bicarbonate solution, once with a 20-mL portion of water, and then dry the organic layer over anhydrous calcium chloride (about 4 g) in a dry Erlenmeyer flask. The mixture can be analyzed by gas chromatography at this point or the unreacted 1-chlorobutane can be removed by fractional distillation (up to bp 85°C) and the pot residue analyzed by gas chromatography.

Cleaning Up   Empty the gas trap and combine the contents with all the aqueous layers and washes. Neutralize the aqueous solution with sodium carbonate and flush the solution down the drain with a large excess of water. The drying agent will be coated with chlorinated product and therefore must be disposed of in the hazardous waste container for solvent-contaminated drying agent. Any unused starting material and product must be placed in the halogenated organic waste container.

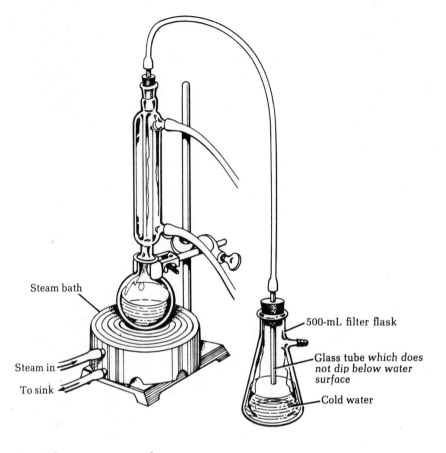

Steam bath

Steam in

To sink

500-mL filter flask

Glass tube *which does not dip below water surface*

Cold water

FIG. 1   Apparatus for chlorination of 1-chlorobutane with SO_2 and HCl gas trap.

Gas Chromatography

See Chapter 12 for information about gas chromatography. A Carbowax column works best, although any other nonpolar phase such as silicone rubber should work as well. With a nonpolar column packing the products are expected to come out in the order of their boiling points. A typical set of operating conditions would be column temperature 100°C, He flow rate 35 mL/min, column size 5-mm dia × 2 m, sample size 5 microliters, attenuation 16.

The molar amounts of each compound present in the reaction mixture are proportional to the areas under the peaks in the chromatogram. Because 1-chlorobutane is present in large excess, let this peak run off the paper, but be sure to keep the four product peaks on the paper. To determine relative peak areas simply cut out the peaks with a pair of scissors and weigh them. This method of peak integration works very well and depends on the uniform thickness of paper, which results in its weight being proportional to its area. Calculate the relative percent of each product molecule and the partial rate factors relative to the primary hydrogens on carbon-4. Compare your results with those for the chlorination of 2-methylbutane, the data for which were given previously.

Distinguishing Between Alkanes and Alkenes

When an olefin such as cyclohexene is allowed to react with bromine at a very low concentration, substitution instead of addition occurs. Although addition of a halogen atom to the double bond occurs readily, the intermediate free radical (**1**) is not very stable and, if it does not encounter another halogen molecule soon, will revert to starting material. If, however, an allylic free radical (**2**) is formed initially, it is much more stable and will survive long enough to react with a halogen molecule, even when the halogen is at low concentration. Thus, low halogen concentration favors allylic substitution, while high halogen concentration favors addition.

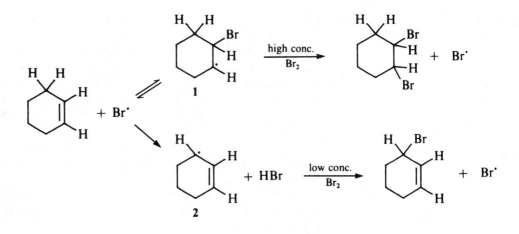

N-Bromosuccinimide (NBS) is a reagent that will continuously generate bromine molecules at a low concentration. This occurs when the HBr molecule from an allylic substitution reacts with NBS:

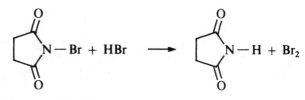

N-Bromosuccinimide

Alkenes react with NBS   The net reaction is one of allylic bromination:

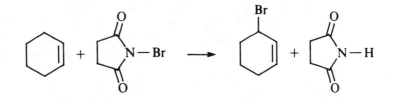

In aqueous solution NBS will react with cyclohexene to form the bromohydrin, a reaction that may or may not involve the intermediate formation of HOBr:

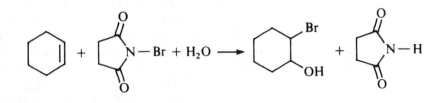

As seen in the preceding experiment, free radical halogenation of alkanes proceeds by substitution to give the chloroalkane and hydrogen chloride gas. Halogenation of an alkene, on the other hand, proceeds by addition across the double bond. These two different modes of reaction can be used to distinguish alkenes from alkanes. Bromine is used as the test reagent. The disappearance of the red bromine color indicates a reaction has taken place. Cyclohexene and dibromocyclohexane are both colorless. If hydrogen bromide has evolved, it can be detected by breathing across the test tube. The hydrogen bromide dissolves in the moist air to give a cloud of hydrobromic acid.

Alkenes decolorize bromine

Bromine at a relatively high concentration in a nonaqueous solution such as dichloromethane can add to an alkene such as cyclohexene through a free radical process:

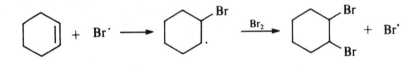

Alkenes react with bromine and evolve HBr

Similarly, it will react with an alkane and will evolve a molecule of hydrogen bromide. The formation of bromine radicals is promoted by light:

$$Br_2 \xrightarrow{h\nu} 2 \cdot \ddot{B}\ddot{r}:$$

$$CH_3CH_2CH_3 + \cdot \ddot{B}\ddot{r}: \longrightarrow HBr + CH_3\dot{C}HCH_3 \xrightarrow{Br_2} CH_3\underset{|}{\overset{Br}{C}}HCH_3 + \cdot \ddot{B}\ddot{r}:$$

In aqueous solution a bromonium ion is formed as the intermediate when bromine reacts with an alkene. This intermediate bromonium ion can react with a molecule of bromine to form a dibromide, and it can react with water to form a bromohydrin. Both reactions give a mixture of products:

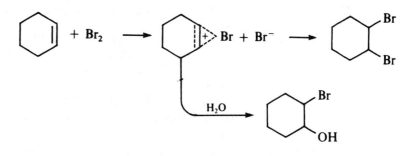

Alkenes decolorize
permanganate

A 10% solution of potassium permanganate is bright purple. When this reagent reacts with an olefin, the purple color disappears and a fine brown suspension of manganese dioxide may be seen. While we will use permanganate simply to distinguish between alkenes and alkanes, it is also an important preparative reagent:

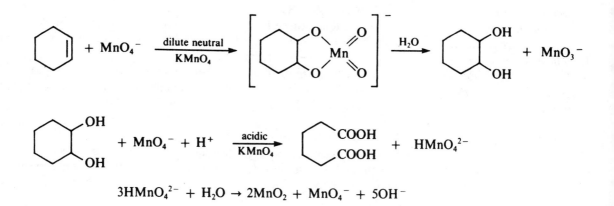

$$3HMnO_4^{2-} + H_2O \rightarrow 2MnO_2 + MnO_4^- + 5OH^-$$

Alkenes dissolve in sulfuric
acid

Alkenes will react with sulfuric acid to form alkyl hydrogen sulfates. In the process the alkene will appear to dissolve in the concentrated sulfuric acid. Alkanes are completely unreactive toward sulfuric acid.

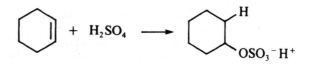

2. Tests for Alkanes and Alkenes

The following tests demonstrate properties characteristic of alkanes and alkenes, provide means of distinguishing between compounds of the two types, and distinguish between pure and impure alkanes. Use your own preparation of cyclohexene (dried over anhydrous sodium sulfate) as a typical alkene, purified 66–75°C ligroin (Eastman Organic Chemicals No. 513) as a typical alkane mixture (the bp of hexane is 69°C), and unpurified ligroin (Eastman No. P513) as an impure alkane. In the following experiments distinguish clearly, as you take notes, between *observations* and *conclusions* based on those observations. Write equations for all positive tests.

These tests can be carried out on a much smaller scale.

(A) Bromine in Nonaqueous Solution. Treat 1.0-mL samples of purified ligroin, unpurified ligroin, and cyclohexene with 5–6 drops of a 3% solution of bromine in dichloromethane. In case decolorization occurs, breathe across the mouth of the tube to see if hydrogen bromide can be detected. If the bromine color persists, illuminate the solution, and if a reaction occurs, test as before for hydrogen bromide.

Bromine (3%) in dichloromethane should be freshly prepared and stored in a brown bottle; test the reagent for decomposition by breathing across the bottle.

(B) Bromine Water. Measure 3 mL of a 3% aqueous solution of bromine into each of three 13 × 100 mm test tubes, then add 1-mL portions of purified ligroin to two of the tubes and 1 mL of cyclohexene to the third. Shake each tube and record the initial results. Put one of the ligroin-containing tubes in the desk out of the light and expose the other to bright sunlight or hold it close to a light bulb. When a change is noted compare the appearance with that of the mixture kept in the dark.

(C) Acid Permanganate Test. To 1-mL portions of purified ligroin, unpurified ligroin, and cyclohexene add a drop of an aqueous solution containing 1% potassium permanganate and 10% sulfuric acid and shake. If the initial portion of reagent is decolorized, add further portions.

CAUTION: Student-prepared cyclohexene may be wet. Dry over anhydrous sodium sulfate before use.

(D) Sulfuric Acid. Cool 1-mL portions of purified ligroin and cyclohexene in ice, treat each with 1 mL of concentrated sulfuric acid, and shake. Observe and interpret the results. Is any reaction apparent? any warming? If the mixture separates into two layers, identify them.

(E) Bromination with Pyridinium Hydrobromide Perbromide ($C_5H_5\overset{+}{N}$ HBr_3)[1] This substance is a crystalline, nonvolatile, odorless complex of high molecular weight (319.84), which, in the presence of a

Note for the instructor

---

1. Crystalline material suitable for small-scale experiments is supplied by Aldrich Chemical Co. Massive crystals commercially available should be recrystallized from acetic acid (4 mL per g). Preparation: Mix 15 mL of pyridine with 30 mL of 48% hydrobromic acid and cool; add 25 g of bromine gradually with swirling, cool, collect the product with use of acetic acid for rinsing and washing. Without drying the solid, crystallize it from 100 mL of acetic acid. Yield of orange needles, 33 g (69%).

bromine acceptor such as an alkene, dissociates to liberate one mole of bromine. For small-scale experiments it is much more convenient and agreeable to measure and use than free bromine.

**Pyridinium hydrobromide
perbromide
MW 319.84**

Add one millimole (320 mg) of the reagent to a 10-mL Erlenmeyer flask and add 2 mL of acetic acid. Swirl the mixture and note that the solid is sparingly soluble. Add one millimole (80 mg) of cyclohexene to the suspension of reagent. Swirl, crush any remaining crystals with a flattened stirring rod, and if after a time the amount of cyclohexene appears insufficient to exhaust the reagent, add a little more. When the solid is all dissolved, dilute with water and note the character of the product. By what property can you be sure that it is the reaction product and not starting material?

(F) Formation of a Bromohydrin.   N-Bromosuccinimide in an aqueous solution will react with an olefin to form a bromohydrin:

**Dioxane
bp 101.5°C, den 1.04
(Dissolves organic
compounds, miscible
with water.)**

Caution: Carcinogen

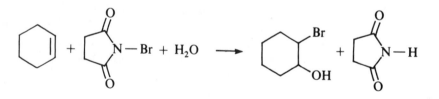

Weigh 178 mg of N-bromosuccinimide, put it into a 13 × 100-mm test tube, and add 0.5 mL of dioxane and 1 millimole (80 mg) of cyclohexene. In another tube chill 0.2 mL of water and add to it 1 millimole of concentrated sulfuric acid. Transfer the cold dilute solution to the first tube with the capillary dropper. Note the result and the nature of the product that separates on dilution with water.

(G) Tests for Unsaturation.   Determine which of the following hydrocarbons are saturated and which are unsaturated or contain unsaturated material. Use any of the above tests that seem appropriate.

Camphene
Pinene, the principal constituent of turpentine
Paraffin oil, a purified petroleum product

Gasoline produced by cracking
Cyclohexane
Rubber (The adhesive Grippit and other rubber cements are solutions
of unvulcanized rubber. Squeeze a drop of it onto a stirring rod and
dissolve it in toluene. For tests with permanganate or with bromine
in dichloromethane use only a drop of the former and just enough of
the latter to produce coloration.)

Cleaning Up   Place the entire test solutions from parts 3(A), 3(B), 3(E),
3(F), and 3(G) in the halogenated organic solvents container. Dilute the
solutions from tests 3(C) and 3(D) with water, place the organic layer in the
organic solvents container and, after neutralization with sodium carbonate,
flush the aqueous layer down the drain.

Questions

1. Draw the structure of the compound formed when cyclohexene dissolves in concentrated sulfuric acid.

2. The reaction of cyclohexene with cold dilute aqueous potassium permanganate gives a compound having the empirical formula $C_6H_{12}O_2$.
What is the structure of this compound? What is the stereochemistry of
the compound?

3. Cyclohexene reacts with a high concentration of bromine to give what
compound? What is its stereochemistry?

4. 1-Hexene is brominated with pyridinium hydrobromide perbromide as
in test (E). The reaction mixture is diluted with water. By what physical
property can you be sure that a reaction product has been produced
and not starting material?

14

Nucleophilic Substitution Reactions of Alkyl Halides

Prelab Exercise: Predict the outcomes of the two sets of experiments to be carried out with the eight halides used in the present experiment.

The alkyl halides, R—X, where X = Cl, Br, I, and sometimes F, play a central role in organic synthesis. They can easily be prepared from, among others, alcohols, alkenes, and industrially, alkanes. In turn, they are the starting materials for the synthesis of a large number of new functional groups. These syntheses are often carried out by nucleophilic substitution reactions in which the halide is replaced by some nucleophile such as cyano, hydroxyl, ether, ester, alkyl—the list is long. As a consequence of the importance of this substitution reaction, it has been studied carefully by employing reactions such as the two used in this experiment. Some of the questions that can be asked include: How does the structure of the alkyl part of the alkyl halide affect the reaction? and What is the effect of changing the nature of the halide, the nature of the solvent, the relative concentrations of the reactants, the temperature of the reaction, or the nature of the nucleophile? In this experiment we shall explore the answers to a few of these questions.

In free radical reactions the covalent bond undergoes homolysis when it breaks

$$R{:}\ddot{\underset{\cdot\cdot}{C}}l{:} \longrightarrow R{\cdot} + {\cdot}\ddot{\underset{\cdot\cdot}{C}}l{:}$$

whereas in ionic reactions it undergoes heterolysis

$$R{:}\ddot{\underset{\cdot\cdot}{C}}l{:} \longrightarrow R^+ + {:}\ddot{\underset{\cdot\cdot}{C}}l{:}^-$$

A carbocation is often formed as a reactive intermediate in these reactions. This carbocation is sp^2 hybridized and trigonal-planar in structure with a

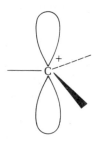

Order of carbocation stability

vacant *p*-orbital. Much experimental evidence of the type obtained in the present experiment indicates that the order of stability of carbocations is

$$
\underset{\overset{\displaystyle |}{R}}{\overset{\displaystyle R}{\underset{\displaystyle |}{R}}}-C^+ > \underset{\overset{\displaystyle |}{H}}{\overset{\displaystyle R}{R}}-C^+ > \underset{\overset{\displaystyle |}{H}}{\overset{\displaystyle H}{R}}-C^+ > \underset{\overset{\displaystyle |}{H}}{\overset{\displaystyle H}{H}}-C^+
$$

The alkyl (R) groups stabilize the positive charge of the carbocation by displacing or releasing electrons toward the positive charge. Delocalization of the charge over several atoms stabilizes the charge.

Many organic reactions occur when a nucleophile (a species with an unshared pair of electrons) reacts with an alkyl halide to replace the halogen with the nucleophile.

Nucleophilic substitution

$$
Nu:^- + R:\ddot{X}: \longrightarrow Nu:R + :\ddot{X}:^-
$$

This substitution reaction can occur in one smooth step

One step

$$
Nu:^- + R:\ddot{X}: \longrightarrow \left[\overset{\delta-}{Nu} \cdots R \cdots \overset{\delta-}{\ddot{X}:} \right] \longrightarrow Nu:R + :\ddot{X}:^-
$$

or it can occur in two discrete steps

Two steps

$$
R:\ddot{X}: \longrightarrow R^+ + :\ddot{X}:^-
$$

$$
Nu:^- + R^+ \longrightarrow Nu:R
$$

depending primarily on the structure of the R group. The nucleophile, Nu:$^-$, can be a substance with a full negative charge, such as $:\ddot{I}:^-$ or $H:\ddot{O}:^-$, or an uncharged molecule with an unshared pair of electrons such as exists on the oxygen atom in water, H—$\ddot{O}$—H. Not all of the halides, $:\ddot{X}:^-$, depart with equal ease in nucleophilic substitution reactions. In this experiment we shall investigate the ease with which the different halogens leave in one of the substitution reactions.

Reaction kinetics

To distinguish between the reaction that occurs as one smooth step and the reaction that occurs as two discrete steps, it is necessary to study the kinetics of the reaction. If the reaction were carried out with several different concentrations of R:$\ddot{X}$: and Nu:$^-$, we could determine if the reaction is bimolecular or unimolecular. In the case of the smooth, one-step reaction, the nucleophile must collide with the alkyl halide. The kinetics of the reaction

$$
Nu:^- + R:\ddot{X}: \longrightarrow \left[\overset{\delta-}{Nu} \cdots R \cdots \overset{\delta-}{\ddot{X}:} \right] \longrightarrow Nu:R + :\ddot{X}:^-
$$

$$
Rate = k\left[Nu:^- \right]\left[R:\ddot{X}: \right]
$$

are found to depend on the concentration of both the nucleophile and the halide. Such a reaction is said to be a bimolecular nucleophilic substitution reaction, S_N2. If the reaction occurs as a two-step process

$$R:\ddot{X}: \xrightarrow{slow} R^+ + :\ddot{X}:^-$$

$$NU:^- + R^+ \xrightarrow{fast} Nu:R$$

$$Rate = k\left[R:\ddot{X}:\right]$$

the rate of the first step, the slow step, depends only on the concentration of the halide, and it is said to be a unimolecular nucleophilic substitution reaction, S_N1.

Racemization: S_N1

The S_N1 reaction proceeds through a planar carbocation. Even if the starting material were chiral, the product would be a 50:50 mixture of enantiomers because the intermediate is planar.

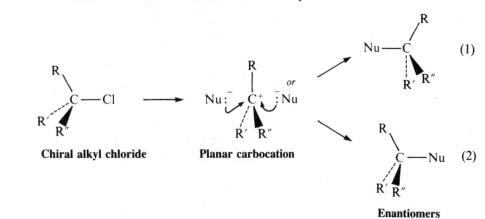

Chiral alkyl chloride   **Planar carbocation**   **Enantiomers**

Inversion: S_N2

The S_N2 reaction occurs with inversion of configuration to give a product of the opposite chirality from the starting material.

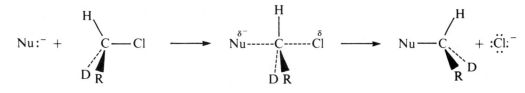

The order of reactivity for *simple* alkyl halides in the S_N2 reaction is

Order of S_N2 reactivity

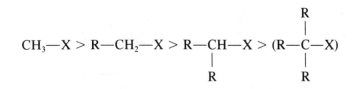

$$CH_3{-}X > R{-}CH_2{-}X > R{-}\underset{\underset{R}{|}}{\overset{\overset{}{}}{C}}H{-}X > (R{-}\underset{\underset{R}{|}}{\overset{\overset{R}{|}}{C}}{-}X)$$

The tertiary halide is in parentheses because it usually does not react by an S_N2 mechanism. The primary factor in this order of reactivity is steric hindrance, i.e., the ease with which the nucleophile can come within bonding distance of the alkyl halide. 2,2-Dimethyl-1-bromopropane

Sterically hindered toward
S_N2

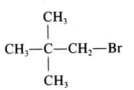

even though it is a primary halide, reacts 100,000 times slower than ethyl bromide, CH_3CH_2Br, because of steric hindrance to attack on the bromine atom in the dimethyl compound.

The primary factor in S_N1 reactivity is the relative stability of the carbocation that is formed. For simple alkyl halides, this means that only tertiary halides react by this mechanism. The tertiary halide must be able to form a planar carbocation. Only slightly less reactive are the allyl carbocations, which derive their great stability from the delocalization of the charge on the carbon by resonance

The allyl carbocation

$$CH_2{=}CH{-}CH_2{-}\ddot{B}r: \longrightarrow CH_2{=}CH{-}\overset{+}{C}H_2 \longleftrightarrow \overset{+}{C}H_2{-}CH{=}CH_2 + :\ddot{B}r:^-$$

Solvent effects

The nature of the solvent has a large effect on the rates of S_N2 reactions. In a solvent with a hydrogen atom attached to an electronegative atom such as oxygen, the protic solvent forms hydrogen bonds to the nucleophile.

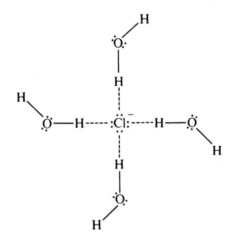

These solvent molecules get in the way during an S_N2 reaction. If the solvent is polar and aprotic, solvation of the nucleophile cannot occur and

*Aprotic: no ionizable
protons*

the S_N2 reaction can occur up to a million times faster. Some common polar, aprotic solvents are

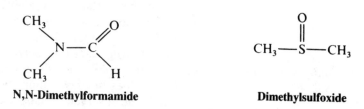

N,N-Dimethylformamide   **Dimethylsulfoxide**

In the S_N1 reaction a polar protic solvent such as water stabilizes the transition state more than it does the reactants, lowering the energy of activation for the reaction and thus increasing the rate, relative to the rate in a nonpolar solvent. Acetic acid, ethanol, and acetone are relatively nonpolar solvents and have lower dielectric constants than the polar solvents water, dimethylsulfoxide, and N,N-dimethylformamide.

The leaving group

The rate of S_N1 and S_N2 reactions depends on the nature of the leaving group, the best leaving groups being the ones that form stable ions. Among the halogens we find that iodide ion, I^-, is the best leaving group as well as the best nucleophile in the S_N2 reaction.

Vinylic and aryl halides

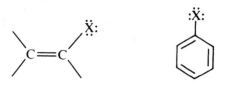

do not normally react by S_N1 or S_N2 reactions because the resulting carbocations

are relatively unstable. The electrons in the nearby double bonds repel the nucleophile, which is either an ion or a polarized neutral species.

Temperature dependence

The rates of both S_N1 and S_N2 reactions depend on the temperature of the reaction. As the temperature increases the kinetic energy of the molecules increases, leading to a greater rate of reaction. The rate of many organic reactions will approximately double when the temperature increases about 10°C.

Experiments

Sodium iodide in acetone reacts by S_N2

In the experiments that follow, eight representative alkyl halides are treated with sodium iodide in acetone and with an ethanolic solution of silver nitrate. Acetone, with a dielectric constant of 21, is a relatively nonpolar solvent that will readily dissolve sodium iodide. The iodide ion is an excellent nucleophile, and the nonpolar solvent, acetone, favors the S_N2 reaction; it does not favor ionization of the alkyl halide. The extent of reaction can be observed because sodium bromide and sodium chloride are not soluble in acetone and precipitate from solution if reaction occurs.

$$Na^+I^- + R—Cl \longrightarrow R—I + NaCl \downarrow$$

$$Na^+I^- + R—Br \longrightarrow R—I + NaBr \downarrow$$

Ethanolic silver nitrate reacts by S_N1

When an alkyl halide is treated with an ethanolic solution of silver nitrate, the silver ion coordinates with an electron pair of the halogen. This weakens the carbon–halogen bond as a molecule of insoluble silver halide is formed, thus promoting an S_N1 reaction of the alkyl halide. The solvent, ethanol, favors ionization of the halide, and the nitrate ion is a very poor nucleophile, so alkyl nitrates do not form by an S_N2 reaction.

$$R—\overset{..}{\underset{..}{X}}: \overset{Ag^+}{\rightleftharpoons} \overset{\delta+}{R} \overset{..}{\underset{..}{X}} \overset{\delta+}{Ag} \longrightarrow R^+ + AgX \downarrow$$

On the basis of the foregoing discussion tertiary halides would be expected to react with silver nitrate most rapidly and primary halides least rapidly.

Procedure

Label eight small containers (reaction tubes, 3-mL centrifuge tubes, 10 × 75-mm test tubes or 1-mL vials) and place 0.1 mL or 100 mg of each of the following halides in the tubes.

$$CH_3CH_2CH_2CH_2Cl$$

1-Chlorobutane
bp 77–78°C

$$CH_3CH_2CH_2CH_2Br$$

1-Bromobutane
bp 100–104°C

$$CH_3CH_2\overset{\overset{\displaystyle Cl}{|}}{C}HCH_3$$

2-Chlorobutane
bp 68–70°C

$$CH_3—\overset{\overset{\displaystyle CH_3}{|}}{\underset{\underset{\displaystyle CH_3}{|}}{C}}—Cl$$

2-Chloro-2-methylpropane
bp 51–52°C

$$CH_3CH=CHCH_2Cl$$

1-Chloro-2-butene mixture
of *cis* and *trans* isomers
bp 63.5°C (*cis*)
68°C (*trans*)

$$CH_3\overset{\overset{\displaystyle CH_3}{|}}{C}HCH_2Cl$$

1-Chloro-2-methylpropane
bp 68–69°C

Chlorobenzene
bp 132°C

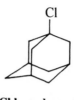

1-Chloroadamantane
mp 165–166°C

To each tube then rapidly add 1 mL of an 18% solution of sodium iodide in acetone, stopper each tube, mix the contents thoroughly, and note the time. Note the time of first appearance of any precipitate. If no reaction occurs within about 5 min place those tubes in a 50°C water bath and watch for any reaction over the next 5 or 6 min.

Empty the tubes, rinse them with ethanol, place the same amount of each of the alkyl halides in each tube as in the first part of the experiment, add 1 mL of 1% ethanolic silver nitrate solution to each tube, mix the contents well, and note the time of addition as well as the time of appearance of the first traces of any precipitate. If a precipitate does not appear in 5 min heat those tubes in a 50°C water bath for 5 to 6 min and watch for any reaction.

To test the effect of solvent on the rate of S_N1 reactivity, compare the time needed for a precipitate to appear when 2-chlorobutane is treated with 1% ethanolic silver nitrate solution (above) and when treated with 1% silver nitrate in a mixture of 50% ethanol and 50% water.

In your analysis of the results from these experiments consider the following for both S_N1 and S_N2 conditions: The nature of the leaving group (Cl vs. Br) in the 1-halobutanes; the effect of structure, i.e., compare simple primary, secondary, and tertiary halides, unhindered primary vs. hindered primary halides, a simple tertiary halide vs. a complex tertiary halide, and an allylic halide vs. a tertiary halide; the effect of solvent polarity on the S_N1 reaction; and the effect of temperature on the reaction.

Cleaning Up   Since all of the test solutions contain halogenated material, all test solutions and washes as well as unused starting materials should be placed in the halogenated organic waste container.

Questions

1. What would be the effect of carrying out the sodium iodide in acetone reaction with the alkyl halides using an iodide solution half as concentrated?

2. The addition of sodium or potassium iodide catalyzes many S_N2 reactions of alkyl chlorides or bromides. Explain.

The S$_N$2 Reaction: 1-Bromobutane

Prelab Exercise: Prepare a detailed flow sheet for the isolation and purification of *n*-butyl bromide. Indicate how each reaction by-product is removed and which layer is expected to contain the product in each separation step.

$$CH_3CH_2CH_2CH_2OH \xrightarrow{\text{NaBr, H}_2\text{SO}_4} CH_3CH_2CH_2CH_2Br + NaHSO_4 + H_2O$$

1-Butanol
bp 118°C
den 0.810
MW 74.12

1-Bromobutane
bp 101.6°C
den 1.275
MW 137.03

Choice of reagents

A primary alkyl bromide can be prepared by heating the corresponding alcohol with (a) constant-boiling hydrobromic acid (47% HBr); (b) an aqueous solution of sodium bromide and excess sulfuric acid, which is an equilibrium mixture containing hydrobromic acid; or (c) with a solution of hydrobromic acid produced by bubbling sulfur dioxide into a suspension of bromine in water. Reagents (b) and (c) contain sulfuric acid at a concentration high enough to dehydrate secondary and tertiary alcohols to undesirable by-products (alkenes and ethers) and hence the HBr method (a) is preferred for preparation of halides of the types R$_2$CHBr and R$_3$CBr. Primary alcohols are more resistant to dehydration and can be converted efficiently to the bromides by the more economical methods (b) and (c), unless they are of such high molecular weight as to lack adequate solubility in the aqueous mixtures. The NaBr-H$_2$SO$_4$ method is preferred to the Br$_2$-SO$_2$ method because of the unpleasant, choking property of sulfur dioxide. The overall equation is given above, along with key properties of the starting material and principal product.

The procedure that follows specifies a certain proportion of 1-butanol, sodium bromide, sulfuric acid, and water; defines the reaction temperature and time; and describes operations to be performed in working up the reaction mixture. The prescription of quantities is based upon considerations of stoichiometry as modified by the results of experimentation. Before undertaking a preparative experiment you should analyze the procedure and calculate the molecular proportions of the reagents. Construction of tables (see p. 178) of properties of starting material, reagents, products, and by-products provides guidance in regulation of temperature and in separa-

Reagents

Reagent	MW	Den	Bp (°C)	Wt used (g)	Moles Theory	Moles Used
n-C$_4$H$_9$OH	74.12	0.810	118	8.0	0.108	0.108
NaBr	102.91	—	—	13.3	0.108	0.129
H$_2$SO$_4$	98.08	1.84	—	20.	0.108	0.200

Product and By-Products

Compound	MW	Den	Bp (°C) Given	Bp (°C) Found	Theoretical Yield Moles	Theoretical Yield g	Found g	Found %
n-C$_4$H$_9$Br	137.03	1.275	101.6		0.11	14.8 (100%)	__g	__%
CH$_3$CH$_2$CH=CH$_2$			−6.3					
C$_4$H$_9$OC$_4$H$_9$			141					

The laboratory notebook

tion and purification of the product and should be entered in the laboratory notebook.

$$NaBr + H_2SO_4 \rightleftharpoons HBr + NaHSO_4$$

One mole of 1-butanol theoretically requires one mole each of sodium bromide and sulfuric acid, but the procedure calls for use of a slight excess of bromide and twice the theoretical amount of acid. Excess acid is used to shift the equilibrium in favor of a high concentration of hydrobromic acid. The amount of sodium bromide taken, arbitrarily set at 1.2 times the theory as an insurance measure, is calculated as follows:

$$\frac{8 \text{ (g of C}_4\text{H}_9\text{OH)}}{74.12 \text{ (MW of C}_4\text{H}_9\text{OH)}} \times 102.91 \text{ (MW of NaBr)} \times 1.2 = 13.3 \text{ g of NaBr}$$

The theoretical yield is 0.11 mole of product, corresponding to the 0.11 mole of butyl alcohol taken; the maximal weight of product is calculated thus:

$$0.11 \text{ (mole of alcohol)} \times 137.03 \text{ (MW of product)} = 14.8 \text{ g 1-bromobutane}$$

The probable by-products are 1-butene, dibutyl ether, and the starting alcohol. The alkene is easily separable by distillation, but the other substances are in the same boiling point range as the product. However, all three possible by-products can be eliminated by extraction with concentrated sulfuric acid.

Experiments

1. Synthesis of 1-Bromobutane

Put 13.3 g of sodium bromide, 15 mL of water, and 10 mL of n-butyl alcohol in a 100-mL round-bottomed flask, cool the mixture in an ice-water bath, and slowly add 11.5 mL of concentrated sulfuric acid with swirling and cooling. Place the flask in an electric flask heater, clamp it securely, and fit it with a short condenser for reflux condensation (Fig. 1). Heat to the boiling point, note the time, and adjust the heat for brisk, steady refluxing. The upper layer that soon separates is the alkyl bromide, since the aqueous solution of inorganic salts has a greater density. Reflux for 45 min, remove the heat, and let the condenser drain for a few minutes (extension of the reaction period to 1 h increases the yield by only 1–2%). Remove the condenser, mount a stillhead in the flask, and set the condenser for downward distillation (see Fig. 5 in Chapter 5) through a bent or vacuum adapter into a 50-mL Erlenmeyer. Distill the mixture, make frequent readings of the temperature, and distill until no more water-insoluble droplets come over, by which time the temperature should have reached 115°C (collect a few drops of distillate in a test tube and see if it is water soluble). The increasing boiling point is due to azeotropic distillation of *n*-butyl bromide with water containing increasing amounts of sulfuric acid, which raises the boiling point.

Pour the distillate into a separatory funnel, shake with about 10 mL of water, and note that *n*-butyl bromide now forms the lower layer. A pink coloration in this layer due to a trace of bromine can be discharged by adding a pinch of sodium bisulfite and shaking again. Drain the lower layer of 1-bromobutane into a clean flask, clean and dry the separatory funnel, and return the 1-bromobutane to it. Then cool 10 mL of concentrated sulfuric acid thoroughly in an ice bath and add the acid to the funnel, shake well, and allow 5 min for separation of the layers. Use care in handling concentrated sulfuric acid. Check to see that the stopcock and stopper don't leak. The relative densities given in the tables presented in the introduction of this experiment identify the two layers; an empirical method of telling the layers apart is to draw off a few drops of the lower layer into a test tube and see whether the material is soluble in water (H$_2$SO$_4$) or insoluble in water (bromobutane). Separate the layers, allow 5 min for further drainage, and separate again. Then wash the 1-bromobutane with 10 mL of 10% sodium hydroxide (den. 1.11) solution to remove traces of acid, separate, and be careful to save the proper layer.

Dry the cloudy 1-bromobutane by adding 1 g of anhydrous calcium chloride with swirling until the liquid clears. Further drying can be effected by transferring the liquid to another flask and adding anhydrous sodium sulfate until the drying agent no longer clumps together. After 5 min decant the dried liquid into a 25-mL flask or filter it through a fluted filter paper, add a boiling stone, distill, and collect material boiling in the range 99–103°C. A typical student yield is in the range 10–12 g. Note the approximate volumes of forerun and residue.

Put the sample in a narrow-mouth bottle of appropriate size; make a neatly printed label giving the name and formula of the product and your name. Press the label onto the bottle under a piece of filter paper and make sure that it is secure. After all the time spent on the preparation, the final product should be worthy of a carefully executed and secured label.

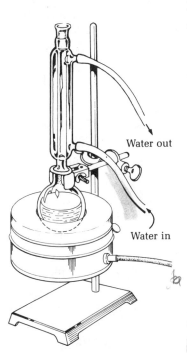

FIG. 1 Refluxing a reaction mixture.

Calcium chloride: removes both water and alcohol from a solution. Not as efficient a drying agent as anhydrous sodium sulfate.

A proper label is important

Check purity by TLC and IR. The refractive index can also be checked.

Cleaning Up Carefully dilute all nonorganic material with water (the reaction pot residue, the sulfuric acid wash, and the sodium hydroxide wash), and combine and neutralize with sodium carbonate before flushing down the drain with excess water. The residue from the distillation of 1-bromobutane goes in the container for halogenated organic solvents. The drying agent, after the solvent is allowed to evaporate from it in the hood, goes in the nonhazardous solid waste container.

Questions

1. What experimental method would you recommend for the preparation of 1-bromooctane? *t*-Butyl bromide?

2. Explain why the crude product is apt to contain certain definite organic impurities.

3. How does each of these impurities react with sulfuric acid when the crude 1-bromobutane is shaken with this reagent?

4. How should the reaction conditions in the present experiment be changed to try to produce 1-chlorobutane?

5. Write a balanced equation for the reaction of sodium bisulfite with bromine.

6. What is the purpose of refluxing the reaction mixture for 30 min? Why not simply boil the mixture in an Erlenmeyer flask?

7. Why is it necessary to remove all water from 1-bromobutane before distilling it?

8. Write reaction mechanisms showing how 1-butene and di-*n*-butyl ether are formed.

9. Why is the resonance of the bromine-bearing carbon atom and the hydroxyl-bearing carbon atom farthest downfield in Figs. 2 and 3?

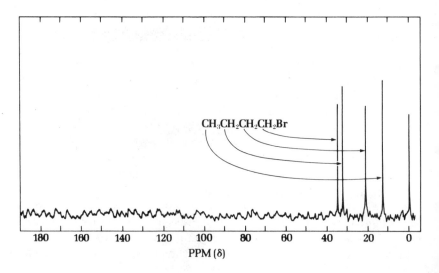

$CH_3CH_2CH_2CH_2Br$

FIG. 2 ^{13}C nmr spectrum of *n*-butyl bromide (22.6 MHz).

PPM (δ)

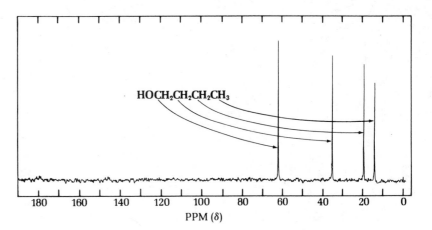

FIG. 3 ^{13}C nmr spectrum of 1-butanol (22.6 MHz).

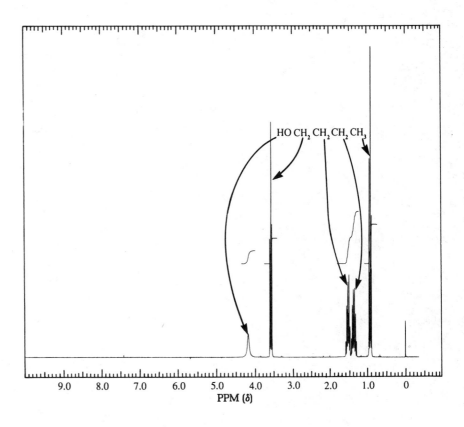

FIG. 4 ^{1}H nmr spectrum of 1-butanol (250 MHz).

16

Liquid Chromatography

Prelab Exercise: Read the instruction manual for the HPLC apparatus you will use, and acquaint yourself with the operating controls for the instrument. Consult reference works on HPLC for examples of solvents and column packings that will effect the separation you would like to carry out.

The newest addition to the chromatographic family of analytical procedures is liquid chromatography, most commonly known as HPLC, which stands for high-performance (or high-pressure) liquid chromatography. This form of chromatography resembles column chromatography (Chapter 10) except that the process involves pumping liquid eluant at high pressure through the column instead of depending on gravity for flow. Rather than evaporating a number of fractions to detect separated products, the experimenter runs the eluant through a detector, most commonly an ultraviolet detector or a refractive index detector. The detector is connected to a chart recorder that gives a record of uv absorbance or refractive index versus mL of eluant—a chromatogram that is very similar to one produced by a gas chromatograph. The uv detector can detect as little as 10^{-10} g of solute, while the refractive index detector can detect 10^{-6} g of solute. The column can be packed with material that separates substances on the basis of their molecular sizes (exclusion chromatography), their ionic charges (ion exchange chromatography), their ability to adsorb to the packing as in ordinary column chromatography (adsorption chromatography), or their ability to partition between a stationary phase attached to the column and a mobile phase that is pumped through the column (partition chromatography). Partition column chromatography is most common in HPLC and will therefore be the focus of this discussion.

The partitioning of a solute between two immiscible solvents is what takes place in a separatory funnel during the process of extraction (Chapter 8). In partition chromatography this extraction process occurs repeatedly as the eluant, containing the mixture to be separated, flows past the fixed organic phase that is attached to beads of silica gel in the column. The advantage of this form of chromatography over gas chromatography is that substances of high molecular weight having very small chemical or structural differences can be separated. For example, insulins from a variety of animals can be separated from one another even though they differ by just one or two amino acids out of 51 and have molecular weights near 6000.

FIG. 1   Silica gel.

Outer Surface

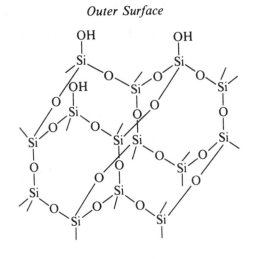

The stationary phase in partition chromatography is usually a long hydrocarbon chain covalently bound to silica gel. Silica gel has the structure shown in Fig. 1. The siloxane bonds ($\sim$Si—O—Si$\sim$) are stable to water and the other solvents used in HPLC between pH 2 and 9. The silanol ($\sim$Si—OH) groups can be bound to long hydrocarbon chains through reactions such as the following:

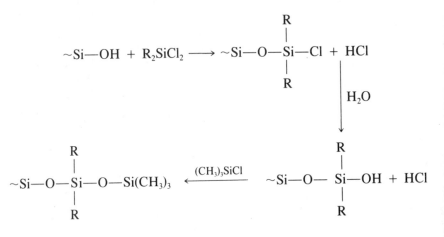

$$R = (CH_2)_{17}CH_3 = \text{octadecyl}$$

The most common R group is the octadecyl group (C-18), which leaves the silica particles coated with hydrocarbon chains. The silica particles are very small ($\sim$40 μ dia. = 4×10^{-3} cm dia.) and very uniform in size. With very small, uniform particles, a molecule in the solvent can rapidly diffuse to the surface of the packing and undergo partitioning between the stationary and mobile phases. Such small-diameter particles present a large surface area to the eluant and offer considerable resistance to the flow of the eluant. Consequently, for a typical column 4 mm in diameter and 25 cm long, the inlet pressure necessary to pump the eluant at a flow rate of 1 mL/ min is often in excess of 1000 psi and can reach as high as 5000 psi. Because the column is packed with such exceedingly fine particles, it is subject to clogging from particulate matter in the samples and solvent. Both must be filtered through very fine membrane filters or centrifuged before being applied to the column.

Because it would be difficult to inject a sample onto the column having a back pressure of >1000 psi at the inlet, a loop injector is employed (Fig. 2). In the load position the sample is injected into the loop, which can be of any capacity, often 0.1 mL. During this time the pump is pushing pure solvent through the valve and column. When the valve is turned, the pump pushes the sample onto the column. Pressure is maintained and the sample goes onto the column in a small volume plug of solution.

A block diagram for a typical high-performance liquid chromatograph is shown in Fig. 3.

An isocratic HPLC is one in which only one solvent or solvent mixture is pumped through the column. An HPLC equipped for gradient elution can automatically mix two or more solvents in changing proportions (a solvent gradient) so that maximum separation of solutes is achieved in the minimum time. Because of the necessity for a pump capable of generating high pressures, a special injection valve, and a uv detector, the cost of HPLC apparatus is several times that for a gas chromatograph. The packed columns cost several hundred dollars each, and the carefully purified solvents cost several times as much as reagent-grade solvents commonly found in the laboratory. In return for this investment, HPLC is proving to be an extremely versatile analytical tool of great sensitivity. HPLC is usually used for analytical purposes only; however, it is possible to collect the eluant, evaporate the solvent, and examine the residue by uv, ir, mass, or

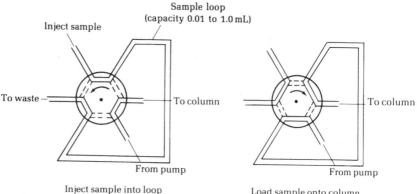

FIG. 2   Rotary loop injector for HPLC.

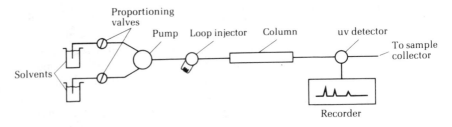

FIG. 3 Gradient elution high-performance liquid chromatograph.

nmr spectroscopy. Up to 10 mg of material can be separated on one injection into an analytical HPLC by overloading the column. Special instruments are manufactured for preparative scale separations. Figure 4 shows a typical separation of a variety of monosubstituted aromatic compounds.

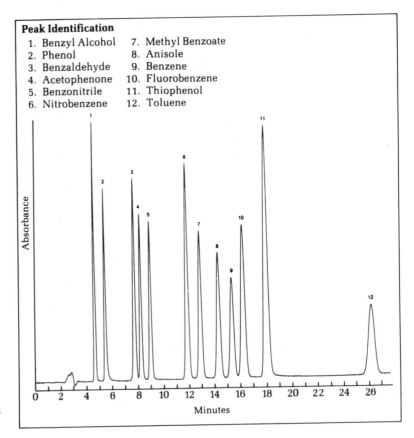

Peak Identification

1. Benzyl Alcohol
2. Phenol
3. Benzaldehyde
4. Acetophenone
5. Benzonitrile
6. Nitrobenzene
7. Methyl Benzoate
8. Anisole
9. Benzene
10. Fluorobenzene
11. Thiophenol
12. Toluene

FIG. 4 Separation of 10 μL of a mixture of monosubstituted aromatic compounds by HPLC employing an octadecyl (C-18) column 4.5-mm dia. × 25 cm long. The mobile phase is 40% acetonitrile/60% H_2O and the flow rate is 1 mL/min. A uv detector operating at 254 nm is used to detect the peaks.

Experiment

Analysis of Mixtures Using Liquid Chromatography

Many of the crude reaction mixtures and extracts encountered in the organic laboratory can be analyzed by HPLC. The most common apparatus is an isocratic (single eluant composition) system employing a C-18 column and an ultraviolet detector operating at 254 nm. Methanol–water mixtures or acetonitrile–water mixtures can be used to elute the compounds. It is absolutely necessary to employ HPLC-grade solvents—methanol, acetonitrile, and water—and to filter each sample through a membrane filter or to centrifuge it at high speed in a clinical centrifuge for 5 min. Employing these precautions ensures that the column will not be blocked. It is good practice to employ a short renewable guard column in front of the main column, as further insurance against blockage.

Use extreme caution to ensure that no insoluble material is injected onto the HPLC column. Do not exceed the recommended pump pressure

Dissolve the sample in exactly the same solvent or solvent mixture being pumped through the column, filter or centrifuge the solution, and inject it into the chromatograph. If the retention time is not long enough or the components are not well resolved, increase the percent of the more polar solvent (water) in the eluant.

HPLC on an octadecyl column can be applied to the analysis of caffeine from tea or cola syrup (Chapter 8—use 20% methanol, 0.8% acetic acid, and 79.2% water as eluant), the acetylation of cholesterol (Chapter 10) pulegone and citronellal from citronellol (Chapter 25), cholesterol from gallstones (Chapter 22), the isolation of eugenol from cloves (Chapter 28—use 10% methanol, 5.4% acetic acid, and 84.6% water as eluant), isolation of lycopene and β-carotene (Chapter 9), and the product obtained from enzymatic reduction of ethyl acetoacetate (Chapter 59).

The ingredients of common pain relievers (acetaminophen, caffeine, salicylamide, aspirin, and salicyclic acid) can be separated using 20% methanol, 0.8% acetic acid, and 79.2% water until the caffeine peak appears. Then the eluant is changed to 20% methanol, 0.6% acetic acid, and 59.4% water using a flow rate of 3 mL/min. If commercial tablets are used, be sure to filter or centrifuge the solution before use in order to remove starch that is used as a binder in the tablets.

Questions

1. Which would you expect to elute first from a C-18 column, eluting with methanol: a C-10 or a C-20 saturated hydrocarbon?

2. Why would you have difficulty detecting saturated hydrocarbons using the uv detector?

3. Since saturated and monounsaturated fatty acids don't absorb uv light, how might they be modified for detection in an HPLC apparatus?

CHAPTER

17

Separation and Purification of the Components of an Analgesic Tablet: Aspirin, Caffeine, and Acetaminophen

Prelab Exercise: Prepare a detailed flow sheet for the separation of aspirin, caffeine, and acetaminophen from an analgesic tablet.

This experiment puts into practice the techniques learned in several previous chapters for separating and purifying, on a very small scale, the components in a common analgesic tablet. It is presumed that thin-layer or high-performance liquid chromatography has already shown that the tablet does indeed contain aspirin, caffeine, and acetaminophen.

Many pharmaceutical tablets are held together with a binder to prevent the components from crumbling on storage or while being swallowed. A close reading of the contents on the package will disclose the nature of the binder. Starch is commonly used, as is microcrystalline cellulose and silica gel. All of these have one property in common: they are insoluble in water and common organic solvents.

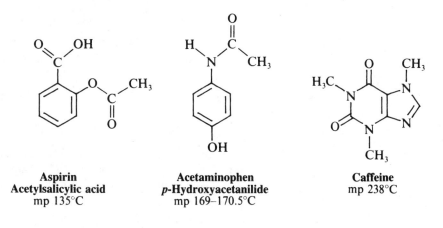

Aspirin	**Acetaminophen**	**Caffeine**
Acetylsalicylic acid	*p*-Hydroxyacetanilide	mp 238°C
mp 135°C	mp 169–170.5°C	

Inspection of the structures of caffeine, acetylsalicylic acid, and acetaminophen reveals that one is a base, one a strong organic acid, and one a weak organic acid. One might be tempted to separate this mixture using exactly the same procedure as employed in Chapter 8 to separate benzoic acid, 2-naphthol, and 1,4-dimethoxybenzene, i.e., dissolve the mixture in dichloromethane; remove the benzoic acid (the strong acid) by reaction with bicarbonate ion, a weak base; then remove the naphthol, the weak acid, by reaction with hydroxide, a strong base. This process would leave the neutral compound, 1,4-dimethoxybenzene, in the dichloromethane solution.

In the present experiment the solubility data (see Table 1) reveal that the weak acid, acetaminophen, is not soluble in ether, chloroform, or dichloromethane, so it cannot be extracted by strong base. We can take advantage of this lack of solubility by dissolving the other two components, caffeine and aspirin, in dichloromethane, and removing the acetaminophen

TABLE 1 Solubilities

	Water	Ethanol	Chloroform	Diethyl ether	Ligroin
Aspirin	0.33 g/100 mL 25°C 1 g/100 mL 37°C	1 g/5 mL	1 g/17 mL	1 g/13 mL	
Acetaminophen	v. sl. sol. cold sol. hot	sol.	sol.	sl. sol.	ins.
Caffeine	1 g/46 mL 25°C 1 g/5.5 mL 80°C 1 g/1.5 mL 100°C	1 g/66 mL 25°C 1 g/22 mL 60°C	1 g/5.5 mL	1 g/530 mL	sl. sol.

by filtration. The binder is also insoluble in dichloromethane, but treatment of the solid mixture with ethanol will dissolve the acetaminophen and not the binder. They can then be separated by filtration and the acetaminophen isolated by evaporation of the ethanol.

This experiment is a test of technique. It is not easy to separate and crystallize a few milligrams of a compound that occurs in a mixture.

Experiment

Extraction and Purification Procedure

In a mortar, grind two Extra Strength Excedrin tablets to a very fine powder. The label states that this analgesic contains 250 mg of aspirin, 250 mg of acetaminophen, and 65 mg of caffeine per tablet. Place this powder in a test tube and add to it 7.5 mL of dichloromethane. Warm the mixture briefly and note that a large part of the material does not dissolve. Filter the mixture into another test tube. This can be done by transferring the slurry to a funnel equipped with a piece of filter paper. Use a Pasteur pipette and complete the transfer with a small portion of dichloromethane. This filtrate is Solution 1.

Transfer the powder on the filter to a test tube, add 4 mL of ethanol, and heat the mixture to boiling (boiling stick). Not all of the material will go into solution. That which does not is the binder. Filter the mixture into a

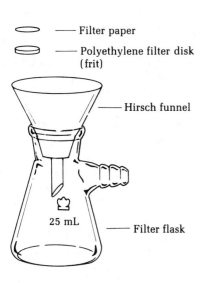

⬭ —— Filter paper

⬮ —— Polyethylene filter disk
       (frit)

—— Hirsch funnel

25 mL

—— Filter flask

FIG. 1   Hirsch funnel with
integral adapter, polyethylene
frit, and 25-mL filter flask.

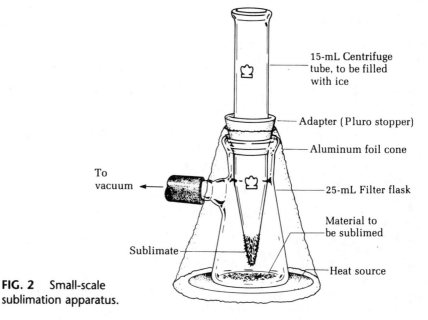

15-mL Centrifuge
tube, to be filled
with ice

Adapter (Pluro stopper)

Aluminum foil cone

To
vacuum ←

25-mL Filter flask

Material to
be sublimed

Sublimate —

Heat source

FIG. 2   Small-scale
sublimation apparatus.

Acetaminophen

tared test tube and complete the transfer and washing using a few drops of
hot ethanol. This is Solution 2.

Evaporate about two-thirds of Solution 2 by boiling off the ethanol
(boiling stick), or better by warming the solution and blowing a stream of air
into the test tube. Heat the residue to boiling (add a boiling stick to prevent
bumping) and add more ethanol if necessary to bring the solid into solution.
Allow the saturated solution to cool slowly to room temperature to deposit
crystals of acetaminophen, which is reported to melt at 169–170.5°C. After
the mixture has cooled to room temperature cool it in ice for several
minutes, remove the solvent with a Pasteur pipette, wash the crystals once
with two drops of ice-cold ethanol, remove the ethanol, and dry the crystals
under aspirator vacuum while heating the tube on a steam bath.

Alternatively, the original ethanol solution is evaporated to dryness
and the residue recrystallized from boiling water. The crystals are best
collected and dried on a Hirsch funnel (Fig. 1). Once the crystals are dry

Check product purity by TLC using 25:1 ethyl acetate/acetic acid to elute the silica gel plates

determine their weight and melting point. Thin-layer chromatographic analysis and the melting points of these crystals and the two other components of this mixture will indicate their purity.

The dichloromethane filtered from the binder and acetaminophen mixture (Solution 1) should contain caffeine and aspirin. These can be separated by extraction either with acid, which will remove the caffeine as a water-soluble salt, or by extraction with base, which will remove the aspirin as a water-soluble salt. We shall use the latter procedure.

To the dichloromethane solution in a test tube add 4 mL of 3M sodium hydroxide solution and shake the mixture thoroughly. Remove the aqueous layer, add 1 mL more water, shake the mixture thoroughly, and again remove the aqueous layer, which is combined with the first aqueous extract.

To the dichloromethane add anhydrous sodium sulfate until the drying agent no longer clumps together. Shake the mixture over a 5- or 10-min period to complete the drying process, then remove the solvent, wash the drying agent with more solvent, and evaporate the combined extracts to dryness under a stream of air to leave crude caffeine.

Caffeine

The caffeine can be purified by sublimation as has been done in the experiment in which it was extracted from tea (Fig. 2) or it can be purified by crystallization. Recrystallize the caffeine by dissolving it in the minimum quantity of 30% ethanol in tetrahydrofuran. It can also be crystallized by dissolving the product in a minimum quantity of hot toluene or acetone and adding to this solution ligroin (hexanes) until the solution is cloudy while at the boiling point. In any case allow the solution to cool slowly to room temperature, then cool the mixture in ice, and remove the solvent from the crystals with a Pasteur pipette. Remove the remainder of the solvent under aspirator vacuum, and determine the weight of the caffeine and its melting point.

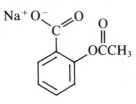

Sodium Acetylsalicylate
(soluble in water)

The aqueous hydroxide extract contains aspirin as the sodium salt of the carboxylic acid. To the aqueous solution add 3M hydrochloric acid dropwise until the solution tests strongly acid to indicator paper, and then add two more drops of acid. This will give a suspension of white acetylsalicylic acid in the aqueous solution. It could be filtered off and recrystallized from boiling water, but this would entail losses in transfer. An easier procedure is to simply heat the aqueous solution that contains the precipitated aspirin and allow it to crystallize on slow cooling.

Aspirin

Add a boiling stick and heat the mixture to boiling, at which time the aspirin should dissolve completely. If it does not, add more water. Long boiling will hydrolyze the aspirin to salicyclic acid, mp 157–159°C. Once completely dissolved the aspirin should be allowed to crystallize slowly as the solution cools to room temperature in an insulated container. Once the tube has reached room temperature it should be cooled in ice for several minutes and then the solvent removed with a Pasteur pipette. The crystals are to be washed with a few drops of ice-cold water and then scraped out onto a piece of filter paper. Squeezing the crystals between sheets of the

Alternative procedure: use the Hirsch funnel (Fig. 8 in Chapter 3) to isolate the aspirin

filter paper will hasten drying. Once they are completely dry, determine the weight of the acetylsalicylic acid and its melting point.

Cleaning Up Place any dichloromethane-containing solutions in the ha- halogenated organic waste container and the other organic liquids in the organic solvents container. The aqueous layers should be diluted and neutralized with sodium carbonate before being flushed down the drain. After it is free of solvent, the sodium sulfate can be placed in the nonhazardous solid waste container.

Questions

1. Write equations showing how caffeine could be extracted from an organic solvent and subsequently isolated.

2. Write equations showing how acetaminophen might be extracted from an organic solvent such as ether, if it were soluble.

3. Write detailed equations showing the mechanism by which aspirin is hydrolyzed in boiling, slightly acidic water.

■CHAPTER■
18

Biosynthesis of Ethanol

Prelab Exercise: List the essential chemical substances, the solvent, and conditions for converting glucose to ethanol.

Fermentation

Human beings have been preparing fermented beverages for more than 5000 years. Materials excavated from Egyptian tombs dating to the third millennium B.C. demonstrate the operations used in making beer and leavened bread. The history of fermentation, whereby sugar is converted to ethanol by the action of yeast, is also a history of chemistry. The word "gas" was coined by van Helmont in 1610 to describe the bubbles produced in fermentation. Leeuwenhoek observed and described the cells of yeast with his newly invented microscope in 1680. Joseph Black in 1754 discovered carbon dioxide and showed it to be a product of fermentation, the burning of charcoal, and respiration. Lavoisier in 1789 showed that sugar gives ethanol and carbon dioxide and made quantitative measurements of the amounts consumed and produced.

Glucose $\rightarrow$ 2 C_2H_5OH + 2 CO_2

Once the mole concept was established, Gay-Lussac in 1815 could show that one mole of glucose gives exactly two moles of ethanol and two moles of carbon dioxide. But the process of fermentation stumped some great chemists. The little-known Kutzing wrote in 1837, on the basis of microscopic observation, "It is obvious that chemists must now strike yeast off the role of chemical compounds, since it is not a compound but an organized body, an organism." On the other side were chemists such as Berzelius, who believed that yeast had a catalytic action, and Liebig, who put forth a "theory of motion of the elements within a compound which caused a disturbance of equilibrium which was communicated to the elements of the substance with which it came in contact thus forming new compounds."

Pasteur's contribution

It remained for Pasteur to show that fermentation was a physiologic action associated with the life processes of yeast. Through his microscope he could see the yeast cells that grew naturally on the surface of grapes. He showed that grape juice carefully extracted from the center of a grape and exposed to clean air would not ferment. In his classic paper of 1857, he described fermentation as the action of a living organism; but because the conversion of glucose to ethanol and carbon dioxide is a balanced equation, other chemists disputed his findings. They searched for the substance in yeast that might cause the reaction. The search lasted for 40 years, eventually ended by a clever experiment by Eduard Büchner. He made a cell-free extract of yeast that would still cause the conversion of sugar to alcohol. This cell-free extract contained the catalysts, which we now call enzymes, that were necessary for fermentation—a discovery that earned him the 1907 Nobel prize. In 1905 Harden discovered that inorganic phosphate added to the enzymes increased the rate of fermentation and was itself consumed. This result led him to eventually isolate fructose 1,6-diphosphate. Clearly the history of biochemistry is intimately associated with the study of alcoholic fermentation.

Ancient peoples discovered many of the essential reactions of alcoholic fermentation completely by accident. That crushed grapes would soon begin to froth and bubble and produce a pleasant beverage is a discovery lost in time. But what of those who lived in colder climates where the grape did not grow? How did they discover that the starch of wheat or barley could be converted to sugar by the enzymes in malt? When grain germinates, enzymes are produced that turn the starch into sugar. The process of malting involves letting the grain start to germinate and then heating and drying the sprouts to stop the process before the enzymes are used up. The color of the malt depends on the temperature of the drying. The darkest is used for stout and porter, the lighter for brown, amber, and pale ale. Because of a discovery some time ago that the resulting beverage did not spoil as rapidly if hops were added, we now also have beer.

Other sources exist for the amylases that catalyze the conversion of starch to glucose. The Peruvian campasinos (peasants) make a drink called "chicha" from masticated wheat, which is dried in small cakes. When water, yeast, and more ground wheat are added, the resulting mixture ferments to a beerlike beverage. The enzyme salivary amylase is the catalyst for this starch-to-glucose conversion.

The baker makes use of fermentation by taking advantage of the gas released to leaven his bread. In the present experiment baker's yeast is used to convert sucrose, ordinary table sugar, into ethanol and carbon dioxide with the aid of some 14 enzymes as catalysts, in addition to adenosine

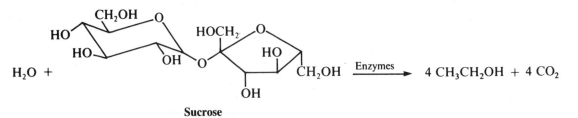

Sucrose

triphosphate (ATP), phosphate ion, thiamine pyrophosphate, magnesium ion, and reduced nicotinamide adenine dinucleotide (NADH), all present in yeast. The fermentation process—known as the Emden-Meyerhof-Parnas scheme—involves the hydrolysis of sucrose to glucose and fructose which, as their phosphates, are cleaved to two three-carbon fragments. These fragments, as their phosphates, eventually are converted to pyruvic acid, which is decarboxylated to give acetaldehyde. Acetaldehyde, in turn, is reduced to ethanol in the final step. Each step requires a specific enzyme as a catalyst and often inorganic ions, such as magnesium and, of course, phosphate. Thirty-one kilocalories of heat are released per mole of glucose consumed in this sequence of anaerobic reactions.

This same sequence of reactions, up to the formation of pyruvic acid, occurs in the human body in times of stress when energy is needed, but not enough oxygen is available for normal aerobic oxidation. The pyruvic acid under these conditions is converted to lactic acid. It is the buildup of lactic acid in the muscles that is partly responsible for the feeling of fatigue.

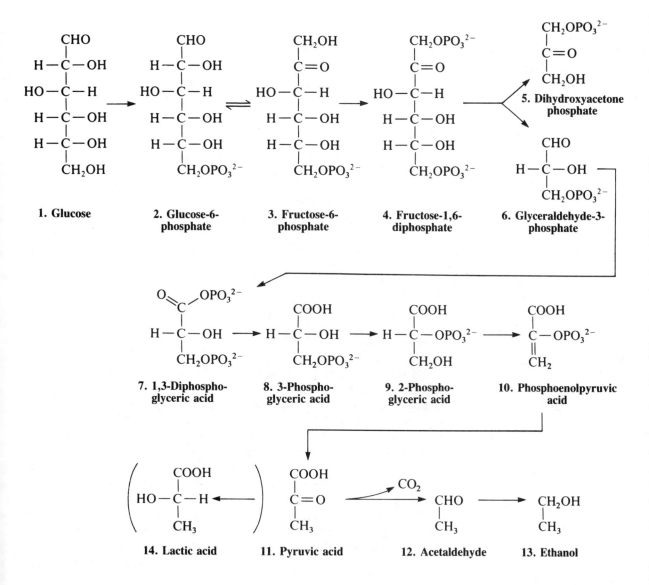

The Emden-Meyerhof-
Parnas scheme

The first step in the sequence is the formation of glucose-6-phosphate (**2**) from glucose (**1**). The reaction requires adenosine triphosphate (ATP), which is converted to the diphosphate (ADP) by catalysis with the enzyme glucokinase, which requires magnesium ion to function. This conversion is one of the reactions in which energy is released. In the living organism this energy can be used to do work; in fermentation it simply creates heat.

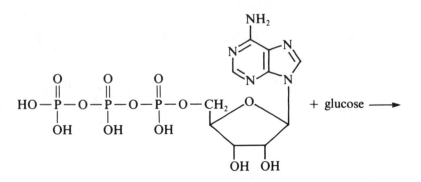

+ glucose ⟶

Adenosine triphosphate (ATP)

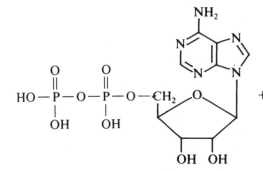

+ glucose-6-phosphate + 9 kcal/mole

Adenosine diphosphate (ADP)

In the next step glucose-6-phosphate (**2**) is converted through the enol of the aldehyde to fructose-6-phosphate (**3**) by the enzyme phosphoglucoisomerase. The fructose monophosphate (**3**) is converted to the diphosphate (**4**) by the action of ATP under the influence of phosphofructokinase with the release of more energy. This diphosphate (**4**) undergoes a reverse aldol reaction catalyzed by aldolase to give dihydroxyacetone phosphate (**5**) and glyceraldehyde-3-phosphate (**6**). The latter two are interconverted by means of triosphosphate isomerase. The aldehyde group of glyceraldehyde phosphate (**6**) is oxidized by nicotinamide adenine dinucleotide (NAD^+) in the presence of another enzyme to a carboxyl group that is phosphorylated with inorganic phosphate. In the next reaction ADP is converted to ATP as **7** loses phosphate to give **8**. A mutase converts the 3-phosphate (**8**) to the 2-phosphate (**9**). An enolase converts **9** to **10**, and a kinase converts **10** to pyruvic acid (**11**). A decarboxylase converts pyruvic acid (**11**) to acetaldehyde (**12**) in the fermentation process. Yeast alcohol dehydrogenase (YAD), a well-studied enzyme, catalyzes the reduction of acetaldehyde to ethanol. The reducing agent is reduced nicotinamide adenine dinucleotide, NADH.

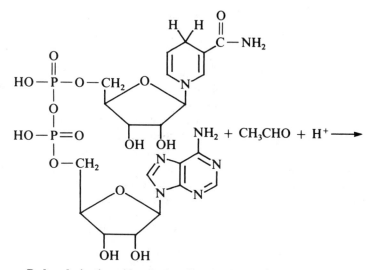

Reduced nicotinamide adenine dinucleotide, NADH

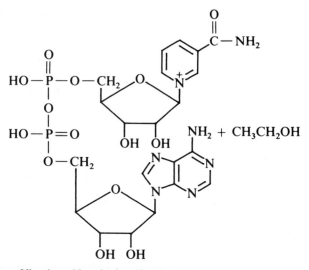

Nicotinamide adenine dinucleotide, NAD⁺

Enzymes are labile

Enzymes are remarkably efficient catalysts, but they are also labile (sensitive) to such factors as heat and cold, changes in pH, and various specific inhibitors. In the first experiment of this chapter you will have an opportunity to observe the biosynthesis of ethanol and to test the effects of various agents on the enzyme system.

This experiment involves the fermentation of ordinary cane sugar using baker's yeast. The resulting dilute solution of ethanol, after removal of the yeast by filtration, can be distilled according to the procedures of Chapter 5.

Experiments

1. Fermentation of Sucrose

Macerate (grind) one-half cake of yeast or half an envelope of dry yeast in 50 mL of water in a beaker, add 0.35 g of disodium hydrogen phosphate, and transfer this slurry to a 500-mL round-bottomed flask. Add a solution of 51.5 g of sucrose in 150 mL of water, and shake to ensure complete mixing. Fit the flask with a one-hole rubber stopper containing a bent glass tube that dips below the surface of a saturated aqueous solution of calcium hydroxide (limewater) in a 6-in. test tube (Fig. 1). The tube in limewater will act as a seal to prevent air and unwanted enzymes from entering the flask, but will allow gas to escape.[1] Place the assembly in a warm spot in your desk (the optimum temperature for the reaction is 35°C) for one week, at which time the evolution of carbon dioxide will have ceased. What is the precipitate in the limewater?

Upon completion of fermentation add 10 g of Celite filter aid (diatomaceous earth, face powder) to the flask, shake vigorously, and filter. Use a 5.5-cm Büchner funnel placed on a neoprene adapter of Filtervac atop a 500-mL filter flask that is attached to the water aspirator through a trap by vacuum tubing (Fig. 2). Since the apparatus is top-heavy, clamp the flask to a ring stand. Moisten the filter paper with water and apply gentle

Protect the fermenting reaction from exposure to oxygen and contaminating materials. Under aerobic conditions acetobacter bacteria can convert ethanol to acetic acid.

Reaction time: 1 week

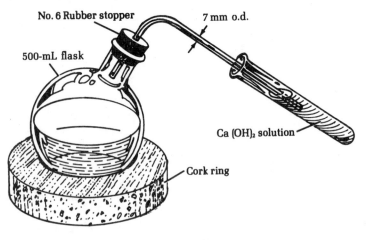

No. 6 Rubber stopper

7 mm o.d.

500-mL flask

Ca (OH)₂ solution

Cork ring

FIG. 1 Apparatus used in the fermentation of sucrose experiment.

---

1. The tube should be about 0.5 cm below the limewater so that limewater will not be sucked back into the flask should the pressure change.

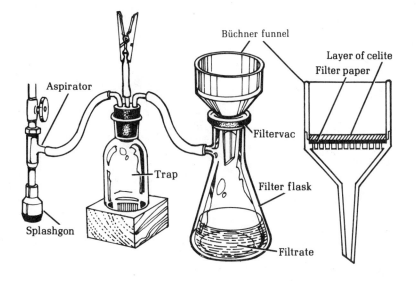

FIG. 2   Vacuum filtration apparatus.

suction (water supply to aspirator turned on full force, clothespin on trap partially closed), and slowly pour the reaction mixture onto the filter. Wash out the flask with a few milliliters of water from your wash bottle and rinse

Filter aid

the filter cake with this water. The filter aid is used to prevent the pores of the filter paper from becoming clogged with cellular debris from the yeast.

The filtrate, which is a dilute solution of ethanol contaminated with bits of cellular material and other organic compounds (acetic acid if you are not careful), is saved in a stoppered flask until it is distilled following the procedure outlined in Chapter 5.

Cleaning Up   Since sucrose, yeast, and ethanol are all natural products, all solutions produced in this experiment contain biodegradable material and can be flushed down the drain after dilution with water. The limewater can be disposed of in the same way. The Celite filter aid can be placed in the nonhazardous solid waste container.

2. Effect of Various Reagents and Conditions on Enzymatic Reactions

Handle the sodium fluoride solution with great care—it is very poisonous if ingested

About 15 min after mixing the yeast, sucrose, and phosphate, remove 20 mL of the mixture and place 4 mL in each of five test tubes. To one tube add 1.0 mL of water, to the next add 1.0 mL of 95% ethanol, to the next add 1.0 mL of 0.5 M sodium fluoride. Heat the next tube for 5 min in a steam bath and cool the next tube for 5 min in ice. Add 10 to 15 drops of mineral oil on top of the reaction mixture in each tube (to exclude air, since the process is anaerobic). Place tubes in a beaker of water at a temperature of 30°C for 15 min, then take them one at a time and connect each to the manometer as shown in Fig. 3. Allow about 30 s for temperature equilibration, then clamp

FIG. 3   Manometric gasometer.

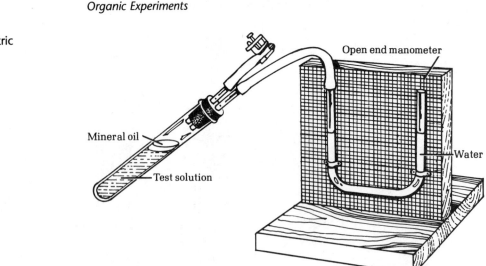

the vent tube and read the manometer. Record the height of the manometer fluid in the open arm of the U-tube every minute for 5 min or until the fluid reaches the top of the manometer. Plot a graph of the height of the manometer fluid against time for each of the four reactions. What conclusions can you draw from the results of these five reactions?

Cleaning Up   Combine all reaction mixtures, remove the mineral oil from the top, and place it in the organic solvents container. The aqueous solutions, after diluting with about 50 volumes of water, can be flushed down the drain. Solutions containing fluoride ion are very toxic and should not be placed in the sewer system but neutralized and treated with excess calcium chloride to precipitate calcium fluoride. The latter is separated by filtration and put in the nonhazardous solid waste container.

Questions

1. Using yeast, can glucose be converted to ethanol? Can fructose be converted to ethanol?

2. Write the equation for the formation of the precipitate formed in the test tube containing calcium hydroxide.

3. In this experiment could 90% ethanol be made by adding more sugar to the fermentation flask?

19

Infrared Spectroscopy

Prelab Exercise: When an infrared (ir) spectrum is run, there is a possibility that the chart paper is not properly placed or that the spectrometer is not mechanically adjusted. Describe how you could calibrate an infrared spectrum.

The presence and also the environment of functional groups in organic molecules can be identified by infrared spectroscopy. Like nuclear magnetic resonance and ultraviolet spectroscopy, infrared spectroscopy is nondestructive. Moreover, the small quantity of sample needed, the speed with which a spectrum can be obtained, the relatively small cost of the spectrometer, and the wide applicability of the method combine to make infrared spectroscopy one of the most useful tools available to the organic chemist.

Infrared radiation, which is electromagnetic radiation of longer wavelength than visible light, is detected not with the eyes but by a feeling of warmth on the skin. When absorbed by molecules, radiation of this wavelength (typically 2.5 to 15 microns), increases the amplitude of vibrations of the chemical bonds joining atoms.

Infrared spectra are measured in units of frequency or wavelength. The wavelength is measured in micrometers, μm, or microns, μ ($1 \mu = 1 \times 10^{-4}$ cm). The positions of absorption bands are measured in frequency units by wavenumbers, $\bar{v}$, which are expressed in reciprocal centimeters, cm^{-1}, corresponding to the number of cycles of the wave in each centimeter.

$$\bar{v} \; cm^{-1} = \frac{10,000}{\mu}$$

Unlike ultraviolet and nmr spectra, infrared spectra are inverted and are not always presented on the same scale. Some spectrometers record the spectra on an ordinate linear in microns, but this compresses the low wavelength region. Other spectrometers present the spectra on a scale linear in reciprocal centimeters, but linear on two different scales, one between 4000 and 2000 cm^{-1}, which spreads out the low wavelength region, and the other, a smaller scale between 200 and 667 cm^{-1}.

To picture the molecular vibrations that interact with infrared light, imagine a molecule as being made up of balls (atoms) connected by springs (bonds). The vibration can be described by Hooke's law from classical mechanics, which says that the frequency of a stretching vibration is directly proportional to the strength of the spring (bond) and inversely proportional to the masses connected by the spring. Thus, we find C—H,

N—H, and O—H bond-stretching vibrations are high-frequency (short wavelength) compared to those of C—C and C—O, because of the low mass of hydrogen compared to that of carbon or oxygen. The bonds connecting carbon to bromine and iodine, atoms of large mass, vibrate so slowly that they are beyond the range of most common infrared spectrometers. A double bond can be regarded as a stiffer, stronger spring, so we find C=C and C=O vibrations at higher frequency than C—C and C—O stretching vibrations. And C≡C and C≡N stretch at even higher frequencies than C=C and C=O (but at lower frequencies than C—H, N—H, and O—H). These frequencies are in keeping with the bond strengths of single (~100 kcal/mole), double (~160 kcal/mole), and triple bonds (~220 kcal/mole).

The stretching vibrations noted above are intense and particularly easy to analyze. A nonlinear molecule of *n* atoms can undergo $3n - 6$ possible modes of vibration, which means cyclohexane with 18 atoms can undergo 48 possible modes of vibration. A single CH_2 group can vibrate in six different ways. Each vibrational mode produces a peak in the spectrum because it corresponds to the absorption of energy at a discrete frequency. These many modes of vibration create a complex spectrum that defies simple analysis, but even in very complex molecules certain functional groups have characteristic frequencies that can easily be recognized. Within these functional groups are the above-mentioned atoms and bonds, C—H, N—H, O—H, C=C, C=O, C≡C, and C≡N. Their absorption frequencies are given in Table 1.

TABLE 1 Characteristic Infrared Absorption Frequencies

	Wavenumber (cm^{-1})	Wavelength (μ)
O—H	3600–3400	2.78–2.94
N—H	3400–3200	2.94–3.12
C—H	3080–2760	3.25–3.62
C≡N	2260–2215	4.42–4.51
C≡C	2150–2100	4.65–4.76
C=O	1815–1650	5.51–6.06
C=C	1660–1600	6.02–6.25
C—O	1050–1200	9.52–8.33

When the frequency of infrared light is the same as the natural vibrational frequency of an interatomic bond, light will be absorbed by the molecule and the amplitude of the bond vibration will increase. The intensity of infrared absorption bands is proportional to the change in dipole moment that a bond undergoes when it stretches. Thus, the most intense bands (peaks) in an infrared spectrum are often from C=O and C—O stretching vibrations, while the C≡C stretching band for a symmetrical acetylene is almost nonexistent because the molecule undergoes no net change of dipole moment when it stretches:

$$\overset{\overset{+\longrightarrow}{|}}{\underset{/}{\overset{\backslash}{C}}}=O \longleftrightarrow \overset{\overset{+\longrightarrow}{|}}{\underset{/}{\overset{\backslash}{C}}}=O \qquad H_3C-C\equiv C-CH_3 \overset{\longleftrightarrow}{} \longleftrightarrow H_2C-C\equiv C-CH_3 \overset{\longleftrightarrow}{}$$

Change in dipole moment   No change in dipole moment

Unlike proton nuclear magnetic resonance spectroscopy, where the area of the peaks is strictly proportional to the numbers of hydrogen atoms causing the peaks, the intensities of infrared peaks are not proportional to the numbers of atoms causing them. And where every peak or group of peaks in an nmr spectrum can be assigned to specific hydrogens in a molecule, the assignment of the majority of peaks in an infrared spectrum is usually not possible. Peaks to the right (longer wavelength) of 1250 cm^{-1} are the result of combinations of vibrations that are characteristic not of individual functional groups, but of the molecule as a whole. This part of the spectrum is often referred to as the "fingerprint region," because it is uniquely characteristic of each molecule. While two organic compounds can have the same melting points or boiling points and can have identical ultraviolet and nmr spectra, they cannot have identical ir spectra. Infrared spectroscopy is thus the final arbiter in deciding whether two compounds are identical.

Analysis of Infrared Spectra

Three rules apply to all analyses: (1) pay most attention to the strongest absorptions, (2) pay more attention to peaks to the left (shorter wavelength) of 1250 cm^{-1}, and (3) pay as much attention to the absence of certain peaks as to the presence of others. The absence of characteristic peaks will definitely exclude certain functional groups. Be wary of weak O—H peaks because water is a common contaminant of many samples. Because KBr is hygroscopic it is often found in the spectra of KBr pellets.

Starting at the left-hand side of the spectrum:

O—H   *3600–3400 cm^{-1}*   The hydroxyl group gives a sharp peak at 3600 cm^{-1} for nonhydrogen-bonded groups and a broad peak at 3400 cm^{-1} for hydrogen-bonded groups. Depending on the concentration of the sample, the relative intensities of these two peaks will vary (see the spectrum of 1-naphthol, Fig. 2 in Chapter 61, where the sharp and broad peaks are just barely resolved). Carboxyl groups give a characteristic very intense and very broad wedge-shaped band, which extends to and obscures the C—H region [see the spectra of benzoic acid, Fig. 2 in Chapter 27, and acetylsalicylic acid (aspirin), Fig. 1 in Chapter 26].

N—H   *3400–3200 cm^{-1}*   A sharp peak at 3400–3200 cm^{-1} is shown by N—H vibrations. The NH$_2$ group usually shows a doublet (see aniline, Fig. 4 in Chapter 41).

C—H   *3300–3000 cm^{-1}* Unsaturated C—H bonds absorb in this region with alkynes at 3300 cm^{-1}, alkenes at 3080–3010 cm^{-1}, and aromatic compounds at 3050 cm^{-1}. Alkanes also absorb over this narrow range. All of these C—H peaks are strong except for the aromatic C—H. The complete absence of peaks to the left of 3000 cm^{-1} indicates no unsaturation while the complete absence of peaks to the right of 3000 cm^{-1} indicates no aliphatic hydrogens. Nmr spectroscopy is the best method for identifying aromatic hydrogens.

C≡N   *2260–2215 cm^{-1}* Although not common this functional group is very easily identified by this infrared peak because only one other peak, that at 2150–2100 cm^{-1} from alkynes, appears near it. It is usually a strong and very sharp peak.

C≡C   *2150–2100 cm^{-1}* Again alkynes are not common functional groups but, if terminal, they give strong and very sharp peaks in this range. If the alkyne is not terminal the peak can be weak or absent in the case of a symmetrical alkyne. No other functional groups give peaks near these for C≡C and C≡N.

C=O   *1870–1825* and *1790–1765 cm^{-1}* Acid anhydrides are also not common functional groups but are easily recognized because they always have two strong peaks in the indicated ranges. See Note below.

1815–1800 cm^{-1} Acid chlorides absorb at this frequency. The presence of halogen is, however, more easily determined by the Beilstein test.

1750–1735 cm^{-1} Esters and lactones give a very strong band in this frequency range. See Note.

1725–1705 cm^{-1} Peaks in this range are very strong and indicative of unsubstituted and unconjugated aldehydes and ketones. See Note.

1700 cm^{-1} Carboxylic acids produce a strong absorption at this frequency. See Note.

1690–1670 cm^{-1} Amides give strong peaks in this frequency range. See Note.

NOTE: The frequencies given for the C=O stretching vibrations for anhydrides, acid chlorides, esters, lactones, aldehydes, ketones, carboxylic acids and amides refer to the open chain or unstrained functional group in a nonconjugated system. If the carbonyl group is conjugated with a double bond or an aromatic ring, the frequency is 30 cm^{-1} less. If it is conjugated to groups on both sides (cross-conjugated), the frequency is 50 cm^{-1} less.

When the carbonyl group is in a ring smaller than six members the frequency is higher by about 25 cm^{-1}, and halogen or oxygen substitution on the carbon adjacent to an aldehyde or ketone carbonyl also increases the frequency by about 25 cm^{-1}. See the detailed list of carbonyl frequencies in Table 2.

C=C	*1660–1625 cm^{-1}* Alkenes appear in this range, but the intensity of the peak is variable and cannot be relied upon. Conjugation shifts the peak to lower frequencies by about 30 cm^{-1}.
	1600 cm^{-1} Aromatic rings give a medium to strong peak near this frequency and also a peak at 1450 cm^{-1}. Conjugated dienes also produce a peak at 1600 cm^{-1}.
—NO$_2$	*1520 and 1350 cm^{-1}* Coupled stretching vibrations of the nitro group give rise to these two intense and easily recognizable bands.
—CH$_3$	*1375 cm^{-1}* A methyl group displays a strong band at this frequency.

$$\begin{array}{c} \diagdown \quad \diagup CH_3 \\ C \\ \diagup \quad \diagdown CH_3 \end{array}$$

1385 and 1365 cm^{-1}   This grouping gives a strong doublet at these two frequencies.

C$_6$H$_5$—O—	*~1200 cm^{-1}*	These carbon-oxygen stretching or bending vibrations give rise to strong bands near the indicated positions, but these may vary. (See the spectrum of acetylsalicylic acid, Fig. 1 in Chapter 26.) Esters often give two strong peaks someplace in this range.
—C—O—	*~1150 cm^{-1}*	
—CH—O—	*~1150 cm^{-1}*	
—CH$_2$—O—	*~1050 cm^{-1}*	

The Double Beam Infrared Spectrometer

The infrared spectrometer

Figure 1 is a schematic representation of a typical double beam, optical null, infrared spectrometer. An electrically heated metal rod serves as the radiation source; the radiation passes through both the sample cell and reference cell, through combs and a beam chopper to the dispersion grating. In some instruments the radiation is dispersed by a prism made of sodium chloride, which is transparent to infrared radiation. Of the electromagnetic radiation frequencies spread out by the grating (or prism), only a small range of frequencies is allowed to pass through the slit to the detector, which is a thermocouple with a very rapid response time.

FIG. 1 Schematic diagram of a double-beam, optical null, infrared spectrometer.

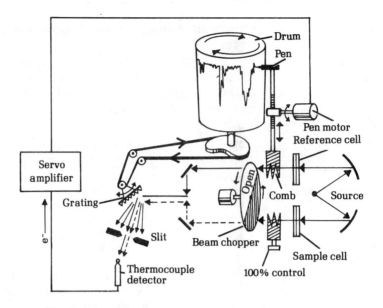

TABLE 2 Characteristic Infrared Carbonyl Stretching Frequencies (chloroform solutions)

	Wavenumber (cm^{-1})	Wavelength (μ)
Aliphatic ketones	1725–1705	5.80–5.87
Acid chlorides	1815–1785	5.51–5.60
α,β-Unsaturated ketones	1685–1666	5.93–6.00
Aryl ketones	1700–1680	5.88–5.95
Cyclobutanones	1775	5.64
Cyclopentanones	1750–1740	5.72–5.75
Cyclohexanones	1725–1705	5.80–5.87
β-Diketones	1640–1540	6.10–6.50
Aliphatic aldehydes	1740–1720	5.75–5.82
α,β-Unsaturated aldehydes	1705–1685	5.80–5.88
Aryl aldehydes	1715–1695	5.83–5.90
Aliphatic acids	1725–1700	5.80–5.88
α,β-Unsaturated acids	1700–1680	5.88–5.95
Aryl acids	1700–1680	5.88–5.95
Aliphatic esters	1740	5.75
α,β-Unsaturated esters	1730–1715	5.78–5.83
Aryl esters	1730–1715	5.78–5.83
Formate esters	1730–1715	5.78–5.83
Vinyl and phenyl acetate	1776	5.63
δ-Lactones	1740	5.75
γ-Lactones	1770	5.65
Acyclic anhydrides (two peaks)	1840–1800 1780–1740	5.44–5.56 5.62–5.75
Primary amides	1694–1650	5.90–6.06
Secondary amides	1700–1670	5.88–6.01
Tertiary amides	1670–1630	5.99–6.14

The spectrometer works on the optical null principle. The detector senses infrared light coming alternately through the substances in the sample and reference cells. If the amount of light is the same from both beams the detector produces a direct current and nothing happens; but if less light comes through the sample beam than the reference beam (because of absorption of radiation by the sample molecules), then the detector senses an alternating current that alternates at the rate the chopper is turning. This current is amplified in the servo amplifier, which activates the pen motor. The pen motor moves the pen down the paper drawing an absorption band and at the same time drives a comb into the reference beam just far enough so that the detector will again sense a null, i.e., no alternating current. The motion of the drum holding the paper is linked to the grating so that as the drum moves the grating moves with it to scan the entire range of frequencies.

Correlation tables

Extensive correlation tables and discussions of characteristic group frequencies can be found in specialized references.[1] As one example, consider the band patterns of toluene, and of *o-*, *m-*, and *p*-xylene, which appear in the frequency range 2000 to 1650 cm^{-1} (Fig. 2). These band patterns are due to changes in the dipole moment accompanying changes in vibrational modes of the aromatic ring and are surprisingly similar to those for monosubstituted and other *o-*, *m-*, and *p*-disubstituted benzenes.

Sample, solvents, and equipment must be dry

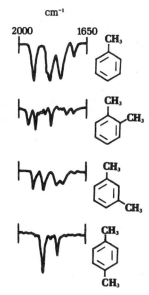

cm^{-1}

2000   1650

FIG. 2   Band patterns of toluene and *o-*, *m-*, and *p*-xylene. These peaks are very weak.

Experimental Aspects

Infrared spectra can be determined on neat (undiluted) liquids, on solutions with an appropriate solvent, and on solids as mulls and KBr pellets. Glass is opaque to infrared radiation; therefore, the sample and reference cells used in infrared spectroscopy are sodium chloride plates. The sodium chloride plates are fragile and can be attacked by moisture. *Handle only by the edges.*

Spectra of Neat Liquids

To run a spectrum of a neat liquid (free of water) remove a demountable cell (Fig. 3) from the desiccator and place a drop of the liquid between the salt plates, press the plates together to remove any air bubbles, and add the top rubber gasket and metal top plate. Next, put on all four of the nuts and *gently* tighten them to apply an even pressure to the top plate. Place the cell in the sample compartment (nearest the front of the spectrometer) and run the spectrum.

Although running a spectrum on a neat liquid is convenient and results in no extraneous bands to interpret, it is not possible to control the path length of the light through the liquid in a demountable cell. A low-viscosity liquid when squeezed between the salt plates may be so thin that the short path length gives peaks that are too weak. A viscous liquid, on the other

---

1. See list of references at end of chapter.

hand, may give peaks that are too intense. A properly run spectrum will have the most intense peak with an absorbance of about 1.0. Unlike the nmr spectrometer, the infrared spectrometer does not usually have a control to adjust the peak intensities; this control is possible only by adjusting the sample concentration.

Another demountable cell is pictured in Fig. 4. The plates are thin wafers of silver chloride, which is transparent to infrared radiation. This cell has advantages over the salt cell in that the silver chloride disks are more resistant to breakage than NaCl plates, they are less expensive, and they are not affected by water. Because silver chloride is photosensitive the wafers must be stored in the dark to prevent them from turning black. Because one side of each wafer is recessed, the thickness of the sample can be varied according to the manner in which the cell is assembled. The disks are cleaned by rinsing them with an organic solvent such as acetone or ethanol and wiping dry with an absorbent paper towel.

A fast, simple alternative: Place a drop of the compound on a round salt plate, add a top plate, and mount the two on the holder pictured in Fig. 4.

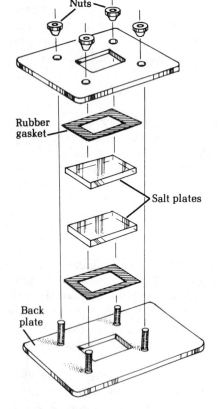

FIG. 3　Exploded view of a demountable salt cell for analyzing the infrared spectra of neat liquids.

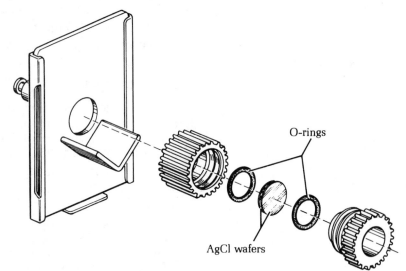

FIG. 4　Demountable silver chloride cell.

Spectra of Solutions

The most widely applicable method of running spectra of solutions involves dissolving an amount of the liquid or solid sample in an appropriate solvent to give a 10% solution. Just as in nmr spectroscopy, the best solvents to use are carbon disulfide and carbon tetrachloride; but because these compounds are not polar enough to dissolve many substances, chloroform is used as a compromise. Unlike nmr solvents, no solvent suitable in infrared spectroscopy is entirely free of absorption bands in the frequency range of interest [Figs. 5(a) and (b)]. In chloroform, for instance, no light passes through the cell between 650 and 800 cm^{-1}. As can be seen from the figures, spectra obtained using carbon disulfide and chloroform cover the entire infrared frequency range. In practice, a base line is run with the same solvent in both cells to ascertain if the cells are clean and matched [Fig. 5(c)]. Often it is necessary to obtain only one spectrum employing one solvent, depending on which region of the spectrum you need to use.

Solvents: CHCl$_3$, CS$_2$, CCl$_4$

CAUTION: Chloroform is toxic; use laboratory hood. Avoid the use of chloroform and carbon disulfide if possible.

Three drops needed to fill cell

Three large drops of solution will fill the usual sealed infrared cell (Fig. 6). A 10% solution of a liquid sample can be approximated by dilution of one drop of the liquid sample with nine of the solvent. Since weights are more difficult to estimate, solid samples should be weighed to obtain a 10% solution.

The infrared cell is filled by inclining it slightly and placing about three drops of the solution in the lower hypodermic port with a capillary dropper. The liquid can be seen rising between the salt plates through the window. In the most common sealed cell, the salt plates are spaced 0.1 mm apart. Make sure that the cell is filled past the window and that no air bubbles are present. Then place the Teflon stopper lightly but firmly in the hypodermic port. Be particularly careful not to spill any of the sample on the outside of the cell windows.

Solvent and sample must be dry. Do not touch or breathe upon NaCl plates.

Fill the reference cell from a clean hypodermic syringe in the same manner as the sample cell and place both cells in the spectrometer, with the sample cell toward the front of the instrument. After running the spectrum, force clean solvent through the sample cell, using a syringe attached to the top port of the cell (Fig. 7). Finally, with the syringe, pull the last bit of solvent from both cells, blow clean, dry, compressed air through the cells to dry them, and store them in a desiccator.

Dispose of waste solvents in the container provided

Cleaning Up   Discard halogenated liquids in the halogenated organic waste container. Other solutions should be placed in the organic solvents container.

Mulls and KBr Disks

Solids insoluble in the usual solvents can be run as mulls or KBr disks. In preparing the mull, the sample is ground to a particle size less than that of the wavelength of light going through the sample (2.5 microns), in order to avoid scattering the light. About 15 to 20 mg of the sample is ground for 3 to

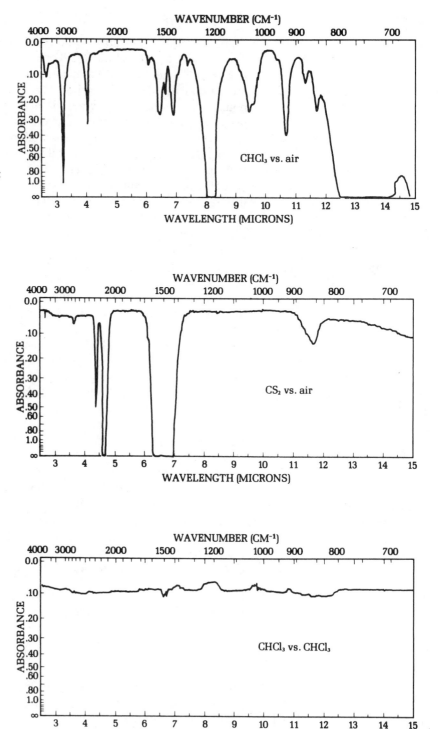

FIG. 5(a)   Spectrum of chloroform in sample cell, air in reference cell. No infrared light passes through chloroform between 1200 and 1250 cm^{-1} and between 650 and 800 cm^{-1}; therefore no information about sample absorption in those regions can be obtained.

CHCl$_3$ vs. air

FIG. 5(b)   Spectrum of carbon disulfide in sample cell, air in reference cell. No infrared light passes through carbon disulfide between 1430 and 1550 cm^{-1}.

CS$_2$ vs. air

FIG. 5(c)   Spectrum of chloroform in both sample and reference cells. A typical baseline.

CHCl$_3$ vs. CHCl$_3$

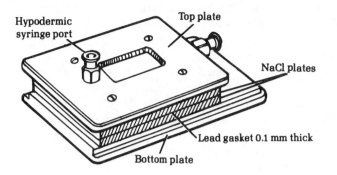

Hypodermic syringe port

Top plate

NaCl plates

Lead gasket 0.1 mm thick

Bottom plate

FIG. 6   Sealed infrared sample cell.

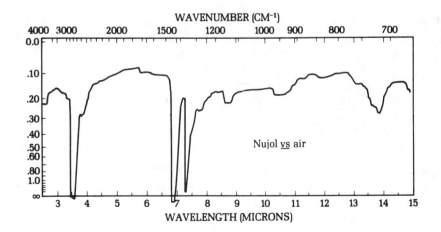

FIG. 7   Flushing the infrared sample cell. The solvent used to dissolve the sample is used in this process.

Sample must be finely ground

A fast, simple alternative: Grind a few mgs of the solid with a drop of tetrachloroethylene between two round salt plates. Mount on the holder shown in Fig. 4.

10 min in an agate mortar until it is spread over the entire inner surface of the mortar and has a caked and glassy appearance. Then, to make a mull, 1 or 2 drops of paraffin oil (Nujol) (Fig. 8) is added, and the sample ground 2 to 5 more minutes. The mull is transferred to the bottom salt plate of a demountable cell (Fig. 3) using a rubber policeman, the top plate added and twisted to distribute the sample evenly and to eliminate all air pockets, and the spectrum run. Since the bands from Nujol obscure certain frequency regions, running another mull using Fluorolube as the mulling agent will allow the entire infrared spectral region to be covered. If the sample has not been ground sufficiently fine, there will be marked loss of transmittance at the short-wavelength end of the spectrum. After running the spectrum, the salt plates are wiped clean with a cloth saturated with an appropriate solvent.

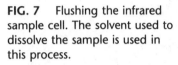

FIG. 8   Infrared spectrum of Nujol (paraffin oil).

Nujol <u>vs</u> air

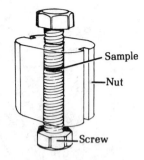

FIG. 9   KBr disk die. Pressure is applied by tightening the machine screws with a wrench.

The spectrum of a solid sample can also be run by incorporating the sample in a KBr disk. This procedure needs only one disk to cover the entire spectral range, since KBr is completely transparent to infrared radiation. Although very little sample is required, making the disk calls for special equipment and time to prepare it. Since KBr is hygroscopic, water is a problem. The sample is first ground as for a mull and 1.5 mg of this is added to 300 mg of spectroscopic grade KBr (previously dried in an oven and stored in a desiccator). The two are gently mixed (not ground) and quickly placed in a 13-mm die and subjected to 14,000–16,000 lb/sq in. pressure for 3–6 min while under vacuum in a specially constructed hydraulic press. A transparent disk is produced, which is removed from the die with tweezers and placed in a special holder, prior to running the spectrum.

A simple, low-cost, small press is illustrated in Fig. 9. The press consists of a large nut and two machine screws. The sample is placed between the two machine screws (which have polished faces), and the screws are tightened with a wrench with the nut held in a vise. The screws are then loosened and removed. The KBr disk is left in the nut, which is then mounted in the spectrometer to run the spectrum. An opaque area in the disk indicates insufficient pressure was applied. Too much pressure can result in crushed disks.

Running the Spectrum

Satisfactory spectra are easily obtained with the lower-cost spectrometers, even those which have only a few controls and require few adjustments. To run a spectrum the paper must be positioned accurately, the pen set between 90% and 100% transmittance with the 100% control (0.0 and 0.05 absorbance), and the speed control set for a fast scan (usually one of about three minutes). The calibration of a given spectrum can be checked by backing up the drum and superimposing a spectrum of a thin polystyrene film. This film, mounted in a cardboard holder that has the frequencies of important peaks printed on it, will be found near most spectrometers. The film is held in the sample beam and parts of the spectrum to be calibrated are rerun. The spectrometer gain (amplification) should be checked frequently and adjusted when necessary. To check the gain, put the pen on the 90% transmittance line with the 100% control. Place your finger in the sample beam so that the pen goes down to 70% T. Then quickly remove your finger. The pen should overshoot the 90% T line by 2%.

Calibration with polystyrene film

Throughout the remainder of this book representative infrared spectra of starting materials and products will be presented and the important bands in each spectrum identified.

Experiment

Unknown Carbonyl Compound

Sample, solvents, and apparatus must be dry

Run the infrared spectrum of an unknown carbonyl compound obtained from the laboratory instructor. Be particularly careful that all apparatus and solvents are completely free of water, which will damage the sodium chloride cell plates. The spectrum can be calibrated by positioning the spectrometer pen at a wavelength of about 6.2 μ without disturbing the paper, and rerunning the spectrum in the region from 6.2 to 6.4 μ while holding the polystyrene calibration film in the sample beam. This will superimpose a sharp calibration peak at 6.246 μ (1601 cm^{-1}) and a less intense peak at 6.317 μ (1583 cm^{-1}) on the spectrum. Determine the frequency of the carbonyl peak and list the possible types of compounds that could correspond to this frequency (Table 2).

References

1. L. J.Bellamy, *The Infrared Spectra of Complex Organic Molecules,* Chapman and Hall, New York, 2nd Ed., 1980.

2. R. T.Conley, *Infrared Spectroscopy,* Allyn and Bacon, Boston, 2nd Ed., 1972.

3. K.Nakanishi and P. H. Solomon, *Infrared Absorption Spectroscopy,* Holden-Day, San Francisco, 1977.

4. R. M.Silverstein, C. G. Bassler, and T. C. Morrill, *Spectrometric Identification of Organic Compounds,* John Wiley, New York, 5th Ed., 1991. (Includes ir, uv, and nmr.)

5. H. A.Szymanski and R. E. Erickson, *Infrared Band Handbook,* Plenum, New York, 2nd Ed., 1970.

6. C. J.Pouchert, *The Aldrich Library of Infrared Spectra,* Aldrich Chemical Co., Milwaukee, 1970.

7. J. W.Cooper, *Spectroscopy Techniques for Organic Chemists,* John Wiley, New York, 1980.

8. G.Socrates, *Infrared Characteristic Group Frequencies,* John Wiley, New York, 1980.

9. J. R.Ferraro and L. J. Basile, eds., *Fourier Transform Infrared Spectroscopy,* Vol. 1, 1973; Vol. 2, 1979; Vol. 3, 1982, Techniques; Vol. 4, 1985, Applications to Chemical Systems, Academic Press, New York.

20

Nuclear Magnetic Resonance Spectroscopy

Prelab Exercise: Outline the preliminary solubility experiments you would carry out on an unknown using inexpensive solvents, before preparing a solution of the compound for nmr spectroscopy.

nmr—Determination of the number, kind, and relative locations of hydrogen atoms (protons) in a molecule

Integration

Chemical shifts, δ

Coupling constant, J

Nuclear magnetic resonance (nmr) spectroscopy is a means of determining the number, kind, and relative locations of certain atoms, principally hydrogen, in molecules. Experimentally, the sample, 0.3 mL of a 20% solution in a 5-mm o.d. glass tube, is placed in the probe of the spectrometer between the faces of a powerful (1.4 Tesla) permanent or electromagnet and irradiated with radiofrequency energy (60 MHz; 60,000,000 Hz for protons). The absorption of radiofrequency energy versus magnetic field strength is plotted by the spectrometer to give a spectrum.

In a typical ^{1}H spectrum (Figs. 1 and 2, ethyl iodide) the relative numbers of hydrogen atoms (protons) in the molecule are determined from the *integral*, the stair-step line over the peaks. The height of the step is proportional to the area under the nmr peak, and in nmr spectroscopy (contrasted with infrared, for instance) the area of each group of peaks is directly proportional to the number of hydrogen atoms causing the peaks. Integrators are part of all nmr spectrometers and running the integral takes no more time than running the spectrum. The different kinds of protons are indicated by their *chemical shifts*.[1] For ethyl iodide the two protons adjacent to the electronegative iodine atom are *downfield* (at lower magnetic field strength) ($\delta = 3.20$ ppm) from the three methyl protons at $\delta = 1.83$ ppm. Tables of chemical shifts for protons in various environments can be found in reference books and are given graphically in Fig. 3.

The relative locations of the five protons in ethyl iodide are indicated by the pattern of peaks on the spectrum. The three peaks indicate methyl protons adjacent to two protons; four peaks indicate methylene protons adjacent to three methyl protons. In general, in molecules of this type, a given set of protons will appear as $n + 1$ peaks if they are adjacent to n protons. The distance between adjacent peaks in the quartet and triplet is the *coupling constant, J*. In this example, $J = 8$ Hz.

Not all nmr spectra are as easily analyzed as the spectrum for ethyl iodide. Consider the one for 3-hexanol (Fig. 4). Twelve protons give rise to an unintelligible group of peaks between 1.0 and 2.2 ppm. It is not clear from a 3-hexanol spectrum which of the two low-field peaks (on the left-

---

1. Chemical shift is a measure of a peak's position relative to the peak of a standard substance (e.g., TMS, tetramethylsilane), which is assigned a chemical shift value of 0.0.

FIG. 1   Proton nmr spectrum of ethyl iodide (60 MHz). The staircase-like line is the integral. In the integral mode of operation the recorder pen moves from left to right and moves vertically a distance proportional to the areas of the peaks over which it passes. Hence, the relative area of the quartet of peaks at 3.20 ppm to the triplet of peaks at 1.83 ppm is given by the relative heights of the integral (4 cm is to 6 cm as 2 is to 3). The relative numbers of hydrogen atoms are proportional to the peak areas (2H and 3H).

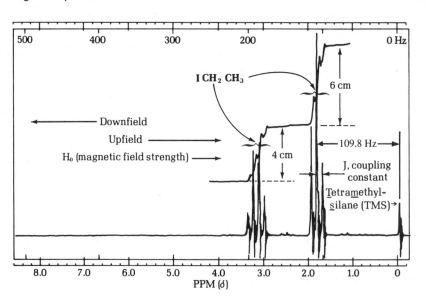

hand side of the spectrum) should be assigned to the hydroxyl proton and which should be assigned to the proton on C-3.

Deuterium atoms give no peaks in the nmr spectrum; thus, the peak in the 3-hexanol spectrum for the proton on C-3 will be evident if the spectrum of the alcohol is one in which the hydroxyl proton of the alcohol has been replaced by deuterium. The hydroxyl proton of 3-hexanol is acidic and will exchange rapidly with the deuterium of D_2O.

Deuterium exchange

Shift reagents

Addition of a few milligrams of a hexacoordinate complex of europium {tris(dipivaloylmethanato)europium(III),$[Eu(dpm)_3]$} to an nmr sample, which contains a Lewis base center (an amine or basic oxygen, such as a hydroxyl group), has a dramatic effect on the spectrum. This lanthanide (soluble in CS_2 and $CDCl_3$) causes large shifts in the positions of peaks arising from the protons near the metal atom in this molecule and is therefore referred to as a shift reagent. It produces the shifts by complexing with the unshared electrons of the hydroxyl oxygen, the amine nitrogen, or other Lewis base centers. As part of the complex the europium atom exerts a large dipolar effect on nearby hydrogens, with resulting changes in the nmr spectrum.

The chemical shift of the triplet is 109.8 Hz from the TMS peak in Fig. 1. Chemical shift are reported in parts per million according to this equation:

$$\delta_i(\text{ppm}) = \frac{\nu_i - \nu_{TMS}}{\nu_0} \times 10^6$$

where δ_i is the chemical shift of proton i, ν_i is the resonance frequency of that proton (e.g., 109.8 Hz), ν_{TMS} is the resonance frequency of TMS (0 Hz), and ν_0 is the operating frequency of the instrument—in this case 60 MHz = 60×10^6 Hz. Thus, the methyl group has a chemical shift, δ_i, of 1.83 ppm.

In Fig. 2 is the spectrum of the same compound run on a 250-MHz spectrometer. The chemical shift of the triplet is now 457.5 Hz from the TMS peak. But, expressed in dimensionless δ units, it is still 1.83 ppm according to the above equation. The distance between individual peaks in the triplet and the quartet is still 8 Hz because the size of the coupling constant does not change with magnetic field strength. High field spectra like this one are easier to analyze than low field (60 MHz) spectra because the amount of peak overlap in complex molecules is much less. In this text you will find both high and low field spectra.

With no shift reagent present, the nmr spectrum of 2-methyl-3-pentanol [Fig. 5(A)] is not readily analyzed. Addition of about 10 mg of Eu(dpm)$_3$ (the shift reagent) to the sample (0.5 mL of a 0.4 M solution) causes very large downfield shifts of peaks owing to protons near the coordination site [Fig. 5(B)].[2] Further additions of 10-mg portions of the shift reagent cause further downfield shifts [Figs. 5(D)–(F)], so that in Fig. 5(G) only the methyl peaks appear within 500 Hz of TMS. Finally, when the mole ratio of Eu(dpm)$_3$ to alcohol is 1 : 1 we find the spectrum shown in Fig. 6. The two protons on C-4 and the two methyls on C-2 are magnetically nonequivalent because they are adjacent to a chiral center (an asymmetric carbon atom, C-3), and therefore each gives a separate set of peaks. With shift reagent added, this spectrum can be analyzed by inspection.

Quantitative information about molecular geometry can be obtained from shifted spectra. The shift induced by the shift reagent, $\Delta H/H$, is related to the distance (r) and the angle (θ), which the proton bears to the europium atom.

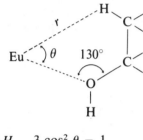

$$\frac{\Delta H}{H} = \frac{3 \cos^2 \theta - 1}{r^3}$$

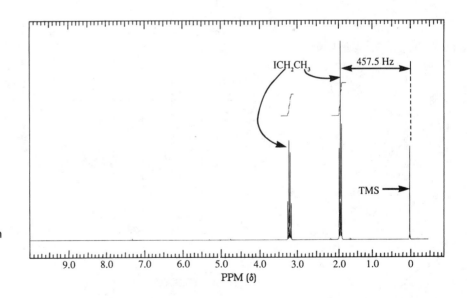

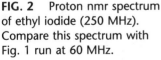

FIG. 2   Proton nmr spectrum of ethyl iodide (250 MHz). Compare this spectrum with Fig. 1 run at 60 MHz.

2. See also K. L. Williamson, D. R. Clutter, R. Emch, M. Alexander, A. E. Burroughs, C. Chua, and M. E. Bogel, *J. Am. Chem. Soc.*, **96**, 1471 (1974).

FIG. 3 ¹H Chemical shifts, ppm from TMS.

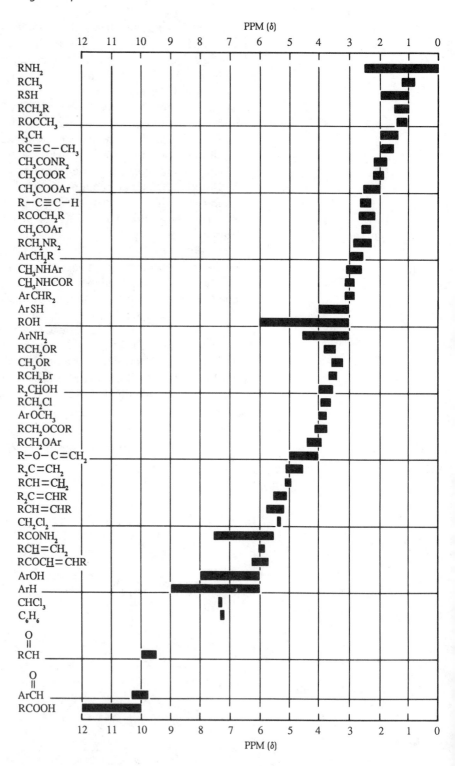

FIG. 4   Proton nmr spectrum of 3-hexanol (60 MHz).

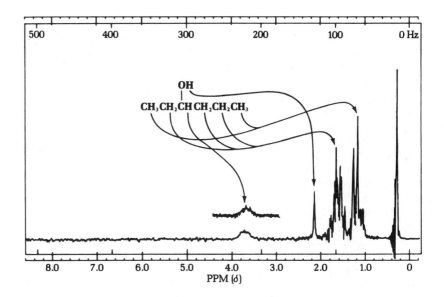

Eu(dpm)$_3$ is a metal chelate (*chele,* Greek, claw). The β-diketone dipivaloylmethane (2,2,6,6-tetramethylheptane-3,5-dione) is a ligand—in this case a bidentate ligand—attached to the europium at two places. Since Eu^{3+} is hexacoordinate, three dipivaloylmethane molecules cluster about this metal atom. However, when a molecule with a basic group like an amine or an alcohol is in solution with Eu(dpm)$_3$ the europium will expand its coordination sphere to complex with this additional molecule. Such a complex is weak and so its nmr spectrum is an average of the complexed and uncomplexed molecule.

Chiral shift reagents will cause differential shifts of enantiomeric protons. See Chapter 64.

Carbon-13 Spectra

Although the most common and least expensive nmr spectrometers are those capable of observing protons, spectrometers capable of observing carbon atoms of mass 13 are becoming prevalent. In many cases ^{13}C spectra are much simpler and easier to interpret than ^{1}H spectra because the spectrum, as usually presented, consists of a single line for each chemically and magnetically distinct carbon atom. The element carbon consists of 98.9% carbon with mass 12 and spin 0 (nmr inactive) and only 1.1% ^{13}C with spin $\frac{1}{2}$ (nmr active). Carbon, with such a low concentration of spin $\frac{1}{2}$ nuclei, gives such a small signal in a conventional nmr spectrometer that special means must be employed to obtain an observable spectrum. In a CW (continuous wave) spectrometer, proton spectra are produced by sweeping the radiofrequency through the spectrum while holding the magnetic field constant, a process that requires anywhere from one to five minutes. If one

CW spectrometer

FIG. 5 The 60-MHz ^{1}H nmr spectrum of 2-methyl-3-pentanol (0.4 M in CS_2) with various amounts of shift reagent present. (A) No shift reagent present. All methyl peaks are superimposed. The peak for the proton adjacent to the hydroxyl group is downfield from the others because it is adjacent to the electronegative oxygen atom. (B) 2-Methyl-3-pentanol + $Eu(dpm)_3$. Mole ratio of $Eu(dpm)_3$ to alcohol = 0.05. The hydroxyl proton peak at 1.6 ppm in spectrum A appears at 6.2 ppm in spectrum B because it is closest to the Eu in the complex formed between $Eu(dpm)_3$ and the alcohol. The next closest proton, the one on the hydroxyl-bearing carbon atom, gives a peak at 4.4 ppm. Peaks due to the three different methyl groups at 1.1–1.5 ppm begin to differentiate. (C) Mole ratio of $Eu(dpm)_3$ to alcohol = 0.1. The hydroxyl proton does not appear in this spectrum because its chemical shift is greater than 8.6 ppm with this much shift reagent present. (D) Mole ratio of $Eu(dpm)_3$ to alcohol = 0.25. Further differentiation of methyl peaks (2.3–3.0 ppm) is evident. (E) Mole ratio of $Eu(dpm)_3$ to alcohol = 0.5. Separate groups of peaks begin to appear in the region 4.2–6.2 ppm. (F) Mole ratio of $Eu(dpm)_3$ to alcohol = 0.7. Three groups of peaks (at 6.8, 7.7, and 8.1 ppm) due to the protons on C-2 and C-4 are evident, and three different methyls are now apparent. (G) Mole ratio of $Eu(dpm)_3$ to alcohol = 0.9. Only the methyl peaks appear on the spectrum. The two doublets come from the methyls attached to C-2 and the triplet comes from the terminal methyl at C-5.

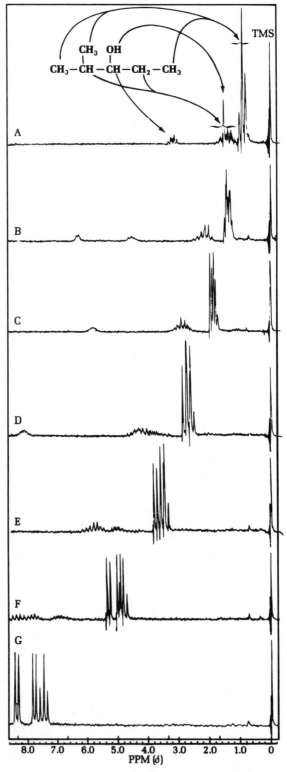

FIG. 6 Proton nmr spectrum of 2-methyl-3-pentanol containing Eu(dpm)₃ (60 MHz). Mole ratio of Eu(dpm)₃ to alcohol = 1.0. Compare this spectrum to those shown in Fig. 5. Protons nearest the hydroxyl group are shifted most. Methyl groups are recorded at reduced spectrum amplitude. Note the large chemical shift difference between the two protons on C-4. The average conformation of the molecule is the one shown and was calculated from the equation on p. 219.

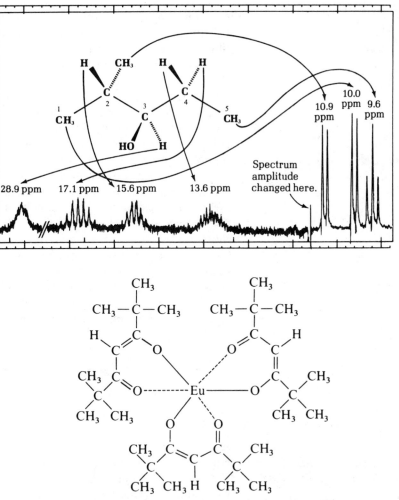

Tris(dipivaloylmethanato)europiumIII, Eu(dpm)₃(III)
MW 701.78, mp 188-189°C

tries to obtain a ¹³C spectrum under these conditions the signal is so small it cannot be distinguished from the random noise in the background. In order to increase the signal-to-noise ratio a number of spectra are averaged in a small computer built into the spectrometer. Since noise is random and the signal coherent, the signal will increase in size and the noise decrease as many spectra are averaged together. Spectra are accumulated rapidly by applying a very short pulse of radiofrequency energy to the sample and then storing the resulting free induction decay signal in digital form in the computer. The free induction decay signal (FID), which takes just 0.6 s to acquire, contains frequency information about all the signals in the spectrum. In a few minutes several hundred FID's can be obtained and averaged in the computer. The FID is converted to a spectrum of conventional

Fourier transform spectrometer

appearance by carrying out a Fourier transform computation on the signal using the spectrometer's computer. These principles are illustrated for very dilute proton spectra in Fig. 7.

Because only one in a hundred carbon atoms have mass 13 the chances of a molecule having two ^{13}C atoms adjacent to one another are small. Consequently, coupling of one carbon with another is not observed. Coupling of the ^{13}C atoms with ^1H atoms leads to excessively complex spectra, so this is ordinarily eliminated by noise decoupling the protons. This decoupling distorts the peaks such that peak areas of carbon spectra are not proportional to the number of carbon atoms present. Unlike proton spectra, which are observed over a range of approximately 15 ppm, carbon spectra occur over a 200-ppm range and peaks from magnetically distinct carbons rarely overlap. For example, each of the 12 carbons of sucrose gives a separate line (Fig. 8). The same factors that control proton chemical shifts are operative for carbon atoms. Carbons of high electron density (e.g., methyl groups) appear upfield near the carbon atoms of tetramethylsilane, the zero of reference, while carbon atoms bearing electron-withdrawing groups or atoms appear downfield. It should not be surprising that carbonyl carbons are found furthest downfield, between 160 and 220 ppm. Figure 9 gives the chemical shift ranges for carbon atoms.

Noise decoupling (margin note)

Experiments

1. Running an nmr Spectrum

Procedure

Sample size: 0.3 mL of 10–20% solution plus TMS (margin note)

A typical ^1H nmr sample is 0.3 to 0.5 mL of a 10–20% solution of a nonviscous liquid or a solid in a proton-free solvent contained in a 5-mm dia. glass tube. The sample tube must be of uniform outside and inside diameter with uniform wall thickness. Test a sample tube by rolling it down a very slightly inclined piece of plate glass. Reject all tubes that roll unevenly.

The ideal solvent, from the nmr standpoint, is carbon tetrachloride. It is proton-free and nonpolar but unfortunately a poor solvent. Carbon disulfide is an excellent compromise. It will, however, react with amines.

Deuterochloroform ($CDCl_3$) is one of the most widely used nmr solvents. Although more expensive than nondeuterated solvents, it will dissolve a wider range of samples than carbon disulfide or carbon tetrachloride. Residual protons in the $CDCl_3$ will always give a peak at 7.27 ppm. Chemical shifts of protons are measured relative to the sharp peak of the protons in tetramethylsilane (taken as 0.0 ppm). Stock solutions of 3–5% tetramethylsilane in carbon disulfide and in deuterochloroform are useful for preparing routine samples.

CS_2 and CCl_4: use only under laboratory hood. CS_2 is toxic and flammable; CCl_4 can cause liver and kidney damage and is a suspected carcinogen.
TMS boils at room temperature. It is expensive and flammable. Handle with care. (margin note)

A wide variety of completely deuterated solvents are commercially available, e.g., deuteroacetone (CD_3COCD_3), deuterodimethylsulfoxide (CD_3SOCD_3), deuterobenzene (C_6D_6), although they are expensive. For highly polar samples a mixture of the expensive deuterodimethylsulfoxide

FIG. 7 ¹H nmr spectra of cortisone acetate, 300 μg/0.3 mL, in CDCl₃. (A) Continuous wave (CW) spectrum, 500-s scan time. (B) Fourier transform (FT) spectrum of the same sample, 250 scans (500 s). The H₂O and CHCl₃ are contaminants.

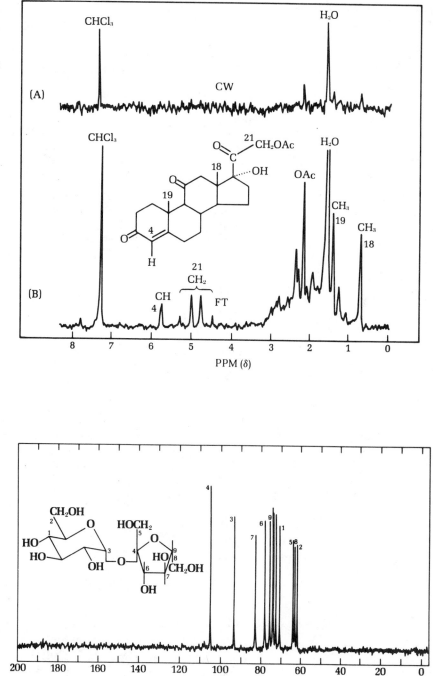

FIG. 8 ¹³C nmr spectrum of sucrose (22.6 MHz). Not all lines have been assigned to individual carbon atoms.

FIG. 9 ^{13}C chemical shifts, ppm from TMS.

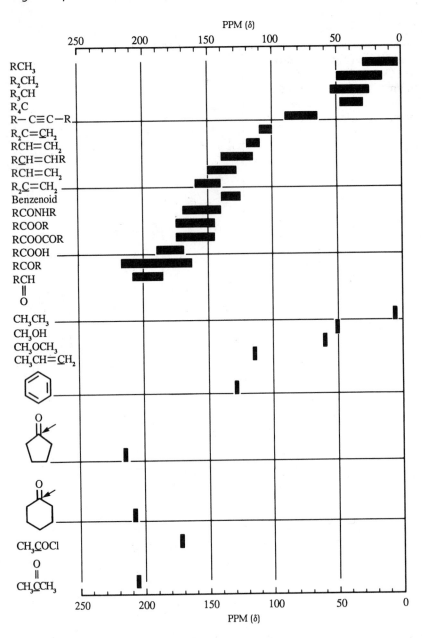

with the less expensive deuterochloroform will often be satisfactory. Water-soluble samples are dissolved in deuterated water containing a water-soluble salt [DSS, $(CH_3)_3SiCH_2CH_2CH_2SO_3^-Na^+$] as a reference substance. The protons on the three methyl groups bound to the silicon in this salt absorb at 0.0 ppm.

Erratic spectra from ferromagnetic impurities; remove by filtration

Solid impurities in nmr samples will cause very erratic spectra. If two successive spectra taken within minutes of each other are not identical, suspect solid impurities, especially ferromagnetic ones. These can be

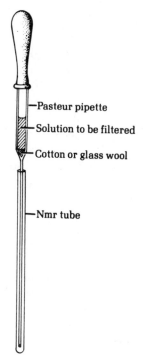

—Pasteur pipette

—Solution to be filtered

—Cotton or glass wool

—Nmr tube

FIG. 10 Micro filter for nmr samples. Solution to be filtered is placed in the top of the Pasteur pipette, rubber bulb put in place, and pressure applied to force sample through the cotton or glass wool into an nmr sample tube.

removed by filtration of the sample through a tightly packed wad of glass wool in a capillary pipette (Fig. 10). If very high resolution spectra (all lines very sharp) are desired, oxygen, a paramagnetic impurity, must be removed by bubbling a fine stream of pure nitrogen through the sample for 60 s. Routine samples do not require this treatment.

The usual nmr sample has a volume of 0.3 mL to 0.7 mL, even though the volume sensed by the spectrometer receiver coils (referred to as the active volume) is much smaller (Fig. 11). To average the magnetic fields produced by the spectrometer within the sample, the tube is spun by an air turbine at 20–40 revolutions per second while taking the spectrum. Too rapid spinning or an insufficient amount of sample will cause the vortex produced by the spinning to penetrate the active volume, giving erratic nonreproducible spectra. A variety of microcells are available for holding and proper positioning of small samples with respect to the receiver coils of the spectrometer (Fig. 12). The vertical positioning of these cells in the spectrometer is critical. If microcells are used, only one or two mg of the sample are needed to give satisfactory spectra, in contrast to the 20–30 mg usually needed for a CW spectrum.

Cleaning Up   Place halogenated solvents and compounds in the halogenated organic waste container. All others go into the organic solvents container.

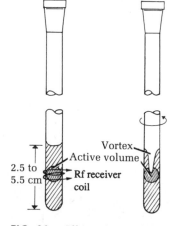

2.5 to 5.5 cm

Vortex
Active volume
Rf receiver coil

FIG. 11 Effect of too rapid spinning or insufficient sample. The active volume is the only part of the sample detected by the spectrometer.

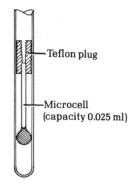

—Teflon plug

—Microcell (capacity 0.025 ml)

FIG. 12 NMR microcell positioned in an nmr tube. The Teflon plug supports the microcell and allows the sample to be centered in the active volume of the spectrometer.

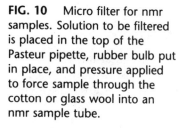

Adjusting the Continuous Wave (CW) Spectrometer

Sweep zero

To be certain the spectrometer is correctly adjusted and working properly, record the spectrum of the standard sample of chloroform and tetramethylsilane (TMS) usually found with the spectrometer. While recording the spectrum from left to right, the $CHCl_3$ peak should be brought to 7.27 ppm and the TMS peak to 0.0 ppm with the sweep zero control. The most important adjustment, the resolution control (also called homogeneity, or Y-control), should be adjusted for each sample so the TMS peak is as high and narrow as possible, with good ringing (Fig. 13). The signal (the peak traced by the spectrometer) should also be properly phased; it will then have the same appearance in both forward and backward scans (Fig. 13).

Resolution: sharp, narrow peaks with good ringing = high resolution

Use scrap paper for all but the final, best spectrum

Small peaks symmetrically placed on each side of a principal peak are artifacts called spinning side bands (Fig. 14). They are recognized as such by changing the spin speed (see again Fig. 14), which causes the spinning side bands to change positions. Two controls on the spectrometer determine the height of a signal as it is recorded on the paper. One, the spectrum amplitude control, increases the size of the signal as well as the baseline noise (the jitter of the pen when no signal is present). The other, the radiofrequency (rf) power control, increases the size of the signal alone, but only to a point, after which saturation occurs (Fig. 15). Applying more than the optimum rf power will cause the peak to become distorted and of low intensity.

Adjusting the Fourier Transform (FT) Spectrometer

Since FT spectrometers lock on the resonance of deuterium to achieve field/frequency stabilization all samples must be dissolved in a solvent containing deuterium. The fact that lock is obtained is registered on a meter

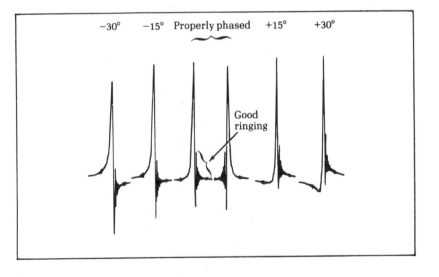

FIG. 13 Effect of phasing on signal shape. The TMS peaks on both forward and backward scans are quite high and narrow, with good ringing and perfect symmetry. Ringing is seen only on CW spectrometers.

FIG. 14 (Left)   Spinning side bands.

FIG. 15 (Right)   Saturation of the nmr signal for chloroform. At the optimum rf power level (0.075–0.10) the signal reaches maximum intensity. Further increase of power causes the signal to become smaller and broader, with poor ringing–saturation. Ringing is seen on CW but not on FT spectrometers.

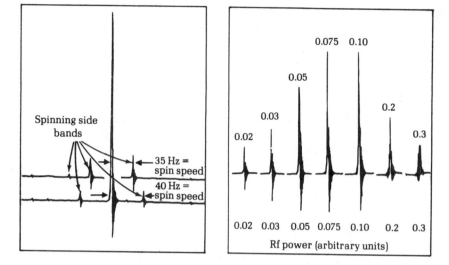

or oscilloscope. If the spectrometer utilizes an electromagnet, then the Y and Curvature controls are adjusted to achieve the highest possible lock signal and thus maximum field homogeneity. If the spectrometer utilizes a superconducting solenoid to produce the magnetic field, then the Y and Y^2 controls are used to maximize the field homogeneity and achieve the highest resolution.

Two-Dimensional NMR Spectroscopy

The large and fast computers associated with Fourier transform spectrometers allow for a series of precisely timed pulses and data accumulations to give a large data matrix that can be subjected to Fourier transformation in two dimensions to produce a two-dimensional nmr spectrum.

One of the most common and useful of these is the COSY (<u>co</u>rrelated <u>s</u>pectroscop<u>y</u>) spectrum (Fig. 16). In this spectrum two ordinary one-dimensional spectra are correlated with each other through spin-spin coupling. The 2D spectrum is a topographic representation, where spots represent peaks. The 1D spectra at the top and side are projections of these peaks. Along the diagonal of the 2D spectrum is a spot for each group of peaks in the molecule.

In Figure 16 is the 2D spectrum of citronellol. See Fig. 24.1 for the 1D spectrum of this compound. Consider the spot A on the diagonal. From a table of chemical shifts it is known that this is a vinyl proton, the single proton on the double bond. From the structure of citronellol we can expect this proton to have a small coupling, over four bonds, to the methyl groups, and a stronger coupling to the protons on carbon-6. In the absence of a 2D spectrum it is not obvious which group of peaks belongs to C-6, but the spot at B correlates with spot C on the diagonal, which is directly below the spot labeled 6 on the spectrum.

FIG. 16   The 2D COSY nmr
spectrum of citronellol.

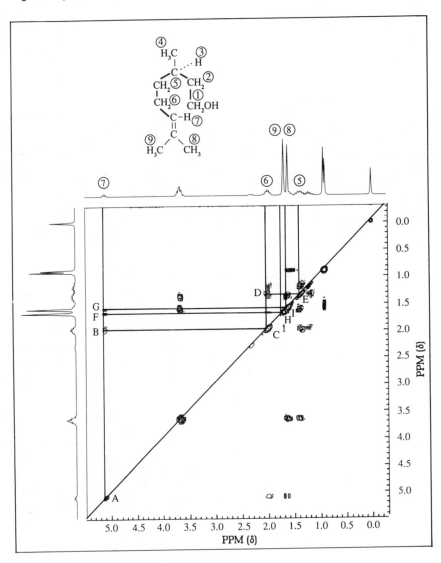

The diagonal spot C, which we have just assigned to C-6, correlates through the off-diagonal spot D with the diagonal spot E, which lies just below the group of peaks labeled 5. Spot A on the diagonal also correlates through spots F and G with spots H and I on the diagonal, which lie directly below methyl peaks 8 and 9. In this way one can determine the complete connectivity pattern of the molecule, seeing which protons are coupled to other protons.

Other 2D experiments allow proton spectra on one axis to be correlated with carbon spectra on the other axis. NOESY (nuclear Overhauser effect spectroscopy) experiments give cross peaks for protons that are near to each other in space but not spin coupled to each other.

2. Identification of Unknown Alcohol or Amine by ^1H nmr

Shift reagents are expensive

Dispose of used solvents in the container provided

Using a stock solution of 4% TMS in carbon disulfide, prepare 0.5 mL of a 0.4 M (or 10%) solution of an unknown alcohol or amine. Filter the solution, if necessary, into a clean, *dry* nmr tube. Set the TMS peak at 0.0 ppm, check the phasing, maximize the resolution, and run a spectrum over a 500-Hz range using a 250-s sweep time. To the unknown solution add about 5 mg of $Eu(dpm)_3$ (the shift reagent), shake thoroughly to dissolve, and run another spectrum. Continue adding $Eu(dpm)_3$ in 10-mg portions until the spectrum is shifted enough for easy analysis. Integrate peaks and groups of peaks if in doubt about their relative areas. To protect the $Eu(dpm)_3$ from moisture store in a desiccator.

Cleaning Up   All samples containing shift reagents go into a hazardous waste container for heavy metals.

Questions

1. Propose a structure or structures consistent with the proton nmr spectrum of Fig. 17. Numbers adjacent to groups of peaks refer to relative peak areas. Account for missing lines.

2. Propose a structure or structures consistent with the proton nmr spectrum of Fig. 18. Numbers adjacent to groups of peaks refer to relative peak areas.

3. Propose a structure or structures consistent with the proton nmr spectrum of Fig. 19.

4. Propose structures for a, b, and c consistent with the carbon-13 nmr spectra of Figs. 20, 21, and 22. These are isomeric alcohols with the empirical formula $C_4H_{10}O$.

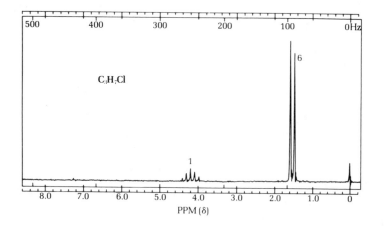

FIG. 17   Proton nmr spectrum (60 MHz), Question 1.

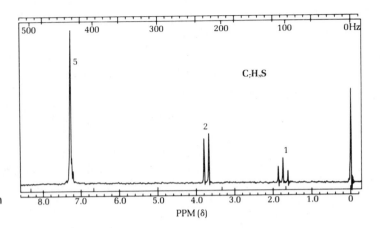

FIG. 18   Proton nmr spectrum (60 MHz), Question 2.

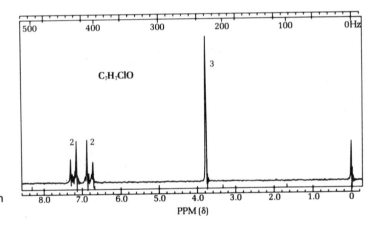

FIG. 19   Proton nmr spectrum (60 MHz), Question 3.

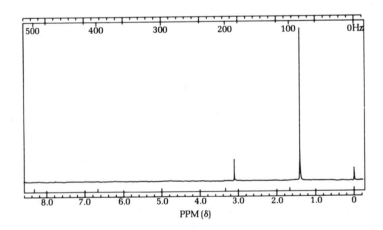

FIG. 20   ¹H nmr spectrum of $C_4H_{10}O$ (90 MHz), Question 4(a).

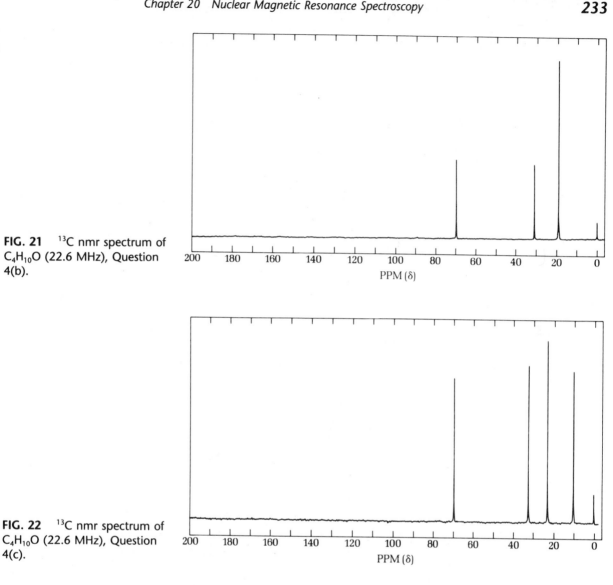

FIG. 21 ^{13}C nmr spectrum of $C_4H_{10}O$ (22.6 MHz), Question 4(b).

FIG. 22 ^{13}C nmr spectrum of $C_4H_{10}O$ (22.6 MHz), Question 4(c).

References

1. L. M. Jackman and S. Sternhell, *Applications of Nuclear Magnetic Resonance Spectroscopy in Organic Chemistry*, Pergamon Press, Elmsford, N.Y., 2nd Ed., 1969.

2. D. Shaw, *Fourier Transform NMR Spectroscopy*, Elsevier, Amsterdam, 1976.

3. T. C. Farrer and E. D. Becker, *Pulse and Fourier Transform NMR*, Academic Press, New York, 1971.

4. M. L. Martin, J.-J. Delpuech, and G. J. Martin, *Practical NMR Spectroscopy,* Heyden & Son, Philadelphia, 1980.

5. E. D. Becker, *High Resolution NMR: Theory and Chemical Applications,* 2nd Ed., Academic Press, New York, 1980.

6. *High Resolution NMR Spectra Catalogs,* Varian Associates, Palo Alto, Vol. 1, 1962, Vol. 2, 1963.

7. J. W. Cooper, *Spectroscopic Techniques for Organic Chemists,* John Wiley, New York, 1980.

8. J. B. Stothers, *Carbon-13 NMR Spectroscopy,* Academic Press, New York, 1972.

9. G. C. Levy, R. Lichter, and G. L. Nelson, *Carbon-13 NMR Spectroscopy,* 2nd Ed., Wiley Interscience, New York, 1980.

10. L. F. Johnson and W. C. Jankowski, *Carbon-13 NMR Spectra,* Wiley Interscience, New York, 1972.

11. R. J. Abraham, J. Fisher, and P. Loftus, *Introduction to NMR Spectroscopy,* Heyden & Son, Philadelphia, 1988.

12. J. W. Akitt, *NMR and Chemistry,* 2nd Ed., Chapman and Hall, London, 1983.

13. O. Jardetzky and G. C. K. Roberts, *NMR in Molecular Biology,* Academic Press, New York, 1981.

14. A. E. Derome, *Modern NMR Techniques for Chemistry Research,* Pergamon Press, New York, 1986.

15. G. E. Martin and A. S. Zektzer, *Two-Dimensional NMR Methods for Establishing Molecular Connectivity,* VCH Publishers, New York, 1988.

16. W. R. Croasman and R. M. K. Carlson, Eds., *Two-Dimensional NMR Spectroscopy,* VCH Publishers, New York, 1987.

21

Ultraviolet Spectroscopy

Prelab Exercise: Predict the appearance of the ultraviolet spectrum of 1-butylamine in acid, of methoxybenzene in acid and base, and of benzoic acid in acid and base.

---

uv—Electronic transitions within molecules

Ultraviolet spectroscopy gives information about electronic transitions within molecules. Whereas absorption of low-energy infrared radiation causes bonds in a molecule to stretch and bend, the absorption of short-wavelength, high-energy ultraviolet radiation causes electrons to move from one energy level to another with energies that are often capable of breaking chemical bonds.

We shall be most concerned with transitions of π-electrons in conjugated and aromatic ring systems. These transitions occur in the wavelength region 200 to 800 nm (nanometers, 10^{-9} meters, formerly known as mμ, millimicrons). Most common ultraviolet spectrometers cover the region 200 to 400 nm as well as the visible spectral region 400 to 800 nm. Below 200 nm air (oxygen) absorbs uv radiation; spectra in that region must therefore be obtained in a vacuum or in an atmosphere of pure nitrogen.

Consider ethylene, even though it absorbs uv radiation in the normally inaccessible region at 163 nm. The double bond in ethylene has two *s* electrons in a σ-molecular orbital and two, less tightly held, *p* electrons in a π-molecular orbital. Two unoccupied, high-energy-level, antibonding orbitals are associated with these orbitals. When ethylene absorbs uv radiation, one electron moves up from the bonding π-molecular orbital to the antibonding π^*-molecular orbital (Fig. 1). As the diagram indicates, this change requires less energy than the excitation of an electron from the σ to the σ^* orbital.

By comparison with infrared spectra and nmr spectra, uv spectra are fairly featureless (Fig. 2). This condition results as molecules in a number of different vibrational states undergo the same electronic transition, to produce a band spectrum instead of a line spectrum.

Band spectra

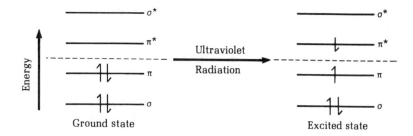

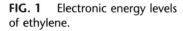

FIG. 1 Electronic energy levels of ethylene.

FIG. 2 The ultraviolet spectrum of cholesta-3,5-diene in ethanol.

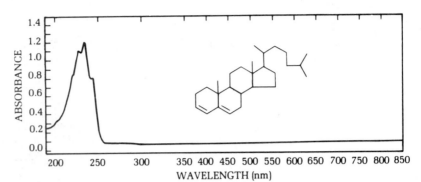

Beer-Lambert Law

Unlike ir spectroscopy, ultraviolet spectroscopy lends itself to precise quantitative analysis of substances. The intensity of an absorption band is usually given by the molar extinction coefficient ϵ, which, according to the Beer-Lambert Law, is equal to the absorbance A, divided by the product of the molar concentration c, and the path length l, in centimeters.

$$\epsilon = \frac{A}{cl}$$

λ_{max}, *wavelength of maximum absorption*

The wavelength of maximum absorption (the tip of the peak) is given by λ_{max}. Because uv spectra are so featureless it is common practice to describe a spectrum like that of cholesta-3,5-diene (Fig. 2) as λ_{max} 234 nm (ϵ = 20,000), and not bother to reproduce the actual spectrum.

ϵ, *extinction coefficient*

The extinction coefficients of conjugated dienes and enones are in the range 10,000–20,000, so only very dilute solutions are needed for spectra. In the example of Fig. 2 the absorbance at the tip of the peak, A, is 1.2, and the path length is the usual 1 cm; so the molar concentration needed for this spectrum is 6×10^{-5} mole per liter.

$$c = \frac{A}{l\epsilon} = \frac{1.2}{20,000} = 6 \times 10^{-5} \text{ mole per liter}$$

which is 0.221 mg per 10 mL of solvent.

Spectro grade solvents

The usual solvents for uv spectroscopy are 95% ethanol, methanol, water, and saturated hydrocarbons such as hexane, trimethylpentane, and isooctane; the three hydrocarbons are often specially purified to remove impurities that absorb in the uv region. Any transparent solvent can be used for spectra in the visible region.

uv cells are expensive; handle with care

Sample cells for spectra in the visible region are made of glass, but uv cells must be of the more expensive fused quartz, since glass absorbs uv radiation. The cells and solvents must be clean and pure, since very little of a substance produces a uv spectrum. A single fingerprint will give a spectrum!

Ethylene has λ_{max} 163 nm (ϵ = 15,000) and butadiene has λ_{max} 217 nm (ϵ = 20,900). As the conjugated system is extended, the wavelength of maximum absorption moves to longer wavelengths (toward the visible region). For example, lycopene with 11 conjugated double bonds has λ_{max} 470 nm (ϵ = 185,000), Fig. 3. Since lycopene absorbs blue visible light at 470 nm the substance appears bright red. It is responsible for the color of tomatoes; its isolation is described in Chapter 9.

Woodward and Fieser Rules for dienes and dienones

The wavelengths of maximum absorption of conjugated dienes and polyenes and conjugated enones and dienones are given by the Woodward and Fieser Rules, Tables 1 and 2.

The application of the rules in the above tables is demonstrated by the spectra of pulegone (1) and carvone (2), Fig. 4, with the calculations given in Tables 4 and 5.

These rules will be applied in a later experiment in which cholesterol is converted into an α,β-unsaturated ketone.

TABLE 1   Rules for the Prediction of λ_{max} for Conjugated Dienes and Polyenes

	Increment (nm)
Parent acyclic diene (butadiene)	217
Parent heteroannular diene	214
Double bond extending the conjugation	30
Alkyl substituent or ring residue	5
Exocyclic location of double bond to any ring	5
Groups: OAc, OR	0
	0
Solvent correction, see Table 3 ( )	
λ_{max} = Total	

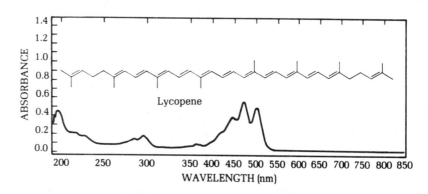

FIG. 3   The ultraviolet-visible spectrum of lycopene in isooctane.

TABLE 2 Rules for the Prediction of λ_{max} for Conjugated Enones and Dienones

$$\overset{\beta}{|}\ \overset{\alpha}{|}\ \overset{R}{|}\qquad\qquad\overset{\delta}{|}\ \overset{\gamma}{|}\ \overset{\beta}{|}\ \overset{\alpha}{|}\ \overset{R}{|}$$

$$\beta-C=C-C=O \quad \text{and} \quad \delta-C=C-C=C-C=O \qquad \text{Increment (nm)}$$

	Increment (nm)
Parent α,β-unsaturated system	215
Double bond extending the conjugation	30
R (alkyl or ring residue), OR, OCOCH$_3$ α	10
β	12
γ, δ, and higher	18
α-Hydroxyl, enolic	35
α-Cl	15
α-Br	23
exo-Location of double bond to any ring	5
Homoannular diene component	39

Solvent correction, see Table 3

$$\lambda_{max}^{EtOH} = \text{Total}$$

TABLE 3 Solvent Correction

Solvent	Factor for Correction to Ethanol
Hexane	+11
Ether	+7
Dioxane	+5
Chloroform	+1
Methanol	0
Ethanol	0
Water	−8

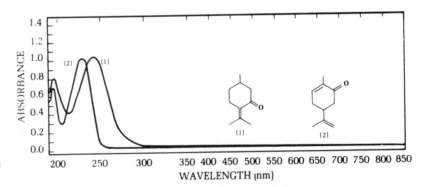

FIG. 4 Ultraviolet spectra of (1) pulegone and (2) carvone in hexane.

TABLE 4   Calculation of λ_{max} for Pulegone (See Fig. 4)

Parent α,β-unsaturated system	215 nm
α-Ring residue, R	10
β-Alkyl group (two methyls)	24
Exocyclic double bond	5
Solvent correction (hexane)	-11

Calcd λ_{max} 245 nm; found 244 nm

TABLE 5   Calculation of λ_{max} for Carvone (See Fig. 4)

Parent α,β-unsaturated system	215 nm
α-Alkyl group (methyl)	10
β-Ring residue	12
Solvent correction (hexane)	-11

Calcd λ_{max} 226 nm; found 229 nm

No simple rules exist for calculation of aromatic ring spectra, but several generalizations can be made. From Fig. 5 it is obvious that as polynuclear aromatic rings are extended linearly, λ_{max} shifts to longer wavelengths.

As alkyl groups are added to benzene, λ_{max} shifts from 255 nm for benzene to 261 nm for toluene to 272 nm for hexamethylbenzene. Substituents bearing nonbonding electrons also cause shifts of λ_{max} to longer wavelengths, e.g., from 255 nm for benzene to 257 nm for chlorobenzene, 270 nm for phenol, and 280 nm for aniline ($\epsilon = 6{,}200$–$8{,}600$). That these effects are the result of interaction of the π-electron system with the nonbonded electrons is seen dramatically in the spectra of vanillin and the derived anion (Fig. 6). Addition of two more nonbonding electrons in the anion causes λ_{max} to shift from 279 nm to 351 nm and ϵ to increase.

Effect of acid and base on λ_{max}

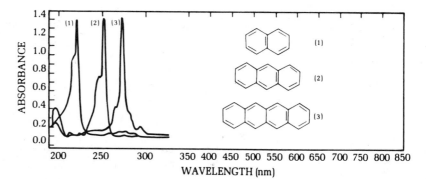

FIG. 5   The ultraviolet spectra of (1) naphthalene, (2) anthracene, and (3) tetracene.

Removing the electrons from the nitrogen of aniline by making the anilinium cation causes λ_{max} to decrease from 280 nm to 254 nm (Fig. 7). These changes of λ_{max} as a function of pH have obvious analytical applications.

Intense bands result from π-π conjugation of double bonds and carbonyl groups with the aromatic ring. Styrene, for example, has λ_{max} 244 nm (ϵ = 12,000) and benzaldehyde λ_{max} 244 nm (ϵ = 15,000).

Experiment

Ultraviolet Spectrum of Unknown Acid, Base, or Neutral Compound

Determine whether an unknown compound obtained from the instructor is acidic, basic, or neutral from the ultraviolet spectra in the presence of acid and base as well as in neutral media.

Cleaning Up   Since uv samples are extremely dilute solutions in ethanol, they can normally be flushed down the drain.

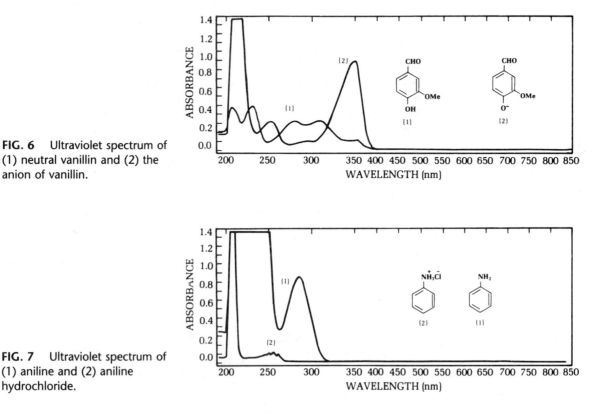

FIG. 6   Ultraviolet spectrum of (1) neutral vanillin and (2) the anion of vanillin.

FIG. 7   Ultraviolet spectrum of (1) aniline and (2) aniline hydrochloride.

Questions

1. Calculate the ultraviolet absorption maximum for 2-cyclohexene-1-one.

2. Calculate the ultraviolet absorption maximum for 3,4,4-trimethyl-2-cyclohexene-1-one.

3. Calculate the ultraviolet absorption maximum for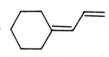

4. What concentration, in g/mL, of a substance with MW 200 should be prepared in order to give an absorbance value A equal to 0.8 if the substance has $\epsilon = 16,000$ and a cell with path length 1 cm is employed?

References

1. A. E. Gillam and E. S. Stern, *An Introduction to Electronic Absorption Spectroscopy in Organic Chemistry,* Edward Arnold, London, 1967.

2. C. N. R. Rao, *Ultraviolet and Visible Spectroscopy,* Butterworths, London, 2nd ed., 1967.

3. H. H. Jaffe and M. Orchin, *Theory and Application of Ultraviolet Spectroscopy,* John Wiley, New York, 1962.

4. R. M. Silverstein, C. G. Bassler, and T. C. Morrill, *Spectrometric Identification of Organic Compounds,* John Wiley, New York, 4th Ed., 1981. (Includes ir, uv, and nmr.)

22

Cholesterol from Human Gallstones

Prelab Exercise: Explain how the process of bromination and debromination of cholesterol can free it from impurities. Write a detailed mechanism for the bromination of cholesterol, taking care to consider the stereochemistry of the product.

In this experiment cholesterol will be isolated from human gallstones. Cholesterol is an unsaturated alcohol containing 27 carbon atoms and 46 hydrogen atoms:

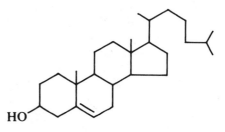

It is a solid, mp 148.5°C, and is insoluble in water but soluble in boiling ethanol and dioxane.

Gallstones

The gallbladder is attached to the undersurface of the liver just below the rib cage. It retains bile produced by the liver and feeds it into the upper part of the small intestine as needed for digestion. Bile consists primarily of bile acids, which are carboxylic acids closely resembling cholesterol and which aid in the digestion of fats by functioning as emulsifying agents. The gallbladder also harbors free cholesterol. If the concentration of cholesterol in the bile exceeds a certain critical level, it will come out of solution and agglomerate into particles that grow to form gallstones. An amateur geologist given a bottle of gallstones to identify once labeled them a "riverbed conglomerate"—and indeed they do resemble stones in color, texture, and hardness. They come in a variety of shapes and colors and can be up to an inch in diameter.

As gallstones collect they irritate the lining of the gallbladder, causing severe pain, nausea, and vomiting. The stones can block the bile duct and lead to fatal complications. The remedy is surgery although, recently, gallstones have been dissolved right in the gallbladder, not with boiling alcohol but with methyl *tert*-butyl ether. The ether is injected directly into the gallbladder through a 1.7-mm tube, which pierces the skin and liver on

Cholesterol: soluble in methyl tert-butyl ether in the gallbladder

its way to the gallbladder. Depending on the number and size of the gallstones, the dissolution process can take from 7 to 18 h.

In the average human, approximately 200 g of cholesterol is concentrated primarily in the spinal cord, brain, and nerve tissue. Insoluble in water and plasma, it is transported in the bloodstream bound to lipoproteins, which are proteins attached to lipids (fats). Recent research has divided these lipoproteins, when centrifuged, into two broad classes—high-density (HDL) and low-density (LDL) lipoproteins. A relatively high concentration of HDL bound to cholesterol seems to cause no problems and in fact is beneficial; but a high ratio of LDL-cholesterol leads to the deposition of cholesterol both in the gallbladder (resulting in gallstones) and on the walls of the arteries (causing a plaque that cuts off blood flow and hastens hardening of the arteries or atherosclerosis).

HDL: high-density lipoproteins (good)
LDL: low-density lipoproteins (bad)

Mounting evidence points to unsaturated fats such as those found in vegetable oils as favoring the HDL-cholesterol bond, while LDL-cholesterol formation is speeded up by saturated fats such as those found in animals. The HDL-cholesterol level goes down with smoking or eating large amounts of sugar. It goes up with regular exercise and with the consumption of *moderate* amounts of alcohol (a glass of wine per day). The 1985 Nobel prize in physiology or medicine went to Michael Brown and Joseph Goldstein for their pioneering work on LDL- and HDL-cholesterol.

1985 Nobel prize: Brown and Goldstein

The average American woman at age 75 has a 50% chance of developing gallstones while for a man of the same age the chance is only half as great. Gallstones and coronary heart disease are also much more common in overweight people. Almost 70% of the women in certain American Indian tribes get gallstones before the age of 30, whereas only 10% of black women are afflicted. Swedes and Finns have gallstones more often than Americans; the problem is almost unknown among the Masai people of East Africa.

The most common treatment at present for gallstones is surgical removal, an operation performed 500,000 times each year in the United States. This is the source of the gallstones used in the present experiment.

Experiment

Cholesterol from Gallstones[1]

Swirl 2 g of crushed gallstones in a 25-mL Erlenmeyer flask with 15 mL of 2-butanone on a hot plate for a few minutes until the solid has disintegrated and the cholesterol has dissolved. Filter the solution while hot, taking precautions to warm the stemless funnel prior to the filtration. Its dirty-yellow appearance is from a brown residue of the bile pigment bilirubin, a metabolite of hemoglobin.

Note for the instructor

1. Obtainable from the department of surgery of a hospital. Wrap the stones in a towel and crush by lightly pounding with a hammer. Do not breathe the dust as the gallstones may not be sterile. (See also *Instructor's Manual for Organic Experiments,* 7th ed. for a procedure for making "artificial gallstones.")

*Crystallization from mixed
solvents—Chapter 5*

If the filtrate is highly colored, add decolorizing charcoal and refilter the hot solution in the usual way (see Chapter 3). This will ordinarily not be necessary. Dilute the filtrate with 10 mL of methanol, reheat the filtrate to the boiling point, add a little water gradually until the solution is saturated with cholesterol at the boiling point, cool the solution in ice, and let the solution stand for crystallization. Collect the crystals, wash, dry, and take the melting point. Typical result: 1.5 g of large, colorless plates or needles, mp 146–147.5°C. Use 1 g for purification and save the rest for color tests.

Bilirubin

$\lambda_{max}^{CHCl_3}$ 450 nm

Cleaning Up   The solvents, 2-butanone and methanol, should be placed in the organic solvents container and the bilirubin in the nonhazardous solid waste container.

Question

1. Cholesterol is an alcohol. Why is it more soluble in organic solvents than in water?

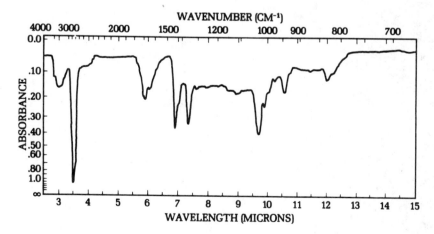

FIG. 1   Infrared spectrum of cholesterol.

23

Bromination and Debromination: Purification of Cholesterol

Prelab Exercise: Make a molecular model of 5α,6β-dibromocholestan-3β-ol and convince yourself that the bromine atoms in this molecule are *trans* and *diaxial*.

The bromination of a double bond is an important and well-understood organic reaction. In the present experiment it is employed for the very practical purpose of purifying crude cholesterol through the process of bromination, crystallization, and then zinc dust debromination.

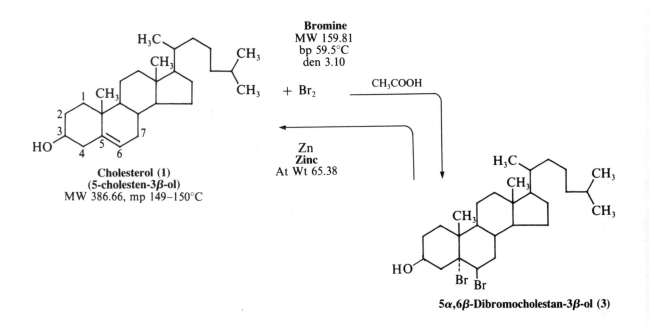

Cholesterol (1)
(5-cholesten-3β-ol)
MW 386.66, mp 149–150°C

Bromine
MW 159.81
bp 59.5°C
den 3.10

$+ Br_2$

CH_3COOH

Zn
Zinc
At Wt 65.38

5α,6β-Dibromocholestan-3β-ol (3)

The reaction involves nucleophilic attack by the alkene on bromine with the formation of a tertiary carbocation that probably has some bromonium ion character resulting from sharing of the nonbonding electrons on bromine with the electron-deficient C-5 carbon. This ion is attacked from the backside by bromide ion to form dibromocholesterol with the bromine atoms in the *trans* and diaxial configuration, the usual result when brominating a cyclohexene.

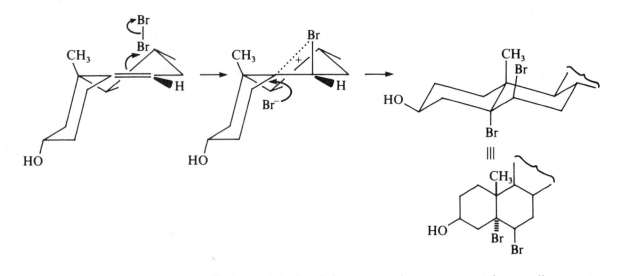

Cholesterol isolated from natural sources contains small amounts (0.1–3%) of 3β-cholestanol, 7-cholesten-3β-ol, and 5,7-cholestadien-3β-ol.[1] These are so very similar to cholesterol in solubility that their removal by crystallization is not feasible. However, complete purification can be accomplished through the sparingly soluble dibromo derivative 5α,6β-dibromocholestan-3β-ol. 3β-Cholestanol is saturated and does not react with bromine; thus, it remains in the mother liquor. 7-Cholesten-3β-ol and 5,7-cholestadien-3β-ol are dehydrogenated by bromine to dienes and trienes, respectively, that likewise remain in the mother liquor and are eliminated along with colored by-products.

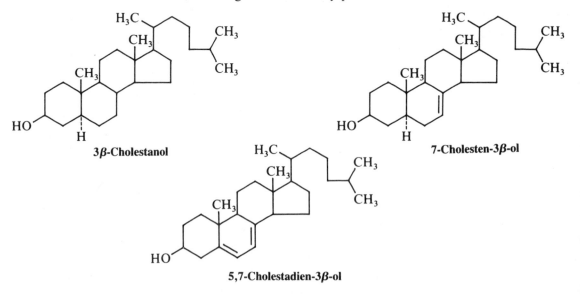

3β-Cholestanol

7-Cholesten-3β-ol

5,7-Cholestadien-3β-ol

1. A fourth companion, cerebrosterol, or 25-hydroxycholesterol, is easily eliminated by crystallization from alcohol.

The cholesterol dibromide that crystallizes from the reaction solution is collected, washed free of the impurities or their dehydrogenation products, and debrominated with zinc dust, with regeneration of cholesterol in pure form. Specific color tests can differentiate between pure cholesterol and tissue cholesterol purified by ordinary methods.

Experiments

1. Bromination of Cholesterol

In a 25-mL Erlenmeyer flask dissolve 1 g of gallstone cholesterol or of commercial cholesterol (content of 7-cholesten-3β-ol about 0.6%) in 7 mL of ether by gentle warming and, with a pipette fitted with a pipetter or with a plastic syringe (Fig. 1), or from a burette in the hood, add 5 mL of a solution of bromine and sodium acetate in acetic acid.[2] Cholesterol dibromide begins to crystallize in a minute or two. Cool in an ice bath and stir the crystalline paste with a stirring rod for about 10 min to ensure complete crystallization, and at the same time cool a mixture of 3 mL of ether and 7 mL of acetic acid in ice. Then collect the crystals on a small suction funnel and wash with the iced ether-acetic acid solution to remove the yellow mother liquor. Finally, wash with a little methanol, continuing to apply suction, and transfer the white solid without drying it (dry weight 1.2 g) to a 50-mL Erlenmeyer flask.

Cleaning Up The filtrate from this reaction contains halogenated material and so must be placed in the halogenated organic waste container.

2. Zinc Dust Debromination

To the flask containing dibromocholesterol add 20 mL of ether, 5 mL of acetic acid, and 0.2 g of zinc dust and swirl.[3] In about 3 min the dibromide dissolves; after 5–10 min swirling, zinc acetate usually separates to form a white precipitate (the dilution sometimes is such that no separation occurs). Stir for 5 min more and then add water by drops (no more than 0.5 mL) until any solid present (zinc acetate) dissolves to make a clear solution. Decant the solution from the zinc into a separatory funnel, and wash the ethereal solution twice with water and then with 10% sodium hydroxide (to remove traces of acetic acid). Then shake the ether solution with an equal volume of saturated sodium chloride solution to reduce the water content, dry the ether with anhydrous sodium sulfate, remove the drying agent, add 10 mL of methanol (and a boiling stone), and evaporate the solution on the steam

FIG. 1 Plastic syringes.

Caution: Be extremely careful to avoid getting the solution of bromine and acetic acid on the skin. Carry out the reaction in the hood and wear disposable gloves for maximum safety.

Aspirator tube

---

2. Weigh a 125-mL Erlenmeyer flask on a balance placed in the hood, add 4.5 g of bromine by a Pasteur pipette (avoid breathing the vapor), and add 50 mL of acetic acid and 0.4 g of sodium acetate (anhydrous).

3. If the reaction is slow add more zinc dust or place the flask in an ultrasonic cleaning bath, which will activate the zinc. The amount specified is adequate if material is taken from a freshly opened bottle, but zinc dust deteriorates on exposure to air.

bath to the point where most of the ether is removed and the purified cholesterol begins to crystallize. Remove the solution from the steam bath, let crystallization proceed at room temperature and then in an ice bath, collect the crystals, and wash them with cold methanol; you should obtain 0.6–0.7 g, mp 149–150°C.

Cleaning Up The unreacted zinc dust can be discarded in the non-hazardous solid waste container after it has been allowed to dry and then sit exposed to the air for about one-half hour on a watch glass. Sometimes the zinc dust at the end of this reaction will get quite hot as it air oxidizes. The aqueous and acetic acid solutions after neutralization can be flushed down the drain, and the organic filtrates containing ether and methanol should be placed in the organic solvents container. After the solvent has evaporated from the sodium sulfate it can be placed in the nonhazardous solid waste container.

Questions

1. What is the purpose of the acetic acid in this reaction?

2. Why might old zinc dust not react in the debromination reaction?

3. Why does not acetate ion attack the intermediate bromonium ion to give the 5-acetoxy-6-bromo compound in the bromination of cholesterol?

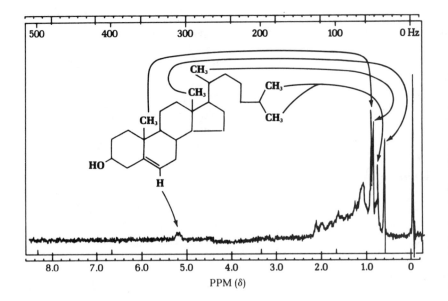

FIG. 2 ^{1}H nmr spectrum of cholesterol (60 MHz).

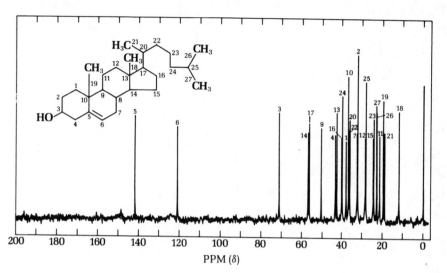

FIG. 3 ^{13}C nmr spectrum of cholesterol (22.6 MHz).

CHAPTER

24

Pulegone from Citronellol: Oxidation with Pyridinium Chlorochromate

Prelab Exercise: Write a balanced equation for the oxidation of citronellol to citronellal.

Oxidizing agents:
$Cr_2O_7^{2-}$ in HOAc, $KMnO_4$, CrO_3 in aq H_2SO_4, Jones reagent,
CrO_3 + pyridine, Collins reagent,
Corey reagents, PCC

E. J. Corey of Harvard, winner of the 1991 Nobel prize in chemistry

A variety of reagents and conditions can be employed to oxidize alcohols to carbonyl compounds. The choice of which reagents and set of conditions to use depends on such factors as the scale of the reaction, the speed of the reaction, anticipated yield, and ease of isolation of the products. Cyclohexanol can be oxidized to cyclohexanone by dichromate in acetic acid. This is much faster than permanganate oxidation of the same alcohol. Cholesterol can be oxidized to 5-cholesten-3-one by either the Jones reagent (chromium trioxide in sulfuric acid and water) or the Collins reagent, $(C_5H_5N)_2CrO_3$, prepared from chromium trioxide in anhydrous pyridine. Both of these reagents have the advantage they they will not cause isomerization of the labile Δ^5 double bond to the Δ^4 position. E. J. Corey has added two new Cr^{6+} reagents to the repertory. Although they might seem similar to the other dichromate reagents, they each have unique advantages. While not nearly as fast as the Collins or Jones reagents, the Corey reagents are characterized by extraordinary ease of product isolation, because one merely removes the reagent by filtration and evaporates the solvent to obtain the product. Both reagents are easily prepared or can be purchased commercially (Aldrich). The reagent used in this experiment, pyridinium chlorochromate, $C_5H_5NH^+ClCrO_3^-$, will oxidize primary alcohols to aldehydes in high yield without oxidizing the aldehyde further to a carboxylic acid.[1] It will oxidize secondary alcohols to ketones, but being somewhat acidic, it will cause rearrangement of 5-cholesten-3-one to 4-cholesten-3-one.

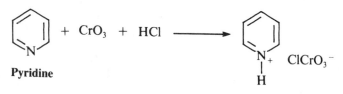

Pyridine

Pyridinium chlorochromate, PCC
MW 215.56

---

1. E. J. Corey and J. W. Suggs, *Tetrahedron Letters*, 2647 (1975).

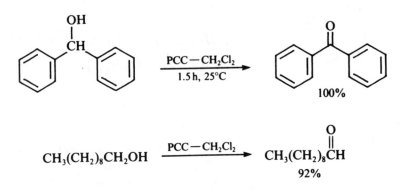

100%

92%

The other Corey reagent, pyridinium dichromate (PDC), $(C_5H_5NH^+)_2Cr_2O_7^{2-}$, dissolved in dimethylformamide (DMF) will oxidize allylic alcohols to α,β-unsaturated aldehydes without oxidizing the aldehyde to the carboxylic acid.[2]

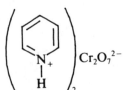

Pyridinium Dichromate, PDC

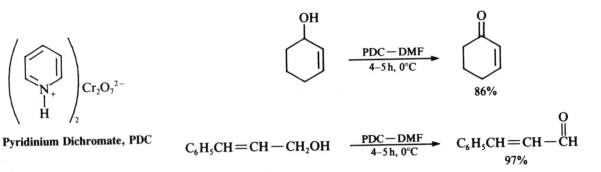

86%

$$C_6H_5CH=CH-CH_2OH \xrightarrow[\text{4-5 h, 0°C}]{PDC-DMF} C_6H_5CH=CH-\overset{\displaystyle O}{\overset{\displaystyle \|}{C}}H$$

97%

If the primary alcohol is not conjugated, then it is oxidized to the carboxylic acid:

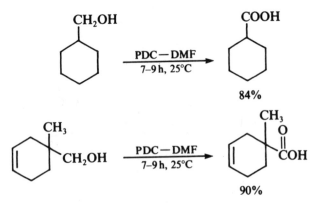

84%

90%

2. E. J. Corey and G. Schmidt, *Tetrahedron Letters*, 399 (1979).

When suspended in CH_2Cl_2, pyridinium dichromate will oxidize primary alcohols to the corresponding aldehydes and no further:

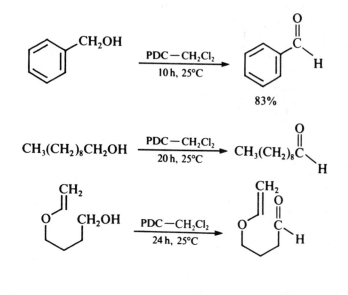

83%

Experiments

$$RCH_2OH \xrightarrow[\text{alumina}]{\text{PCC on}} RCHO$$

In the present experiment pyridinium chlorochromate (PCC) is used to oxidize citronellol, a constituent of rose and geranium oil, to the corresponding aldehyde. If PDC in CH_2Cl_2 were used, the reaction would stop at this point and citronellal could be isolated in 92% yield. Or if sodium acetate (a buffer) were added to the PCC reaction, the oxidation would also stop at the aldehyde stage, in 82% yield. It has been found that PCC deposited on alumina will effect the same oxidation in less time and even higher yield.[3]

If the buffer is omitted, the pyridinium chlorochromate is acidic enough to cause the intermediate citronellal to cyclize to isopulegols (four possible isomers). These secondary alcohols are oxidized to two isomeric isopulegones. On treatment with base, the double bond will migrate to give pulegone, as shown on the following page.

The course of the reaction can easily be followed by thin-layer chromatography.[4] The isolation of the oxidation products, citronellal and isopulegone, could not be simpler: the reduced reagent is removed by filtration and the solvent is evaporated. In the case of the second experiment the reaction can be stopped once isopulegone is formed, or the isopulegone can be isomerized to pulegone as a separate reaction.

3. Y.-S. Cheng, W.-L. Liu and S. Chen, *Synthesis*, 223 (1980).

4. E. J. Corey and J. W. Suggs, *J. Org. Chem.*, **41**, 380 (1976). TLC R_f values: citronellol, 0.17; citronellal, 0.65; isopulegols, 0.27 and 0.35; isopulegone, 0.41; pulegone, 0.36, CH_2Cl_2 developer.

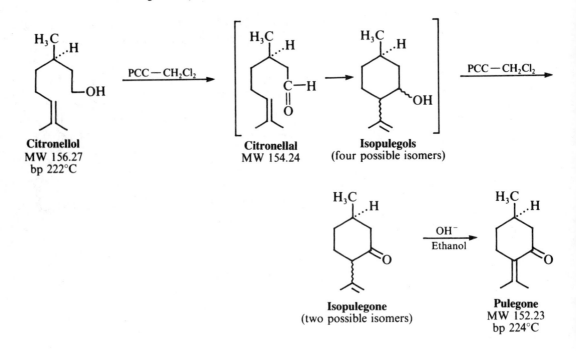

Citronellol
MW 156.27
bp 222°C

Citronellal
MW 154.24

Isopulegols
(four possible isomers)

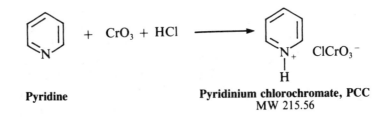

Isopulegone
(two possible isomers)

Pulegone
MW 152.23
bp 224°C

1. Preparation of Pyridinium Chlorochromate, PCC

Pyridine

Pyridinium chlorochromate, PCC
MW 215.56

Dissolve 12 g of chromium trioxide in 22 mL of 6 N hydrochloric acid, and add 9.5 g of pyridine during a 10-min period while maintaining the temperature at 45°C. Cool the mixture to 0°C and collect the crystalline yellow-orange pyridinium chlorochromate on a sintered glass funnel and dry it for 1 h *in vacuo*. The compound is not hygroscopic and is stable at room temperature.

Cleaning Up   The filtrate should have pH < 3 and then be treated with a 50% excess of sodium bisulfite to reduce the orange dichromate ion to the green chromic ion. This solution should then be made basic with ammonium hydroxide to precipitate chromium as the hydroxide. This precipitate is collected on a filter paper, which is placed in the hazardous solid waste container for heavy metals. The filtrate from this latter treatment can go down the drain.

2. Preparation of PCC-Alumina Reagent

To prepare PCC on alumina, cool the reaction mixture described above to 10°C until the product crystallizes; then reheat the solution to 40°C to dissolve the solid. To the resulting solution add 100 g of alumina with stirring at 40°C. The solvent is removed on a rotary evaporator, and the orange solid is dried in vacuum for 2 h at room temperature. If stored under vacuum in the dark, the reagent will retain its activity for several weeks.

Cleaning Up   The used alumina, since it contains adsorbed hexavalent chromium ion, should be placed in the hazardous solid waste container for Cr^{6+} contaminated alumina. Place dichloromethane in the halogenated waste container.

CAUTION: *The dust (but not the solutions) of Cr^{6+} is a cancer suspect agent.*

3. Citronellal from Citronellol Using Pyridinium Chlorochromate on Alumina

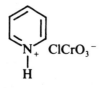

Pyridinium chlorochromate
MW 215.56

PCC on alumina (7.5g, 6.1 mmol) is added to a flask containing a solution of citronellol (0.60 g, 3.8 mmol) in 10 mL of *n*-hexane. After stirring or shaking for up to 3 h (follow the course of the reaction by TLC), remove the solid by filtration, wash it with three 10-mL portions of ether, and remove the solvents from the filtrate by distillation or evaporation. The last trace of solvent can be removed under vacuum. (See Fig. 9 in Chapter 3.) The residue should be pure citronellal, bp 90°C at 14 mm. Check its purity by TLC and infrared spectroscopy. A large number of other primary and secondary alcohols can be oxidized to aldehydes and ketones using this same procedure.

Cleaning Up   The used alumina, since it contains adsorbed hexavalent chromium ion, should be placed in the hazardous solid waste container. Organic material goes in the organic solvents container.

4. Isopulegone from Citronellol Using Pyridinium Chlorochromate in CH_2Cl_2

To 16 g of pyridinium chlorochromate suspended in 100 mL of dichloromethane add 4.0 g of citronellol. The best yields are obtained if the mixture is stirred at room temperature for 36 h or more using a magnetic stirrer. Alternatively, let the mixture sit at room temperature for a week with occasional shaking. The chromium salts are removed by vacuum filtration and washed on the filter with 15 mL of dichloromethane. Remove the solvent by evaporation to give isopulegone. Follow the course of the reaction and analyze the product by TLC on silica gel plastic sheets, developing with dichloromethane.

Cleaning Up   Dichloromethane should be placed in the halogenated organic waste container. The chromium salts should be dissolved in hydrochloric acid, diluted to <5%, reduced, and precipitated as described above under the preparation of pyridinium chlorochromate.

5. Pulegone from Isopulegone

Crush and then dissolve one sodium hydroxide pellet under 30 mL of ethanol in a 50-mL Erlenmeyer flask, add 4 g of isopulegone, and warm the mixture on the steam bath for one hour. Evaporate most of the ethanol on the steam bath; then add 10 mL of cold water and extract the product with 20 mL of ether. Wash the ether extract with about 5 mL of water, followed by 5 mL of saturated sodium chloride solution, and then dry the ether over anhydrous sodium sulfate. Decant the ether into an Erlenmeyer flask, wash the sodium sulfate with a few milliliters more ether, and evaporate the ether to leave crude pulegone. Pulegone can be distilled at atmospheric pressure (bp 224°C) but is best distilled at reduced pressure (bp 103°C/17 mm).

Ultraviolet spectra
α,β-unsaturated ketone

Pulegone occurs naturally in pennyroyal oil and has a pleasant odor somewhere between that of peppermint and camphor. It is an α,β-unsaturated ketone and therefore should have λ_{max} 254 nm as calculated by application of the Fieser and Woodward Rules. (See p. 237.) The actual value is 253.3 nm. The starting material is not a conjugated ketone and so should not have intense uv absorption. Compare the two compounds by dissolving 1 mg of each in 50 mL of ethanol and determine their ultraviolet spectra. If the final product is not purified by vacuum distillation, this entire experiment can be scaled down by a factor of four or more.

Cleaning Up   Combine the aqueous layers, neutralize with dilute hydrochloric acid, and after dilution flush down the drain. Dry the sodium sulfate in the hood and then place it in the nonhazardous waste container. Combine all organic material and place it in the organic solvents container.

Questions

1. Draw the chair conformations of the four possible isopulegols.

2. Assign as many peaks as possible to specific protons in the ¹H nmr spectra of citronellol (Fig. 1) and citronellal (Fig. 2).

3. Give the mechanism for the base-catalyzed isomerization of isopulegone to pulegone.

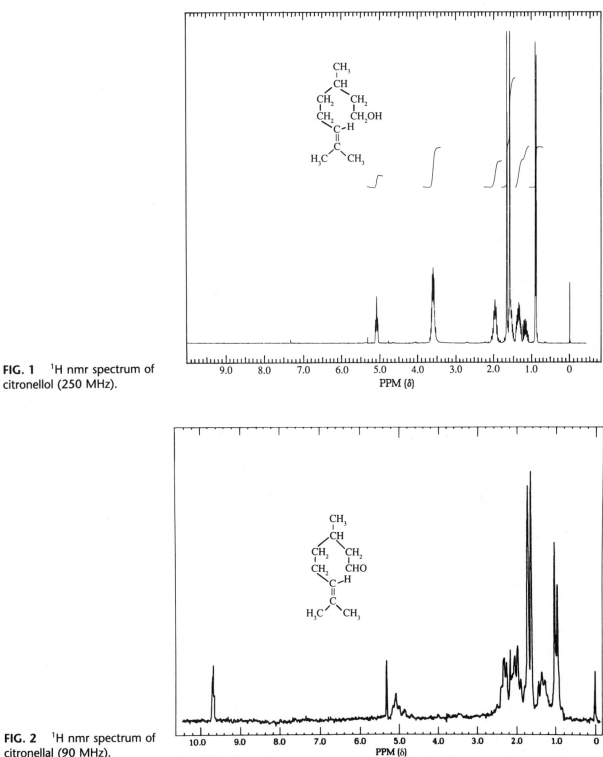

FIG. 1   ¹H nmr spectrum of citronellol (250 MHz).

FIG. 2   ¹H nmr spectrum of citronellal (90 MHz).

Oxidation: Cyclohexanol to Cyclohexanone; Cyclohexanone to Adipic Acid

Prelab Exercise: Write balanced equations for the dichromate and hypochlorite oxidations of cyclohexanol to cyclohexanone and for the permanganate oxidation of cyclohexanone to adipic acid.

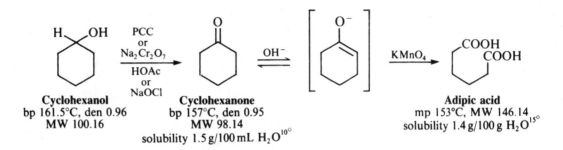

Cyclohexanol
bp 161.5°C, den 0.96
MW 100.16

Cyclohexanone
bp 157°C, den 0.95
MW 98.14
solubility 1.5 g/100 mL $H_2O^{10°}$

Adipic acid
mp 153°C, MW 146.14
solubility 1.4 g/100 g $H_2O^{15°}$

The oxidation of a secondary alcohol to a ketone is done by a very large number of oxidizing agents, including sodium dichromate, pyridinium chlorochromate, and sodium hypochlorite (household bleach). The ketone can be oxidized further to the dicarboxylic acid giving adipic acid. Both of these oxidations can be carried out by permanganate ion to give the diacid. Nitric acid is a powerful oxidizing agent that can oxidize cyclohexane, cyclohexene, cyclohexanol, or cyclohexanone to adipic acid.

Dichromate oxidation

The mechanism of oxidation of an alcohol to a ketone by dichromate appears to be the following:

$$H_2O + Cr_2O_7^{2-} \rightleftharpoons 2\ HCrO_4^{-}$$

Permanganate oxidation

A number of intermediate valence states of chromium are involved in this reaction as orange Cr^{6+} ultimately is reduced to green Cr^{3+}. The course of the oxidation can be followed by these color changes.

The oxidation of a ketone by permanganate to the dicarboxylic acid takes place through the enol form of the ketone. The reaction can be followed as the bright purple permanganate solution reacts to give a brown precipitate of manganese dioxide. A possible mechanism for this reaction is the following:

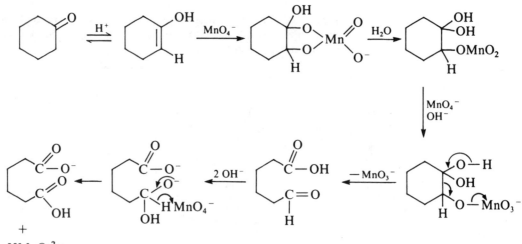

$$3\ HMnO_4{}^{2-} + H_2O \longrightarrow 2\ MnO_2 + MnO_4{}^- + 5\ OH^-$$

Nitric acid oxidation

The balanced equation for the nitric acid oxidation of cyclohexanone to adipic acid is:

In this reaction nitric acid is reduced to nitric oxide.

In the following experiments cyclohexanol is oxidized to cyclohexanone using pyridinium chlorochromate in dichloromethane. The progress of the reaction can be followed by thin-layer chromatography. On a larger scale this reaction would be carried out using sodium dichromate in acetic acid because the reagents are less expensive, the reaction is faster, and much less solvent is required.

Cr(VI) is probably the most widely used and versatile laboratory oxidizing agent and is used in a number of different forms to carry out selective oxidations in this text. But from an environmental standpoint, it is far from ideal. Inhalation of the dust from insoluble Cr(VI) compounds can lead to cancer of the respiratory system. The product of the reaction (Cr(III)) should not be flushed down the drain because it is toxic to aquatic life at extremely low concentrations, so, as stated in the Cleaning Up section of the dichromate oxidation experiment, the Cr(III) must be precipitated as insoluble $Cr(OH)_3$, and this material dealt with as a hazardous waste.

Sodium hypochlorite oxidation

There is an alternative oxidant for secondary alcohols that is just as efficient and much safer from an environmental standpoint: 5.25% (0.75 M) sodium hypochlorite solution available in the grocery store as household bleach.[1] The mechanism of the reaction is not clear. It is not a free radical reaction; the reaction is much faster in acid than in base; elemental chlorine is presumably the oxidant; and hypochlorous acid must be present. It may form an intermediate alkyl hypochlorite ester, which, by an E_2 elimination, gives the ketone and chloride ion.

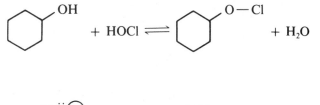

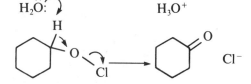

Excess hypochlorite is easily destroyed with bisulfite; the final product is chloride ion, much less toxic to the environment than Cr(III).

Experiments

1. Cyclohexanone from Cyclohexanol

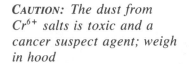

CAUTION: The dust from Cr^{6+} salts is toxic and a cancer suspect agent; weigh in hood

In a 125-mL Erlenmeyer flask dissolve 15 g of sodium dichromate dihydrate in 25 mL of acetic acid by swirling the mixture on the hot plate and then cool the solution with ice to 15°C. In a second Erlenmeyer flask chill a mixture of 15.0 g of cyclohexanol and 10 mL of acetic acid in ice. After the first solution is cooled to 15°C, transfer the thermometer and adjust the tempera-

1. J. Mohrig, D. M. Nienhuis, C. F. Linck, C. Van Zoeren, and B. G. Fox, *J. Chem. Ed.*, **62**, 519 (1985).

ture in the second flask to 15°C. Wipe the flask containing the dichromate solution, pour the solution into the cyclohexanol-acetic acid mixture, rinse the flask with a little solvent (acetic acid), note the time, and take the initially light orange solution from the ice bath, but keep the ice bath ready for use when required.[2] The exothermic reaction that is soon evident can get

Reaction time 45 min

out of hand unless controlled. When the temperature rises to 60°C cool in ice just enough to prevent a further rise and then, by intermittent brief cooling, keep the temperature close to 60°C for 15 min. No further cooling is needed, but the flask should be swirled occasionally and the temperature watched. The usual maximal temperature is 65°C (25–30 min). When the temperature begins to drop and the solution becomes pure green, the reaction is over. Allow 5–10 min more reaction time and then pour the green solution into a 250-mL round-bottomed flask, rinse the Erlenmeyer flask with 100 mL of water, and add the solution to the flask for steam distillation (Chapter 6) of the product (Fig. 3 in Chapter 6). Distill as long as any oil passes over with the water and, because cyclohexanone is appreciably soluble in water, continue somewhat beyond this point (about 80 mL will be collected).

Alternatively, instead of setting up the apparatus for steam distillation, simply add a boiling chip to the 250-mL flask and distill 40 mL of liquid, cool the flask slightly, add 40 mL of water to it, and distill 40 mL more. Note the temperature during the distillation. This is a steam distillation in which steam is generated *in situ* rather than from an outside source.

Isolation of Cyclohexanone from Steam Distillate

Cyclohexanone is fairly soluble in water. Dissolving inorganic salts such as potassium carbonate or sodium chloride in the aqueous layer will decrease the solubility of cyclohexanone such that it can be completely extracted with ether. This process is known as "salting out."

To salt out the cyclohexanone, add to the distillate 0.2 g of sodium chloride per milliliter of water present and swirl to dissolve the salt. Then pour the mixture into a separatory funnel, rinse the flask with ether, add more ether to a total volume of 25–30 mL, shake, and draw off the water layer. Then wash the ether layer with 25 mL of 10% sodium hydroxide solution to remove acetic acid, test a drop of the wash liquor to make sure it contains excess alkali, and draw off the aqueous layer.

---

2. When the acetic acid solutions of cyclohexanol and dichromate were mixed at 25°C rather than at 15°C the yield of crude cyclohexanone was only 6.9 g. A clue to the evident importance of the initial temperature is suggested by an experiment in which the cyclohexanol was dissolved in 12.5 mL of benzene instead of 10 mL of acetic acid and the two solutions were mixed at 15°C. Within a few minutes orange-yellow crystals separated and soon filled the flask; the substance probably is the chromate ester, $(C_6H_{11}O)_2CrO_2$. When the crystal magma was let stand at room temperature the crystals soon dissolved, exothermic oxidation proceeded, and cyclohexanone was formed in high yield. Perhaps a low initial temperature ensures complete conversion of the alcohol into the chromate ester before side reactions set in.

To dry the ether, which contains dissolved water, shake the ether layer with an equal volume of saturated aqueous sodium chloride solution. Draw off the aqueous layer, pour the ether out of the neck of the separatory funnel into an Erlenmeyer flask, add about 5 g of anhydrous sodium sulfate, and complete final drying of the ether solution by occasional swirling of the solution over a 5-min period. Remove the drying agent by decantation or gravity filtration into a tared Erlenmeyer flask and rinse the flask that contained the drying agent, the sodium sulfate, and the funnel with ether. Add a boiling chip to the ether solution and evaporate the ether on the steam bath under an aspirator tube (Fig. 5 in Chapter 8). Cool the contents of the flask to room temperature, evacuate the crude cyclohexanone under aspirator vacuum to remove final traces of ether (Fig. 3 in Chapter 10), and weigh the product. Yield is 11–12.5 g.

The crude cyclohexanone can be purified by simple distillation or used directly in the following experiment.

Cleaning Up   To the residue from the steam distillation add sodium bisulfite to destroy any excess dichromate ion, neutralize the solution, dilute it with a large quantity of water, and pour it down the drain if local laws allow. If not, collect the precipitate of chromium hydroxide that forms when the solution is just slightly basic and dispose of the filter paper and precipitate in the hazardous waste container designated for heavy metals. The filtrate can be flushed down the drain. Combine water layers from the extraction process, neutralize with dilute hydrochloric acid, and dipose of the solution down the drain. Any ether should be placed in the organic solvents container, and the drying agent, if completely dry, can be placed in the nonhazardous waste container. If it still has organic solvent on it, then it must be placed in the hazardous waste container. It is *much* more expensive to dispose of hazardous waste.

The use of Celite will speed up the filtration

2. Hypochlorite Oxidation of Cyclohexanol

Into a 250-mL Erlenmeyer flask in the hood place 8 mL (0.075 mole) of cyclohexanol and 4 mL of acetic acid. Introduce a thermometer and slowly add to the flask with swirling 115 mL of a commercial household bleach such as Clorox (usually 5.25% by weight NaOCl, which is 0.75 molar). This can be added from a separatory funnel clamped to a ring stand or from another Erlenmeyer flask. Take care not to come in contact with the reagent. During the addition, keep the temperature in the range of 40–50°C. Have an ice bath available in case the temperature goes above 50°C, but do not allow the temperature to go below 40°C because oxidation will be incomplete. The addition should take about 15–20 min. Swirl the reaction mixture periodically for the next 20 min to complete the reaction.

Since the exact concentration of the hypochlorite in the bleach depends on its age, it is necessary to analyze the reaction mixture for unreacted hypochlorite and to reduce excess to chloride ion with bisulfite.

Add a drop of the reaction mixture to a piece of starch-iodide paper. Any unreacted hypochlorite will cause the appearance of the blue starch-triiodide complex. Add 1-mL portions of saturated sodium bisulfite solution to the reaction mixture until the starch-iodide test is negative.

Add a few drops of thymol blue indicator solution to the mixture, and then slowly add with swirling 15–20 mL of 6 M sodium hydroxide solution until the mixture is neutral as indicated by the blue color.

Note the temperature of the vapor during this distillation

Transfer the reaction mixture to a 250-mL round-bottomed flask, add a boiling chip, and set up the apparatus for simple distillation. Distillation under these conditions is a steam distillation (see Chapter 6) in which the steam is generated *in situ*. Continue the distillation until no more cyclohexanone comes over with the water (40 mL of distillate should be collected).

See the section entitled ''Isolation of Cyclohexanone from Steam Distillate'' in the previous experiment to complete the reaction sequence.

Cleaning Up   The residue from the steam distillation contains only chloride, sodium, and acetate ions and can therefore be flushed down the drain.

3. Adipic Acid

The reaction to prepare adipic acid is conducted with 10.0 g of cyclohexanone, 30.5 g of potassium permanganate, and amounts of water and alkali that can be adjusted to provide an attended reaction period of one-half hour, procedure (a), or an unattended overnight reaction, procedure (b).

Choice of controlling the temperature for 30 min or running an unattended overnight reaction

(a) For the short-term reaction, mix the cyclohexanone and permanganate with 250 mL of water in a 500-mL Erlenmeyer flask, adjust the temperature to 30°C, note that there is no spontaneous temperature rise, and then add 2 mL of 10% sodium hydroxide solution. A temperature rise is soon registered by the thermometer. It may be of interest to determine the temperature at which you can just detect warmth, by holding the flask in the palm of the hand and by touching the flask to your cheek. When the temperature reaches 45°C (15 min) slow the oxidation process by brief ice-cooling and keep the temperature at 45°C for 20 min. Wait for a slight further rise (47°C) and an eventual drop in temperature (25 min), and then heat the mixture by swirling it over a flame to complete the oxidation and to coagulate the precipitated manganese dioxide. Make a spot test by withdrawing reaction mixture on the tip of a stirring rod and touching it to a filter paper; permanganate, if present, will appear in a ring around the spot of manganese dioxide. If permanganate is still present, add small amounts of sodium bisulfite until the spot test is negative. Then filter the mixture by suction on an 11-cm Büchner funnel, wash the brown precipitate well with water, add a boiling chip, and evaporate the filtrate over a flame from a large beaker to a volume of 70 mL. If the solution is not clear and colorless, clarify it with decolorizing charcoal and evaporate again to 70 mL. Acidify the hot solution with concentrated hydrochloric acid to pH 1–2, add 10 mL acid in excess, and let the solution stand to crystallize. Collect the crystals on a small Büchner funnel, wash them with a very small quantity of *cold* water, press the crystals between sheets of filter paper to remove excess water, and set them aside to dry. A typical yield of adipic acid, mp 152–153°C, is 6.9 g.

(b) In the alternative procedure the weights of cyclohexanone and permanganate are the same but the amount of water is doubled (500 mL) to moderate the reaction and make temperature control unnecessary. All the permanganate must be dissolved before the reaction is begun. Heat the flask and swirl the contents vigorously. Test for undissolved permanganate with a glass rod. After adjusting the temperature of the solution to 30°C, 10 mL of 10% sodium hydroxide is added and the mixture is swirled briefly and let stand overnight (maximum temperature 45–46°C). The work-up is the same as in procedure (a) and a typical yield of adipic acid is 8.3 g.

A longer oxidizing time does no harm

Cleaning Up   Place the manganese dioxide precipitate in the hazardous waste container for heavy metals. Neutralize the aqueous solution with sodium carbonate and flush it down the drain.

Questions

1. In the oxidation of cyclohexanol to cyclohexanone, what purpose does the acetic acid serve?

2. Explain the order of the chemical shifts of the carbon atoms in the ^{13}C spectra of cyclohexanone (Fig. 1) and adipic acid (Fig. 2).

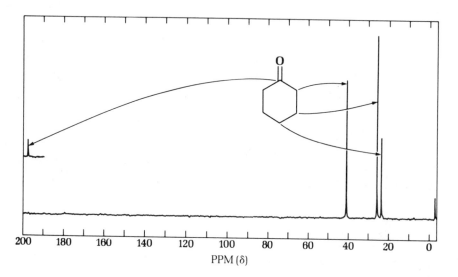

FIG. 1 ^{13}C nmr spectrum of cyclohexanone (22.6 MHz).

PPM (δ)

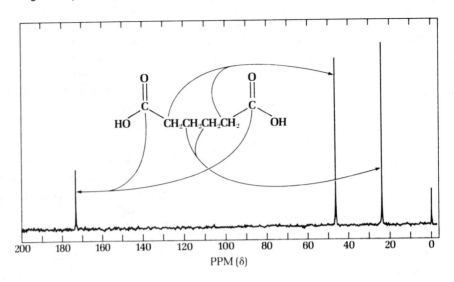

FIG. 2 ^{13}C nmr spectrum of
adipic acid (22.6 MHz).

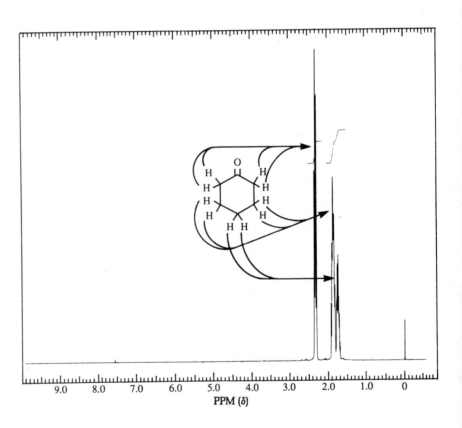

FIG. 3 ^{1}H nmr spectrum of
cyclohexanone (250 MHz).

26

Acetylsalicylic Acid (Aspirin)

Prelab Exercise: Write detailed mechanisms showing how pyridine and sulfuric acid catalyze the formation of acetylsalicylic acid.

Aspirin is among the most fascinating and versatile drugs known to medicine, and it is among the oldest—the first known use of an aspirinlike preparation can be traced to ancient Greece and Rome. Salicigen, an extract of willow and poplar bark, has been used as a pain reliever (analgesic) for centuries. In the middle of the last century it was found that salicigen is a glycoside formed from a molecule of salicylic acid and a sugar molecule. Salicylic acid is easily synthesized on a large scale by heating sodium phenoxide with carbon dioxide at 150°C under slight pressure (the Kolbe synthesis):

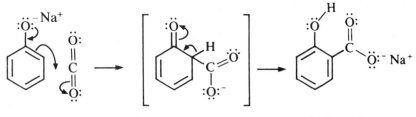

Sodium salicylate

But unfortunately salicylic acid attacks the mucous membranes of the mouth and esophagus and causes gastric pain that may be worse than the discomfort it was meant to cure. Felix Hoffmann, a chemist for Friedrich Bayer, a German dye company, reasoned that the corrosive nature of salicylic acid could be altered by addition of an acetyl group; and in 1893 the Bayer Company obtained a patent on acetylsalicylic acid, despite the fact that it had been synthesized some forty years previously by Charles Gerhardt. Bayer coined the name Aspirin for their new product to reflect its acetyl nature and its natural occurrence in the Spiraea plant. Over the years they have allowed the term aspirin to fall into the public domain so it is no longer capitalized. The manufacturers of Coke and Sanka work hard to prevent a similar fate befalling their products.

In 1904 the head of Bayer, Carl Duisberg, decided to emulate John D. Rockefeller's Standard Oil Company and formed an "interessen gemeinschaft" (I.G.) of the dye industry (Farbenindustrie). This cartel completely dominated the world dye industry before World War I and it continued to prosper between the wars even though some of their assets were seized and sold after World War I. After World War I an American company, Sterling Drug, bought the rights to aspirin. The company's Glenbrook Laboratories division still is the major manufacturer of aspirin in the United States (Bayer Aspirin).

Because of their involvement at Auschwitz the top management of IG Farbenindustrie was tried and convicted at the Nuremberg trials after World War II and the cartel broken into three large branches—Bayer, Hoechst, and BASF (Badische Anilin and Sodafabrik)—each of which does more business than DuPont, the largest American chemical company.

By law all drugs sold in the United States must meet purity standards set by the Food and Drug Administration, and so all aspirin is essentially the same. Each 5-grain tablet contains 0.325 g of acetylsalicylic acid held together with a binder. The remarkable difference in price for aspirin is primarily a reflection of the advertising budget of the company that sells it.

Aspirin is an analgesic (painkiller), an antipyretic (fever reducer), and an anti-inflammatory agent. It is the premier drug for reducing fever, a role for which it is uniquely suited. As an anti-inflammatory, it has become the most widely effective treatment for arthritis. Patients suffering from arthritis must take so much aspirin (several grams per day) that gastric problems may result. For this reason aspirin is often combined with a buffering agent. Bufferin is an example of such a preparation.

The ability of aspirin to diminish inflammation is apparently due to its inhibition of the synthesis of prostaglandins, a group of C-20 molecules that enhance inflammation. Aspirin alters the oxygenase activity of prostaglandin synthetase by moving the acetyl group to a terminal amine group of the enzyme.

Each day in the United States 80 million aspirin tablets are consumed.

If aspirin were a new invention, the U.S. Food and Drug Administration (FDA) would place many hurdles in the path of its approval. It has been implicated, for example, in Reyes syndrome, a brain disorder that strikes children and young people under 18. It has an effect on platelets, which play a vital role in blood clotting. In newborn babies and their mothers, aspirin can lead to uncontrolled bleeding and problems of circulation for the baby—even brain hemorrhage in extreme cases. This same effect can be turned into an advantage, however. Heart specialists urge potential stroke victims to take aspirin regularly to inhibit clotting in their arteries, and it has recently been shown that one-half tablet per day will help prevent heart attacks in healthy men.

Aspirin is found in more than 100 common medications, including Alka-Seltzer, Anacin ("contains the pain reliever doctors recommend most"), Coricidin, Excedrin, Midol, and Vanquish. Despite its side effects, aspirin remains the safest, cheapest, and most effective nonprescription drug. It is made commercially, employing the same synthesis used here.

The mechanism for the acetylation of salicylic acid is as follows:

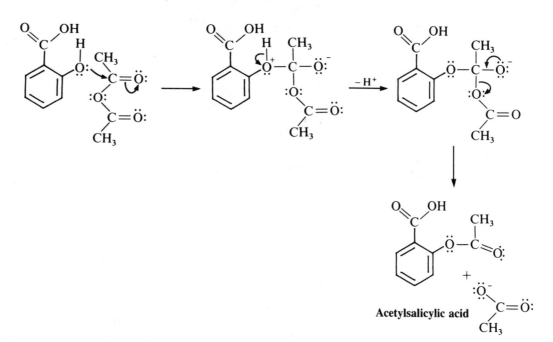

Experiment

Synthesis of Acetylsalicylic Acid (Aspirin)

Measure acetic anhydride in the hood as it is very irritating to breathe

Acetylation catalysts
  1. $CH_3COO^-Na^+$

  2.

  3. $BF_3 \cdot (C_2H_5)_2O$
  4. H_2SO_4

Pyridine has a bad odor and boron trifluoride etherate and sulfuric acid are corrosive; work in hood

Note for the instructor

Place 1 g of salicylic acid in each of four 13 × 100-mm test tubes and add to each tube 2 mL of acetic anhydride. To the first tube add 0.2 g of anhydrous sodium acetate, note the time, stir with a thermometer, and record the time required for a 4°C rise in temperature. Replace the thermometer and continue to stir occasionally while starting the next acetylation. Obtain a clean thermometer, put it in the second tube, add 5 drops of pyridine, observe as before, and compare with the first results. To the third and fourth tubes add 5 drops of boron trifluoride etherate[1] and 5 drops of concentrated sulfuric acid, respectively. What is the order of activity of the four catalysts as judged by the rates of the reactions?

Put all tubes in hot water (beaker) for 5 min to dissolve solid material and complete the reactions, and then pour all the solutions into a 125-mL Erlenmeyer flask containing 50 mL of water and rinse the tubes with water. Swirl to aid hydrolysis of excess acetic anhydride and then cool thoroughly in ice, scratch the side of the flask with a stirring rod to induce crystallization, and collect the crystalline solid; yield is 4 g.

Acetylsalicyclic acid melts with decomposition at temperatures reported from 128 to 137°C. It can be crystallized by dissolving it in ether,

---

1. Commercial reagent if dark should be redistilled (bp 126°C, water-white).

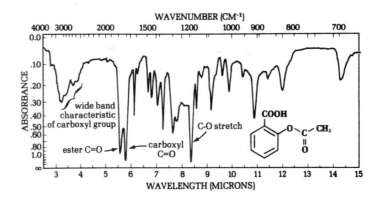

FIG. 1 Infrared spectrum of acetylsalicylic acid (aspirin) in $CHCl_3$.

adding an equal volume of petroleum ether, and letting the solution stand undisturbed in an ice bath.

Test the solubility of your sample in toluene and in hot water and note the peculiar character of the aqueous solution when it is cooled and when it is then rubbed against the tube with a stirring rod. Note also that the substance dissolves in cold sodium bicarbonate solution and is precipitated by addition of an acid. Compare a tablet of commercial aspirin with your sample. Test the solubility of the tablet in water and in toluene and observe if it dissolves completely. Compare its behavior when heated in a melting point capillary with the behavior of your sample. If an impurity is found, it is probably some substance used as binder for the tablets. Is it organic or inorganic? To interpret your results, consider the mechanism whereby salicylic acid is acetylated.

Ether and petroleum ether are very flammable

Propose mechanisms to interpret your results

Note that acetic acid is eliminated during the reaction. What effect would sodium acetate have? How might boron fluoride etherate or sulfuric acid affect the nucleophilic attack of the phenolic oxygen on acetic anhydride? With what might the base, pyridine, associate?

Cleaning Up   Combine the aqueous filtrates, dilute with water, and flush the solution down the drain.

Questions

1. Hydrochloric acid is about as strong a mineral acid as sulfuric acid. Why would it not be a satisfactory catalyst in this reaction?

2. How do you account for the smell of vinegar when an old bottle of aspirin is opened?

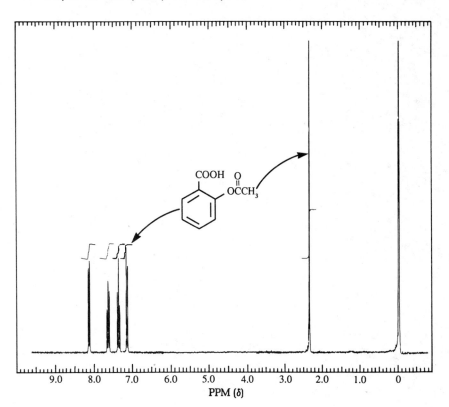

FIG. 2   ¹H nmr spectrum of acetylsalicylic acid (aspirin) (250 MHz). The COOH proton does not appear on this spectrum.

27

Esterification

Prelab Exercise: Give the detailed mechanism for the acid catalyzed hydrolysis of methyl benzoate.

The ester group

$$R-\overset{\overset{\displaystyle O}{\|}}{C}-O-R'$$

Flavors and fragrances

is an important functional group that can be synthesized in a number of different ways. The low molecular weight esters have very pleasant odors and indeed are the major components of the flavor and odor components of a number of fruits. Although the natural flavor may contain nearly a hundred different compounds, single esters approximate the natural odors and are often used in the food industry for artificial flavors and fragrances (Table 1).

Esters can be prepared by the reaction of a carboxylic acid with an alcohol in the presence of a catalyst such as concentrated sulfuric acid, hydrogen chloride, *p*-toluenesulfonic acid, or the acid form of an ion exchange resin:

$$\underset{\textbf{Acetic acid}}{CH_3\overset{\overset{\displaystyle O}{\|}}{C}-OH} + \underset{\textbf{Methanol}}{CH_3OH} \overset{H+}{\rightleftharpoons} \underset{\textbf{Methyl acetate}}{CH_3\overset{\overset{\displaystyle O}{\|}}{C}-OCH_3} + H_2O$$

Fischer esterification

This Fischer esterification reaction reaches equilibrium after a few hours of refluxing. The position of the equilibrium can be shifted by adding more of the acid or of the alcohol, depending on cost or availability. The mechanism of the reaction involves initial protonation of the carboxyl group, attack by the nucleophilic hydroxyl, a proton transfer, and loss of water followed by loss of the catalyzing proton to give the ester. Because each of these steps is completely reversible, this process is also, in reverse, the mechanism for the hydrolysis of an ester:

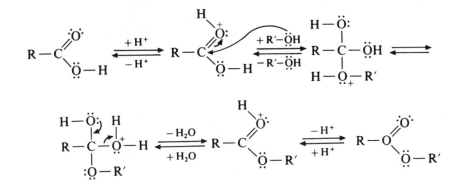

TABLE 1 Fragrances and Boiling Points of Esters

Ester	Formula	Bp (°C)	Fragrance
Isobutyl formate	$\overset{O}{\overset{\|}{HC}}-OCH_2\overset{CH_3}{\overset{\|}{CH}}CH_3$	98.4	Raspberry
n-Propyl acetate	$CH_3\overset{O}{\overset{\|}{C}}-OCH_2CH_2CH_3$	101.7	Pear
Methyl butyrate	$CH_3CH_2CH_2\overset{O}{\overset{\|}{C}}-OCH_3$	102.3	Apple
Ethyl butyrate	$CH_3CH_2CH_2\overset{O}{\overset{\|}{C}}-OCH_2CH_3$	121	Pineapple
Isobutyl propionate	$CH_3CH_2\overset{O}{\overset{\|}{C}}-OCH_2\overset{CH_3}{\overset{\|}{CH}}CH_3$	136.8	Rum
Isoamyl acetate	$CH_3\overset{O}{\overset{\|}{C}}-OCH_2CH_2\overset{CH_3}{\overset{\|}{CH}}CH_3$	142	Banana
Benzyl acetate	$CH_3\overset{O}{\overset{\|}{C}}-OCH_2$ ⬡	206	Peach
Octyl acetate	$CH_3\overset{O}{\overset{\|}{C}}-OCH_2(CH_2)_6CH_3$	210	Orange
Methyl salicylate	⬡$\overset{O}{\overset{\|}{C}}-OCH_3$ with OH	222	Wintergreen

Other methods are available for the synthesis of esters, most of them more expensive but readily carried out on a small scale. For example, alcohols react with anhydrides and with acid chlorides:

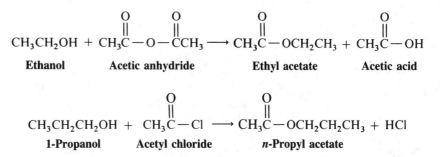

Ethanol   **Acetic anhydride**   **Ethyl acetate**   **Acetic acid**

$$CH_3CH_2CH_2OH + CH_3\overset{O}{\overset{\|}{C}}-Cl \longrightarrow CH_3\overset{O}{\overset{\|}{C}}-OCH_2CH_2CH_3 + HCl$$

1-Propanol   **Acetyl chloride**   ***n*-Propyl acetate**

In the latter reaction an organic base such as pyridine is usually added to react with the hydrogen chloride.

A number of other methods can be used to synthesize the ester group. Among these are the addition of 2-methylpropene to an acid to form *t*-butyl esters, the addition of ketene to make acetates, and the reaction of a silver salt with an alkyl halide:

Other ester syntheses

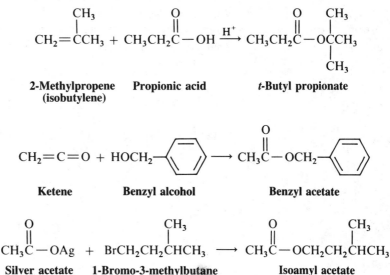

2-Methylpropene   Propionic acid   *t*-Butyl propionate
(isobutylene)

Ketene   Benzyl alcohol   Benzyl acetate

Silver acetate   1-Bromo-3-methylbutane   Isoamyl acetate

As noted above, Fischer esterification is an equilibrium process. Consider the reaction of acetic acid with 1-butanol to give *n*-butyl acetate:

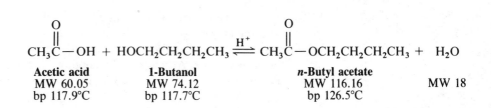

Acetic acid	**1-Butanol**	***n*-Butyl acetate**	
MW 60.05	MW 74.12	MW 116.16	MW 18
bp 117.9°C	bp 117.7°C	bp 126.5°C	

The equilibrium constant is:

$$K_{eq} = \frac{[n\text{-BuOAc}][H_2O]}{[n\text{-BuOH}][HOAc]}$$

For primary alcohols reacting with unhindered carboxylic acids, $K_{eq} \approx 4$. If equal quantities of 1-butanol and acetic acid are allowed to react, at equilibrium the theoretical yield of ester is only 67%. To upset the equilibrium we can, by Le Châtelier's principle, increase the concentration of either the alcohol or acid, as noted above. If either one is doubled the theoretical yield increases to 85%. When one is tripled it goes to 90%. But note that in the example cited the boiling point of the relatively nonpolar ester is only about 8°C higher than the boiling points of the polar acetic acid and 1-butanol, so a difficult separation problem exists if either starting material is increased in concentration and the product isolated by distillation.

Another way to upset the equilibrium is to remove water. This can be done by adding to the reaction mixture molecular sieves (an artificial zeolite), which preferentially adsorb water. Most other drying agents, such as anhydrous sodium sulfate, will not remove water at the temperatures used to make esters. A third way to upset the equilibrium is to preferentially remove the water as an azeotrope (a constant-boiling mixture of water and an organic liquid).

Azeotropic distillation

Experiments

Methyl Benzoate by Fischer Esterification

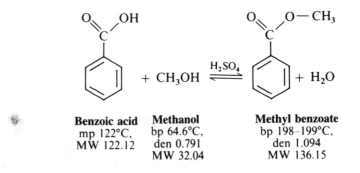

Benzoic acid
mp 122°C,
MW 122.12

Methanol
bp 64.6°C,
den 0.791
MW 32.04

Methyl benzoate
bp 198–199°C,
den 1.094
MW 136.15

Place 10.0 g of benzoic acid and 25 mL of methanol in a 125-mL round-bottomed flask, cool the mixture in ice, pour 3 mL of concentrated sulfuric acid *slowly and carefully down the walls of the flask,* and then swirl to mix the components. Attach a reflux condenser, add a boiling chip, and reflux the mixture gently for 1 h. See Fig. 1 for a drawing of the apparatus. Cool the

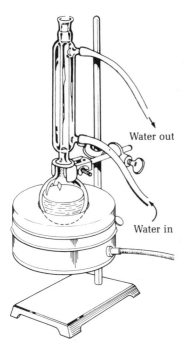

FIG. 1 Macroscale apparatus for refluxing a reaction mixture.

Water out

Water in

Tared: previously weighed

solution, decant it into a separatory funnel containing 50 mL of water, and rinse the flask with 35 mL of practical grade (not anhydrous) ether. Add this ether to the separatory funnel, shake thoroughly, and drain off the water layer, which contains the sulfuric acid and the bulk of the methanol. Wash the ether in the separatory funnel with 25 mL of water followed by 25 mL of 5% sodium bicarbonate to remove unreacted benzoic acid. Again shake, with frequent release of pressure by inverting the separatory funnel and opening the stopcock, until no further reaction is apparent; then drain off the bicarbonate layer into a beaker. If this aqueous material is made strongly acidic with hydrochloric acid, unreacted benzoic acid may be observed. Wash the ether layer in the separatory funnel with saturated sodium chloride solution and dry the solution over anhydrous sodium sulfate in an Erlenmeyer flask. Add sufficient anhydrous sodium sulfate so that it no longer clumps together on the bottom of the flask. After 10 min decant the dry ether solution into a flask, wash the drying agent with an additional 5 mL of ether, and decant again.

Remove the ether by simple distillation or by evaporation on the steam bath under an aspirator tube. See Fig. 5 in Chapter 5 or Fig. 9 in Chapter 3 or use a rotary evaporator (Fig. 7 in Chapter 10). When evaporation ceases, add 2–3 g of anhydrous sodium sulfate to the residual oil and heat for about 5 min longer. Then decant the methyl benzoate into a 50-mL round-bottomed flask, attach a stillhead, dry out the ordinary condenser and use it without water circulating in the jacket, and distill. The boiling point of the ester is so high (199°C) that a water-cooled condenser is liable to crack. Use a tared 25-mL Erlenmeyer as the receiver and collect material boiling above 190°C. A typical student yield is about 7 g. See Chapter 36 for the nitration of methyl benzoate.

Cleaning Up Pour the sulfuric acid layer into water, combining it with the bicarbonate layer, neutralize it with sodium carbonate, and flush the solution down the drain with much water. The saturated sodium chloride layer can also be flushed down the drain. If the sodium sulfate is free of ether and methyl benzoate, it can be placed in the nonhazardous solid waste container; otherwise it must go into the hazardous waste container. Ether goes into the organic solvents container, along with the pot residues from the final distillation.

Questions

1. In the preparation of methyl benzoate what is the purpose of (a) washing the organic layer with sodium bicarbonate solution? (b) washing the organic layer with saturated sodium chloride solution? (c) treating the organic layer with anhydrous sodium sulfate?

2. Assign the resonances in Fig. 4 to specific protons in methyl benzoate.

3. Figures 5 and 6 each have two resonances that are very small. What do the carbons causing these peaks have in common?

4. Throughout this chapter the esters have been given their more common trivial names. Name the esters in Table 1 according to the IUPAC system of nomenclature.

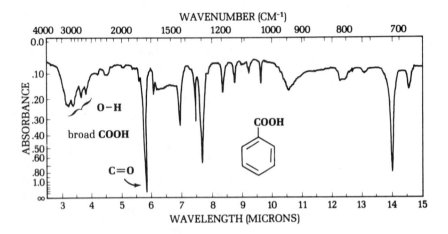

FIG. 2   Infrared spectrum of benzoic acid in CS_2.

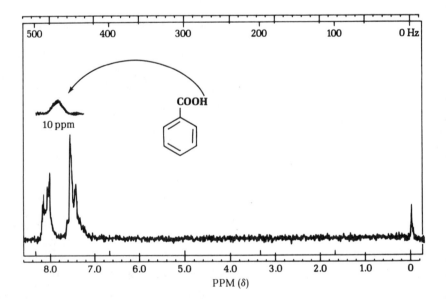

FIG. 3   ¹H nmr spectrum of benzoic acid (90 MHz).

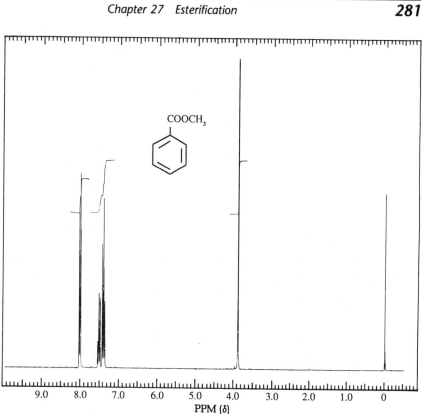

FIG. 4 ¹H nmr spectrum of methyl benzoate (250 MHz).

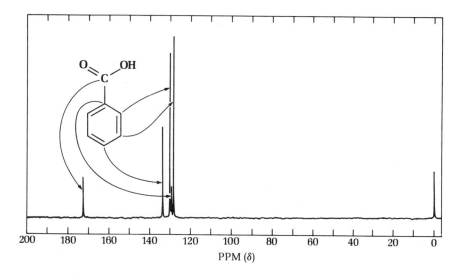

FIG. 5 ¹³C nmr spectrum of benzoic acid (22.6 MHz).

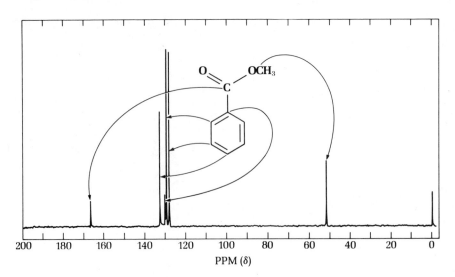

FIG. 6 ^{13}C nmr spectrum of methyl benzoate (22.6 MHz).

Diels-Alder Reaction

Prelab Exercise: Describe in detail the laboratory operations, reagents, and solvents you would employ to prepare:

Otto Diels and his pupil Kurt Alder received the Nobel Prize in 1950 for their discovery and work on the reaction that bears their names. Its great usefulness lies in its high yield and high stereospecificity. A cycloaddition reaction, it involves the 1,4-addition of a conjugated diene in the s-*cis*-conformation to an alkene in which two new sigma bonds are formed from two pi bonds.

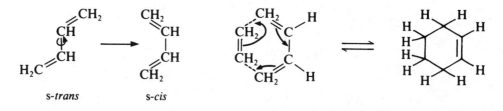

s-*trans*   s-*cis*

The adduct is a six-membered ring alkene. The diene can have the two conjugated bonds contained within a ring system as with cyclopentadiene or cyclohexadiene, or the molecule can be an acyclic diene that must be in the *cis* conformation about the single bond before reaction can occur.

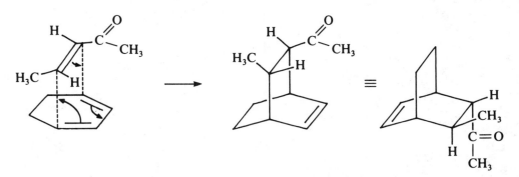

The reaction works best when there is a marked difference between the electron densities in the diene and the alkene with which it reacts, the dienophile. Usually the dienophile has electron-attracting groups attached to it while the diene is electron rich, e.g., as in the reaction of methyl vinyl ketone with 1,3-butadiene.

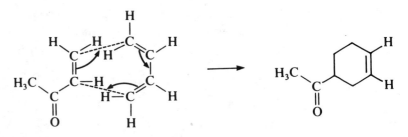

Methyl vinyl ketone   1,3-Butadiene

Retention of the configurations of the reactants in the products implies that both new sigma bonds are formed almost simultaneously. If not, then the intermediate with a single new bond could rotate about that bond before the second sigma bond is formed, thus destroying the stereospecificity of the reaction.

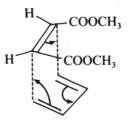

Dimethyl maleate
+
1,3-Butadiene

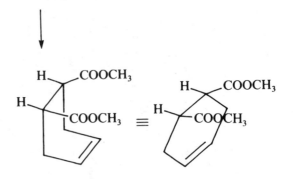

cis-isomer

This does not happen:

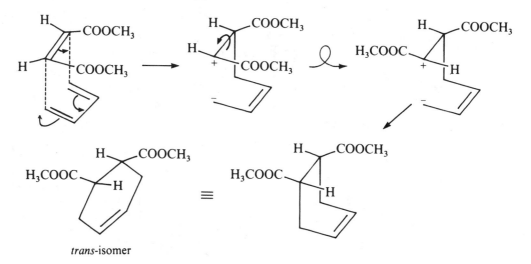

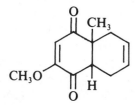

trans-isomer

A highly stereospecific reaction

This reaction is not polar in that no charged intermediates are formed. Neither is it radical because no unpaired electrons are involved. It is instead known as a concerted reaction, or one in which several bonds in the transition state are simultaneously made and broken. When a cyclic diene and a cyclic dienophile react with each other as in the present reaction, more than one stereoisomer may be formed. The isomer that predominates is the one which involves maximum overlap of pi electrons in the transition state. The transition state for the formation of the *endo*-isomer in the present reaction involves a sandwich with the diene directly above the dienophile. To form the *exo*-isomer the diene and dienophile would need to be arranged in a stair-step fashion.

The Diels-Alder reaction has been used extensively in the synthesis of complex natural products because it is possible to exploit the formation of a number of chiral centers in one reaction and also the regioselectivity of the reaction. For example, the first step in R. B. Woodward's synthesis of cortisone was the formation of a Diels-Alder adduct.

Woodward's Diels-Alder adduct

But the reaction is also subject to steric hindrance, especially when the difference between the electron-withdrawing and donating characters of the two reactants is not great. When Woodward tried to synthesize cantharidin, the active ingredient in Spanish fly, by the Diels-Alder condensation of furan with dimethylmaleic anhydride, the reaction did not work. The reaction possesses $-\Delta V^*$ (it proceeds with a net decrease in volume). High pressure should overcome this problem, but this reaction will not proceed even at 600,000 lb/in^2. A closely related reaction will proceed at 300,000 lb/in^2 and has been used to synthesize this molecule.[1] Cantharidin is a powerful vesicant (blister-former).

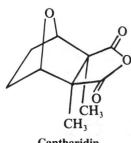

Cantharidin

1. W. G. Dauben, C. R. Kessel and K. H. Takemura, *J. Am. Chem. Soc.*, **102**, 6893 (1980).

Maximum overlap
of pi electrons

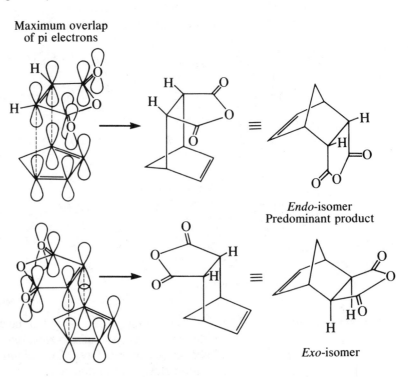

Endo-isomer
Predominant product

Exo-isomer

    Woodward and Roald Hoffmann, Nobel prize winners for their work, formulated the theoretical rules involving the correlation of orbital symmetry, which govern the Diels-Alder and other electrocyclic reactions.

    Cyclopentadiene is obtained from the light oil from coal tar distillation but exists as the stable dimer, dicyclopentadiene, which is the Diels-Alder adduct from two molecules of the diene. Thus, generation of cyclopentadiene by pyrolysis of the dimer represents a reverse Diels-Alder reaction. See Figs. 1 and 2 for nmr and infrared spectra of dicyclopentadiene. In the Diels-Alder addition of cyclopentadiene and maleic anhydride the two molecules approach each other in the orientation shown in the drawing above, as this orientation provides maximal overlap of π-bonds of the two reactants and favors formation of an initial π-complex and then the final *endo*-product. Dicyclopentadiene also has the *endo*-configuration.

$$\xrightleftharpoons[\text{room temperature}]{\sim 160°C}$$

Dicyclopentadiene
den 0.98
MW 132.20

Cyclopentadiene
bp 41°C, den 0.80
MW 66.10

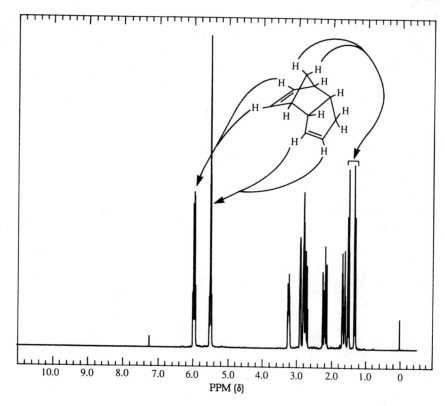

FIG. 1 ¹H nmr spectrum of
dicyclopentadiene (250 MHz).

PPM (δ)

Experiments

Gas chromatography reveals that cyclopentadiene is 8% dimerized in 4 h
and 50% dimerized in 24 h at room temperature. It should be kept on ice and
used as soon as possible after being prepared.

1. Cracking of Dicyclopentadiene

Measure 20 mL of dicyclopentadiene into a 100 mL flask and arrange for
fractional distillation into an ice-cooled receiver (Fig. 1 in Chapter 11). Heat
the dimer with a Bunsen burner until it refluxes briskly and at such a rate
that the monomeric diene begins to distill in about 5 min and soon reaches a
steady boiling point in the range 40–42°C. Apply heat continuously to
promote rapid distillation without exceeding the boiling point of 42°C.
Distillation for 45 min should provide the 12 mL of cyclopentadiene required for two preparations of the adduct; continued distillation for another
half hour gives a total of about 20 mL of monomer.

Reverse Diels-Alder

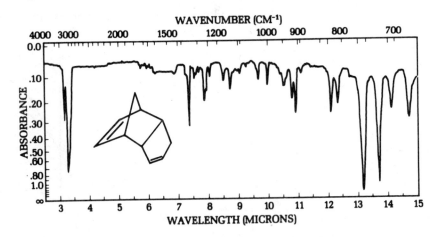

FIG. 2   Infrared spectrum of dicyclopentadiene.

Check old dicyclopentadiene for peroxides

Cleaning Up   Pour the pot residue of dicyclopentadiene and unused cyclopentadiene into the recovered dicyclopentadiene container. This recovered material can, despite its appearance, be cracked in the future to give cyclopentadiene. If the pot residue is not to be recycled, place it in the organic solvents container. If calcium chloride is used to dry the cyclopentadiene allow the organic material to evaporate from the drying agent and then place it in the nonhazardous solid waste container.

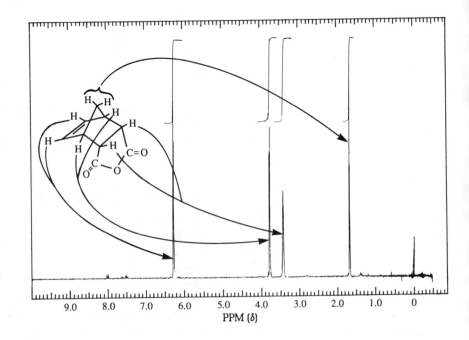

FIG. 3   ¹H nmr spectrum of *cis*-norbornene-5,6-*endo*-dicarboxylic anhydride (250 MHz).

2. *cis*-Norbornene-5,6-*endo*-dicarboxylic Anhydride

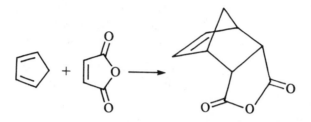

Maleic anhydride
mp 53°C, MW 98.06

cis-**Norbornene-5,6-*endo*-
dicarboxylic anhydride**
mp 165°C, MW 164.16

Place 6 g of maleic anhydride in a 125-mL Erlenmeyer flask and dissolve the anhydride in 20 mL of ethyl acetate by heating on a hot plate or steam bath. Add 20 mL of ligroin, bp 60–90°C, cool the solution thoroughly in an ice-water bath, and leave it in the bath (some anhydride may crystallize).

The distilled cyclopentadiene may be slightly cloudy, because of the condensation of moisture in the cooled receiver and water in the starting material. Add about 1 g of calcium chloride to remove the moisture. It will redimerize; use it immediately. Measure 6 mL of dry cyclopentadiene, and add it to the ice-cold solution of maleic anhydride. Swirl the solution in the ice bath for a few minutes until the exothermic reaction is over and the

Cyclopentadiene is flammable

Rapid addition at 0°C

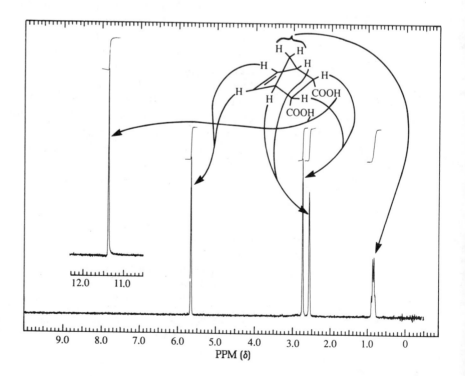

FIG. 4   ¹H nmr spectrum of *cis*-norbornene-5,6-*endo*-dicarboxylic acid (250 MHz).

adduct separates as a white solid. Then heat the mixture on a hot plate or steam bath until the solid is all dissolved.[2] If you let the solution stand undisturbed, you will be rewarded with a beautiful display of crystal formation. The anhydride crystallizes in long spars, mp 164–165°C; a typical yield is 8.2 g.[3]

Cleaning Up   Place the crystallization solvent mixture in the organic solvents container. It contains a very small quantity of the product.

3. *cis*-Norbornene-5,6-*endo*-dicarboxylic Acid

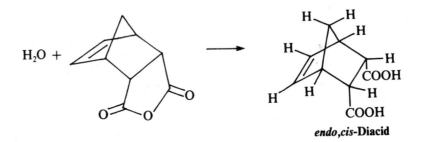

endo,cis-**Diacid**

For preparation of the *endo,cis*-diacid, place 4.0 g of bicyclic anhydride and 50 mL of distilled water in a 125-mL Erlenmeyer flask, grasp this with a clamp, swirl the flask over a hot plate, and bring the contents to the boiling point, at which point the solid partly dissolves and partly melts. Continue to heat until the oil is all dissolved and let the solution stand undisturbed. Because the diacid has a strong tendency to remain in supersaturated solution, allow half an hour or more for the solution to cool to room temperature and then drop in a boiling stone or touch the surface of the liquid once or twice with a stirring rod. Observe the stone and its surroundings carefully, waiting several minutes before applying the more effective method of making one scratch with a stirring rod on the inner wall of the flask at the air-liquid interface. Let crystallization proceed spontaneously

---

2. In case moisture has gotten into the system, a little of the corresponding diacid may remain undissolved at this point and should be removed by filtration of the hot solution.

3. The student need not work up the mother liquor but may be interested in learning the result. Concentration of the solution to a small volume is not satisfactory because of the presence of dicyclopentadiene, formed by dimerization of excess monomer; the dimer has high solvent power. Hence the bulk of the solvent is evaporated on the steam bath, the flask is connected to the water pump with a rubber stopper and glass tube and heated under vacuum on the steam bath until dicyclopentadiene is removed and the residue solidifies. Crystallization from 1:1 ethyl acetate-ligroin affords 1.3 g of adduct, mp 156–158°C; total yield is 95%.

The temperature of decomposition is variable

to give large needles, then cool the solution in ice and collect the product. Yield is about 4 g; mp 180–185°C, dec (anhydride formation).[4]

Cleaning Up   The aqueous filtrate from the crystallization contains a very small quantity of the diacid. It can be flushed down the drain.

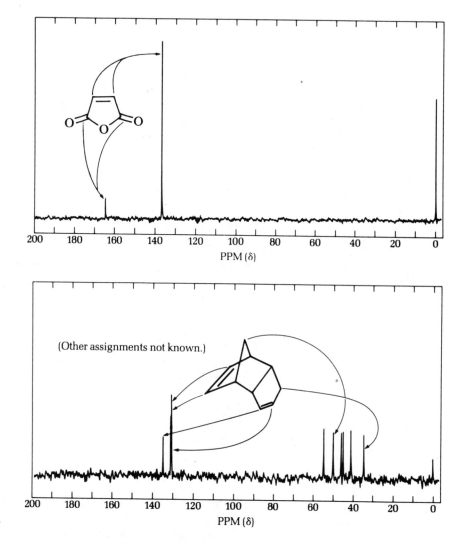

FIG. 5   ¹³C nmr spectrum of maleic anhydride.

FIG. 6   ¹³C nmr spectrum of dicyclopentadiene (22.6 MHz).

4. The *endo,cis*-diacid is stable to alkali but can be isomerized to the *trans*-diacid (mp 192°C) by conversion to the dimethyl ester (3 g of acid, 10 mL methanol, 0.5 mL concentrated H_2SO_4; reflux 1 h). This ester is equilibrated with sodium ethoxide in refluxing ethanol for three days and saponified. For an account of a related epimerization and discussion of the mechanism, see J. Meinwald and P. G. Gassman, *J. Am. Chem. Soc.*, **82**, 5445 (1960). See also K. L. Williamson, Y.-F. Li, R. Lacko, and C. H. Youn, *J. Am. Chem. Soc.*, **91**, 6129 (1969) and K. L. Williamson and Y.-F. Li, *J. Am. Chem. Soc.*, **92**, 7654 (1970).

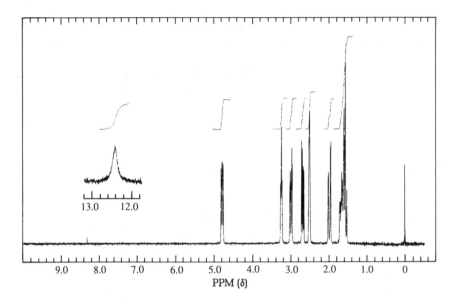

FIG. 7 ¹H nmr spectrum of compound X, prepared from *endo,cis*-diacid. Low-field peak represents one proton (250 MHz).

4. Synthesis of Compound X[5]

For preparation of X, place 1 g of the *endo,cis*-diacid and 5 mL of concentrated sulfuric acid in a 50-mL Erlenmeyer and heat gently on the hot plate for a minute or two until the crystals are all dissolved. Then cool in an ice bath, add a small piece of ice, swirl to dissolve, and add further ice until the volume is about 20 mL. Heat to the boiling point and let the solution simmer on the hot plate for 5 min. Cool well in ice, scratch the flask (see Chapter 3) to induce crystallization, and allow for some delay in complete separation. Collect, wash with water, and crystallize from water. Compound X (about 0.7 g) forms large prisms, mp 203°C.

Cleaning Up   Dilute the aqueous filtrate with water, neutralize it with sodium carbonate, and flush the resulting solution down the drain. It contains a small quantity of Compound X.

5. Structure Determination of X

To determine the formula for X try to answer the following questions: What intermediate is formed when the diacid dissolves in concentrated sulfuric acid? Why is the nmr spectrum of X so much more complex than the diacid and anhydride spectra? What functional group is missing from X that is seen in Figs. 3 and 4? Write formulas for possible structures of X and devise tests to distinguish among them. Compound X is an *isomer* of the starting diacid.

---

5. Introduced by James A. Deyrup.

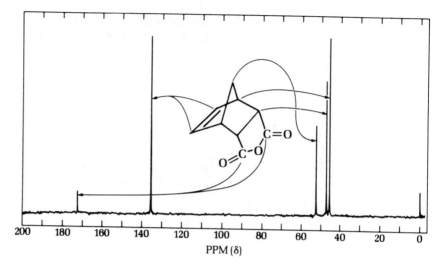

FIG. 8 ^{13}C nmr spectrum of
cis-norbornene-5,6-*endo*-
dicarboxylic anhydride
(22.6 MHz).

Questions

1. In the cracking of dicyclopentadiene, why is it necessary to distill the product very slowly?

2. Draw the products of the following reactions:

(a)

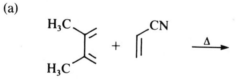

(b)

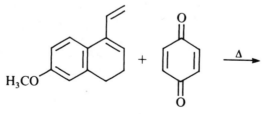

(c)

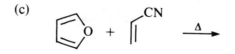

3. What starting material would be necessary to prepare the following compound by the Diels-Alder reaction?

4. If the Diels-Alder reaction between dimethylmaleic anhydride and furan had worked would cantharidin have been formed?

Ferrocene [Bis(cyclopentadienyl)iron]

Prelab Exercise: Propose a detailed outline of the procedure for the synthesis of ferrocene, paying particular attention to the time required for each step.

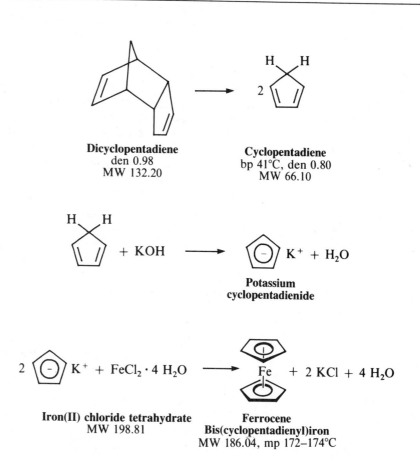

Dicyclopentadiene
den 0.98
MW 132.20

Cyclopentadiene
bp 41°C, den 0.80
MW 66.10

Potassium cyclopentadienide

Iron(II) chloride tetrahydrate
MW 198.81

Ferrocene Bis(cyclopentadienyl)iron
MW 186.04, mp 172–174°C

The Grignard reagent is a classical organometallic compound. The magnesium ion in Group IIA of the periodic table needs to lose two and only two electrons to achieve the inert gas configuration. This metal has a strong tendency to form ionic bonds by electron transfer:

$$RBr + Mg \longrightarrow \overset{\delta^-}{R}{-}\overset{\delta^+}{MgBr}$$

Among the transition elements the situation is not so simple. Consider the bonding between iron and carbon monoxide in $Fe(CO)_5$:

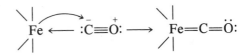

The pair of electrons on the carbon atom is shared with iron to form a σ bond between the carbon and iron. The π bond between iron and carbon is formed from a pair of electrons in the *d*-orbital of iron. The π bond is thus formed by the overlap of a *d* orbital of iron with the *p*-π bond of the carbonyl group. This mutual sharing of electrons results in a relatively nonpolar bond.

Iron has 6 electrons in the 3*d* orbital, 2 in the 4*s*, and none in the 4*p* orbital. The inert gas configuration requires 18 electrons—ten 3*d*, two 4*s*, and six 4*p* electrons. Iron pentacarbonyl enters this configuration by accepting two electrons from each of the five carbonyl groups, a total of 18 electrons. Back-bonding of the *d*-π type distributes the excess electrons among the five carbon monoxide molecules.

Early attempts to form σ-bonded derivatives linking alkyl carbon atoms to iron were unsuccessful, but P. L. Pauson in 1951 succeeded in preparing a very stable substance, ferrocene, $C_{10}H_{10}Fe$, by reacting two moles of cyclopentadienylmagnesium bromide with anhydrous ferrous chloride. Another group of chemists—Wilkinson, Rosenblum, Whiting, and Woodward—recognized that the properties of ferrocene (remarkable stability to water, acids, and air and its ease of sublimation) could only be explained if it had the structure depicted and that the bonding of the ferrous iron with its six electrons must involve all twelve of the π-electrons on the two cyclopentadiene rings, with a stable 18-electron inert gas structure as the result.

In the present experiment ferrocene is prepared by reaction of the anion of cyclopentadiene with iron(II) chloride. Abstraction of one of the acidic allylic protons of cyclopentadiene with base gives the aromatic cyclopentadienyl anion. It is considered aromatic because it conforms to the Hückel rule in having $4n + 2\pi$ electrons (where *n* is 1). Two molecules of this anion will react with iron(II) to give ferrocene, the most common member of the class of metal-organic compounds referred to as metallocenes. In this centrosymmetric sandwich-type π complex, all carbon atoms are equidistant from the iron atom, and the two cyclopentadienyl rings rotate more or less freely with respect to each other. The extraordinary stability of ferrocene (stable to 500°C) can be attributed to the sharing of the 12 π electrons of the two cyclopentadienyl rings with the six outer shell electrons of iron(II) to give the iron a stable 18-electron inert gas configuration. Ferrocene is soluble in organic solvents, can be dissolved in concentrated sulfuric acid and recovered unchanged, and is resistant to other acids and bases as well (in the absence of oxygen). This behavior is

Ferrocene: soluble in organic solvents, stable to 500°C

consistent with that of an aromatic compound; ferrocene is found to undergo electrophilic aromatic substitution reactions with ease.

Cyclopentadiene readily dimerizes at room temperature by a Diels-Alder reaction to give dicyclopentadiene. This dimer can be "cracked" by heating (an example of the reversibility of the Diels-Alder reaction) to give low-boiling cyclopentadiene. In most syntheses of ferrocene the anion of cyclopentadiene is prepared by reaction of the diene with metallic sodium. Subsequently, this anion is allowed to react with anhydrous iron(II) chloride. In the present experiment the anion is generated using powdered potassium hydroxide, which functions as both a base and a dehydrating agent.

The anion of cyclopentadiene rapidly decomposes in air, and iron(II) chloride, although reasonably stable in the solid state, is readily oxidized to the iron(III) (ferric) state in solution. Consequently this reaction must be carried out in the absence of oxygen, accomplished by bubbling nitrogen gas through the solutions to displace dissolved oxygen and to flush air from the apparatus. In research laboratories rather elaborate apparatus is used to carry out an experiment in the absence of oxygen. In the present experiment, because no gases are evolved, no heating is necessary, and the reaction is only mildly exothermic, very simple apparatus is used.

Experiments

1. Synthesis of Ferrocene

CAUTION: Potassium hydroxide is extremely corrosive and hygroscopic. Immediately wash any spilled powder or solutions from the skin and wipe up all spills. Keep containers tightly closed. Work in the hood.

CAUTION: Dimethoxyethane can form peroxides. Discard 90 days after opening bottle because of peroxide formation.

Following the procedure described in Chapter 28, prepare 6 mL of cyclopentadiene. It need not be dry. While this distillation is taking place, rapidly weigh 25 g of finely powdered potassium hydroxide[1] into a 125-mL Erlenmeyer flask, add 60 mL of dimethoxyethane ($CH_3OCH_2CH_2OCH_3$), and immediately cool the mixture in an ice bath. Swirl the mixture in the ice bath for a minute or two, then bubble nitrogen through the solution for about 2 min. Quickly cork the flask and shake the mixture to dislodge the cake of potassium hydroxide from the bottom of the flask and to dissolve as much of the base as possible (much will remain undissolved).

Grind 7 g of iron(II) chloride tetrahydrate to a fine powder and then add 6.5 g of the green salt to 25 mL of dimethyl sulfoxide (DMSO) in a 50-mL Erlenmeyer flask. Pass nitrogen through the DMSO mixture for about 2 min, cork the flask, and shake it vigorously to dissolve all the iron(II) chloride. Gentle warming of the flask on a steam bath may be necessary to dissolve the last traces of iron(II) chloride. Transfer the solution rapidly to a

---

1. Potassium hydroxide is easily ground to a fine powder in 75-g batches in one minute employing an ordinary food blender (e.g., Waring, Osterizer). The finely powdered base is transferred in a hood to a bottle with a tightly fitting cap. If a blender is unobtainable, crush and then grind 27 g of potassium hydroxide pellets in a large mortar and quickly weigh 25 g of the resulting powder into the 125-mL Erlenmeyer flask.

60-mL separatory funnel equipped with a cork to fit the 25-mL Erlenmeyer flask, flush air from the funnel with a stream of nitrogen, and stopper it.

Transfer 5.5 mL of the freshly distilled cyclopentadiene to the slurry of potassium hydroxide in dimethoxyethane. Shake the flask vigorously and note the color change as the potassium cyclopentadienide is formed. After waiting about 5 min for the anion to form, replace the cork on the Erlenmeyer flask with the separatory funnel quickly (to avoid admission of air to the flask). See Fig. 1. Add the iron(II) chloride solution to the base dropwise over a period of 20 min with vigorous swirling and shaking. Dislodge the potassium hydroxide should it cake on the bottom of the flask. The shaking will allow nitrogen to pass from the Erlenmeyer flask into the separatory funnel as the solution leaves the funnel.[2] Continue to shake and swirl the solution for 10 min after all the iron(II) chloride is added, then pour the dark slurry onto a mixture of 90 mL of 6 M hydrochloric acid and 100 g of ice in a 500-mL beaker. Stir the contents of the beaker thoroughly to dissolve and neutralize all the potassium hydroxide. Collect the crystalline orange ferrocene on a Büchner funnel, wash the crystals with water, press out excess water, and allow the product to dry on a watch glass overnight.

Recrystallize the ferrocene from methanol or, better, from ligroin. It is also very easily sublimed. In a hood place about 0.5 g of crude ferrocene on a watch glass on a hot plate set to about 150°C. Invert a glass funnel over the

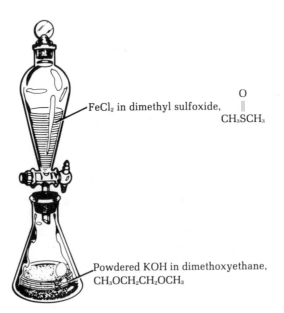

FeCl$_2$ in dimethyl sulfoxide, $\overset{\text{O}}{\underset{\|}{\text{CH}_3\text{SCH}_3}}$

Powdered KOH in dimethoxyethane, $\text{CH}_3\text{OCH}_2\text{CH}_2\text{OCH}_3$

FIG. 1　Apparatus for ferrocene synthesis.

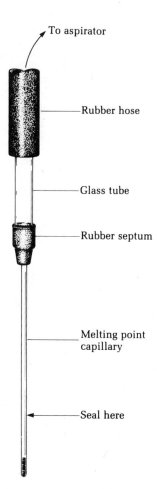

To aspirator

Rubber hose

Glass tube

Rubber septum

Melting point
capillary

Seal here

FIG. 2   Evacuation of a
melting point capillary prior to
sealing.

watch glass. Ferrocene will sublime in about one hour, leaving nonvolatile impurities behind. Pure ferrocene melts at 172–174°C. Determine the melting point in an evacuated capillary (Fig. 2) since the product sublimes at the melting point. Compare the melting points of your sublimed and recrystallized materials.

Cleaning Up   The filtrate from the reaction mixture should be slightly acidic. Neutralize it with sodium carbonate, dilute it with water, and flush it down the drain. Place any unused cyclopentadiene in the recovered dicyclopentadiene or the organic solvents container. If the ferrocene has been crystallized from methanol or ligroin, place the mother liquor in the organic solvents container.

Questions

1. If ferrocene is stable to air and all of the reagents are stable to air before the reaction begins, why must air be so carefully excluded from this reaction?

2. What special properties do the solvents dimethoxyethane and dimethyl sulfoxide have compared to diethyl ether, for example, that make them particularly suited for this reaction?

3. What is there about ferrocene that allows it to sublime easily where many other compounds do not?

30

Aldehydes and Ketones

Prelab Exercise: Outline a logical series of experiments designed to identify an unknown aldehyde or ketone with the least effort. Consider the time required to complete each identification reaction.

The carbonyl group occupies a central place in organic chemistry. Aldehydes and ketones—compounds such as formaldehyde, acetaldehyde, acetone, and 2-butanone—are very important industrial chemicals used by themselves and as starting materials for a host of other substances. For example, in 1985 8.2 *billion* lb of formaldehyde-containing plastics were produced in the United States.

The carbonyl carbon is sp^2 hybridized, the bond angles between adjacent groups are 120°, and the four atoms R, R′, C, and O lie in one plane:

The electronegative oxygen polarizes the carbon-oxygen bond, rendering the carbon electron deficient and hence subject to nucleophilic substitution.

$$\overset{\diagdown}{\underset{\diagup}{C}}=\ddot{O}: \quad\longleftrightarrow\quad \overset{\diagdown}{\underset{\diagup}{\overset{+}{C}}}-\ddot{O}:^{-} \quad \text{or} \quad \overset{\diagdown}{\underset{\diagup}{\overset{\delta+}{C}}}\cdots\overset{\delta-}{\ddot{O}}:$$

Geometry of the carbonyl group

Attack on the sp^2 hybridized carbon occurs via the π-electron cloud above or below the plane of the carbonyl group:

Reactions of the Carbonyl Group

Many reactions of carbonyl groups are acid-catalyzed. The acid attacks the electronegative oxygen, which bears a partial negative charge, to create a carbocation that subsequently reacts with the nucleophile:

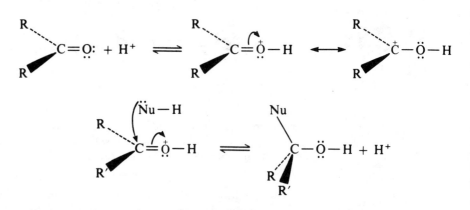

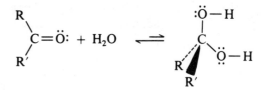

The strength of the nucleophile and the structure of the carbonyl compound determine whether the equilibrium lies on the side of the carbonyl compound or the tetrahedral adduct. Water, a weak nucleophile, does not usually add to the carbonyl group to form a stable compound:

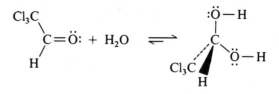

A stable hydrate: chloral hydrate

but in the special case of trichloroacetaldehyde the electron-withdrawing trichloromethyl group allows a stable hydrate to form:

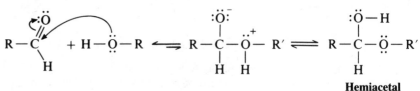

The compound so formed, chloral hydrate, was discovered by Liebig in 1832 and was introduced as one of the first sedatives and hypnotics (sleep-inducing substances) in 1869. It is now most commonly encountered in detective fiction as a "Mickey Finn" or "knockout drops."

Reaction with an alcohol. Hemiacetals

In an analogous manner, an aldehyde or ketone can react with an alcohol. The product, a hemiacetal or hemiketal, is usually not stable, but in the case of certain cyclic hemiacetals the product can be isolated. Glucose is an example of a stable hemiacetal.

$$R-C\overset{:\ddot{O}}{\underset{H}{\diagup}} + H-\ddot{O}-R \rightleftharpoons R-\overset{:\ddot{O}:^-}{\underset{\underset{H}{|}}{\overset{|}{C}}}-\overset{+}{\underset{\underset{H}{|}}{\ddot{O}}}- R' \rightleftharpoons R-\overset{:\ddot{O}-H}{\underset{\underset{H}{|}}{\overset{|}{C}}}-\ddot{O}-R'$$

Hemiacetal
usually not stable

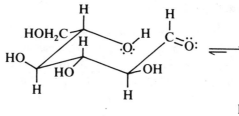
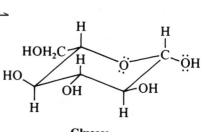

Glucose
a stable cyclic hemiacetal

Bisulfite Addition

The bisulfite ion is a strong nucleophile but a weak acid. It will attack the unhindered carbonyl group of an aldehyde or methyl ketone to form an addition product:

$$\underset{H}{\overset{R}{C}}=\ddot{O}: \;+\; :SO_3H^- \underset{Na^+}{\rightleftharpoons} R-\underset{\underset{H}{|}}{\overset{\overset{:O:^-\;Na^+}{|}}{C}}-SO_3H \longrightarrow R-\underset{\underset{H}{|}}{\overset{\overset{:\ddot{O}-H}{|}}{C}}-SO_3^-Na^+$$

Since these bisulfite addition compounds are ionic water-soluble compounds and can be formed in up to 90% yield, they serve as a useful means of separating aldehydes and methyl ketones from mixtures of organic compounds. At high sodium bisulfite concentrations these adducts crystallize and can be isolated by filtration. The aldehyde or ketone can be regenerated by adding either a strong acid or base:

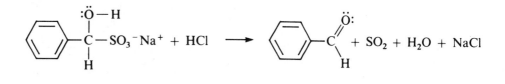

$$CH_3CH_2-\underset{\underset{CH_3}{|}}{\overset{\overset{:\ddot{O}-H}{|}}{C}}-SO_3^-Na^+ + NaOH \longrightarrow CH_3CH_2-\overset{\overset{\ddot{O}:}{\|}}{C}\diagdown_{CH_3} + Na_2SO_3 + H_2O$$

Cyanide Addition

A similar reaction occurs between aldehydes and ketones and hydrogen cyanide, which, like bisulfite, is a weak acid but a strong nucleophile. The reaction is hazardous to carry out because of the toxicity of cyanide, but the cyanohydrins are useful synthetic intermediates:

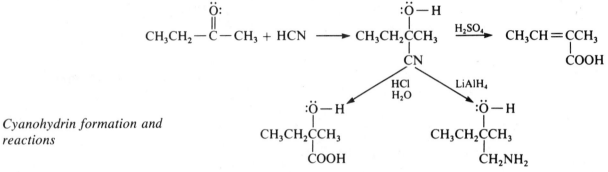

Cyanohydrin formation and reactions

Amines are good nucleophiles and readily add to the carbonyl group:

$$\text{R}-\overset{\displaystyle :\ddot{\text{O}}}{\underset{\displaystyle \text{H}}{\text{C}}} + \text{H}_2\ddot{\text{N}}\text{R}' \rightleftharpoons \text{R}-\overset{\displaystyle :\ddot{\text{O}}:^-}{\underset{\displaystyle \text{H}}{\overset{\displaystyle |}{\underset{\displaystyle |}{\text{C}}}}}-\overset{+}{\text{N}}\text{H}_2\text{R}' \rightleftharpoons \text{R}-\overset{\displaystyle :\ddot{\text{O}}-\text{H}}{\underset{\displaystyle \text{H}}{\overset{\displaystyle |}{\underset{\displaystyle |}{\text{C}}}}}-\ddot{\text{N}}\text{HR}'$$

The reaction is strongly dependent on the pH. In acid the amine is protonated ($\text{RN}\overset{+}{\text{H}}_3$) and is no longer a nucleophile. In strong base there are no protons available to catalyze the reaction. But in weak acid solution (pH 4–6) the equilibrium between acid and base (a) is such that protons are available to protonate the carbonyl (b) and yet there is free amine present to react with the protonated carbonyl (c):

(a) $\text{CH}_3\ddot{\text{N}}\text{H}_2 + \text{HCl} \rightleftharpoons \text{CH}_3\overset{+}{\text{N}}\text{H}_3 + \text{Cl}^-$

(b) $\text{CH}_3-\overset{\displaystyle :\ddot{\text{O}}}{\underset{\displaystyle \text{H}}{\text{C}}} + \text{HCl} \rightleftharpoons \left[\text{CH}_3-\overset{\displaystyle \overset{+}{\ddot{\text{O}}}-\text{H}}{\underset{\displaystyle \text{H}}{\text{C}}} \longleftrightarrow \text{CH}_3-\overset{\displaystyle :\ddot{\text{O}}-\text{H}}{\underset{\displaystyle \text{H}}{\overset{+}{\text{C}}}} \right] + \text{Cl}^-$

(c) $\text{CH}_3-\overset{\displaystyle \overset{+}{\ddot{\text{O}}}-\text{H}}{\underset{\displaystyle \text{H}}{\text{C}}} + \text{CH}_3\ddot{\text{N}}\text{H}_2 \rightleftharpoons \text{CH}_3-\overset{\displaystyle :\ddot{\text{O}}-\text{H}}{\underset{\displaystyle \text{H}}{\overset{\displaystyle |}{\underset{\displaystyle |}{\text{C}}}}}-\overset{+}{\text{N}}\text{H}_2\text{CH}_3 \overset{-\text{H}^+}{\rightleftharpoons} \text{CH}_3-\overset{\displaystyle :\ddot{\text{O}}-\text{H}}{\underset{\displaystyle \text{H}}{\overset{\displaystyle |}{\underset{\displaystyle |}{\text{C}}}}}-\overset{\displaystyle |}{\underset{\displaystyle \text{H}}{\ddot{\text{N}}}}-\text{CH}_3$

Schiff Bases

The intermediate hydroxyamino form of the adduct is not stable and spontaneously dehydrates under the mildly acidic conditions of the reaction to give an imine, commonly referred to as a Schiff base:

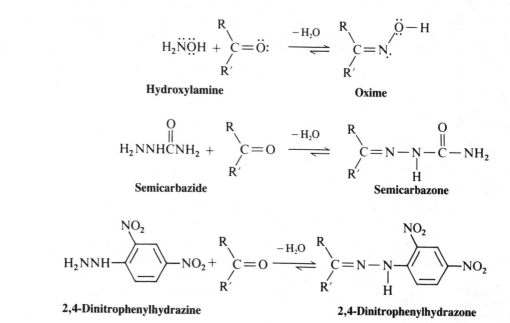

Imine or Schiff base formation

The biosynthesis of most amino acids proceeds through Schiff base intermediates.

Oximes, Semicarbazones, and 2,4-Dinitrophenylhydrazones

Three rather special amines form useful stable imines:

These imines are solids and are useful for the characterization of aldehydes and ketones. For example, infrared and nmr spectroscopy may indicate that a certain unknown is acetaldehyde. It is difficult to determine the boiling point of a few milligrams of a liquid, but if it can be converted to a solid derivative the mp *can* be determined with that amount. The 2,4-dinitrophenylhydrazones are usually the derivatives of choice because they are nicely crystalline compounds with well-defined melting or decomposition points and they increase the molecular weight by 180. Ten milligrams of acetaldehyde will give 51 mg of 2,4-DNP:

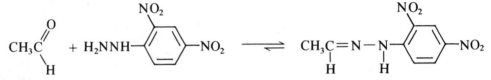

Acetaldehyde	2,4-Dinitrophenylhydrazine	Acetaldehyde 2,4-dinitrophenylhydrazone
MW 44.05	MW 198.14	MW 224.19
bp 20.8°C	mp 196°C	mp 168.5°C

Tollens' Reagent

Before the advent of nmr and ir spectroscopy the chemist was often called upon to identify aldehydes and ketones by purely chemical means. Aldehydes can be distinguished chemically from ketones by their ease of oxidation to carboxylic acids. The oxidizing agent, an ammoniacal solution of silver nitrate, Tollens' reagent, is reduced to metallic silver, which is deposited on the inside of a test tube as a silver mirror.

$$2\ Ag(NH_3)_2OH + R-C\!\!\begin{array}{c}O\\ \diagup\\ \diagdown\\ H\end{array} \longrightarrow 2\ Ag + R-C\!\!\begin{array}{c}O\\ \diagup\\ \diagdown\\ O^-\end{array} NH_4^+ + H_2O + 3\ NH_3$$

Schiff's Reagent

Another way to distinguish aldehydes from ketones is to use Schiff's reagent. This is a solution of the red dye Basic Fuchsin, which is rendered colorless on treatment with sulfur dioxide. In the presence of an aldehyde the colorless solution turns magenta.

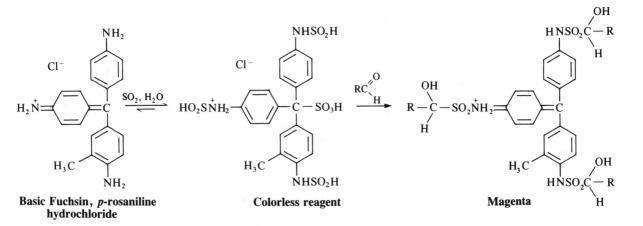

Basic Fuchsin, *p*-rosaniline hydrochloride   **Colorless reagent**   **Magenta**

Iodoform Test

Methyl Ketone

Methyl ketones can be distinguished from other ketones by the iodoform test. The methyl ketone is treated with iodine in a basic solution. Introduction of the first iodine atom increases the acidity of the remaining methyl protons, so halogenation stops only when the triiodo compound has been produced. The base then allows the relatively stable triiodomethyl carbanion to leave and a subsequent proton transfer gives iodoform, a yellow crystalline solid of mp 119–123°C. The test is also positive for fragments easily oxidized to methyl ketones, such as CH_3CHOH- and ethanol. Acetaldehyde also gives a positive test because it is both a methyl ketone and an aldehyde.

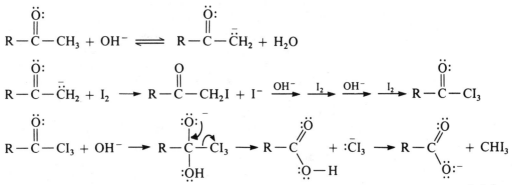

Iodoform
mp 123°C

Experiments

1. Unknowns

You will be given an unknown that may be any of the aldehydes or ketones listed in Table 1. At least one derivative of the unknown is to be submitted to the instructor; but if you first do the bisulfite and iodoform characterizing tests, the results may suggest derivatives whose melting points will be particularly revealing.

Carry out three tests:
Known positive
Known negative
Unknown

In conducting the following tests you should perform three tests simultaneously: on a compound known to give a positive test, on a compound known to give a negative test, and on the unknown. In this way you will be able to determine whether the reagents are working as they should as well as interpret a positive or a negative test.

2. 2,4-Dinitrophenylhydrazones

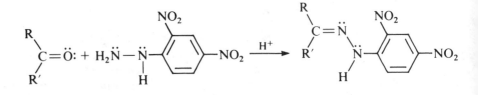

To 10 mL of the stock solution[1] of 2,4-dinitrophenylhydrazine in phosphoric acid add about 0.1 g of the compound to be tested. Ten milliliters of the 0.1 M solution contains 1 millimole (0.001 mole) of the reagent. If the compound to be tested has a molecular weight of 100 then 0.1 g is 1 millimole. Warm the reaction mixture for a few minutes in a water bath and then let crystallization proceed. Collect the product by suction filtration (Fig. 1), wash the crystals with a large amount of water to remove all phosphoric acid, press a piece of moist litmus paper on to the crystals, and if they are acidic wash them with more water. Press the product as dry as possible between sheets of filter paper and recrystallize from ethanol. Occasionally a high-molecular-weight derivative won't dissolve in a reasonable quantity (20 mL) of ethanol. In that case cool the hot suspension and isolate the crystals by suction filtration. The boiling ethanol treatment removes impurities so that an accurate melting point can be obtained on the isolated material.

An alternative procedure is applicable when the 2,4-dinitrophenylhydrazone is known to be sparingly soluble in ethanol. Measure 1 millimole (0.198 g) of crystalline 2,4-dinitrophenylhydrazine into a 125-mL Erlenmeyer flask, add 30 mL of 95% ethanol, digest on the steam bath until all

---

1. Dissolve 2.0 g of 2,4-dinitrophenylhydrazine in 50 mL of 85% phosphoric acid by heating, cool, add 50 mL of 95% ethanol, cool again, and clarify by suction filtration from a trace of residue.

TABLE 1 Melting Points of Derivatives of Some Aldehydes and Ketones

Compound[c]	Formula	MW	Den	Water solubility	Melting Points (°C)		
					Phenyl-hydrazone	2,4-DNP	Semi-carbazone
Acetone	CH_3COCH_3	58.08	0.79		42	126	187
n-Butanal	$CH_3CH_2CH_2CHO$	72.10	0.82	4 g/100 g	Oil	123	95(106)[a]
Diethyl ketone	$CH_3CH_2COCH_2CH_3$	86.13	0.81	4.7 g/100 g	Oil	156	138
Furfural	$C_4H_3O \cdot CHO$	96.08	1.16	9 g/100 g	97	212(230)[a]	202
Benzaldehyde	C_6H_5CHO	106.12	1.05	Insol.	158	237	222
Heptane-2,6-dione	$CH_3COCH_2CH_2CH_2COCH_3$	114.14	0.97	∞	120[b]	257[b]	224[b]
2-Heptanone	$CH_3(CH_2)_4COCH_3$	114.18	0.83	Insol.	Oil	89	123
3-Heptanone	$CH_3(CH_2)_3COCH_2CH_3$	114.18		Insol.	Oil	81	101
n-Heptanal	$n\text{-}C_6H_{13}CHO$	114.18	0.82	Insol.	Oil	108	109
Acetophenone	$C_6H_5COCH_3$	120.66	1.03	Insol.	105	238	198
2-Octanone	$CH_3(CH_2)_5COCH_3$	128.21	0.82	Insol.	Oil	58	122
Cinnamaldehyde	$C_6H_5CH=CHCHO$	132.15	1.10	Insol.	168	255	215
Propiophenone	$C_6H_5COCH_2CH_3$	134.17	1.01	Insol.	about 48°	191	182

a. Both melting points have been found, depending on crystalline form of derivative.
b. Monoderivative or diderivative.
c. See Tables 3 and 14 in Chapter 70 for additional aldehydes and ketones.

particles of solid are dissolved, and then add 1 millimole of the compound to be tested and continue warming. If there is no immediate change, add, from a Pasteur pipette, 6–8 drops of concentrated hydrochloric acid as catalyst and note the result. Warm for a few minutes, then cool and collect the product. This procedure would be used for an aldehyde like cinnamaldehyde ($C_6H_5CH=CHCHO$).

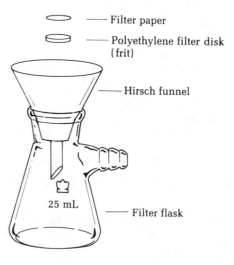

— Filter paper
— Polyethylene filter disk (frit)
— Hirsch funnel
25 mL
— Filter flask

FIG. 1 Hirsch funnel filtration apparatus.

The alternative procedure strikingly demonstrates the catalytic effect of hydrochloric acid, but it is not applicable to a substance like diethyl ketone, whose 2,4-dinitrophenylhydrazone is much too soluble to crystallize from the large volume of ethanol. The first procedure is obviously the one to use for an unknown.

Cleaning Up   The filtrate from the preparation of the 2,4-dinitrophenylhydrazone should have very little 2,4-dinitrophenylhydrazine in it, so after dilution with water and neutralization with sodium carbonate it can be flushed down the drain. Similarly the mother liquor from crystallization of the phenylhydrazone should have very little product in it and so should be diluted and flushed down the drain. If solid material is detected, it should be collected by suction filtration, the filtrate flushed down the drain, and the filter paper placed in the solid hazardous waste container since hydrazines are toxic.

3. Semicarbazones

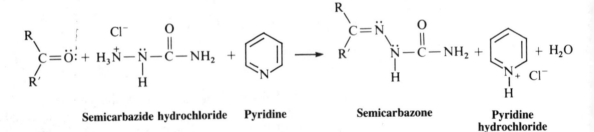

Semicarbazide hydrochloride   **Pyridine**   **Semicarbazone**   **Pyridine hydrochloride**

Semicarbazide (mp 96°C) is not very stable in the free form and is used as the crystalline hydrochloride (mp 173°C). Since this salt is insoluble in methanol or ethanol and does not react readily with typical carbonyl compounds in alcohol-water mixtures, a basic reagent, pyridine, is added to liberate free semicarbazide.

To 0.5 mL of the stock solution[2] of semicarbazide hydrochloride, which contains 1 millimole of the reagent, add 1 millimole of the compound to be tested and enough methanol (1 mL) to produce a clear solution; then add 10 drops of pyridine (a twofold excess) and warm the solution gently on the steam bath for a few minutes. Cool the solution slowly to room temperature. It may be necessary to scratch the inside of the test tube in order to induce crystallization. Cool the tube in ice, collect the product by suction filtration, and wash it with water followed by a small amount of cold methanol. Recrystallize the product from methanol, ethanol, or ethanol/water.

---

2. Prepare a stock solution by dissolving 1.11 g of semicarbazide hydrochloride in 5 mL of water; 0.5 mL of this solution contains 1 millimole of reagent.

Cleaning Up   Combine the filtrate from the reaction and the mother liquor from the crystallization, dilute with water, make very slightly acidic with dilute hydrochloric acid, and flush the mixture down the drain.

4. Tollens' Test

Test for aldehydes

$$R-C{\overset{O}{\underset{H}{}}} + 2\ Ag(NH_3)_2OH \longrightarrow 2\ Ag + RCOO^-NH_4^+ + H_2O + 3\ NH_3$$

Clean 4 or 5 test tubes by adding a few milliliters of 10% sodium hydroxide solution to each and heating them in a water bath while preparing Tollens' reagent.

To 2.0 mL of 5% silver nitrate solution, add 1.0 mL of 10% sodium hydroxide in a test tube. To the gray precipitate of silver oxide, Ag_2O, add 0.5 mL of a 2.8% ammonia solution (10 mL of concentrated ammonium hydroxide diluted to 100 mL). Stopper the tube and shake it. Repeat the process until *almost* all of the precipitate dissolves (3.0 mL of ammonia at most); then dilute the solution to 10 mL. Empty the test tubes of sodium hydroxide solution, rinse them, and add 1 mL of Tollen's reagent to each. Add one drop (no more) of the substance to be tested by allowing it to run down the inside of the inclined test tube. Set the tubes aside for a few minutes without agitating the contents. If no reaction occurs, warm the mixture briefly on a water bath. As a known aldehyde try one drop of a 0.1 M solution of glucose. A more typical aldehyde to test is benzaldehyde.

At the end of the reaction promptly destroy excess Tollens' reagent with nitric acid: It can form an explosive fulminate on standing. Nitric acid can also be used to remove silver mirrors from the test tubes.

Cleaning Up   Place all solutions used in this experiment in a beaker (unused ammonium hydroxide, sodium hydroxide solution used to clean out the tubes, Tollens' reagent from all tubes). Remove any silver mirrors from reaction tubes with a few drops of nitric acid, which is added to the beaker. Make the mixture acidic with nitric acid to destroy unreacted Tollens' reagent, then neutralize the solution with sodium carbonate and add some sodium chloride solution to precipitate silver chloride (about 40 mg). The whole mixture can be flushed down the drain or the silver chloride collected by suction filtration and the filtrate flushed down the drain. The silver chloride would go in the nonhazardous solid waste container.

5. Schiff's Test

Very sensitive test for aldehydes

Add three drops of the unknown to 2 mL of Schiff's reagent.[3] A magenta color will appear within 10 min with aldehydes. As in all of these tests, compare the colors produced by a known aldehyde, a known ketone, and the unknown compound.

Cleaning Up   Neutralize the solution with sodium carbonate and flush it down the drain. The amount of *p*-rosaniline in this mixture is negligible (1 mg).

6. Iodoform Test

$$R-\overset{\overset{\displaystyle :\ddot{O}}{\|}}{C}-CH_3 + 3\ I_2 + 4\ OH^- \longrightarrow R-C\overset{\displaystyle \ddot{O}}{\underset{\displaystyle \ddot{O}:^-}{\diagup}} + CHI_3 + 3\ :\ddot{\underset{..}{I}}:^- + 3\ H_2O$$

Test for methyl ketones

The reagent contains iodine in potassium iodide solution[4] at a concentration such that 2 mL of solution, on reaction with excess methyl ketone, will yield 174 mg of iodoform. If the substance to be tested is water-soluble, dissolve four drops of a liquid or an estimated 50 mg of a solid in 2 mL of water in a 20 × 150-mm test tube; add 2 mL of 10% sodium hydroxide and then slowly add 3 mL of the iodine solution. In a positive test the brown color of the reagent disappears and yellow iodoform separates. If the substance to be tested is insoluble in water, dissolve it in 2 mL of 1,2-dimethoxyethane, proceed as above, and at the end dilute with 10 mL of water.

Suggested test substances are hexane-2,5-dione (water soluble), *n*-butyraldehyde (water soluble), and acetophenone (water insoluble).

Iodoform can be recognized by its odor and yellow color and, more securely, from the melting point (119–123°C). The substance can be isolated by suction filtration of the test suspension or by adding 0.5 mL of dichloromethane, shaking the stoppered test tube to extract the iodoform into the small lower layer, withdrawing the clear part of this layer with a Pasteur pipette, and evaporating it in a small tube on the steam bath. The crude solid is crystallized from methanol-water.

Cleaning Up   Combine all reaction mixtures in a beaker, add a few drops of acetone to destroy any unreacted iodine in potassium iodide reagent, remove the iodoform by suction filtration, and place it in the halogenated organic waste container. The filtrate can be flushed down the drain after neutralization (if necessary).

3. Schiff's reagent is prepared by dissolving 0.1 g Basic Fuchsin (*p*-rosaniline hydrochloride) in 100 mL of water and then adding 4 mL of a saturated aqueous solution of sodium bisulfite. After 1 h add 2 mL of concentrated hydrochloric acid.

4. Dissolve 25 g of iodine in a solution of 50 g of potassium iodide in 200 mL of water.

7. Bisulfite Test

Forms with unhindered carbonyls

$$\underset{H}{\overset{R}{\diagdown}}C=\ddot{O}: \ +\ Na^{+}SO_{3}H^{-} \ \rightleftharpoons \ R-\underset{H}{\overset{:\ddot{O}H}{\underset{|}{\overset{|}{C}}}}-SO_{3}^{-}Na^{+}$$

Put 1 mL of the stock solution[5] into a 13 × 100-mm test tube and add five drops of the substance to be tested. Shake each tube during the next 10 min and note the results. A positive test will result from aldehydes, unhindered cyclic ketones such as cyclohexanone, and unhindered methyl ketones.

If the bisulfite test is applied to a liquid or solid that is very sparingly soluble in water, formation of the addition product is facilitated by adding a small amount of methanol before the addition to the bisulfite solution.

Cleaning Up   Dilute the bisulfite solution or any bisulfite addition products (they will dissociate) with a large volume of water and flush the mixture down the drain. The amount of organic material being discarded is negligible.

8. NMR and IR Spectroscopy

A peak at 9.6–10 ppm is highly characteristic of aldehydes because almost no other peaks appear in this region. On some spectrometers an offset will be needed to detect this region. Similarly a sharp singlet at 2.2 ppm is very characteristic of methyl ketones; but beware of contamination of the sample by acetone, which is often used to clean glassware.

Infrared spectroscopy is extremely useful in analyzing all carbonyl-containing compounds, including aldehydes and ketones. See the extensive discussion in Chapter 19.

Questions

1. What is the purpose of making derivatives of unknowns?

2. Why are 2,4-dinitrophenylhydrazones better derivatives then phenyl-hydrazones?

3. Using chemical tests, how would you distinguish among 2-pentanone, 3-pentanone, and pentanal?

4. Draw the structure of a compound, C_5H_8O, which gives a positive iodoform test and does not decolorize permanganate.

---

5. Prepare a stock solution from 50 g of sodium bisulfite dissolved in 200 mL of water with brief swirling.

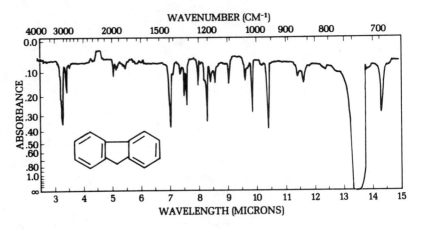

FIG. 2   Infrared spectrum of fluorene in CS_2.

5. Draw the structure of a compound, C_5H_8O, which gives a positive Tollens' test and does not react with bromine in dichloromethane.

6. Draw the structure of a compound, C_5H_8O, which reacts with phenylhydrazine, decolorizes bromine in dichloromethane, and does not give a positive iodoform test.

7. Draw the structure of two geometric isomers, C_5H_8O, which give a positive iodoform test.

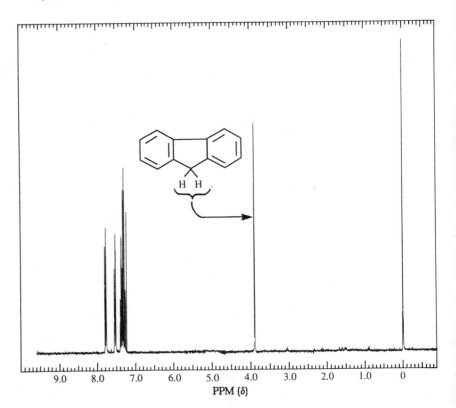

FIG. 3   ¹H nmr spectrum of fluorene (250 MHz).

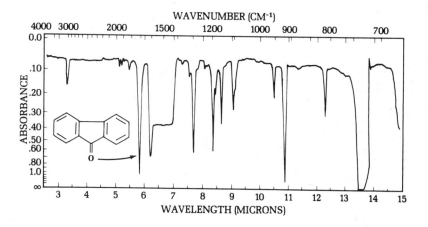

FIG. 4   Infrared spectrum of fluorenone.

8. Assign the various peaks in the ^1H nmr spectrum of 2-butanone to specific protons in the molecule (Fig. 6).

9. Assign the various peaks in the ^1H nmr spectrum of crotonaldehyde to specific protons in the molecule (Fig. 7).

10. What vibrations cause the peaks at about 3.4 μ (3000 cm^{-1}) in the infrared spectrum of fluorene (Fig. 2)?

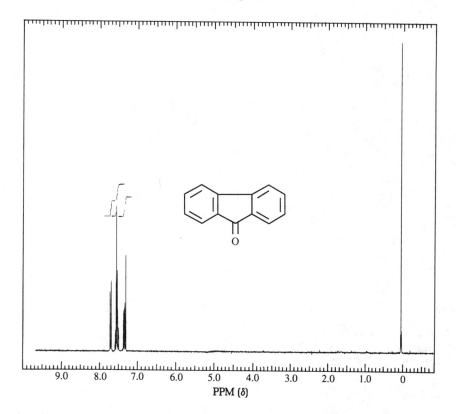

FIG. 5 ^1H nmr spectrum of fluorenone (250 MHz).

Organic Experiments

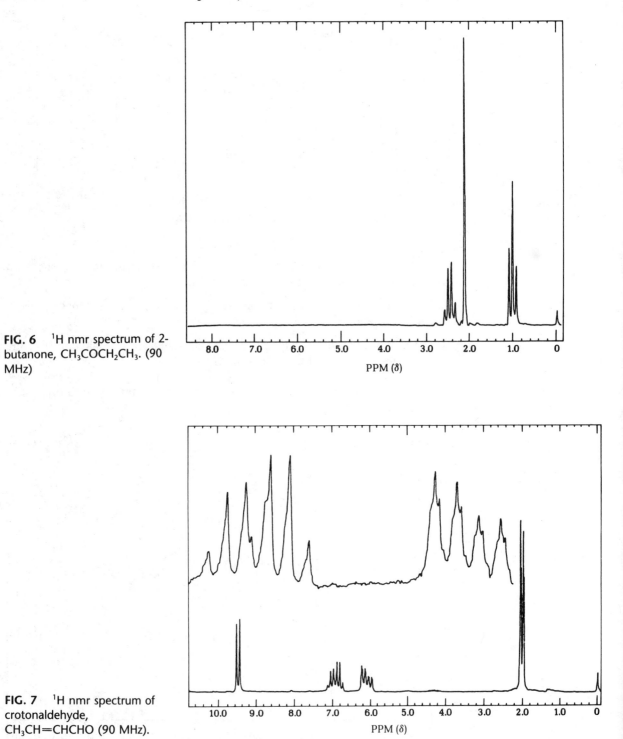

FIG. 6 ¹H nmr spectrum of 2-butanone, $CH_3COCH_2CH_3$. (90 MHz)

FIG. 7 ¹H nmr spectrum of crotonaldehyde, $CH_3CH=CHCHO$ (90 MHz).

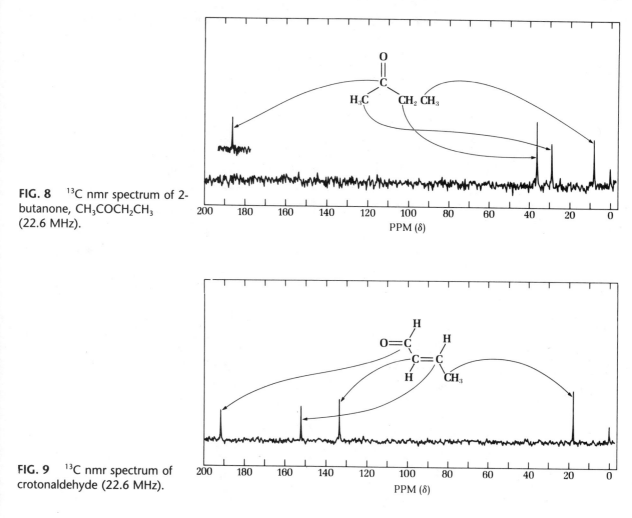

FIG. 8 ^{13}C nmr spectrum of 2-butanone, $CH_3COCH_2CH_3$ (22.6 MHz).

FIG. 9 ^{13}C nmr spectrum of crotonaldehyde (22.6 MHz).

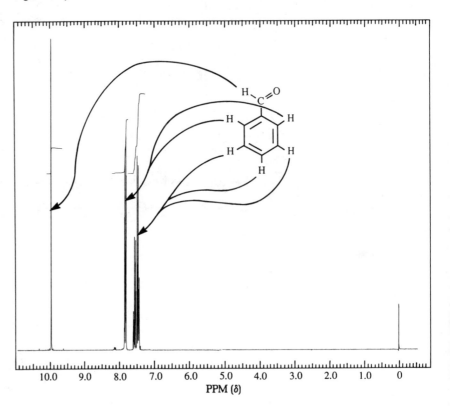

FIG. 10 ¹H nmr spectrum of
benzaldehyde (250 MHz).

31

Grignard Synthesis of Triphenylmethanol and Benzoic Acid

Prelab Exercise: Prepare a flow sheet for the preparation of triphenylmethanol. Through a knowledge of the physical properties of the solvents, reactants, and products, show how the products are purified. Indicate which layer in separations should contain the product.

In 1912 Victor Grignard received the Nobel prize in chemistry for his work on the reaction that bears his name, a carbon-carbon bond-forming reaction by which almost any alcohol may be formed from appropriate alkyl halides and carbonyl compounds. The Grignard reagent is easily formed by reaction of an alkyl halide, in particular a bromide, with magnesium metal in anhydrous ether. Although the reaction can be written and thought of as simply

$$R-Br + Mg \longrightarrow R-Mg-Br$$

it appears that the structure of the material in solution is rather more complex. There is evidence that dialkylmagnesium is present

$$2\ R-Mg-Br \rightleftharpoons R-Mg-R + MgBr_2$$

Structure of the Grignard reagent

and that the magnesium atoms, which have the capacity to accept two electron pairs from donor molecules to achieve a four-coordinated state, are solvated by the unshared pairs of electrons on ether:

$$\begin{array}{c} Et \diagdown \underset{\cdot\cdot}{\overset{\cdot\cdot}{O}} \diagup Et \\ R-Mg-Br \\ Et \diagup \underset{\cdot\cdot}{\overset{\cdot\cdot}{O}} \diagdown Et \end{array}$$

A strong base and strong nucleophile

The Grignard reagent is a strong base and a strong nucleophile. As a base it will react with all protons that are more acidic than those found on alkenes and alkanes. Thus, Grignard reagents react readily with water, alcohols, amines, thiols, etc., to regenerate the alkane:

$$R—Mg—Br + H_2O \longrightarrow R—H + MgBrOH$$

$$R—Mg—Br + R'OH \longrightarrow R—H + MgBrOR'$$

$$R—Mg—Br + R'NH_2 \longrightarrow R—H + MgBrNHR'$$

The starting material for preparing the Grignard reagent can contain no acidic protons. The reactants and apparatus must all be completely dry; otherwise the reaction will not start. If proper precautions are taken, however, the reaction proceeds smoothly.

The magnesium metal, in the form of turnings, has a coat of oxide on the outside. A fresh surface can be exposed by crushing a piece under the absolutely dry ether in the presence of the organic halide. Reaction will begin at the exposed surface, as evidenced by a slight turbidity in the solution and evolution of bubbles. Once the exothermic reaction starts it proceeds easily, the magnesium dissolves, and a solution of the Grignard reagent is formed. The solution is often turbid and gray due to impurities in the magnesium. The reagent is not isolated but reacted immediately with, most often, an appropriate carbonyl compound:

$$R — Mg — Br + R' — \overset{\overset{\ddot{O}:}{\|}}{C} — R'' \longrightarrow R' — \overset{\overset{:\ddot{O}:^- MgBr^+}{|}}{\underset{\underset{R}{|}}{C}} — R''$$

to give, in another exothermic reaction, the magnesium alkoxide. In a simple acid-base reaction this alkoxide is reacted with acidified ice water to give the covalent, ether-soluble alcohol and the ionic water-soluble magnesium salt:

$$R' — \overset{\overset{:\ddot{O}:^- MgBr^+}{|}}{\underset{\underset{R}{|}}{C}} — R'' \quad + H^+Cl^- \longrightarrow R' — \overset{\overset{:\ddot{O} — H}{|}}{\underset{\underset{R}{|}}{C}} — R'' \quad + \quad MgBrCl$$

The great versatility of this reaction lies in the wide range of reactants that undergo the reaction. Thirteen representative reactions are shown on the following page.

A versatile reagent

In every case except reaction 1 the intermediate alkoxide must be hydrolyzed to give the product. The reaction with oxygen (reaction 2) is usually not a problem because the ether vapor over the reagent protects it from attack by oxygen, but this reaction is one reason why the reagent cannot usually be stored without special precautions. Reactions 6 and 7 with ketones and aldehydes giving respectively tertiary and secondary

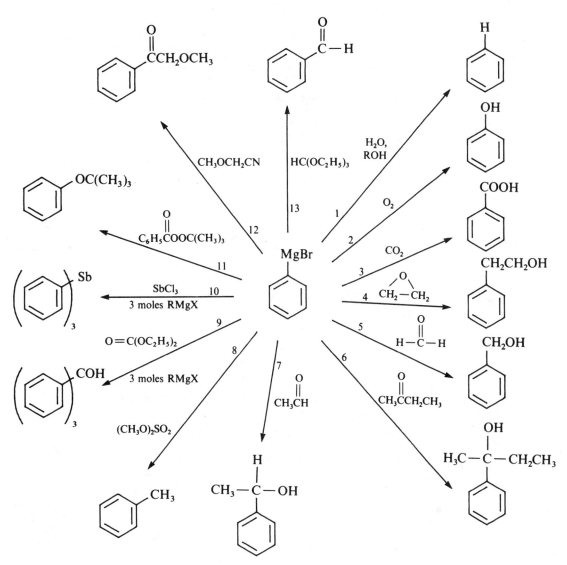

alcohols are among the most common. Reactions 8–12 are not nearly so common.

In the experiment we shall carry out another common type of Grignard reaction, the formation of a tertiary alcohol from two moles of the reagent and one of an ester. The ester employed is the methyl benzoate synthesized in Chapter 27. The initially formed product is unstable and decomposes to a ketone, which, being more reactive than an ester, immediately reacts with more Grignard reagent:

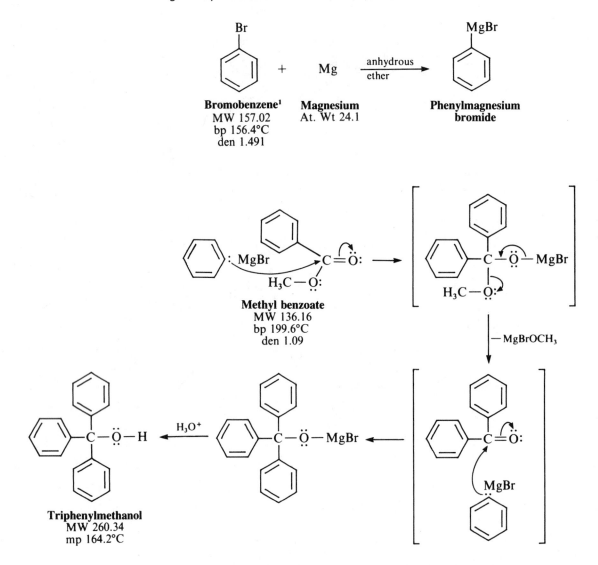

Bromobenzene[1]
MW 157.02
bp 156.4°C
den 1.491

Magnesium
At. Wt 24.1

**Phenylmagnesium
bromide**

Methyl benzoate
MW 136.16
bp 199.6°C
den 1.09

Triphenylmethanol
MW 260.34
mp 164.2°C

The primary impurity in the present experiment is biphenyl, formed by the reaction of phenylmagnesium bromide with unreacted bromobenzene. The most effective way to lessen this side reaction is to add the bromobenzene slowly to the reaction mixture so it will react with the magnesium and not be present in high concentration to react with previously formed Grignard reagent. The impurity is easily eliminated since it is much more soluble in hydrocarbon solvents than triphenylmethanol.

1. See Fig. 3 for ^{13}C nmr spectrum.

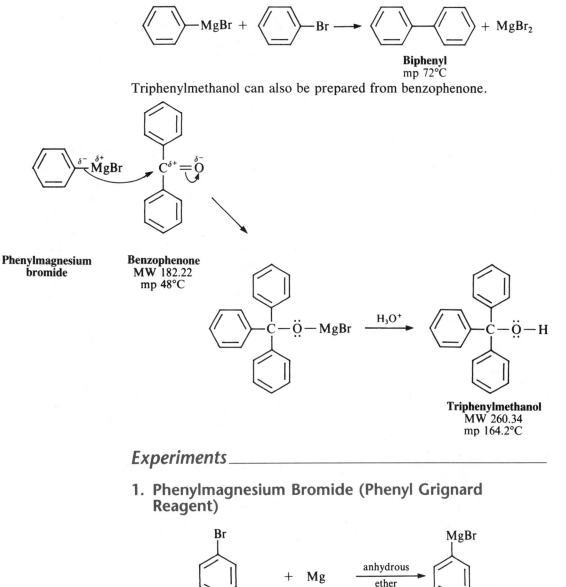

Triphenylmethanol can also be prepared from benzophenone.

Phenylmagnesium bromide

Benzophenone
MW 182.22
mp 48°C

Triphenylmethanol
MW 260.34
mp 164.2°C

Experiments

1. Phenylmagnesium Bromide (Phenyl Grignard Reagent)

Bromobenzene
MW 157.02
bp 156°C
den 1.491

Magnesium
At Wt
24.31

Phenylmagnesium bromide
not isolated, used *in situ*

All apparatus and reagents must be absolutely dry. The Grignard reagent is prepared in a dry 100-mL round-bottomed flask fitted with a long reflux condenser. A calcium chloride drying tube inserted in a cork that will fit either the flask or the top of the condenser is also made ready (Fig. 1). The flask, condenser, and magnesium (2 g = 0.082 mole of magnesium turnings) should be as dry as possible to begin with, and then should be dried in a 110°C oven for at least 35 min. Alternatively, the magnesium is placed in the flask, the calcium chloride tube is attached directly, and the flask is heated gently but thoroughly with a cool luminous flame. Do not overheat the magnesium. It will become deactivated through oxidation or, if strongly overheated, can burn. The flask on cooling pulls dry air through the calcium chloride. Cool to room temperature before proceeding! **Extinguish all flames!** Ether vapor is denser than air and can travel along bench tops and into sinks. Use care.

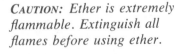

CAUTION: Ether is extremely flammable. Extinguish all flames before using ether.

Specially dried ether is required

Make an ice bath ready in case control of the reaction becomes necessary, although this is usually not the case. Remove the drying tube and fit it to the top of the condenser. Then pour into the flask through the condenser 15 mL of *absolute* ether (absolutely dry, anhydrous) and 9 mL (13.5 g = 0.086 mole) of bromobenzene. Be sure the graduated cylinders used to measure the ether and bromobenzene are absolutely dry. (More ether is to be added as soon as the reaction starts, but at the outset the concentration of bromobenzene is kept high to promote easy starting.) If there is no immediate sign of reaction, insert a *dry* stirring rod with a flattened end and crush a piece of magnesium firmly against the bottom of

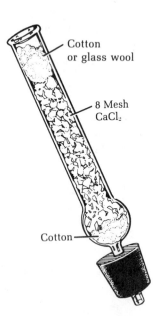

FIG. 1   Calcium chloride drying tube fitted with a rubber stopper. Store for future use with cork in top, pipette bulb on bottom.

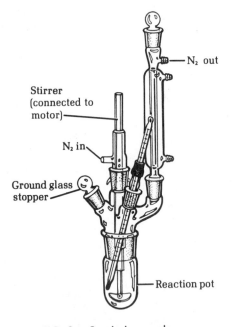

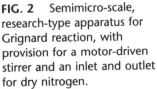

FIG. 2   Semimicro-scale, research-type apparatus for Grignard reaction, with provision for a motor-driven stirrer and an inlet and outlet for dry nitrogen.

the flask under the surface of the liquid, giving a twisting motion to the rod. When this is done properly the liquid becomes slightly cloudy, and ebullition commences at the surface of the compressed metal. Be careful not to punch a hole in the bottom of the flask. Attach the condenser at once, swirl the flask to provide fresh surfaces for contact, and, as soon as you are sure that the reaction has started, add 25 mL more absolute ether through the top of the condenser before spontaneous boiling becomes too vigorous (replace the drying tube). Note the volume of ether in the flask. Cool in ice if necessary to slow the reaction but do not overcool the mixture; the reaction can be stopped by too much cooling. Any difficulty in initiating the reaction can be dealt with by trying the following expedients in succession.

Starting the Grignard reaction

1. Warm the flask with your hands or a beaker of warm water. Then see if boiling continues when the flask (condenser attached) is removed from the warmth.
2. Try further mashing of the metal with a stirring rod.

Ether can be kept anhydrous by storing over Linde 5A molecular sieves. Discard after 90 days because of peroxide formation.

3. Add a tiny crystal of iodine as a starter (in this case the ethereal solution of the final reaction product should be washed with sodium bisulfite solution to remove the yellow color).
4. Add a few drops of a solution of phenylmagnesium bromide or of methylmagnesium iodide (which can be made in a test tube).
5. Start afresh, taking greater care with the dryness of apparatus and reagents, and sublime a crystal or two of iodine on the surface of the magnesium to generate Gattermann's "activated magnesium" before beginning the reaction again.

Use minimum steam to avoid condensation on outside of the condenser

Once the reaction begins, spontaneous boiling in the diluted mixture may be slow or become slow. If so, mount the flask and condenser on the steam bath (one clamp supporting the condenser suffices) and reflux gently until the magnesium has disintegrated and the solution has acquired a cloudy or brownish appearance. The reaction is complete when only a few small remnants of metal (or metal contaminants) remain. Check to see that the volume of ether has not decreased. If it has, add more anhydrous ether. Since the solution of Grignard reagent deteriorates on standing, the next step should be started at once.

2. Triphenylmethanol

Mix 5 g (0.037 mole) of methyl benzoate and 15 mL of absolute ether in a separatory funnel, cool the flask containing phenylmagnesium bromide solution briefly in an ice bath, remove the drying tube, and insert the stem of the separatory funnel into the top of the condenser. Run in the methyl benzoate solution *slowly* with only such cooling as is required to control the mildly exothermic reaction, which affords an intermediate addition compound that separates as a white solid. Replace the calcium chloride tube, swirl the flask until it is at room temperature and the reaction has subsided.

This is a suitable stopping point

The reaction is then completed by either refluxing the mixture for one-half hour, or stoppering the flask with the calcium chloride tube and letting the mixture stand overnight (subsequent refluxing is then unnecessary).[2]

Pour the reaction mixture into a 250-mL Erlenmeyer flask containing 50 mL of 10% sulfuric acid and about 25 g of ice and use both ordinary ether and 10% sulfuric acid to rinse the flask. Swirl well to promote hydrolysis of the addition compound; basic magnesium salts are converted into water-soluble neutral salts and triphenylmethanol is distributed into the ether layer. An additional amount of ether (ordinary) may be required. Pour the mixture into a separatory funnel (rinse the flask with ether), shake, and draw off the aqueous layer. Shake the ether solution with 10% sulfuric acid to further remove magnesium salts, and wash with saturated sodium chloride solution to remove water that has dissolved in the ether. The amounts of liquid used in these washing operations are not critical. In general, an amount of wash liquid equal to one-third of the ether volume is adequate.

Saturated aqueous sodium chloride solution removes water from ether

To effect final drying of the ether solution, pour the ether layer out of the neck of the separatory funnel into an Erlenmeyer flask, add about 5 g of granular anhydrous sodium sulfate, swirl the flask from time to time, and after 5 min remove the drying agent by gravity filtration through a filter paper held in a funnel into a tared Erlenmeyer flask. Rinse the drying agent with a small amount of ether. Add 25 mL of 66–77°C ligroin and concentrate the ether-ligroin solutions (steam bath) in an Erlenmeyer flask under an aspirator tube (see Fig. 5 in Chapter 8). Evaporate slowly until crystals of triphenylcarbinol just begin to separate and then let crystallization proceed, first at room temperature and then at 0°C. The product should be colorless and should melt not lower than 160°C. Concentration of the mother liquor may yield a second crop of crystals. A typical student yield is 5.0 g. Evaporate the mother liquors to dryness and save the residue for later isolation of the components by chromatography.

Analyze the first crop of triphenylmethanol and the residue from the evaporation of the mother liquors by thin-layer chromatography. Dissolve equal quantities of the two solids (a few crystals) and also biphenyl in equal quantities of dichloromethane (1 or 2 drops). Using a microcapillary, spot equal quantities of material on silica gel TLC plates (Eastman No. 13181) and develop the plates in an appropriate solvent system. Try 1:3 dichloromethane–petroleum ether first and adjust the relative quantities of solvent as needed. The spots can be seen by examining the TLC plate under a fluorescent lamp or by treating the TLC plate with iodine vapor. From this analysis decide how pure each of the solids is and whether it would be worthwhile to attempt to isolate more triphenylmethanol from the mother liquors.

Dispose of recovered and waste solvents in the appropriate containers

Turn in the product in a vial labeled with your name, the name of the compound, its melting point, and the overall percent yield from benzoic acid.

---

2. A rule of thumb for organic reactions: a 10°C rise in temperature will double the rate of the reaction.

Cleaning Up   Combine all aqueous layers, dilute with a large quantity of water, and flush the slightly acidic solution down the drain. The ether/ligroin mother liquor from the crystallization goes in the organic solvents container. The thin-layer chromatography developer, which contains dichloromethane, is placed in the halogenated organic waste container. Calcium chloride from the drying tube should be dissolved in water and flushed down the drain.

Questions

1. Why does rapid addition of bromobenzene to magnesium favor the formation of the undesirable by-product, biphenyl, over phenylmagnesium bromide?

2. Triphenylmethanol can also be prepared from the reaction of ethyl benzoate with phenylmagnesium bromide and by the reaction of diethylcarbonate

$$(C_2H_5OCOC_2H_5)$$

with phenylmagnesium bromide. Write stepwise reaction mechanisms for these two reactions.

3. If the ethyl benzoate used to prepare triphenylmethanol is wet, what by-product is formed?

4. Exactly what weight of dry ice is needed to react with 2 mmoles of phenylmagnesium bromide?

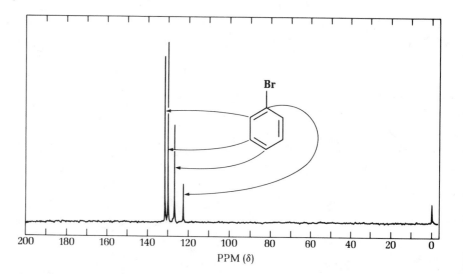

FIG. 3 ^{13}C nmr spectrum of bromobenzene (22.6 MHz).

PPM (δ)

5. In the synthesis of benzoic acid, benzene is often detected as an impurity. How does this come about?

6. The benzoic acid could have been extracted from the ether layer using sodium bicarbonate solution. Give equations showing how this might be done and how the benzoic acid would be regenerated. What practical reason makes this extraction less desirable than sodium hydroxide extraction?

CHAPTER

32

Reactions of Triphenylmethyl Carbocation, Carbanion, and Radical

Stable carbocation, carbanion, and radical

Prelab Exercise: Write all of the resonance structures of the triphenylmethyl carbocation.

Triphenylmethanol, prepared in the experiment in Chapter 31, has played an interesting part in the history of organic chemistry. It was converted to the first stable carbocation and the first stable free radical. In this experiment triphenylmethanol is easily converted to the triphenylmethyl (trityl) carbocation, carbanion, and radical. Each of these is stabilized by ten contributing resonance forms and consequently is unusually stable. Because of their long conjugated systems, these forms absorb radiation in the visible region of the spectrum and thus can be detected visually.

Experiments

The Triphenylmethyl Carbocation, the Trityl Carbocation

The reactions of triphenylmethanol are dominated by the ease with which it dissociates to form the relatively stable triphenylmethyl carbocation. When colorless triphenylmethanol is dissolved in concentrated sulfuric acid, an orange-yellow solution results that gives a fourfold depression of the melting point of sulfuric acid, meaning that four moles of ions are produced. If the triphenylmethanol simply were protonated only two moles of ions would result.

$$(C_6H_5)_3COH + 2\ H_2SO_4 \rightleftharpoons (C_6H_5)_3C^+ + H_3O^+ + 2\ HSO_4^-$$

$$(C_6H_5)_3COH + H_2SO_4 \overset{\times}{\rightleftharpoons} (C_6H_5)_3COH_2^+ + HSO_4^-$$

The central carbon atom in the carbocation is sp^2 hybridized and thus the three carbons attached to it are coplanar and disposed at angles of 120°:

However, the three phenyl groups, because of steric hindrance, cannot lie in one plane. Therefore, the carbocation is propeller-shaped:

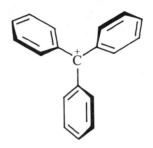

Triphenylmethanol is a tertiary alcohol and undergoes, as expected, S_N1 reactions. The intermediate cation, however, is stable enough to be seen in sulfuric acid solution as a red-brown to yellow solution. Upon dissolution in concentrated sulfuric acid, the hydroxyl is protonated; then the OH_2^+ portion is lost as H_2O (which is itself protonated) leaving the carbocation. The bisulfate ion is a very weak nucleophile and does not compete with methanol in the formation of the product, trityl methyl ether.

1. Trityl Methyl Ether

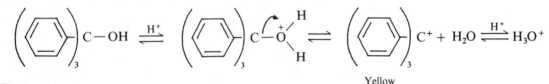

Triphenylmethanol
MW 260.34, mp 163°C

Yellow

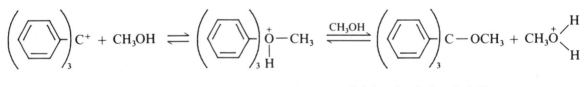

Triphenylmethyl methyl ether
Trityl methyl ether
MW 274.37, mp 96°C

In a 10-mL Erlenmeyer flask, place 0.5 g of triphenylmethanol, grind the crystals to a fine powder with a glass stirring rod, and add 5 mL of concentrated sulfuric acid. Continue to stir the mixture to dissolve all of the alcohol. Using a Pasteur pipette, transfer the sulfuric acid solution to 30 mL of ice-cold methanol in a 50-mL Erlenmeyer flask. Use some of the cold methanol to rinse out the first tube. Induce crystallization if necessary by cooling the solution, adding a seed crystal, or scratching the solution with a glass stirring rod at the liquid-air interface. Collect the product by filtration on the Hirsch funnel. Don't use filter paper on top of the polyethylene frit. This strongly acid solution may attack the paper. Wash the crystals well with water and squeeze them between sheets of filter paper to aid drying. Determine the weight of the crude material, calculate the percent yield, and save a sample for melting point determination. Recrystallize the product from boiling methanol and determine the melting point of the purified trityl methyl ether.

Cleaning Up   Dilute the filtrate with water, neutralize with sodium carbonate, and flush down the drain.

2. Triphenylmethyl Bromide, Trityl Bromide

In this experiment the triphenylmethanol is dissolved in a good ionizing solvent, acetic acid, and allowed to react with a strong acid and nucleophile, hydrobromic acid. The intermediate carbocation reacts immediately with bromide ion.

Reactions that proceed through the carbocation

$$\left(\text{Ph}\right)_3 C-OH + HBr \rightleftharpoons \left(\text{Ph}\right)_3 C-\overset{+}{O}\overset{H}{\underset{H}{<}} \rightleftharpoons \left(\text{Ph}\right)_3 C^+ + H_2O$$

Br⁻

$$\Big\Updownarrow Br^-$$

Triphenylmethyl bromide
Trityl bromide
MW 323.24, mp 154°C

$$\left(\text{Ph}\right)_3 C-Br$$

   Dissolve 0.5 g of triphenylmethanol in 10 mL of warm acetic acid, add 1.0 mL of 47% hydrobromic acid, and heat the mixture for 5 min on the steam bath or in a beaker of boiling water. Cool it in ice, collect the product on the Hirsch funnel, wash the product with water and ligroin, allow it to dry, determine the weight, and calculate the percent yield. Recrystallize the product from ligroin and compare the melting points of the recrystallized and crude material. The compound crystallizes slowly; allow adequate time for crystals to form. Use the Beilstein test (Chapter 70) to test for halogen.

Cleaning Up   Dilute the filtrate from the reaction with water, neutralize with sodium carbonate, and flush down the drain. Ligroin mother liquor from the crystallization goes in the halogenated organic solvents container.

3. Triphenylmethyl Iodide, Trityl Iodide?

In a reaction very similar to the preparation of the bromide, the iodide might be prepared. Bisulfite is added to react with any iodine formed.

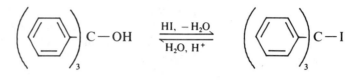

Triphenylmethyl iodide
Trityl iodide
MW 370.22, mp 183°C

Dissolve 0.5 g of triphenylmethanol in 10 mL of warm acetic acid, add 1.0 mL of 47% hydriodic acid, heat the mixture for 1 hr on the steam bath, cool it, and add it to a solution of 0.5 g of sodium bisulfite dissolved in 10 mL of water in a 25-mL Erlenmeyer flask. Collect the product on the Hirsch funnel, wash it with water, press out as much water as possible, and recrystallize the crude, moist product from methanol (about 15 mL). Determine the weight of the dry product, calculate the percent yield, and determine the melting point. Run the Beilstein test. What has been produced and why?

Cleaning Up   Dilute the filtrate from the reaction with water, neutralize with sodium carbonate, and flush down the drain. The methanol from the crystallization should be placed in the organic solvents container.

4. Triphenylmethyl Fluoborate and the Tropilium Ion

Tropilium ion—an aromatic ion

Reaction of triphenylmethanol with fluoboric acid in the presence of acetic anhydride generates the stable salt, trityl fluoborate. The fluoboric acid protonates the triphenylmethanol, which loses the elements of water in an equilibrium reaction. The water reacts with the acetic anhydride to form acetic acid and thus drives the reaction to completion.

The salt so formed is the fluoborate of the triphenyl carbocation; it is a powerful base. It can be isolated, but in this case will be used in situ to react with the hydrocarbon, cycloheptatriene.

This hydrocarbon is more basic than most hydrocarbons and will lose a proton to the triphenylmethyl carbocation to give triphenylmethane and the cycloheptatrienide carbocation, the tropilium ion. To demonstrate that hydride ion transfer has taken place, isolate triphenylmethane.

The Hückel rule: $4n + 2\pi$ electrons

The tropilium ion has a planar structure, each carbon bears a single proton, and the ion contains 6π electrons—it is aromatic, a characteristic that can be confirmed by nuclear magnetic resonance spectroscopy.

The fluoborate group can be displaced by the iodide ion to prepare tropilium iodide. You can see that the iodide ion is ionic by watching the reaction of the aqueous solution of this ion with silver ion.

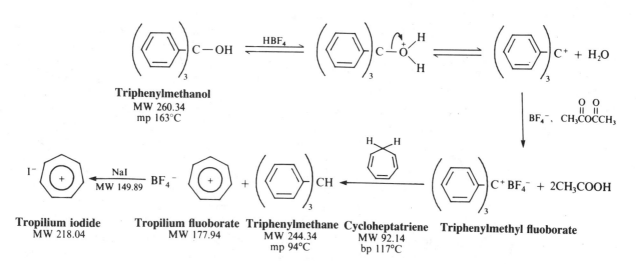

Triphenylmethanol
MW 260.34
mp 163°C

Tropilium iodide
MW 218.04

Tropilium fluoborate
MW 177.94

Triphenylmethane
MW 244.34
mp 94°C

Cycloheptatriene
MW 92.14
bp 117°C

Triphenylmethyl fluoborate

In a 25-mL Erlenmeyer flask place 8.75 mL of acetic anhydride, cool the tube, and add 0.44 g of fluoboric acid. Add 1 g of triphenylmethanol with thorough stirring. Warm the mixture to give a homogeneous dark solution of the triphenylmethyl fluoborate, then add 0.4 g of cycloheptatriene. The color of the trityl cation should fade during this reaction and the tropilium fluoborate begin to precipitate. Add 10 mL of anhydrous ether to the reaction tube, stir the contents well while cooling on ice, and collect the product by filtration on the Hirsch funnel. Wash the product with 10 mL of dry ether and then dry the product between sheets of filter paper.

To the filtrate add 3 M sodium hydroxide solution and shake the flask to allow all of the acetic anhydride and fluoboric acid to react with the base. Test the aqueous layer with indicator paper to ascertain that neutralization is complete and then transfer the mixture to a reaction tube and draw off the aqueous layer. Wash the ether once with 10 mL of water, once with 10 mL of saturated sodium chloride solution, and dry the ether over anhydrous sodium sulfate, adding the drying agent until it no longer clumps together.

Transfer the ether to a tared reaction tube and evaporate the solvent to leave crude triphenylmethane. Remove a sample for melting point determination and recrystallize the residue from an appropriate solvent, determined by experimentation. Prove to yourself that the compound isolated is indeed triphenylmethane. Obtain an infrared spectrum and a nuclear magnetic resonance spectrum.

Structure proof: nmr

To determine the nmr spectrum of the tropilium fluoborate collected on the Hirsch funnel, dissolve about 50 mg of the product in 0.3 mL of deuterated dimethyl sulfoxide that contains 1% tetramethylsilane as a reference. Compare the nmr spectrum obtained with that of the starting material, cycloheptatriene. The spectrum of the latter can be obtained in deuterochloroform, again using tetramethylsilane as the reference compound.

The tropilium fluoborate can be converted to tropilium iodide by dissolving the borate in the minimum amount of hot, but not boiling, water (a few drops) and adding to this solution 1.25 mL of a saturated solution of sodium iodide. Cool the mixture in ice, remove the solvent with a pipette, and wash the crystals with 2.5 mL of ice-cold methanol. Scrape most of the crystals onto a piece of filter paper and allow them to dry before determining the weight. To the crystalline residue in the test tube add a few drops of water; warm the tube, if necessary, to dissolve the tropilium iodide; and then add a drop of 2% aqueous silver nitrate solution and note the result.

Cleaning Up   Solutions that contain the fluoborate ion (the neutralized filtrate, the nmr sample, the filtrate from the iodide preparation) should be treated with aqueous $CaCl_2$ to precipitate insoluble CaF_2, which is removed by filtration and placed in the nonhazardous solid waste container. The filtrate can be flushed down the drain. Allow the ether to evaporate from the sodium sulfate, and then place it in the nonhazardous solid waste container. Recrystallization solvent goes in the organic solvents container.

The Trityl Carbanion

The triphenylmethyl carbanion, the trityl anion, can be generated by the reaction of triphenylmethane with the very powerful base, *n*-butyllithium. The reaction generates the blood-red lithium triphenylmethide and butane. The triphenylmethyl anion reacts much as a Grignard reagent does. In the present experiment it reacts with carbon dioxide to give triphenylacetic acid after acidification. Avoid an excess of *n*-butyllithium; on reaction with carbon dioxide, it gives the vile-smelling pentanoic acid.

Triphenylmethane
MW 244.34
mp 94°C

n-**Butyllithium**
MW 64.06

Trityl carbanion

CO_2, H^+

Triphenylacetic acid
MW 288.35, mp 270–273°C

Triphenylmethyl, a Stable Free Radical

In the early part of the nineteenth century many attempts were made to prepare methyl, ethyl, and similar radicals in a free state, as sodium had been prepared from sodium chloride. Many well-known chemists tried: Gay-Lussac's CN turned out to be cyanogen, $(CN)_2$; Bunsen's cacodyl from $(CH_3)_2AsCl$ proved to be $(CH_3)_2As-As(CH_3)_2$; and the Kolbe electrolysis gave $CH_3CH_2CH_2CH_3$ instead of CH_3CH_2. Moses Gomberg at the University of Michigan prepared the first free radical in 1900. He had prepared triphenylmethane and chlorinated it to give triphenylmethyl chloride, which he hoped to couple to form hexaphenylethane:

Gomberg: The first free radical

Triphenylmethyl chloride   **Hexaphenylethane**

The solid that he obtained instead was a high-melting, sparingly soluble, white solid, which turned out on analysis to be not a hydrocarbon but instead an oxygen-containing compound $(C_{38}H_{30}O_2)$. Repeating the experiment in the absence of air, he obtained a yellow solution that, on evaporation in the absence of air, deposited crystals of a colorless hydrocarbon that was remarkably reactive. It readily reacted with oxygen, bromine, chlorine, and iodine. On dissolution the white hydrocarbon gave the yellow solution in the absence of air; the hydrocarbon was deposited when he evaporated the solution once more. Gomberg interpreted these results as follows: "The experimental evidence . . . forces me to the conclusion that we have to deal here with a free radical, triphenylmethyl, $(C_6H_5)_3C\cdot$. The action of zinc results, as it seems to me, in a mere abstraction of the halogen:

$$2(C_6H_5)_3C-Cl + Zn \longrightarrow 2(C_6H_5)_3C\cdot + ZnCl_2$$

Now as a result of the removal of the halogen atom from triphenylchloromethane, the fourth valence of the methane is bound either to take up the complicated group $(C_6H_5)_3C\cdot$ or remain as such, with carbon as trivalent. Apparently the latter is what happens."

For a long time after Gomberg first carried out this reaction in 1900 it was assumed the radical dimerized to form hexaphenylethane:

$$2 (C_6H_5)_3C\cdot \longrightarrow (C_6H_5)_3C-C(C_6H_5)_3$$

But in 1968 nmr and uv evidence showed that the radical is in equilibrium with a different substance:

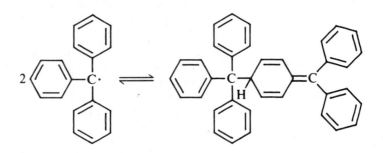

Hexaphenylethane has not yet been synthesized, presumably because of steric hindrance.

In this experiment the trityl radical is prepared in much the same fashion as Gomberg used, by the reaction of trityl bromide with zinc in the absence of oxygen. The yellow solution will then deliberately be exposed to air to give the peroxide.

Trityl radical

$$\Big|O_2$$

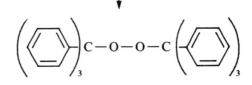

Triphenylmethyl peroxide
MW 518.67, mp 186°C

5. Synthesis of Triphenylmethyl Peroxide

In a small test tube dissolve 100 mg of trityl bromide or chloride in 0.5 mL of toluene. Material prepared in Experiment 2 above may be used; it need not be recrystallized. Cap the tube with a septum, insert an empty needle, and flush the tube with nitrogen while shaking the contents. Add 0.2 g of fresh zinc dust as quickly as possible, then flush the tube with nitrogen once more. Shake the tube vigorously for about 10 min, and note the appearance of the reaction mixture. Using a Pasteur pipette transfer the solution, but

The trityl radical

not the zinc, to a 30-mL beaker. Roll the solution around the inside of the beaker to give it maximum exposure to the air and after a few minutes collect the peroxide on the Hirsch funnel. Wash the solid with a little cold toluene, allow it to dry in the air, and determine the melting point. It is reported to melt at 186°C.

Cleaning Up   Transfer the mixture of zinc and zinc bromide to the hazardous waste container.

A Puzzle

Carry out both of the following reactions. Compare the products formed. Run necessary solubility and simple qualitative tests. Interpret your results and propose structures for the compounds produced in each of the reactions.

6. Synthesis of Compound 1

Malonic acid:
HOOCCH₂COOH

Into an Erlenmeyer flask place 1 g of triphenylmethanol and 2 g of malonic acid. Grind the two solids together with a glass stirring rod and then heat the flask at 160°C for 7 min. Use an aspirator tube mounted at the mouth of the tube to carry away undesirable fumes. Allow the tube to cool and then dissolve the contents in about 2 mL of toluene. Dilute the solution with 10 mL of 60–80°C ligroin, and after cooling the tube in ice isolate the crystals, wash them with a little cold ligroin, and once they are dry determine their weight and melting point.

Cleaning Up   Place the toluene/ligroin mother liquor in the organic solvents container.

7. Synthesis of Compound 2

Chloroacetic acid
ClCH₂COOH

In a 25-mL Erlenmeyer flask dissolve 0.5 g of triphenylmethanol in 10 mL of acetic acid and then add to this solution 2 mL of an acetic acid solution containing 5% of chloroacetic acid and 1% of sulfuric acid. Heat the mixture for 5 min, add about 5–6 mL of water to produce a hot saturated solution, and let the tube cool slowly to room temperature. Collect the product by filtration, and wash the crystals with 1:1 methanol–water. Once they are dry, determine the weight and melting point and then perform tests on Compounds 1 and 2 to deduce their identities.

Cleaning Up   Dilute the filtrate with water, neutralize with sodium carbonate, and flush down the drain.

Questions

1. Give the mechanism for the free radical chlorination of triphenylmethane.

2. Is the propeller-shaped triphenylmethyl carbocation a chiral species?

3. Without carrying out the experiments, speculate on the structures of Compounds 1 and 2 made in Experiments 6 and 7.

4. Is the production of triphenylmethyl peroxide a chain reaction?

5. What product would you expect from the reaction of carbon tetrachloride, benzene, and aluminum chloride?

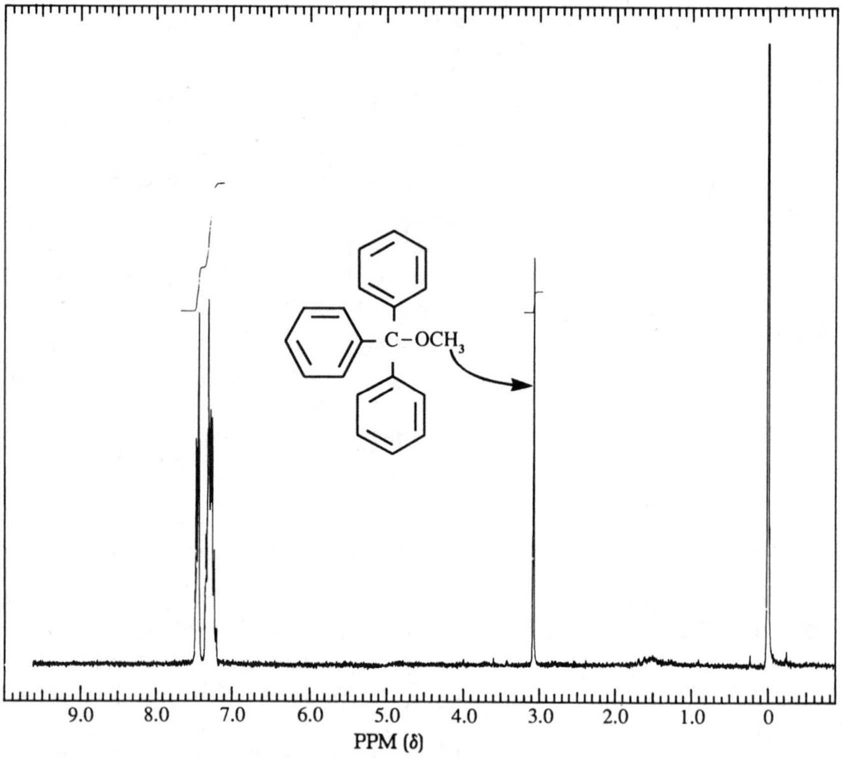

FIG. 1 ^{1}H nmr spectrum of triphenylmethyl methyl ether (trityl methyl ether) (250 MHz).

33

Dibenzalacetone by the Aldol Condensation

Prelab Exercise: Calculate the volumes of benzaldehyde and acetone needed for this reaction, taking into account the densities of the liquids and the number of moles of each required.

$$2 \underset{\substack{\text{Benzaldehyde} \\ \text{MW 106.13, bp 178°C} \\ \text{den 1.04}}}{\boxed{}\text{CHO}} + \underset{\substack{\text{Acetone} \\ \text{(2-Propanone)} \\ \text{MW 58.08, bp 56°C} \\ \text{den 0.790}}}{\text{CH}_3\text{CCH}_3} \xrightarrow[\text{MW 40.01}]{\text{NaOH}} \underset{\substack{\text{Dibenzalacetone} \\ \text{(1,5-Diphenyl-1,4-pentadien-3-one)} \\ \text{MW 234.30, mp 110–111°C}}}{\text{—CH=CHCCH=CH—}}$$

The reaction of an aldehyde with a ketone employing sodium hydroxide as the base is an example of a mixed aldol condensation reaction, the Claisen-Schmidt reaction. Dibenzalacetone is readily prepared by condensation of acetone with two equivalents of benzaldehyde. The aldehyde carbonyl is more reactive than that of the ketone and therefore reacts rapidly with the anion of the ketone to give a β-hydroxyketone, which easily undergoes base-catalyzed dehydration. Depending on the relative quantities of the reactants, the reaction can give either mono- or dibenzalacetone.

In the present experiment sufficient ethanol is present as solvent to readily dissolve the starting material, benzaldehyde, and also the intermediate, benzalacetone. The benzalacetone, once formed, can then easily react with another mole of benzaldehyde to give the product, dibenzalacetone. The detailed mechanism for the formation of benzalacetone is:

$$\text{CH}_3\text{CCH}_3 + \text{OH}^- \rightleftharpoons \left[\text{CH}_3\text{C}-\text{CH}_2{}^- \leftrightarrow \text{CH}_3\text{C}=\text{CH}_2 \right] + \text{H}_2\text{O}$$

$$\text{CH}_3\text{CCH}_2 \cdots \text{C} \underset{\text{O}}{\overset{\text{H}}{\big\langle}} \rightleftharpoons \text{CH}_3\text{CCH}_2\text{C}-\underset{\text{O}^-}{\overset{\text{H}}{\big\langle}} \overset{\text{H}_2\text{O}}{\rightleftharpoons} \text{CH}_3\text{CCH}_2\text{C}-\underset{\text{OH}}{\overset{\text{H}}{\big\langle}} + \text{OH}^-$$

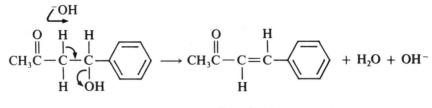

Benzalacetone
4-Phenyl-3-butene-2-one
mp 42°C

Experiment

1. Synthesis of Dibenzalacetone

If sodium hydroxide gets on the skin, wash until the skin no longer has a "soapy" feeling

Mix 0.05 mole of benzaldehyde with the theoretical quantity of acetone, add one-half the mixture to a solution of 5 g of sodium hydroxide dissolved in 50 mL of water and 40 mL of ethanol at room temperature (<25°C). After 15 min add the remainder of the aldehyde-ketone mixture and rinse the container with a little ethanol to complete the transfer. After one-half hour, during which time the mixture is swirled frequently, collect the product by suction filtration on a Büchner funnel. Break the suction and carefully pour 100 mL of water on the product. Reapply the vacuum. Repeat this process three times in order to remove all traces of sodium hydroxide. Finally, press the product as dry as possible on the filter using a cork, then press it between sheets of filter paper to remove as much water as possible. Save a small sample for melting point determination and then recrystallize the product from ethanol using about 10 mL of ethanol for each 4 g of dibenzalacetone. Pure dibenzalacetone melts at 110–111°C, and the yield after recrystallization should be about 4 g.

Cleaning Up   Dilute the filtrate from the reaction mixture with water and neutralize it with dilute hydrochloric acid before flushing down the drain. The ethanol filtrate from the crystallization should be placed in the organic solvents container.

Questions

1. Why is it important to maintain equivalent proportions of reagents in this reaction?

2. What side products do you expect in this reaction? How are they removed?

3. Write formulas to show the possible geometric isomers of dibenzalacetone. Which isomer would you expect to be most stable? Why?

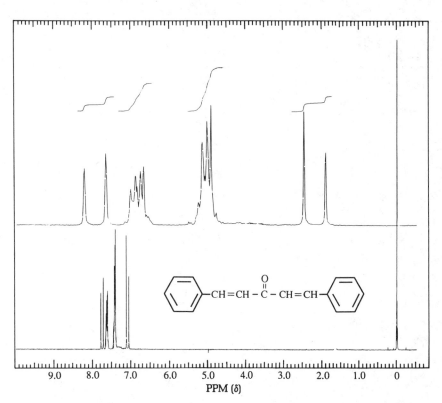

FIG. 1 ^{1}H nmr spectrum of dibenzalacetone (250 MHz). (An expanded spectrum appears above the normal spectrum.)

4. What evidence do you have that your product consists of a single geometric isomer or a mixture of isomers? Does the melting point give such information?

5. From the ^{1}H nmr spectrum of dibenzalacetone (Fig. 1) can you deduce what geometric isomer(s) is (are) formed?

6. How would you change the above procedure if you wished to synthesize benzalacetone, $C_6H_5CH=CHCOCH_3$? benzalacetophenone, $C_6H_5CH=CHCOC_6H_5$?

34

Wittig-Horner Reaction

Prelab Exercise: Account for the fact that the Wittig-Horner reaction of cinnamaldehyde gives almost exclusively the E,E~butadiene with very little contaminating E,Z-product.

The Wittig reaction affords an invaluable method for the conversion of a carbonyl compound to an olefin, for example the conversion of benzaldehyde and *n*-butyl bromide to a mixture of *cis*- and *trans*-1-phenyl-2-*n*-propylethylene.

$(C_6H_5)_3P + CH_3CH_2CH_2CH_2Br$
Triphenylphosphine   *n*-Butyl bromide

Br^-
$(C_6H_5)_3\overset{+}{P}CH_2CH_2CH_2CH_3$
Triphenylbutylphosphonium bromide

$\downarrow$ RLi

$(C_6H_5)_3\overset{+}{P}—\overset{-}{C}HCH_2CH_2CH_3 \longleftrightarrow (C_6H_5)_3P=CHCH_2CH_2CH_3 + RH + LiBr$

Wittig reagent, an ylide

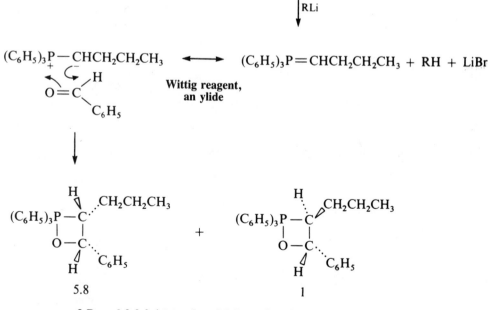

5.8   1

3-Propyl-2,2,2,4-tetraphenyl-1,2-oxaphosphetane

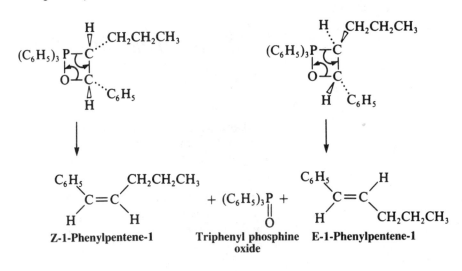

Z-1-Phenylpentene-1   **Triphenyl phosphine oxide**   **E-1-Phenylpentene-1**

Because the active reagent, an ylide, is unstable, it is generated in the presence of the carbonyl compound by dehydrohalogenation of an alkyltriphenylphosphonium bromide with phenyllithium in dry ether in a nitrogen atmosphere. The existence of the four-membered ring intermediate was proved by nmr in 1984.[1]

When the halogen compound employed in the first step has an activated halogen atom ($RCH=CHCH_2X$, $C_6H_5CH_2X$, XCH_2CO_2H) a simpler procedure known as the Horner phosphonate modification of the Wittig reaction is applicable. When benzyl chloride is heated with triethyl phosphite, ethyl chloride is eliminated from the initially formed phosphonium chloride with the production of diethyl benzylphosphonate. This phospho-

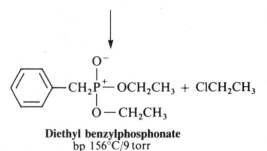

Benzyl chloride	**Triethyl phosphite**	**Triethylbenzylphosphonium chloride**
bp 179°C, MW 126.59	bp 156°C, MW 166.16	
den 1.10	den 0.94	

Diethyl benzylphosphonate
bp 156°C/9 torr

1. B. E. Maryanoff, A. B. Reitz, and M. S. Mutter, *J. Am. Chem. Soc.*, **106**, 1873 (1984).

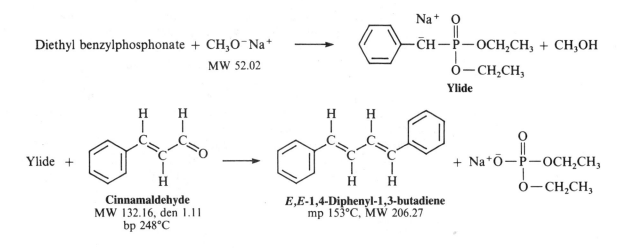

Diethyl benzylphosphonate + CH$_3$O$^-$Na$^+$
MW 52.02

Ylide

Ylide +

Cinnamaldehyde
MW 132.16, den 1.11
bp 248°C

E,E-1,4-Diphenyl-1,3-butadiene
mp 153°C, MW 206.27

Simplified Wittig reaction

nate is stable, but in the presence of a strong base such as the sodium methoxide used here it condenses with a carbonyl component in the same way that a Wittig ylide condenses. Thus, it reacts with benzaldehyde to give *E*-stilbene and with cinnamaldehyde to give *E,E*-1,4-diphenyl-1,3-butadiene.

CAUTION: Handle benzyl chloride in the hood. Severe lachrymator and irritant of the respiratory tract. Take care to keep organophosphorus compounds off the skin.

Use freshly prepared or opened sodium methoxide; keep bottle closed; avoid skin contact with dimethylformamide

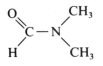

Dimethylformamide
MW 73.10, bp 153°C

A highly polar solvent capable of dissolving ionic compounds such as sodium methoxide, yet miscible with water

Experiment

Synthesis of 1,4-Diphenyl-1,3-butadiene

With the aid of pipettes and a pipetter, measure into a 25 × 150 mm test tube 5 mL of benzyl chloride (α-chlorotoluene) and 7.7 mL of triethyl phosphite. Add a boiling stone, insert a cold finger condenser, and with a flask heater reflux the liquid gently for 1 h. Alternatively, carry out the reaction in a 25-mL round-bottomed flask equipped with a reflux condenser. (Elimination of ethyl chloride starts at about 130°C, and in the time period specified the temperature of the liquid rises to 190–200°C.) Let the phosphonate ester cool to room temperature, pour it into a 125-mL Erlenmeyer flask containing 2.4 g of sodium methoxide, and add 40 mL of dimethylformamide, using a part of this solvent to rinse the test tube. Swirl the flask vigorously in a water-ice bath to thoroughly chill the contents and continue swirling while running in 5 mL of cinnamaldehyde by pipette. The mixture soon turns deep red and then crystalline hydrocarbon starts to separate. When there is no further change (about 2 min) remove the flask from the cooling bath and let it stand at room temperature for about 10 min. Then add 20 mL of water and 10 mL of methanol, swirl vigorously to dislodge crystals, and finally collect the product on a suction funnel using the red mother liquor to wash the flask. Wash the product with water until the red color of the product is all replaced by yellow. Then wash with methanol to remove the yellow impurity, and continue until the wash liquor is colorless. The yield of the crude faintly yellow hydrocarbon, mp

150–151°C, is 5.7 g. This material is satisfactory for use in Chapter 35 (1.5 g required). A good solvent for crystallization of the remainder of the product is methylcyclohexane (bp 101°C, 10 mL per g; use more if the solution requires filtration). Pure *E,E*-1,4-diphenyl-1,3-butadiene melts at 153°C.

Cleaning Up   The filtrate and washings from this reaction are dark, oily, and smell bad. The mixture contains dimethylformamide and could contain traces of all starting materials. Keep the volume as small as possible and place it in the hazardous waste container for organophosphorus compounds. If methylcyclohexane was used for recrystallization, place the mother liquor in the organic solvents container.

Questions

1. Show how 1,4-diphenyl-1,3-butadiene might be synthesized from benzaldehyde and an appropriate halogenated compound.

2. Explain why the methyl groups of trimethyl phosphite give two peaks in the ^1H nmr spectrum (Figure 1).

3. Write the equation for the reaction between sodium methoxide and moist air.

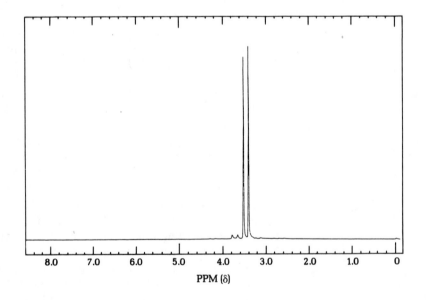

FIG. 1 ^1H nmr spectrum of trimethyl phosphite (90 MHz).

35

p-Terphenyl by the Diels-Alder Reaction

Prelab Exercise: Why does the dicarboxylate, **5**, undergo double decarboxylation to give terphenyl, while the diacid formed from the reaction of cyclopentadiene with maleic anhydride (Chapter 24) does not undergo decarboxylation under the same reaction conditions?

Diels-Alder reaction

E-E-1,4-Diphenyl-1,3-butadiene (**1**) is most stable in the transoid form, but at a suitably elevated temperature the cisoid form present in the equilibrium adds to dimethyl acetylenedicarboxylate (**2**) to give dimethyl 1,4-diphenyl-1,4-dihydrophthalate (**3**). This low-melting ester is obtained as an oil and, when warmed briefly with methanolic potassium hydroxide, is isomerized to the high-melting *E*-ester (**4**). The free *E*-acid can be obtained in 86% yield by refluxing the suspension of (**3**) in methanol for 4 h; but in the recommended procedure the isomerized ester is collected, washed to remove dark mother liquor, and hydrolyzed by brief heating with potassium hydroxide in a high-boiling solvent. The final step, an oxidative decarboxylation, is rapid and nearly quantitative. It probably involves reaction of the oxidant with the dianion (**5**) with removal of two electrons and formation of a diradical, which loses carbon dioxide with formation of *p*-terphenyl (**6**).

Experiment

Synthesis of p-Terphenyl

CAUTION: *Handle dimethyl acetylenedicarboxylate with care.*

CH₃OCH₂CH₂OCH₂
|
CH₃OCH₂CH₂OCH₂

Triethylene glycol dimethyl ether (triglyme)
Miscible with water. bp 222°C

Reaction time: 30 min; reflux over a flame

Place 1.5 g of *E,E*-1,4-diphenyl-1,3-butadiene (from the Wittig reaction, Chapter 34) and 1.0 mL (1.1 g) of dimethyl acetylenedicarboxylate (*caution, skin irritant*[1]) in a 25 × 150-mm test tube and rinse down the walls with 5 mL of triethylene glycol dimethyl ether (triglyme) (bp 222°C). Clamp the test tube in a vertical position, introduce a cold finger condenser, and reflux the mixture gently for 30 min. Alternatively, carry out the experiment in a 25-mL round-bottomed flask equipped with a reflux condenser. Cool the yellowish solution under the tap, pour into a separatory funnel, and rinse out the reaction vessel with a total of about 50 mL of ether. Extract twice with water (50–75 mL portions) to remove the high-boiling solvent, shake the ethereal solution with saturated sodium chloride solution, and dry the ether layer over anhydrous sodium sulfate. Filter or decant the ether solution into a tared 125-mL Erlenmeyer flask and evaporate the

1. This ester is a powerful lachrymator (tear producer) and vesicant (blistering agent) and should be dispensed from a bottle provided with a pipette and a pipetter. Even a trace of ester on the skin should be washed off promptly with methanol, followed by soap and water.

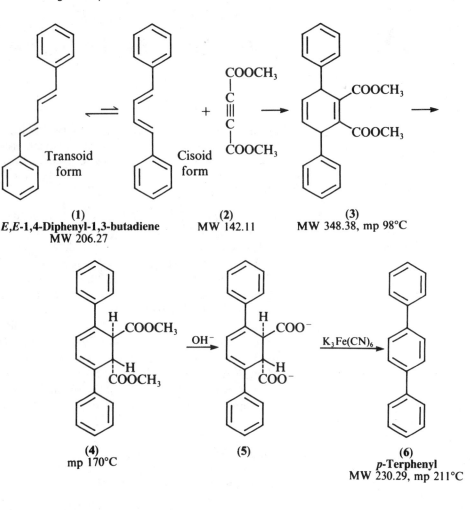

(1)
E,E-1,4-Diphenyl-1,3-butadiene
MW 206.27

(2)
MW 142.11

(3)
MW 348.38, mp 98°C

(4)
mp 170°C

(5)

(6)
p-Terphenyl
MW 230.29, mp 211°C

CAUTION: Extinguish all flames before working with ether

Potassium hydroxide is easily powdered in a Waring blender.

ether on the steam bath, eventually with evacuation at the aspirator, until the weight of yellow oil is constant; yield is about 2.5–2.8 g.

While evaporation is in progress, dissolve 0.5 g of potassium hydroxide (about 5 pellets) in 10 mL of methanol by heating and swirling; the process is greatly hastened by crushing the lumps with a stirring rod with a flattened head. Crystallization of the yellow oil containing (3) can be initiated by cooling and scratching; this provides assurance that the reaction has proceeded properly. Pour in the methanolic potassium hydroxide and heat with swirling on the hot plate for about 1 min until a stiff paste of crystals of the isomerized ester (4) appears. Cool, thin the mixture with methanol, collect the product and wash it free of dark mother liquor, and spread it thinly on a paper for rapid drying. The yield of pure, white ester (4) is 1.7–1.8 g. Solutions in methanol are strongly fluorescent.

Place the ester (4) in a 25 × 150-mm test tube, add 0.7 g of potassium hydroxide (7–8 pellets), and pour in 5 mL of triethylene glycol. Stir the mixture with a thermometer and heat, raising the temperature to 140°C in

Rapid hydrolysis of the hindered ester (4)

the course of about 5 min. By intermittent heating, keep the temperature close to 140°C for 5 min longer and then cool the mixture under the tap. Pour into a 125-mL Erlenmeyer flask and rinse the tube with about 50 mL of water. Heat to boiling and, in case there is a small precipitate or the solution is cloudy, add a little pelletized Norit, swirl, and filter the alkaline solution by gravity. Then add 3.4 g of potassium ferricyanide and heat on the hot plate with swirling for about 5 min to dissolve the oxidant and to coagulate the white precipitate which soon separates. The product can be air-dried overnight or dried to constant weight by heating in an evacuated Erlenmeyer flask on the steam bath. The yield of colorless *p*-terphenyl, mp 209–210°C, is 0.7–0.8 g.

Cleaning Up   The aqueous layer from the reaction after dilution can be flushed down the drain. Allow the ether to evaporate from the sodium sulfate and then discard it in the nonhazardous solid waste container. The aqueous reaction mixture after dilution with water is neutralized with dilute hydrochloric acid and then flushed down the drain.

Questions

1. What is the driving force for the isomerization of **3** to **4**?

2. Why is the *trans* and not the *cis* diester **4** formed in the isomerization of **3** to **4**?

3. Why does hydrolysis of **4** in methanol require 4 h whereas hydrolysis in triethylene glycol requires only 10 min?

36

Nitration of Methyl Benzoate

Prelab Exercise: Draw the complete mechanism for the nitration of methyl benzoate. Show the resonance forms that make the methyl ester group a *meta* director and deactivator of the aromatic ring.

Benzene and somewhat less reactive aromatic compounds such as methyl benzoate can be nitrated with a mixture of nitric and sulfuric acids, which ionizes completely to generate the nitronium and hydronium ions:

$$HNO_3 + 2H_2SO_4 \rightleftharpoons NO_2^+ + 2HSO_4^- + H_3O^+$$

Hot concentrated nitric acid is also a good oxidizing agent. For example, benzoin is oxidized easily to benzil (Chapter 50). Activated aromatic compounds such as amines and phenols can be nitrated using just concentrated nitric acid:

$$2\ HNO_3 \rightleftharpoons NO_2^+ + NO_3^- + H_2O$$
$$ 95\% 5\%$$

1,3,5-Trinitrobenzene cannot be prepared by nitration of *m*-dinitrobenzene, even with the use of heat, concentrated sulfuric acid, and fuming nitric acid because the two nitro groups strongly deactivate the benzene ring.

In the present experiment sulfuric acid serves as the solvent:

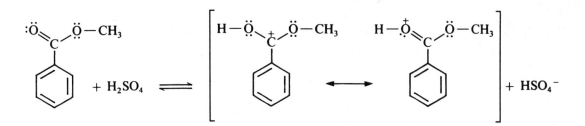

and nitration occurs at the *meta* position because of the partial positive charges residing at the *ortho* and *para* positions.

Experiment

Nitration of Methyl Benzoate

$$HNO_3 + 2H_2SO_4 \rightleftharpoons NO_2^+ + 2HSO_4^- + H_3O^+$$

Nitronium ion   Hydronium ion

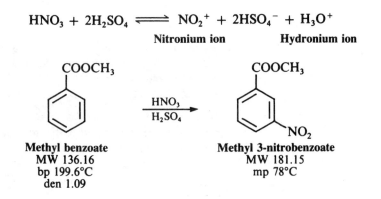

Methyl benzoate
MW 136.16
bp 199.6°C
den 1.09

Methyl 3-nitrobenzoate
MW 181.15
mp 78°C

CAUTION: Use care in handling concentrated sulfuric and nitric acids

In a 125-mL Erlenmeyer flask cool 12 mL of concentrated sulfuric acid to 0°C and then add 6.1 g of methyl benzoate.[1] Again cool the mixture to 0–10°C. Now add dropwise, using a Pasteur pipette, a cooled mixture of 4 mL of concentrated sulfuric acid and 4 mL of concentrated nitric acid. During the addition of the acids, swirl the mixture frequently and maintain the temperature of the reaction mixture in the range of 5–15°C.

When all the nitric acid has been added, warm the mixture to room temperature and after 15 min pour it on 50 g of cracked ice in a 250-mL beaker. Isolate the solid product by suction filtration using a small Büchner funnel and wash well with water, then with two 10-mL portions of ice-cold methanol. A small sample is saved for a melting point determination. The remainder is weighed and crystallized from an equal weight of methanol. The crude product should be obtained in about 80% yield and with a mp of 74–76°C. The recrystallized product should have a mp of 78°C.

Cleaning Up   Dilute the filtrate from the reaction with water, neutralize with sodium carbonate, and flush down the drain. The methanol from the crystallization should be placed in the organic solvents container.

Questions

1. Why does methyl benzoate dissolve in concentrated sulfuric acid? Write an equation showing the ions that are produced.

2. What would you expect the structure of the dinitro ester to be? Consider the directing effects of the ester and the first nitro group upon the addition of the second nitro group.

---

1. Methyl benzoate prepared in Chapter 27 can be used.

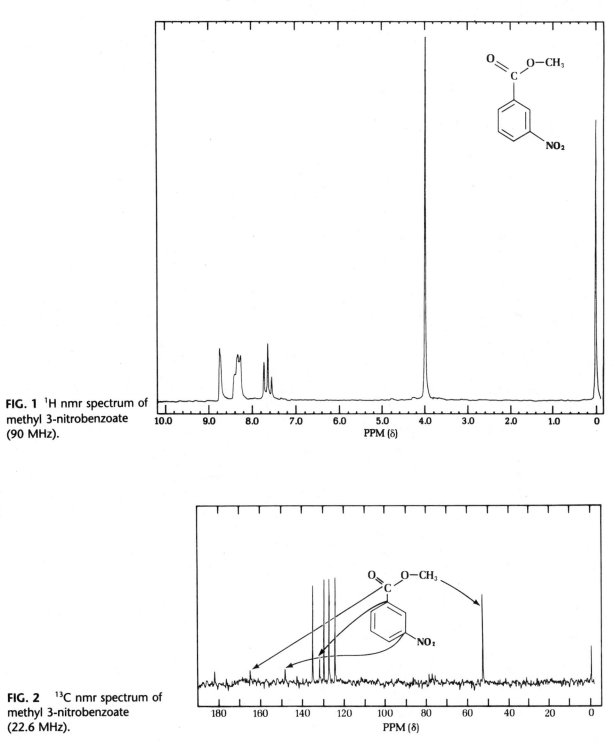

FIG. 1 ¹H nmr spectrum of methyl 3-nitrobenzoate (90 MHz).

FIG. 2 ¹³C nmr spectrum of methyl 3-nitrobenzoate (22.6 MHz).

37

Friedel-Crafts Alkylation of Benzene and Dimethoxybenzene; Host-Guest Chemistry

Prelab Exercise: Prepare a flow sheet for each reaction, indicating how the catalysts and unreacted starting materials are removed from the reaction mixture.

Friedel-Crafts alkylation of aromatic rings most often employs an alkyl halide and a strong Lewis acid catalyst. Some of the catalysts that can be used, in order of decreasing activity, are the halides of Al, Sb, Fe, Ti, Sn, Bi, and Zn. Although useful, the reaction has several limitations. The aromatic ring must be unsubstituted or bear activating groups, and because the product, an alkylated aromatic molecule, is more reactive than the starting material, multiple substitution usually occurs. Furthermore, primary halides will rearrange under the reaction conditions:

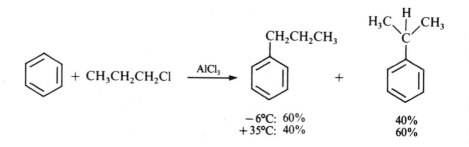

	$-6°C$: 60%	40%
	$+35°C$: 40%	60%

In the present reaction a tertiary halide and the most powerful Friedel-Crafts catalyst, $AlCl_3$, are allowed to react with benzene. The initially formed *t*-butylbenzene is a liquid while the product, 1,4-di-*t*-butylbenzene, which has a symmetrical structure, is a beautifully crystalline solid. The alkylation reaction probably proceeds through the carbocation under the conditions of the present experiment:

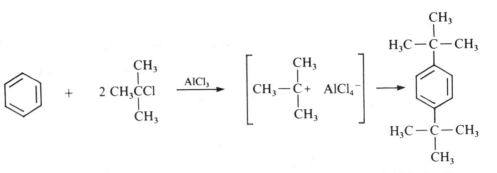

Benzene
MW 78.11, den 0.88
bp 80°C

2-Chloro-2-methylpropane
(*t*-Butyl chloride)
MW 92.57, den 0.85
bp 51°C

1,4-Di-*t*-butylbenzene
MW 190.32, mp 77–79°C
bp 167°C

The reaction is reversible. If 1,4-di-*t*-butylbenzene is allowed to react with *t*-butyl chloride and aluminum chloride (1.3 moles) at 0–5°C, 1,3-di-*t*-butylbenzene, 1,3,5-tri-*t*-butylbenzene, and unchanged starting material are found in the reaction mixture. Thus, the mother liquor from crystallization of 1,4-di-*t*-butylbenzene in the present experiment probably contains *t*-butylbenzene, the desired 1,4-di-product, the 1,3-di-isomer, and 1,3,5-tri-*t*-butylbenzene.

Host-guest chemistry

Although the mother liquor probably contains a mixture of several components, the 1,4-di-*t*-butylbenzene can be isolated easily as an inclusion complex. Inclusion complexes are examples of host-guest chemistry. The molecule thiourea, NH_2CSNH_2, the host, has the interesting property of crystallizing in a helical crystal lattice that has a cylindrical hole in it. The guest molecule can reside in this hole if it is the correct size. It is not bound to the host, and there are often a nonintegral number (on the average) of host molecules per guest. The inclusion complex of thiourea and 1,4-di-*t*-butylbenzene crystallizes very nicely from a mixture of the other hydrocarbons and thus more of the product can be obtained. Because thiourea is very soluble in water the product is recovered from the complex by shaking it with a mixture of ether and water. The complex immediately decomposes and the product dissolves in the ether layer, from which it can be recovered.

Compare the length of the 1,4-di-*t*-butylbenzene molecule with the length of various *n*-alkanes and predict the host/guest ratio for a given alkane. You can then check your prediction experimentally.

Experiments

1. 1,4-Di-*t*-butylbenzene

Measure in the hood 20 mL of 2-chloro-2-methylpropane (*t*-butyl chloride) and 10 mL of benzene in a 125-mL filter flask equipped with a one-holed

CAUTION: *Benzene is a carcinogen. Handle in the hood, do not breathe vapors, or allow liquid to come in contact with the skin.*

Reaction time: about 15 min

Aluminum chloride dust is extremely hygroscopic and irritating. It hydrolyzes to hydrogen chloride on contact with moisture. Clean up spilled material immediately.

Add powdered anhydrous sodium sulfate until it no longer clumps together

Spontaneous crystallization gives beautiful needles or plates

CAUTION: *Thiourea is a carcinogen. Handle the solid in a hood. Do not breathe dust.*

Thiourea
MW 76.12

Inclusion complex starts to crystallize in 10 min

rubber stopper fitted with a thermometer. Place the flask in an ice-water bath to cool. Weigh 1 g of fresh aluminum chloride onto a creased paper, and scrape it with a small spatula into a 10 × 75-mm test tube; close the tube at once with a cork.[1] Connect the side arm of the flask to the aspirator (preferably one made of plastic) and operate it at a rate sufficient to carry away hydrogen chloride formed in the reaction, or make a trap for the hydrogen chloride similar to that shown in Fig. 1 in Chapter 13. Cool the liquid to 0–3°C, add about one quarter of the aluminum chloride, replace the thermometer, and swirl the flask vigorously in the ice bath. After an induction period of about 2 min a vigorous reaction sets in, with bubbling and liberation of hydrogen chloride. Add the remainder of the catalyst in three portions at intervals of about 2 min. Toward the end, the reaction product begins to separate as a white solid. When this occurs, remove the flask from the bath and let stand at room temperature for 5 min. Add ice and water to the reaction mixture, allow most of the ice to melt, and then add ether for extraction of the product, stirring with a rod or spatula to help bring the solid into solution. Transfer the solution to a separatory funnel and shake; draw off the lower layer and wash the upper ether layer with water and then with a saturated sodium chloride solution. Dry the ether solution over anhydrous sodium sulfate for 5 min, filter the solution to remove the drying agent, remove the ether by evaporation on the steam bath, and evacuate the flask using the aspirator to remove traces of solvent until the weight is constant; yield of crude product should be 15 g.

The oily product should solidify on cooling. For crystallization, dissolve the product in 20 mL of hot methanol and let the solution come to room temperature without disturbance. If you are in a hurry, lift the flask without swirling; place it in an ice-water bath and observe the result. After thorough cooling at 0°C, collect the product and rinse the flask and product with a little ice-cold methanol. The yield of 1,4-di-*t*-butylbenzene from the first crop is 8.2–8.6 g of satisfactory material. Save the product for the next step as well as the mother liquor, in case you later wish to work it up for a second crop.

In a 25-mL Erlenmeyer flask dissolve 5 g of thiourea (See margin note) and 3 g of 1,4-di-*t*-butylbenzene in 50 mL of warm methanol (break up lumps with a flattened stirring rod) and let the solution stand for crystallization of the complex, which occurs with ice cooling. Collect the crystals and rinse with a little methanol and dry to constant weight; yield is 5.8 g. Bottle a small sample, determine carefully the weight of the remaining complex, and place the material in a separatory funnel along with about 25 mL each of water and ether. Shake until the crystals disappear, draw off the aqueous layer containing thiourea, wash the ether layer with saturated sodium chloride, and dry the ether layer over anhydrous sodium sulfate. Remove the drying agent by filtration; collect the filtrate in a tared 125-mL Erlenmeyer. Evaporate and evacuate as before, making sure the

---

1. Alternative scheme: Put a wax pencil mark on the test tube 37 mm from the bottom and fill the tube with aluminum chloride to this mark.

weight of hydrocarbon is constant before you record it. Calculate the number of molecules of thiourea per molecule of hydrocarbon (probably *not* an integral number).

Work-up of mother liquor

To work up the mother liquor from the crystallization from methanol, first evaporate the solvent. Note that the residual oil does not solidify on ice cooling. Next, dissolve the oil, together with 5 g of thiourea, in 50 mL of methanol, collect the inclusion complex that crystallizes (3.2 g), and recover 1,4-di-*t*-butylbenzene from the complex as before (0.8 g before crystallization).

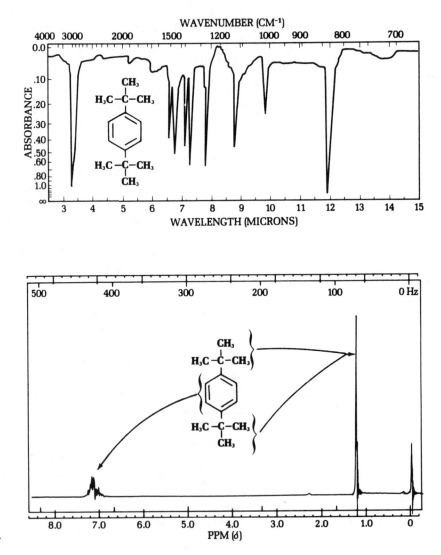

FIG. 1   Infrared spectrum of 1,4-di-*t*-butylbenzene.

FIG. 2   ¹H nmr spectrum of 1,4-di-*t*-butylbenzene (60 MHz).

Cleaning Up   Place any unused *t*-butyl chloride in the halogenated organic waste container, any unused benzene in the hazardous waste container for benzene. Any unused aluminum chloride should be mixed thoroughly with a large excess of sodium carbonate and the solid mixture added to a large volume of water before being flushed down the drain. The combined aqueous layers from the reaction should be neutralized with sodium carbonate and then flushed down the drain. Methanol from the crystallization is to be placed in the organic solvents container.

2. 1,4-Di-*t*-Butyl-2,5-Dimethoxybenzene

This experiment illustrates the Friedel-Crafts alkylation of an activated benzene molecule with a tertiary alcohol in the presence of sulfuric acid as the Lewis acid catalyst. As in the reaction of benzene and *t*-butyl chloride, the substitution involves attack by the electrophilic trimethylcarbocation.

$$CH_3-\overset{\overset{\displaystyle CH_3}{|}}{\underset{\underset{\displaystyle CH_3}{|}}{C}}{}^+$$

Trimethylcarbocation

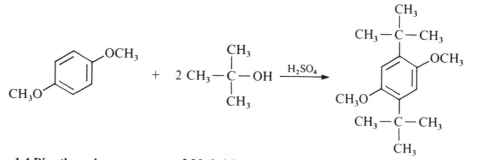

| 1,4-Dimethyoxybenzene (Hydroquinone dimethyl ether) MW 138.16, mp 57°C | 2-Methyl-2-propanol (*t*-Butyl alcohol) MW 74.12, den 0.79 mp 25.5°C, bp 82.8°C | 1,4-Di-*t*-butyl-2,5-dimethoxybenzene MW 250.37, mp 104–105°C |

CAUTION: H_2SO_4 + SO_3 is *fuming sulfuric acid. Highly corrosive to human tissue. Reacts violently with water. Measure these reagents in the hood.*

Reaction time about 12 min

Place 6 g of 1,4-dimethoxybenzene (hydroquinone dimethyl ether) in a 125-mL Erlenmeyer flask, add 10 mL of *t*-butyl alcohol and 20 mL of acetic acid, and put the flask in an ice-water bath to cool. Measure 10 mL of concentrated sulfuric acid into a 50-mL Erlenmeyer flask, add 10 mL of 30% fuming sulfuric acid or 20 mL of concentrated sulfuric acid, and put the flask, properly supported, in the ice bath to cool. For good thermal contact the ice bath should be an ice–water mixture. Put a thermometer in the larger flask and swirl in the ice bath until the temperature is in the range 0–3°C, and remove the thermometer (solid, if present, will dissolve later). Don't use the thermometer as a stirring rod. Clamp a small separatory funnel in a position to deliver into the 125-mL Erlenmeyer flask so that the flask can remain in the ice-water bath, wipe the smaller flask dry, and pour the chilled sulfuric acid solution into the funnel. While swirling the 125-mL flask in the ice bath, run in the chilled sulfuric acid by rapid drops during the course of 4–7 min.

By this time considerable solid reaction product should have separated, and insertion of a thermometer should show that the temperature is in the range 15–20°C. Swirl the mixture while maintaining the temperature at about 20–25°C for 5 min more and then cool in ice. Add ice to the mixture to dilute the sulfuric acid, then add water to nearly fill the flask, cool, and collect the product on a Büchner funnel with suction. It is good practice to clamp the filter flask so it does not tip over. Apply only very gentle suction at first to avoid breaking the filter paper, which is weakened by the strong sulfuric acid solution. Wash liberally with water and then turn on the suction to full force. Press down the filter cake with a spatula and let drain well. Meanwhile, cool a 30-mL portion of methanol for washing to remove a little oil and a yellow impurity. Release the suction, cover the filter cake with a third of the chilled methanol, and then apply suction. Repeat the washing a second and a third time.

Since air-drying of the crude reaction product takes time, the following short procedure is suggested: Place the moist material in a 125-mL Erlenmeyer flask, add a little dichloromethane (10–15 mL) to dissolve the organic material, and note the appearance of aqueous droplets. Add enough anhydrous sodium sulfate to the flask so that the drying agent no longer clumps together, let drying proceed for 10 min, and then remove the drying agent by gravity filtration or careful decantation into another 125-mL Erlenmeyer flask. Add 30 mL of methanol (bp 65°C) to the solution and start evaporation (on the steam bath in the hood) to eliminate the dichloromethane (bp 41°C). When the volume is estimated to be about 30 mL, let the solution stand for crystallization. When crystallization is complete, cool in ice and collect.

From an environmental standpoint it would be better to eliminate the solvents by simple distillation using the apparatus depicted in Fig. 5 in Chapter 5. Leave 30 mL in the flask, allow the mixture to cool slowly, and large crystals will form. Collect the product on a small Büchner funnel. The yield of large plates of pure 1,4-di-*t*-butyl-2,5-dimethoxybenzene is 4–5 g.

Antics of Growing Crystals

R. D. Stolow of Tufts University reported[2] that growing crystals of the di-*t*-butyldimethoxy compound change shape in a dramatic manner: thin plates curl and roll up and then uncurl so suddenly that they propel themselves for a distance of several centimeters. If you do not observe this phenomenon during crystallization of a small sample, you may be interested in consulting the papers cited and pooling your sample with others for trial on a large scale. The solvent mixture recommended by the Tufts workers for observation of the phenomenon is 9.7 mL of acetic acid and 1.4 mL of water per gram of product.

---

2. R. D. Stolow and J. W. Larsen, *Chemistry and Industry*, 449 (1963). See also J. M. Blatchly and N. H. Hartshorne, *Transactions of the Faraday Society*, **62**, 513 (1966).

Figures 3 and 4 present the infrared and nmr spectra of the starting hydroquinone dimethyl ether. Can you predict the appearance of the nmr spectrum of the product?

Cleaning Up   Combine the aqueous layer and methanol washes and crystallization mother liquor, dilute with water, neutralize with sodium carbonate, and flush down the drain. Any spilled sulfuric or fuming sulfuric acid should be covered with a large excess of solid sodium carbonate and the mixture added to water before being flushed down the drain. Dichloromethane mother liquor from the crystallization is placed in the halogenated organic waste container. After the solvent evaporates from the drying agent it is placed in the nonhazardous solid waste container.

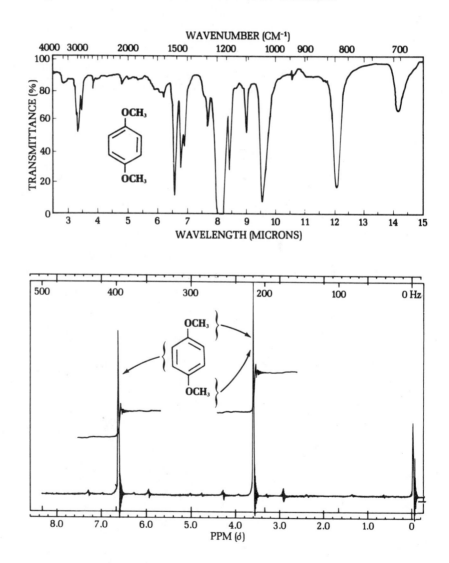

FIG. 3   Infrared spectrum of 1,4-dimethoxybenzene.

FIG. 4   ¹H nmr spectrum of 1,4-dimethoxybenzene (60 MHz).

Questions

1. Explain why the reaction of 1,4-di-*t*-butylbenzene with *t*-butyl chloride and aluminum chloride gives 1,3,5-tri-*t*-butylbenzene.

2. Why must aluminum chloride be protected from exposure to the air?

3. Why does fuming sulfuric acid react violently with water?

4. Draw a detailed mechanism for the formation of *t*-butyl-2,5-dimethoxybenzene.

5. Why is the 1,4 isomer, 1,4-di-*t*-butyl-2,5-dimethoxybenzene, the major product in the alkylation of dimethoxybenzene? Would you expect either of the following compounds to be formed as side products: 1,3-di-*t*-butyl-2,5-dimethoxybenzene; 1,4-dimethoxy-2,3-di-*t*-butylbenzene? Why or why not?

6. Suggest two other compounds that might be used in place of *t*-butyl alcohol to form 1,4-di-*t*-butyl-2,5-dimethoxybenzene.

38

Friedel-Crafts Acylation of Ferrocene: Acetylferrocene

Prelab Exercise: How many possible isomers could exist for diacetylferrocene?

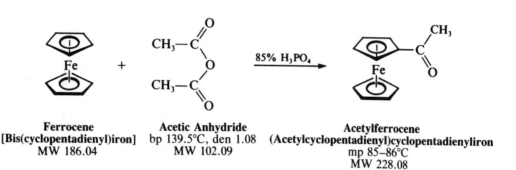

Ferrocene	**Acetic Anhydride**	**Acetylferrocene**
[Bis(cyclopentadienyl)iron]	bp 139.5°C, den 1.08	(Acetylcyclopentadienyl)cyclopentadienyliron
MW 186.04	MW 102.09	mp 85–86°C
		MW 228.08

The Friedel-Crafts acylation of benzene requires aluminum chloride as the catalyst, but ferrocene, which has been referred to as a "superaromatic" compound, can be acylated under much milder conditions with phosphoric acid as catalyst. Since the acetyl group is a deactivating substituent, the addition of a second acetyl group, which requires more vigorous conditions, will occur in the nonacetylated cyclopentadienyl ring to give 1,1′-diacetylferrocene. Because ferrocene gives just one monoacetyl derivative and just one diacetyl derivative, it was assigned an unusual sandwich structure.

Acetylferrocene and ferrocene (both highly colored) are easily separated by column chromatography. See Chapter 10.

Experiment

Acetylferrocene

Both acetic anhydride and sulfuric acid are corrosive to tissue; handle both with care and wipe up any spills immediately

In a 25-mL round-bottomed flask place 3.0 g of ferrocene, 10.0 mL of acetic anhydride, and 2.0 mL of 85% phosphoric acid. Equip the flask with a reflux condenser and a calcium chloride drying tube. Warm the flask gently on the steam bath with swirling to dissolve the ferrocene, then heat strongly for 10 min more. Pour the reaction mixture onto 50 g of crushed ice in a 400-mL beaker and rinse the flask with 10 mL of ice water. Stir the mixture for a few minutes with a glass rod, add 75 mL of 10% sodium hydroxide solution (the solution should still be acidic), then add solid sodium bicarbonate (be careful of foaming) until the remaining acid has been neutralized. Stir and crush all lumps, allow the mixture to stand for 20 min, then collect the product by suction filtration. Press the crude material as dry as possible between sheets of filter paper, save a few crystals for thin-layer chromatography, transfer the remainder to an Erlenmeyer flask, and add 40 mL of hexane or ligroin to the flask. Boil the solvent on the steam bath for a few minutes, then decant the dark orange solution into another Erlenmeyer flask, leaving a gummy residue of polymeric material. Treat the solution with decolorizing charcoal and filter it, through a fluted filter paper placed in a warm stemless funnel, into an appropriately sized Erlenmeyer flask. Evaporate the solvent (use an aspirator tube, Fig. 5 in Chapter 8) until the volume is about 20 mL. Set the flask aside to cool slowly to room temperature. Beautiful rosettes of dark orange-red needles of acetylferrocene will form. After the product has been cooled in ice, collect it on a Büchner funnel and wash the crystals with a small quantity of cold solvent. Pure acetylferrocene has mp 84–85°C. Your yield should be about 1.8 g.

Hexane and ligroin are flammable. No flames!

Analyze your product by thin-layer chromatography. Dissolve very small samples of pure ferrocene, the crude reaction mixture, and recrystallized acetylferrocene, each in a few drops of toluene; spot the three solutions with microcapillaries on silica gel plates; and develop the chromatogram with 30:1 toluene–absolute ethanol. Visualize the spots under a uv lamp if the silica gel has a fluorescent indicator or by adsorption of iodine vapor. Do you detect unreacted ferrocene in the reaction mixture and/or a spot that might be attributed to diacetylferrocene?

Cleaning Up   The reaction mixture filtrate can be flushed down the drain. Unused chromatography, recrystallization, and thin-layer chromatography solvents should be placed in the organic solvents container. The alumina from the column should be placed in the hood to allow the petroleum ether and ether to evaporate from it. Once free of organic solvents, it can be placed in the nonhazardous solid waste container. The decolorizing charcoal, once free of solvent, can also be placed in the nonhazardous solid waste container.

Question

What is the structure of the intermediate species that attacks ferrocene to form acetylferrocene? What other organic molecule is formed?

39

Alkylation of Mesitylene

Prelab Exercise: How much formic acid is consumed in this reaction?

The reaction of an alkyl halide with an aromatic molecule in the presence of aluminum chloride, the Friedel-Crafts reaction, proceeds through the formation of an intermediate carbocation:

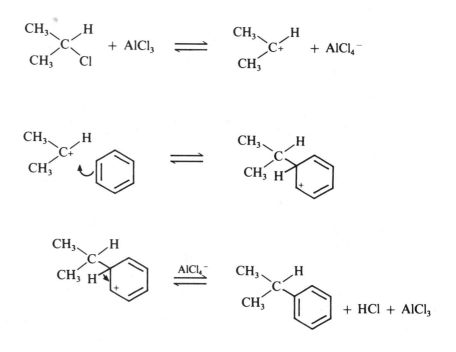

Friedel and Crafts (who later became the president of MIT) discovered this reaction in 1879, but seven years before, Baeyer and his colleagues carried out very similar reactions using aldehydes as the alkylating agent and strong acids as catalysts. These reactions, like the Friedel-Crafts reaction, proceed through carbocation intermediates. Consider the synthesis of DDT (1,1,1-trichloro-2,2-di(*p*-chlorophenyl)ethane)):

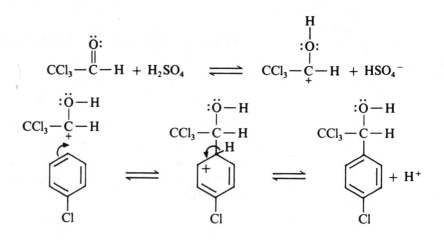

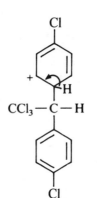 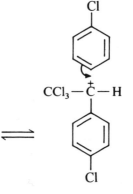 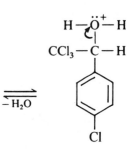

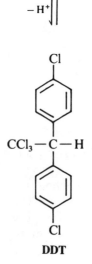

DDT

Trichloroacetaldehyde (chloral) forms a carbocation on reaction with concentrated sulfuric acid. This reacts primarily at the *para*-position of chlorobenzene; and the intermediate alcohol, being benzylic, in the presence of acid readily forms a new carbocation, which in turn attacks another molecule of chlorobenzene. Even though synthesized in 1872, the remarkable insecticidal properties of DDT were not recognized until about 1940. It took another 25 years to realize that this compound, which is resistant to normal biochemical degradation, was building up in rivers, lakes, and streams and causing long-term environmental damage to wildlife. It is now outlawed in many parts of the world.

In the present experiment, discovered by Baeyer in 1872, formaldehyde is allowed to react with mesitylene in the presence of formic acid. The sequence of reactions is very similar to those that form DDT:

Formaldehyde   **Formic acid**

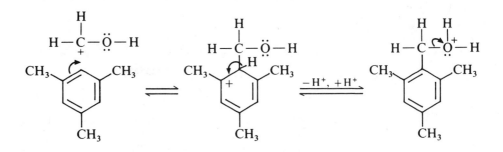

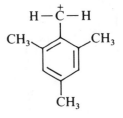

Benzylic carbocation

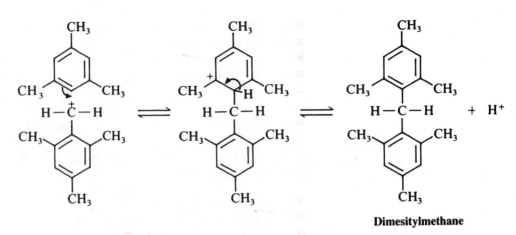

Dimesitylmethane

When aluminum chloride, a much more powerful catalyst, is used in this reaction, the methyl groups on the mesitylene rearrange and disproportionate to form a number of products, including polymeric material. The strongly activated ring of phenol reacts with formaldehyde at the *ortho-* and *para*-positions to form a polymer, Bakelite (see Chapter 67, Polymers).

A convenient form of formaldehyde to use in a reaction of this type is paraldehyde, a polymer that readily decomposes to formaldehyde:

$$+ \ddot{O} - CH_2 - \ddot{O} - CH_2 - \ddot{O} \rightarrow_n \xrightarrow{\text{heat}} CH_2 = \ddot{O} :$$

Paraldehyde   **Formaldehyde**

$$+ CH_2 - O \rightarrow_n \longrightarrow CH_2 = O + \overset{O}{\underset{||}{HCOH}} +$$

| Paraldehyde (Paraformaldehyde) mp 163–165°C (dec) | Formaldehyde MW 30.03 | 95% Formic acid MW 46.03 mp 8.5°C | Mesitylene MW 120.20 bp 162–164°C |

Dimesitylmethane
MW 252.41

Experiment

Dimesitylmethane

To 0.75 g of paraformaldehyde in a 50-mL round-bottomed flask add, in the hood, 4.5 mL of 95% formic acid. Fit the flask with a reflux condenser (see Fig. 1 in Chapter 15) and bring the mixture to a boil on the electrically heated sand bath. After about 5 min most of the paraformaldehyde will have dissolved. Add through the condenser 10.0 mL of mesitylene. Reflux the reaction mixture for 2 h, cool the mixture to room temperature and then in ice. Remove the product by vacuum filtration in the hood, taking great care not to come into contact with the residual formic acid. Wash the crystals with water, aqueous sodium carbonate solution and again with water, and squeeze them between sheets of filter paper to dry. Determine the weight of the crude product and save a few crystals for a melting point determination. Recrystallize the product. A suggested solvent is a mixture of 7.5 mL of toluene and 1 mL of methanol, which is adequate for 5 g of product. Obtain the melting points of the crude and recrystallized product. Calculate the percent yield and turn in the pure product.

CAUTION: Formic acid is corrosive to tissue. Avoid all contact with it or its vapors. Should any come in contact with your skin, wash it off immediately with a large quantity of water. Carrying out this reaction in the hood protects not only against formic acid but also against formaldehyde, a suspected carcinogen formed transitorily in the reaction.

Cleaning Up   At the end of this reaction there should be no formaldehyde remaining in the reaction mixture. Should it be necessary to destroy formaldehyde it should be diluted with water and 7 mL of household bleach added to oxidize 100 mg of paraformaldehyde. After 20 min it can be flushed down the drain. The formic acid solvent from the reaction and aqueous washings should be combined, neutralized with sodium carbonate, and flushed down the drain. Mother liquor from recrystallization is placed in the organic solvents container.

Question

The intermediate benzylic carbocation is stabilized by resonance. Draw the contributing resonance structures.

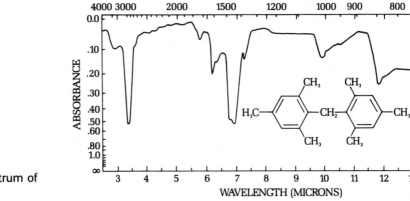

FIG. 1   Infrared spectrum of dimesitylmethane.

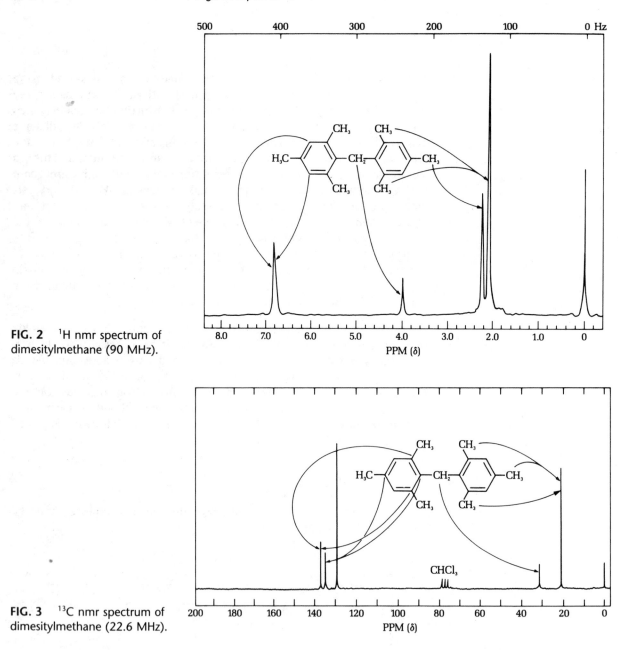

FIG. 2   ¹H nmr spectrum of dimesitylmethane (90 MHz).

FIG. 3   ¹³C nmr spectrum of dimesitylmethane (22.6 MHz).

40

Amines

Prelab Exercise: Explain why an ordinary amide from a primary amine does not dissolve in alkali while the corresponding sulfonamide of the amine does dissolve in alkali.

Amines are weak bases and can be regarded as organic substitution products of ammonia. Just as ammonia reacts with acids to form the ammonium ion, so amines react with acid to form the organoammonium ion:

$$\ddot{N}H_3 + H_3\ddot{O}^+ \rightleftharpoons \overset{+}{N}H_4 + H_2\ddot{O}:$$

Ammonia   **Ammonium ion**

$$CH_3\ddot{N}H_2 + H_3\ddot{O}^+ \rightleftharpoons CH_3\overset{+}{N}H_3 + H_2\ddot{O}:$$

Methylamine   **Methyl-ammonium ion**

When an amine dissolves in water the following equilibrium is established:

$$R\ddot{N}H_2 + H_2O \rightleftharpoons R\overset{+}{N}H_3 + OH^-$$

and from this the basicity constant can be defined as

$$K_b = \frac{[RNH_3{}^+][OH^-]}{[RNH_2]}$$

Strong bases have the larger values of K_b, meaning the amine has a greater tendency to accept a proton from water to increase the concentration of $RNH_3{}^+$ and OH^-.

The basicity constant for ammonia is 1.8×10^{-5}.

$$\ddot{N}H_3 + H_2O \rightleftharpoons NH_4{}^+ + OH^-$$

$$K_b = 1.8 \times 10^{-5} = \frac{[NH_4{}^+][OH^-]}{[NH_3]}$$

Alkyl amines such as methylamine, CH_3NH_2, or *n*-propylamine, $CH_3CH_2CH_2NH_2$, are stronger bases than ammonia because the alkyl groups are electron donors and increase the effective electron density on nitrogen. Aromatic amines, on the other hand, are weaker bases than ammonia. Delocalization of the unshared pair of electrons on nitrogen onto the aromatic ring means the electrons are not available to be shared with an acidic proton.

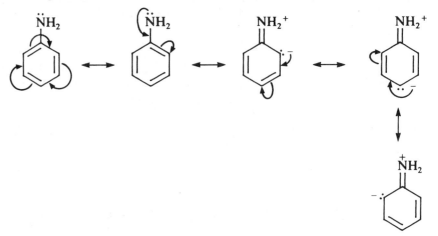

Low-molecular-weight amines have a powerful fishy odor. Slightly higher-molecular-weight diamines have names suggestive of their odors: $H_2N(CH_2)_5NH_2$, cadaverine, and $H_2N(CH_2)_4NH_2$ putrescine. The lower-molecular-weight amines with up to about five carbon atoms are soluble in water. Many amines, especially the liquid aromatic amines, undergo light-catalyzed free radical air oxidation reactions to give a large variety of highly colored decomposition products. The higher-molecular-weight amines that are insoluble in water will dissolve in acid to form ionic amine salts. This reaction is useful for both characterization and for separation purposes. The ionic amine salt on treatment with base will regenerate the amine.

Amides are not basic

Nitriles, $R-C \equiv N$, are not basic and neither are amides or substituted amides. The adjacent electron-withdrawing carbonyl oxygen effectively removes the unshared pair of electrons on nitrogen:

The object of the present experiment is to identify an unknown amine or amine salt. Procedures for solubility tests are given in Section 1.

Section 2 presents the Hinsberg test, a test for distinguishing between primary, secondary, and tertiary amines; Section 3 gives procedures for preparation of solid derivatives for melting point characterizations; and Section 4 gives spectral characteristics. Apply the procedures to known substances along with the unknown.

Experiments

1. Solubility

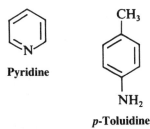

Pyridine

p-Toluidine

Use caution in smelling unknowns and do it only once

If necessary, this experiment can be scaled up by a factor of 2 or 3.
Substances to be tested:

Aniline, $C_6H_5NH_2$, bp 184°C
p-Toluidine, $CH_3C_6H_4NH_2$, mp 43°C
Pyridine, C_5H_5N, bp 115°C (a tertiary amine)
Methylamine hydrochloride, dec. (salt of CH_3NH_2, bp -6.7°C)
Aniline hydrochloride, dec. (salt of $C_6H_5NH_2$)
Aniline sulfate, dec.

Amines are often contaminated with dark oxidation products. The amines can easily be purified by a small scale distillation.

First, see if the substance has a fishy, ammonia-like odor; if so, it probably is an amine of low molecular weight. Then test the solubility in water by putting 1 drop if a liquid or an estimated 10 mg if a solid into a reaction tube or a centrifuge tube, adding 0.1 mL of water (a 5-mm column) and first seeing if the substance dissolves in the cold. If the substance is a solid, rub it well with a stirring rod and break up any lumps before drawing a conclusion.

If the substance is *readily soluble in cold water* and if the odor is suggestive of an amine, test the solution with pH paper and further determine if the odor disappears on addition of a few drops of 10% hydrochloric acid. If the properties are more like those of a salt, add a few drops of 10% sodium hydroxide solution. If the solution remains clear, addition of a little sodium chloride may cause separation of a liquid or solid amine.

If the substance is not soluble in cold water, see if it will dissolve on heating; be careful not to mistake the melting of a substance for dissolving. If it *dissolves in hot water*, add a few drops of 10% alkali and see if an amine precipitates. (If you are in doubt as to whether a salt has dissolved partially or not at all, pour off the supernatant liquid and make it basic.)

If the substance is *insoluble in hot water*, add 10% hydrochloric acid, heat if necessary, and see if it dissolves. If so, make the solution basic and see if an amine precipitates.

Cleaning Up   All solutions should be combined, made basic to free the amines, extracted with an organic solvent such as ligroin, the aqueous layer flushed down the drain, and the organic layer placed in the organic solvents container.

2. Hinsberg Test

If necessary, this experiment can be doubled in scale.
The procedure for distinguishing amines with benzenesulfonyl chlor-

ide is to be run in parallel on the following substances:

Aniline, $C_6H_5NH_2$ (bp 184°C)
N-Methylaniline, $C_6H_5NHCH_3$ (bp 194°C)
Triethylamine, $(CH_3CH_2)_3N$ (bp 90°C)

Primary and secondary amines react in the presence of alkali with benzenesulfonyl chloride, $C_6H_5SO_2Cl$, to give sulfonamides.

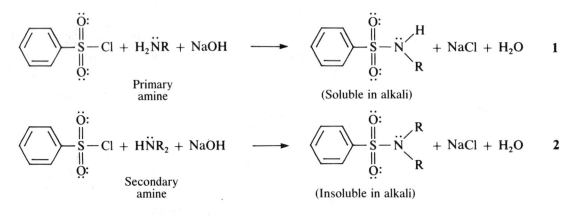

The sulfonamides are distinguishable because the derivative from a primary amine has an acidic hydrogen, which renders the product soluble in alkali (reaction 1); whereas the sulfonamide from a secondary amine is insoluble (reaction 2). Tertiary amines lack the necessary acidic hydrogen for formation of benzenesulfonyl derivatives.

Procedure

In a small test tube or centrifuge tube add 50 mg of the amine, 200 mg of benzenesulfonyl chloride, and 1 mL of methanol. Over the hot sand bath or a steam bath heat the mixture to just below the boiling point, cool, and add 2 mL of 6 M sodium hydroxide. Shake the mixture for 5 min and then allow the tube to stand for 10 more min with occasional shaking. If the odor of benzenesulfonyl chloride is detected, warm the mixture to hydrolyze it. Cool the mixture and acidify it by adding 6 M hydrochloric acid, dropwise, and with stirring. If a precipitate is seen at this point the amine is either primary or secondary. If no precipitate is seen, the amine is tertiary.

If a precipitate is present remove it by filtration on the Hirsch funnel, wash it with 2 mL of water, and transfer it to a reaction tube. Add 2.5 mL of 1.5 M sodium hydroxide solution and warm the mixture to 50°C. Shake the tube vigorously for 2 min. If the precipitate dissolves, the amine is primary. If it does not dissolve it is secondary. The sulfonamide of a primary amine can be recovered by acidifying the alkaline solution. Once dry these sulfonamides can be used to characterize the amine by their melting points.

Cleaning Up   The slightly acidic filtrate can be diluted and then flushed down the drain if the unknown is primary or secondary. If the unknown is tertiary, make the solution basic, extract the amine with ligroin, place the ligroin layer in the organic solvents container, and flush the aqueous layer down the drain.

3. Solid Derivatives

Acetyl derivatives of primary and secondary amines are usually solids suitable for melting point characterization and are readily prepared by reaction with acetic anhydride, even in the presence of water. Benzoyl and benzenesulfonyl derivatives are made by reaction of the amine with the appropriate acid chloride in the presence of alkali, as in Section 2 (the benzenesulfonamides of aniline and of *N*-methylaniline melt at 110°C and 79°C, respectively).

Solid derivatives suitable for characterization of tertiary amines are the methiodides and picrates:

$$R_3N \ + \ CH_3I \ \longrightarrow \ R_3\overset{+}{N}CH_3 \ \overset{I^-}{}$$

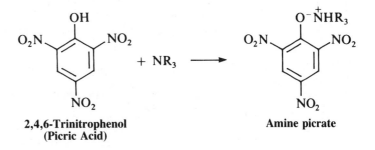

2,4,6-Trinitrophenol
(Picric Acid)   **Amine picrate**

Typical derivatives are to be prepared, and although determination of melting points is not necessary because the values are given, the products should be saved for possible identification of unknowns.

Acetylation of Aniline with Acetic Anhydride

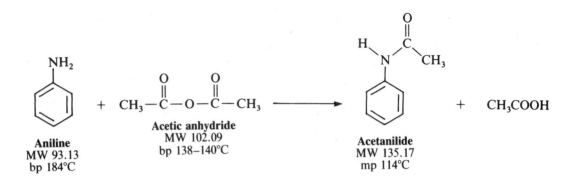

In a dry, small test tube place 46 mg (or 5 drops) of freshly distilled aniline, 51 mg (or 5 drops) of pure acetic anhydride, and a small boiling chip. Reflux the mixture for about 8 min, then add water to the mixture dropwise with heating until all of the product is in solution. This will require very little water. Allow the mixture to cool spontaneously to room temperature and then cool the mixture in ice. Remove the solvent and recrystallize the crude material again from boiling water. Isolate the crystals in the same way, cool the tube in ice water, and wash the crystals with a drop or two of ice-cold acetone. Remove the acetone with a Pasteur pipette and dry the product in the test tube. Once dry determine the weight, calculate the percent yield, and determine the melting point. With patience this reaction and the recrystallizations can be carried out in a melting point tube on one-tenth the indicated quantities of material using essentially the same techniques, and, if deemed necessary, it can be carried out on ten to twenty times as much material. On a larger scale, exercise caution when adding water to the reaction mixture because the exothermic hydrolysis of acetic anhydride is slow at room temperature but can heat up and get out of control.

Acetylation of Aniline Hydrochloride with Acetic Anhydride

If necessary, this experiment can be carried out on ten to twenty times as much material.

Dissolve 0.26 g of aniline hydrochloride in 2.5 mL of water, add 0.21 g of acetic anhydride followed immediately by 0.25 g of sodium acetate. Warm the mixture, cool it to room temperature, and then in ice. Recrystallize the product from water and wash it with acetone as described above. See Chapter 41 for mechanism.

Cleaning Up   The slightly acidic filtrate can be diluted and then flushed down the drain if the unknown is primary or secondary. If the unknown is tertiary, make the solution basic, extract the amine with ligroin, place the ligroin layer in the organic solvents container, and flush the aqueous layer down the drain.

Formation of an Amine Picrate from Picric Acid and Triethylamine

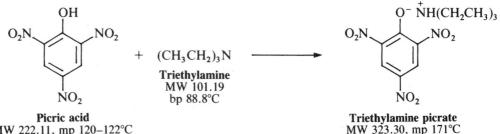

Picric acid
MW 222.11, mp 120–122°C

+ $(CH_3CH_2)_3N$

Triethylamine
MW 101.19
bp 88.8°C

Triethylamine picrate
MW 323.30, mp 171°C

CAUTION: Picric acid can explode if allowed to become dry. Use only the moist reagent (35% water) and do not allow the reagent to dry out. Your instructor may provide a stock solution containing 3 g of moist picric acid in 25 mL of methanol.

If necessary, this reaction can be doubled in scale.

Dissolve 30 mg of moist (35% water) picric acid in 0.25 mL of methanol and to the warm solution add 10.1 mg of triethylamine and let the solution stand to deposit crystals of the picrate, an amine salt that has a characteristic melting point. This picrate melts at 171°C.

Cleaning Up   The filtrate from this reaction and the product as well can be disposed of by dilution with a large volume of water and flushing down the drain. Larger quantities of moist picric acid (1 g) can be reduced with tin and hydrochloric acid to the corresponding triaminophenol.[1] Dry picric acid is said to be hazardous and should not be handled.

4. The NMR and IR Spectra of Amines

The proton bound to nitrogen can appear between 0.6 and 7.0 ppm on the nmr spectrum, the position depending upon solvent, concentration, and structure of the amine. The peak is sometimes extremely broad owing to slow exchange and interaction of the proton with the electric quadrupole of the nitrogen. If addition of a drop of D_2O to the sample causes the peak to disappear, this is evidence for an amine hydrogen; but alcohols, phenols, and enols will also exhibit this exchange behavior. See Fig. 1 for the nmr spectrum of aniline, in which the amine hydrogens appear as a sharp peak at 3.3 ppm. Infrared spectroscopy can also be very useful for identification purposes. Primary amines, both aromatic and aliphatic, show a weak doublet between 3300 and 3500 cm^{-1} and a strong absorption between 1560 and 1640 cm^{-1} due to NH bending (Fig. 2). Secondary amines show a single peak between 3310 and 3450 cm^{-1}. Tertiary amines have no useful infrared absorptions. In Chapter 21 the characteristic ultraviolet absorption shifts of aromatic amines in the presence and absence of acids were discussed.

1. *Hazardous Chemicals Information and Disposal Guide,* M. A. Armour, L. M. Browne, and G. L. Weir, p. 198.

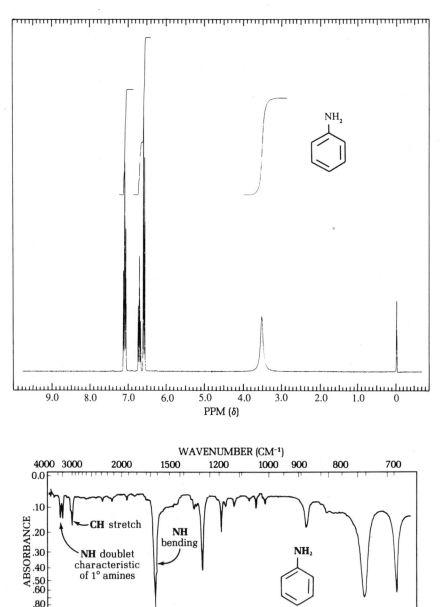

FIG. 1   The ¹H nmr spectrum of aniline (250 MHz). See Figure 1 in Chapter 41 for 60 MHz spectrum.

FIG. 2   Infrared spectrum of aniline in CS_2.

5. Unknowns (see Tables 5 and 6 in Chapter 70)

Determine first if the unknown is an amine or an amine salt and then determine whether the amine is primary, secondary, or tertiary. Complete identification of your unknown may be required.

Questions

1. How could you most easily distinguish between samples of 2-amino-naphthalene and acetanilide?

2. Would you expect the reaction product from benzenesulfonyl chloride and ammonia to be soluble or insoluble in alkali?

3. Is it safe to conclude that a substance is a tertiary amine because it forms a picrate?

4. Why is it usually true that amines that are insoluble in water are odorless?

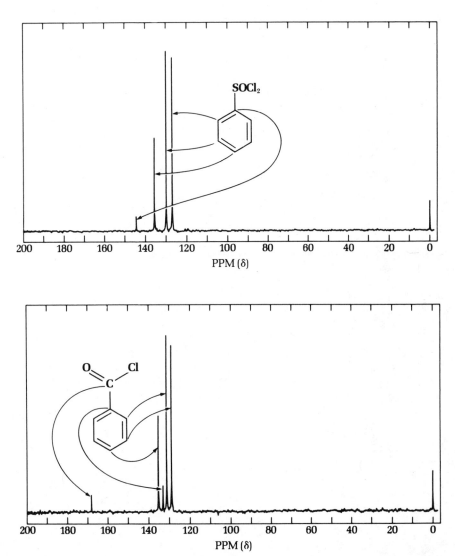

FIG. 3 ^{13}C nmr spectrum of benzenesulfonyl chloride (22.6 MHz).

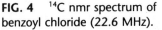

FIG. 4 ^{14}C nmr spectrum of benzoyl chloride (22.6 MHz).

5. Technical dimethylaniline contains traces of aniline and of methylaniline. Suggest a method for elimination of these impurities.

6. How would you prepare aniline from aniline hydrochloride?

7. Write a balanced equation for the reaction of benzenesulfonyl chloride with sodium hydroxide solution in the absence of an amine. What solubility would you expect the product to have in acid and base?

41

Sulfanilamide from Nitrobenzene

Prelab Exercise: Prepare detailed flow sheets for the three reactions. Look up the solubilities of all reagents and indicate in separation steps in which layer you would expect to find the desired product. Calculate the theoretical amount of ammonium hydroxide needed to react with 5 g of *p*-acetaminobenzenesulfonyl chloride to form sulfanilamide.

Paul Ehrlich, the father of immunology and chemotherapy, discovered Salvarsan, an arsenical "magic bullet" (a favorite phrase of his) used to treat syphilis. He hypothesized at the beginning of this century that it might be possible to find a dye that would selectively stain, or dye, a bacterial cell and thus destroy it. In 1932 I.G. Farbenindustrie patented a new azo dye, Prontosil, which they put through routine testing for chemotherapeutic activity when it was noted it had particular affinity for protein fibers like silk.

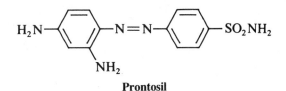

Prontosil

The dye was found to be effective against streptococcal infections in mice, but somewhat surprisingly, ineffective *in vitro* (outside the living animal). A number of other dyes were tested, but only those having the

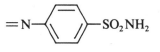

group were effective. French workers hypothesized that the antibacterial activity had nothing to do with the identity of the compounds as dyes, but rather with the reduction of the dyes in the body to *p*-aminobenzenesulfonamide, known commonly as sulfanilamide.

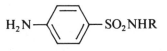

Sulfanilamide   R＝H

381

On the basis of this hypothesis sulfanilamide was tested and found to be the active substance.

Because sulfanilamide had been synthesized in 1908, its manufacture could not be protected by patents, so the new drug and thousands of its derivatives were rapidly synthesized and tested. When the R group in sulfanilamide is replaced with a heterocyclic ring system—pyridine, thiazole, diazine, merazine, etc.—the sulfa drug so produced is often faster-acting or less toxic than sulfanilamide. Although they have been supplanted for the most part by antibiotics of microbial origin, these drugs still find wide application in chemotherapy.

Unlike that of most drugs, the mode of action of the sulfa drugs is now completely understood. Bacteria must synthesize folic acid in order to grow. Higher animals, like man, do not synthesize this vitamin and hence must acquire it in their food. Sulfanilamide inhibits the formation of folic acid, stopping the growth of bacteria; and because the synthesis of folic acid does not occur in man, only the bacteria are affected.

A closer look at these events reveals that bacteria synthesize folic acid using several enzymes, including one called dihydropteroate synthetase, which catalyzes the attachment of *p*-aminobenzoic acid to a pteridine ring system. When sulfanilamide is present it competes with the *p*-aminobenzoic acid (note the structural similarity) for the active site on the enzyme. This activity makes it a competitive inhibitor. Once this site is occupied on the enzyme, folic acid synthesis stops and bacterial growth stops. Folic acid can also be synthesized in the laboratory.[1]

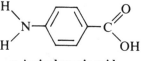

p-**Aminobenzoic acid**

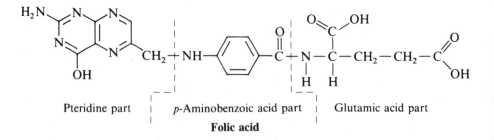

Pteridine part | *p*-Aminobenzoic acid part | Glutamic acid part

Folic acid

This experiment is a multistep synthesis of sulfanilamide starting with nitrobenzene. Nitrobenzene is reduced with tin and hydrochloric acid to

1. L. T. Plante, K. L. Williamson, and E. J. Pastore, "Methods in Enzymology," Vol. 66, D. B. McCormick and L. D. Wright, eds., Academic Press, New York, 1980, p. 533.

give the anilinium ion, which is converted to aniline with base:

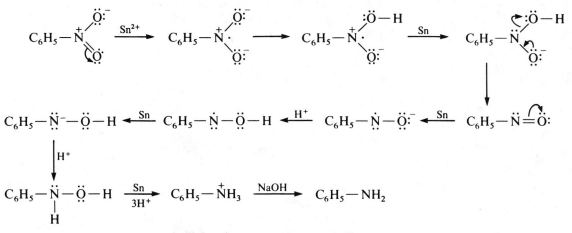

$$C_6H_5-\overset{\underset{\displaystyle |}{H}}{\underset{\displaystyle}{N}}-\overset{..}{\underset{..}{O}}-H \xrightarrow[3H^+]{Sn} C_6H_5-\overset{+}{N}H_3 \xrightarrow{NaOH} C_6H_5-NH_2$$

Anilinium ion   **Aniline**

The acetylation of aniline in aqueous solution to give acetanilide serves to protect the amine group from reaction with chlorosulfonic acid. The acetylation takes place readily in aqueous solution. Aniline reacts with acid to give the water-soluble anilinium ion:

$$C_6H_5-\overset{..}{N}H_2 + H^+ \longrightarrow C_6H_5-\overset{+}{N}H_3$$

Aniline   **Anilinium ion**

The anilinium ion reacts with acetate ion to set up an equilibrium that liberates a small quantity of aniline:

$$C_6H_5-\overset{+}{N}H_3 + CH_3-\overset{\displaystyle \overset{..}{O}:}{\underset{\displaystyle \overset{..}{O}:}{C}} \rightleftharpoons CH_5-\overset{..}{N}H_2 + CH_3-\overset{\displaystyle \overset{..}{O}:}{\underset{\displaystyle \overset{..}{O}-H}{C}}$$

This aniline reacts with acetic anhydride to give water-insoluble acetanilide:

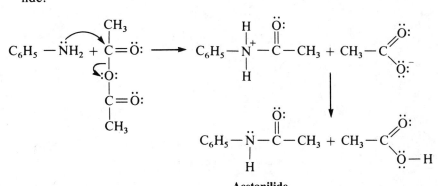

Acetanilide

This upsets the equilibrium, releasing more aniline, which can then react with acetic anhydride to give more acetanilide.

Acetanilide reacts with chlorosulfonic acid in an electrophilic aromatic substitution:

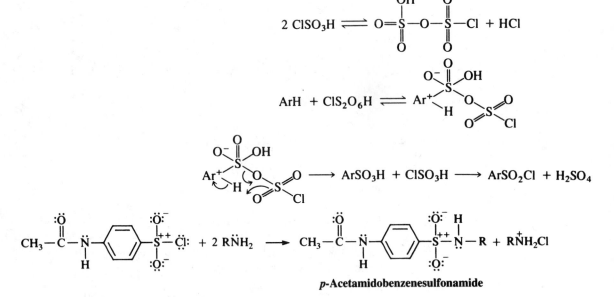

The protecting amide group is removed from the *p*-acetamidobenzenesulfonamide by acid hydrolysis. The amide group is more easily hydrolyzed than the sulfonamide group:

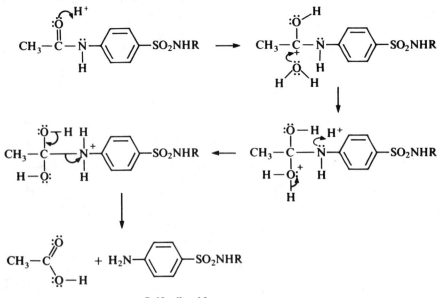

Sulfanilamide

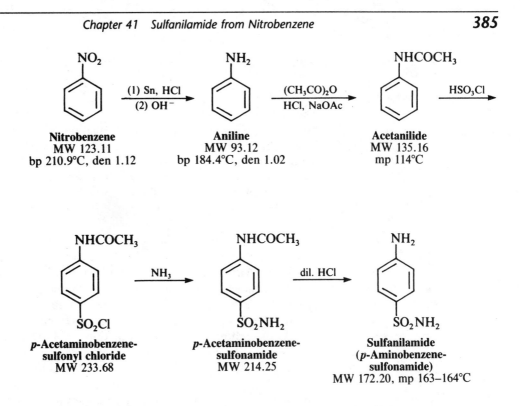

In the present experiment nitrobenzene is reduced to aniline by tin and hydrochloric acid. A double salt with tin having the formula $(C_6H_5NH_3)_2SnCl_4$ separates partially during the reaction, and at the end it is decomposed by addition of excess alkali, which converts the tin into water-soluble stannite or stannate (Na_2SnO_2 or Na_2SnO_3). The aniline liberated is separated from inorganic salts and the insoluble impurities derived from the tin by steam distillation and is then dried, distilled, and acetylated with acetic anhydride in aqueous solution. Treatment of the resulting acetanilide with excess chlorosulfonic acid effects substitution of the chlorosulfonyl group and affords *p*-acetaminobenzenesulfonyl chloride. The alternative route to this intermediate via sulfanilic acid is unsatisfactory, because sulfanilic acid being dipolar is difficult to acetylate. In both processes the amino group must be protected by acetylation to permit formation of the acid chloride group. The next step in the synthesis is ammonolysis of the sulfonyl chloride and the terminal step is removal of the protective acetyl group.

Use the total product obtained at each step as starting material for the next step and adjust the amounts of reagents accordingly. Keep a record of your working time. Aim for a high overall yield of pure final product in the shortest possible time. Study the procedures carefully before coming to the laboratory so that your work will be efficient. A combination of consecutive steps that avoids a needless isolation saves time and increases the yield.

Experiments

1. Preparation of Aniline

Measure nitrobenzene (toxic) in the hood. Do not breathe the vapors.

The reduction of the nitrobenzene is carried out in a 500-mL round-bottomed flask suitable for steam distillation of the reaction product. Put 25 g of granulated tin and 12.0 g of nitrobenzene in the flask, make an ice-water bath ready, add 55 mL of concentrated hydrochloric acid, insert a thermometer, and swirl well to promote reaction in the three-phase system.

Handle aniline with care. It may be a carcinogen.

Let the mixture react until the temperature reaches 60°C and then cool briefly in ice just enough to prevent a rise above 60°C, so the reaction will not get out of hand. Continue to swirl, cool as required, and maintain the temperature in the range 55–60°C for 15 min. Remove the thermometer and rinse it with water, fit the flask with a reflux condenser, and heat on the steam bath with frequent swirling until droplets of nitrobenzene are absent from the condenser, and the color due to an intermediate reduction product is gone (about 15 min). During this period dissolve 40 g of sodium hydroxide in 100 mL of water and cool to room temperature.

Handle the hot solution of sodium hydroxide with care; it is very corrosive.

Suitable point of interruption

At the end of the reduction reaction, cool the acid solution in ice (to prevent volatilization of aniline) during gradual addition of the solution of alkali. This alkali neutralizes the aniline hydrochloride, releasing aniline, which will now be volatile in steam. Attach a stillhead with steam-inlet tube, condenser, adapter, and receiving Erlenmeyer flask (Fig. 3 in Chapter 6); heat the flask with a microburner to prevent the flask from filling with water from condensed steam, and proceed to steam distill. Since aniline is fairly soluble in water (3.6 g/100 g$^{18°}$) distillation should be continued somewhat beyond the point where the distillate has lost its original turbidity (50–60 mL more). Make an accurate estimate of the volume of distillate by filling a second flask with water to the level of liquid in the receiver and measuring the volume of water.

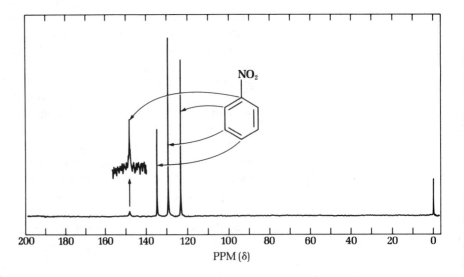

FIG. 1 ^{13}C nmr spectrum of nitrobenzene (22.6 MHz).

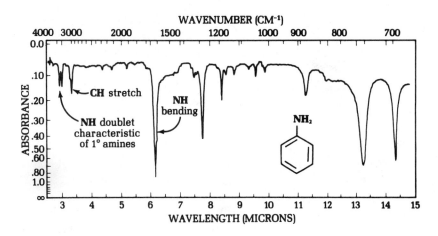

FIG. 2   Infrared spectrum of aniline.

Isolation of Aniline — Alternative Choices

Salting out aniline

At this point aniline can be isolated. You could reduce the solubility of aniline by dissolving in the steam distillate 0.2 g of sodium chloride per milliliter, extract the aniline with 2–3 portions of dichloromethane, dry the extract, distill the dichloromethane (bp 41°C), and then distill the aniline (bp 184°C). Or the aniline can be converted directly to acetanilide. The procedure calls for pure aniline, but note that the first step is to dissolve the aniline in water and hydrochloric acid. Your steam distillate is a mixture of aniline and water, both of which have been distilled. Are they not both water-white and presumably pure? Hence, an attractive procedure would be to assume that the steam distillate contains the theoretical amount of aniline and to add to it, in turn, appropriate amounts of hydrochloric acid, acetic anhydride, and sodium acetate, calculated from the quantities given in Experiment 2.

Cleaning Up   Filter the pot residue from the steam distillation through a layer of Celite filter aid on a Büchner funnel. Discard the tin salts in the nonhazardous solid waste container. The filtrate, after neutralization with hydrochloric acid, can be flushed down the drain.

2. Acetanilide

Acetylation in Aqueous Solution

Handle aniline and acetic anhydride under hood; avoid contact with skin.

Dissolve 5.0 g (0.054 mole) of aniline in 135 mL of water and 4.5 mL (0.054 mole) of concentrated hydrochloric acid, and, if the solution is colored, filter it by suction through a pad of decolorizing charcoal. Measure out 6.2 mL (0.065 mole) of acetic anhydride, and also prepare a solution of 5.3 g (0.065 mole) of anhydrous sodium acetate in 30 mL of water. Add the acetic anhydride to the solution of aniline hydrochloride with stirring and at once add the sodium acetate solution. Stir, cool in ice, and collect the product. It should be colorless and the mp close to 114°C. Since the acetanilide *must be*

FIG. 3   Drying of acetanilide under reduced pressure. Heat flask on steam bath.

completely dry for use in the next step, it is advisable to put the material in a tared 125-mL Erlenmeyer flask and to heat this on the steam bath under evacuation until the weight is constant. (See Fig. 3.)

Cleaning Up   The aqueous filtrate can be flushed down the drain with a large excess of water.

3. Sulfanilamide

The chlorosulfonation of acetanilide in the preparation of sulfanilamide is conducted without solvent in the 125-mL Erlenmeyer flask used for drying the precipitated acetanilide from Procedure 2. Because the reaction is most easily controlled when the acetanilide is in the form of a hard cake, the dried solid is melted by heating the flask over a hot plate; as the melt cools, the flask is swirled to distribute the material as it solidifies over the lower walls of the flask. Let the flask cool while making provision for trapping the

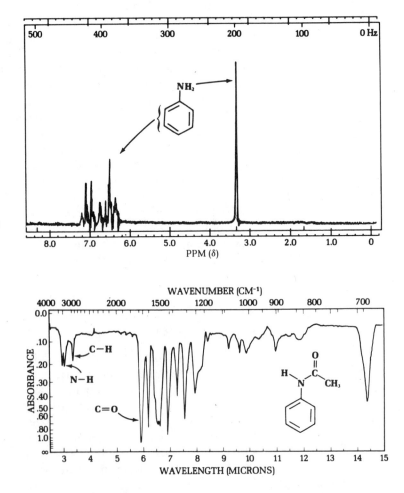

FIG. 4   ¹H nmr spectrum of aniline (60 MHz). See Fig. 1 in Chapter 40 for 250 MHz spectrum.

FIG. 5   Infrared spectrum of acetanilide in CHCl₃.

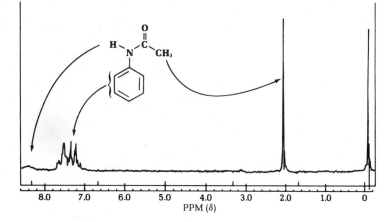

FIG. 6 ¹H nmr spectrum of acetanilide. The amide proton shows a characteristically broad peak (60 MHz).

hydrogen chloride evolved in the chlorosulfonation. Fit the Erlenmeyer with a stopper connected by a section of rubber tubing to a glass tube fitted with a stopper into the neck of a 250-mL filter flask half-filled with water. The tube should be about 1 cm above the surface of the water and *must not dip into the water*. See Fig. 7. Cool the flask containing the acetanilide thoroughly in an ice-water bath, and for 5.0 g of acetanilide measure 12.5 mL of chlorosulfonic acid in a graduate (supplied with the reagent and kept away from water). Add the reagent in 1–2 mL portions with a capillary dropping tube, and connect the flask to the gas trap. The flask is now removed from the ice bath and swirled until a part of the solid has dissolved and the evolution of hydrogen chloride is proceeding rapidly. Occasional

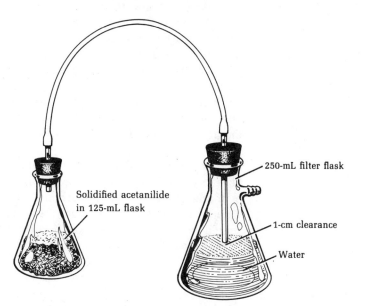

Solidified acetanilide in 125-mL flask

250-mL filter flask

1-cm clearance

Water

FIG. 7 Chlorosulfonation apparatus fitted with HCl gas trap.

$$HSO_3Cl + H_2O \rightarrow$$
$$H_2SO_4 + HCl$$

cooling in ice may be required to prevent too brisk a reaction. In 5–10 min the reaction subsides and only a few lumps of acetanilide remain undissolved. When this point has been reached, heat the mixture on the steam bath for 10 min to complete the reaction, cool the flask under the tap, and deliver the oil by drops with a capillary dropper while stirring it into 75 mL of ice water contained in a beaker cooled in an ice bath (hood). Use extreme caution when adding the oil to ice water and when rinsing out any containers that have held chlorosulfonic acid. Rinse the flask with cold water and stir the precipitated *p*-acetaminobenzenesulfonyl chloride for a few minutes until an even suspension of granular white solid is obtained. Collect and wash the solid on a Büchner funnel. After pressing and draining the filter

Do not let the mixture stand before addition of ammonia.

cake, transfer the solid to the rinsed reaction flask, add (for 5 g of aniline) 15 mL of concentrated aqueous ammonia solution and 15 mL of water, and heat the mixture over a flame with occasional swirling (hood). Maintain the temperature of the mixture just below the boiling point for 5 min. During this treatment a change can be noted as the sulfonyl chloride undergoes transformation to a more pasty suspension of the amide. Cool the suspension well in an ice bath, collect the *p*-acetaminobenzenesulfonamide by suction filtration, press the cake on the funnel, and allow it to drain thoroughly. Any excess water will unduly dilute the acid used in the next step.

Transfer the still moist amide to the well-drained reaction flask, add 5 mL of concentrated hydrochloric acid and 10 mL of water (for 5 g of aniline), boil the mixture gently until the solid has all dissolved (5–10 min), and then continue the heating at the boiling point for 10 min longer (do not evaporate to dryness). The solution when cooled to room temperature should deposit no solid amide, but, if it is deposited, heating should be continued for a further period. The cooled solution of sulfanilamide hydrochloride is shaken with decolorizing charcoal and filtered by suction. Place the solution in a beaker and cautiously add an aqueous solution of 5 g of sodium bicarbonate with stirring to neutralize the hydrochloride. After the foam has subsided, test the suspension with litmus, and, if it is still acidic, add more bicarbonate until the neutral point is reached. Cool thoroughly in ice and collect the granular, white precipitate of sulfanilamide. The crude product (mp 161–163°C) on crystallization from alcohol or water affords pure sulfanilamide, mp 163–164°C, with about 90% recovery. Determine the mp and calculate the overall yield from the starting material.

Cleaning Up   Add the water from the gas trap to the combined aqueous filtrates from all reactions and neutralize the solution by adding either 10% hydrochloric acid or sodium carbonate. Flush the neutral solution down the drain with a large excess of water. Any spilled drops of chlorosulfonic acid should be covered with sodium carbonate, the powder collected in a beaker, dissolved in water, and flushed down the drain.

Questions

1. Why is an acetyl group added to aniline (making acetanilide) and then removed to regenerate the amine group in sulfanilamide?

2. What happens when chlorosulfonic acid comes in contact with water?

3. Acetic anhydride, like any anhydride, reacts with water to form a carboxylic acid. How then is it possible to carry out an acetylation in aqueous solution? What is the purpose of the hydrochloric acid and the sodium acetate in this reaction?

4. What happens when *p*-acetaminobenzenesulfonyl chloride is allowed to stand for some time in contact with water?

42

The Sandmeyer Reaction: 4-Chlorotoluene and 2-Iodobenzoic Acid

Prelab Exercise: Outline the steps necessary to prepare 4-bromotoluene, 4-iodotoluene, and 4-fluorotoluene from benzene.

The Sandmeyer reaction is a versatile means of replacing the amine group of a primary aromatic amine with a number of different substituents:

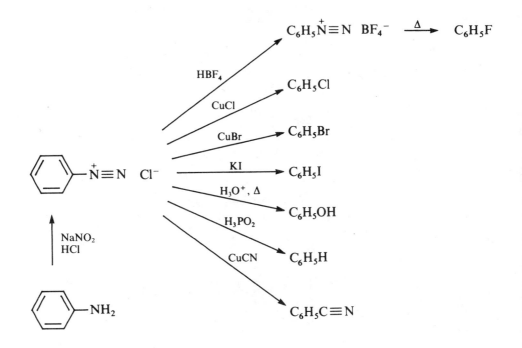

The diazonium salt is formed by the reaction of nitrous acid with the amine in acid solution. Nitrous acid is not stable and must be prepared *in situ*; in strong acid it dissociates to form nitroso ions, ^{+}NO, which attack the nitrogen of the amine. The intermediate so formed loses a proton, rearranges, and finally loses water to form the resonance-stabilized diazonium ion.

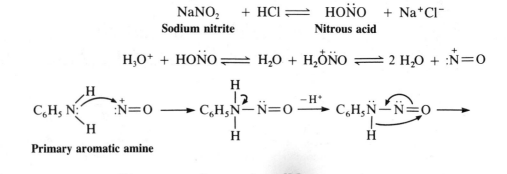

$$NaNO_2 \ + \ HCl \rightleftharpoons \ HO\ddot{N}O \ + \ Na^+Cl^-$$

Sodium nitrite   Nitrous acid

$$H_3O^+ + HO\ddot{N}O \rightleftharpoons H_2O + H_2\overset{+}{\ddot{O}}NO \rightleftharpoons 2\,H_2O + :\overset{+}{N}=O$$

Primary aromatic amine

$$C_6H_5-\ddot{N}=\ddot{N}-OH \overset{H^+}{\rightleftharpoons} C_6H_5-\overset{\frown}{N}\overset{..}{=}\overset{..}{N}-\overset{+}{O}H_2 \overset{-H_2O}{\rightleftharpoons} [C_6H_5-\overset{..}{N}\equiv N \leftrightarrow C_6H_5-\ddot{N}=\overset{+}{\ddot{N}}]$$

Diazonium ion

The diazonium ion is reasonably stable in aqueous solution at 0°C; on warming up it will form phenol, as seen on the previous page. A versatile functional group, it will undergo all of the reactions depicted above as well as couple to aromatic rings activated with substituents such as amino and hydroxyl groups to form the huge class of azo dyes (see Chapter 66).

Benzenediazonium   N,N-Dimethylaniline   A diazo compound
chloride

Diazonium salts are not ordinarily isolated, because the dry solid is explosive.

Two Sandmeyer Reactions

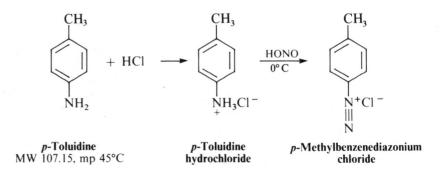

p-Toluidine   *p*-Toluidine   *p*-Methylbenzenediazonium
MW 107.15, mp 45°C   hydrochloride   chloride

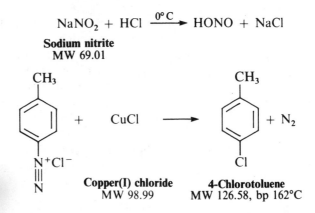

$$NaNO_2 + HCl \xrightarrow{0°C} HONO + NaCl$$

Sodium nitrite
MW 69.01

Copper(I) chloride
MW 98.99

4-Chlorotoluene
MW 126.58, bp 162°C

CAUTION: p-Toluidine, like many aromatic amines, is highly toxic. o-Toluidine is not only highly toxic but also a cancer suspect agent.

The first experiment is a synthesis of 4-chlorotoluene. *p*-Toluidine is dissolved in the required amount of hydrochloric acid, two more equivalents of acid are added, and the mixture cooled in ice to produce a paste of the crystalline amine hydrochloride. When this salt is treated at 0–5°C with one equivalent of sodium nitrite, nitrous acid is liberated and reacts to produce the diazonium salt. The excess hydrochloric acid beyond the two equivalents required to form the amine hydrochloride and react with sodium nitrite maintains acidity sufficient to prevent formation of the diazoamino compound and rearrangement of the diazonium salt.

Copper(I) chloride is made by reduction of copper(II) sulfate with sodium sulfite (which is produced as required from the cheaper sodium bisulfite). The white solid is left covered with the reducing solution for protection against air oxidation until it is to be used and then dissolved in hydrochloric acid. On addition of the diazonium salt solution, a complex forms and rapidly decomposes to give *p*-chlorotoluene and nitrogen. The mixture is very discolored, but steam distillation leaves most of the impurities and all salts behind and gives material substantially pure except for the presence of a trace of yellow pigment, which can be eliminated by distillation of the dried oil.

The third experiment employs potassium iodide, which reacts with a diazonium salt to give an aryl iodide.

Experiments

1. Copper(I) Chloride Solution

$$2CuSO_4 \cdot 5H_2O + 4NaCl + NaHSO_3 + NaOH \rightarrow 2CuCl + 3Na_2SO_4 + 2HCl + 10H_2O$$

MW 249.71   MW 58.45   MW 104.97   MW 40.01   MW 99.02

In a 500-mL round-bottomed flask (to be used later for steam distillation) dissolve 30 g of copper(II) sulfate crystals (CuSO$_4 \cdot$ 5H$_2$O) in 100 mL of water by boiling and then add 10 g of sodium chloride, which may give a

NaHSO$_3$,
Sodium bisulfite, Sodium hydrogen sulfite not Na$_2$S$_2$O$_4$

One may stop here

small precipitate of basic copper(II) chloride. Prepare a solution of sodium sulfite from 7 g of sodium bisulfite, 4.5 g of sodium hydroxide, and 50 mL of water, and add this, not too rapidly, to the hot copper(II) sulfate solution (rinse flask and neck). Shake well and put the flask in a pan of cold water in a slanting position favorable for decantation and let the mixture stand to cool and settle during the diazotization. When you are ready to use the copper(I) chloride, decant the supernatant liquid, wash the white solid once with water by decantation, and dissolve the solid in 45 mL of concentrated hydrochloric acid. The solution is susceptible to air oxidation and should not stand for an appreciable time before use.

2. Diazotization of *p*-Toluidine

CAUTION: p-Toluidine is toxic. Handle with care.

Put 11.0 g of *p*-toluidine and 15 mL of water in a 125-mL Erlenmeyer flask. Measure 25 mL of concentrated hydrochloric acid and add 10 mL of it to the flask. Heat over a hot plate and swirl to dissolve the amine and hence ensure that it is all converted into the hydrochloride. Add the remaining acid and cool thoroughly in an ice bath and let the flask stand in the bath while preparing a solution of 7 g of sodium nitrite in 20 mL of water. To maintain a temperature of 0–5°C during diazotization, add a few pieces of ice to the amine hydrochloride suspension and add more later as the first ones melt. Pour in the nitrite solution in portions during 5 min with swirling in the ice bath. The solid should dissolve to a clear solution of the diazonium salt. After 3–4 min test for excess nitrous acid: dip a stirring rod in the solution, touch off the drop on the wall of the flask, put the rod in a small test tube, and add a few drops of water. Then insert a strip of starch-iodide paper; an instantaneous deep blue color due to a starch-iodine complex indicates the desirable presence of a slight excess of nitrous acid. (The sample tested is diluted with water because strong hydrochloric acid alone produces the same color on starch-iodide paper, after a slight induction period.) Leave the solution in the ice bath.

3. Sandmeyer Reaction: 4-Chlorotoluene

Complete the preparation of copper(I) chloride solution, cool it in the ice bath, pour in the solution of diazonium chloride through a long-stemmed funnel, and rinse the flask. Swirl occasionally at room temperature for 10 min and observe initial separation of a complex of the two components and its decomposition with liberation of nitrogen and separation of an oil. Arrange for steam distillation (Fig. 3 in Chapter 6) or generate steam *in situ* by simply boiling the flask contents with a Thermowell using the apparatus for simple distillation (Fig. 5 in Chapter 5) (add more water during the distillation). Do not start the distillation until bubbling in the mixture has practically ceased and an oily layer has separated. Then steam distill (see Chapter 6) and note that 4-chlorotoluene, although lighter than the solution

Dispose of copper salts and
solutions in the container
provided

CAUTION: *Use aspirator tube*

of inorganic salts in which it was produced, is heavier (den 1.07) than water. Extract the distillate with a little ether, wash the extract with 3 M sodium hydroxide solution to remove any *p*-cresol present, then wash with saturated sodium chloride solution; dry the ether solution over about 5 g of anhydrous sodium sulfate and filter or decant it into a tared flask, evaporate the ether, and determine the yield and percentage yield of product (your yield should be about 9 g). Pure *p*-chlorotoluene, the infrared and nmr spectra of which are shown in Figs. 1–3, is obtained by simple distillation of this crude product. Analyze your crude product by TLC and infrared spectroscopy. Is it pure?

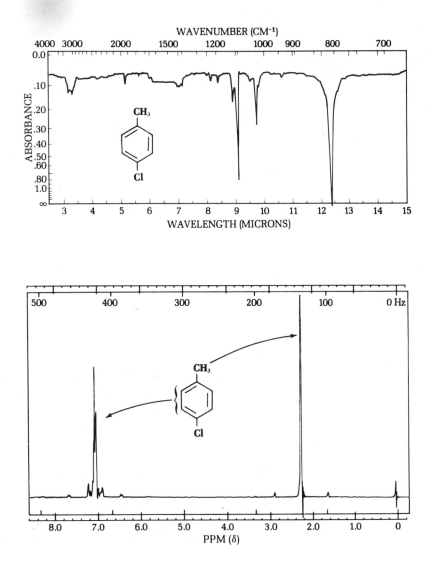

FIG. 1   Infrared spectrum of 4-chlorotoluene in CS_2.

FIG. 2   ¹H nmr spectrum of *p*-chlorotoluene (60 MHz).

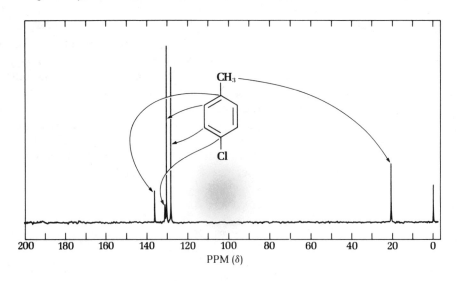

FIG. 3 ¹³C nmr spectrum of
p-chlorotoluene (22.6 MHz).

Cleaning Up   Combine the pot residue from the steam distillation with
the aqueous washings, neutralize with sodium carbonate, dilute with water,
and flush down the drain. Allow the ether to evaporate from the sodium
sulfate in the hood and then place it in the nonhazardous solid waste
container.

4. 2-Iodobenzoic Acid

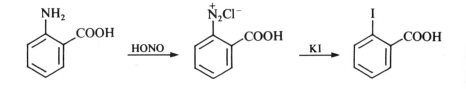

A 500-mL round-bottomed flask containing 13.7 g of anthranilic acid,
100 mL of water, and 25 mL of concentrated hydrochloric acid is heated
until the solid is dissolved. The mixture is then cooled in ice while bubbling
in nitrogen to displace the air. When the temperature reaches 0–5°C a
solution of 7.1 g of sodium nitrite is added slowly. After 5 min a solution of
17 g of potassium iodide in 25 mL of water is added, when a brown complex
partially separates. The mixture is let stand without cooling for 5 min
(under nitrogen) and then warmed to 40°C, at which point a vigorous
reaction ensues (gas evolution, separation of a tan solid). After reacting for
10 min the mixture is heated on the steam bath for 10 min and then cooled in
ice. A pinch of sodium bisulfite is added to destroy any iodine present and
the granular tan product collected and washed with water. The still moist
product is dissolved in 70 mL of 95% ethanol, 35 mL of hot water is added,

and the brown solution is treated with decolorizing charcoal, filtered, diluted at the boiling point with 35–40 mL of water, and let stand. 2-Iodobenzoic acid separates in large, slightly yellow needles of satisfactory purity (mp 164°C) for the experiment; yield is approximately 17 g (71%).

Cleaning Up   The reaction mixture, filtrate and mother liquor from the crystallization are combined, neutralized with sodium carbonate, and flushed down the drain with a large excess of water. Norit is placed in the nonhazardous solid waste container.

Questions

1. Nitric acid is generated by the action of sulfuric acid on sodium nitrate. Nitrous acid is prepared by the action of hydrochloric acid on sodium nitrite. Why is nitrous acid prepared *in situ*, rather than obtained from the reagent shelf?

2. What by-product would be obtained in high yield if the diazotization of *p*-toluidine were carried out at 30°C instead of 0–5°C?

3. How would 4-bromoaniline be prepared from benzene?

Malonic Ester Synthesis: Synthesis of a Barbiturate

Prelab Exercise: Which barbiturate discussed in this chapter cannot be synthesized by the acetoacetic ester reaction?

Barbiturates are central nervous system depressants used as hypnotic drugs and anesthetics. They are all derivatives of barbituric acid (R=R'=H), which has no sedative properties. It is called an acid because the carbonyl groups render the imide hydrogens acidic:

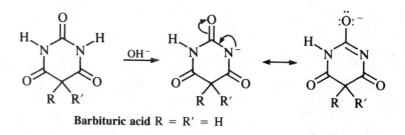

Barbituric acid R = R' = H

Barbituric acid was first synthesized in 1864 by Adolph von Baeyer. It apparently was named at a tavern on St. Barbara's day and is derived from urea. At the turn of the century the great chemist Emil Fischer synthesized the first hypnotic (sleep-inducing) barbiturate, the 5,5-diethyl derivative, at the direction of von Mering. Von Mering, who made the seminal discovery that removal of the pancreas causes diabetes, named the new derivative of barbituric acid Veronal because he regarded Verona as the most restful city on earth.

Barbiturates are the most widely used sleeping pills and are classified according to their duration of action. Because it is quite easy to put almost any conceivable R group onto diethyl malonate, several thousand derivatives of barbituric acid have been synthesized. Studies of these derivatives have shown that as the alkyl chain, R, gets longer or as double bonds are introduced into the chain, the duration of action and the time of onset of action decreases. Maximum sedation occurs when the alkyl chains contain five or six carbons, as found in amobarbital (R=ethyl, R'=3-methylbutyl) and pentobarbital (R=ethyl, R'=1-methylbutyl).

Veronal

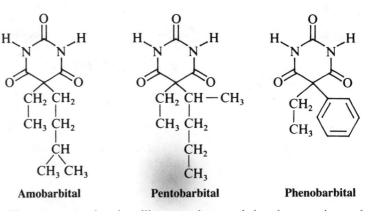

Amobarbital   **Pentobarbital**   **Phenobarbital**

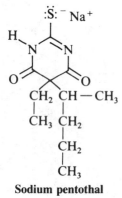

Sodium pentothal

These two molecules illustrate how subtle changes in molecular structure can affect action. Amobarbital requires 30 min to take effect and sedation lasts for 5–6 h, while pentobarbital takes effect in 15 min and sedation lasts only 2–3 h. Phenobarbital (R=ethyl, R'=phenyl), on the other hand, requires over an hour to take effect, but sedation lasts for 6–10 h. When the alkyl chains are made much longer the sedative properties decrease and the substances become anticonvulsants, which are used to treat epileptic seizures. If the alkyl group is too long or is substituted at one of the two nitrogens, convulsants are produced.

The usual dosage is 10–50 mg per pill. Continuous use of barbiturates leads to physiologic dependence (addiction) and withdrawal symptoms are just like those experienced by a heroin addict. Replacing the oxygen atom at carbon-2 with sulfur results in the compound pentothal. Given intravenously as the sodium salt, it is a fast-acting general anesthetic. Like all general anesthetics, the details of its mode of action are unknown. In low, subanesthetic doses sodium pentothal reduces inhibitions and the will to resist and thus functions as the so-called "truth serum," apparently because it takes less mental effort to tell the truth than to prevaricate.

Most barbiturates are made from diethyl malonate. The methylene protons between the two carbonyl groups are acidic and will give a highly stabilized enolate anion.

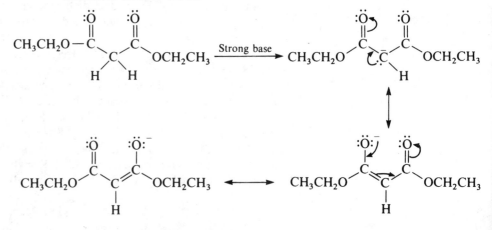

The acidic protons can be removed with a strong base, most often sodium ethoxide in dry ethanol. In the present experiment carbonate functions as the strong base because, in the absence of water, it is not solvated. In association with the tricaprylmethylammonium ion, it is soluble in the organic phase and can react with the diethyl malonate.

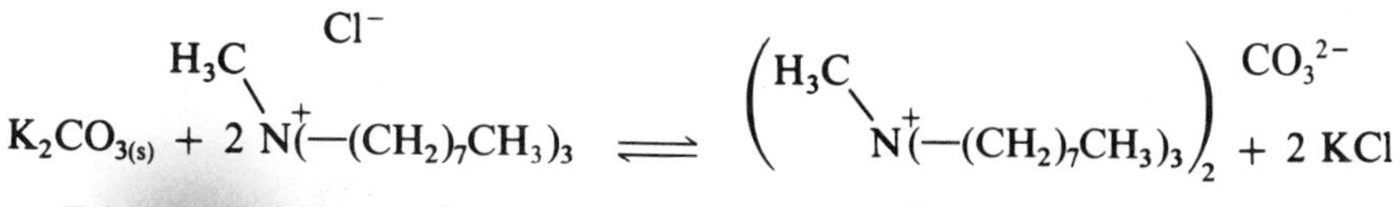

$$K_2CO_{3(s)} + 2 \; \underset{\substack{\textbf{Tricaprylmethylammonium} \\ \textbf{chloride} \\ \text{Phase transfer catalyst}}}{\overset{\overset{\displaystyle H_3C \diagdown \quad Cl^-}{N^+\!(\!-\!(CH_2)_7CH_3)_3}}{}} \; \rightleftharpoons \; \left(\overset{H_3C \diagdown}{\underset{N^+\!(\!-\!(CH_2)_7CH_3)_3}{}} \right)_2 \overset{CO_3^{2-}}{} + 2\,KCl$$

Soluble in organic phase

The enolate anion of diethyl malonate can be alkylated by an S_N2 displacement of bromide to give diethyl *n*-butylmalonate:

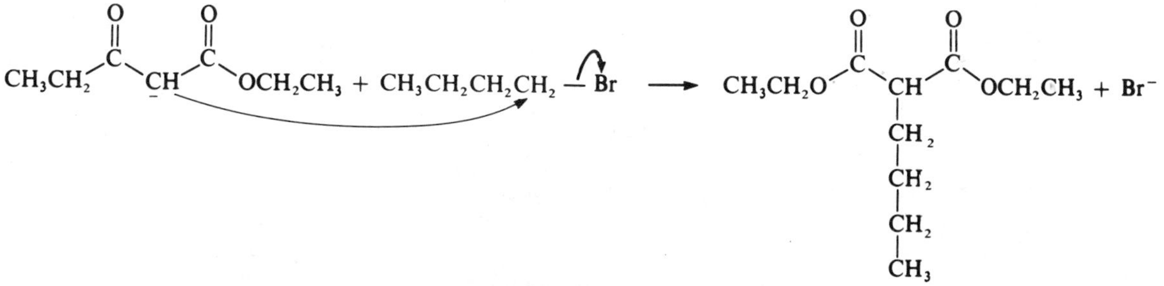

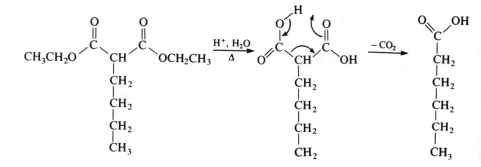

and because the product still contains one acidic proton, the process can be repeated using the same or a different alkyl halide. The product above could be hydrolyzed and decarboxylated to give a carboxylic acid:

In this experiment the diethyl *n*-butylmalonate is allowed to react with urea in the presence of a strong base, sodium ethoxide, to give the barbiturate:

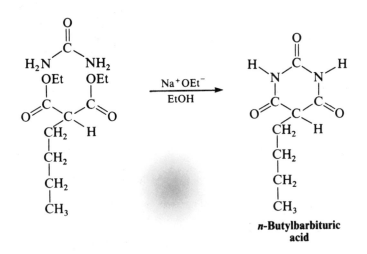

n-**Butylbarbituric acid**

Barbiturates with only one alkyl substituent are rapidly degraded by the body and therefore are physiologically inactive.

Experiments

1. Diethyl *n*-Butylmalonate

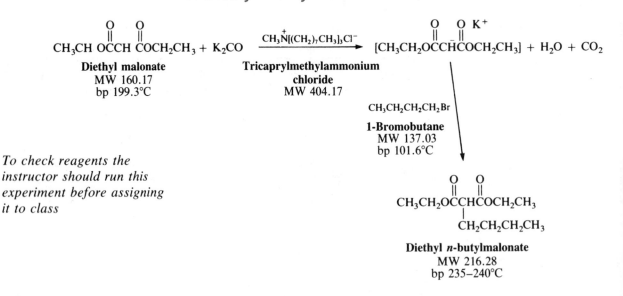

To check reagents the instructor should run this experiment before assigning it to class

CAUTION: Tricaprylmethyl-ammonium chloride (Aliquat 336) is a toxic irritant. Handle with care.

Into a 50-mL round-bottomed flask weigh 0.8 g of tricaprylmethyl-ammonium chloride, 8.00 g of diethyl malonate, 7.53 g of 1-bromobutane, and 8.3 g of anhydrous potassium carbonate. Attach a water-cooled reflux condenser and reflux the reaction mixture for 1.5 h (Fig. 1). Allow the

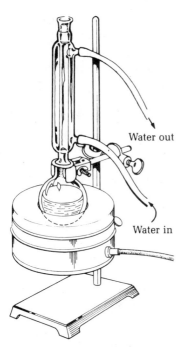

FIG. 1   Macroscale reflux apparatus.

CAUTION: Handle sodium metal with care. It will react violently with water and can cause fire or an explosion. Avoid all contact with moisture in any form.

mixture to cool somewhat and then transfer it, using 30 mL of water to rinse out the flask, to a small separatory funnel. Shake the mixture thoroughly and draw off the aqueous layer after the layers have separated. Place the organic layer in a 50-mL Erlenmeyer flask and extract the aqueous layer with two 10-mL portions of diethyl ether. Dry the combined extracts over anhydrous sodium sulfate (about 10 g), decant the liquid into a 100-mL round-bottomed flask, and evaporate the solvent on the rotary evaporator. Distill the residue from a 25-mL round-bottomed flask, collecting the material that boils above 215°C. Use the apparatus pictured in Fig. 5.5. The product, diethyl *n*-butylmalonate, is reported to boil at 235–240°C.

Cleaning Up   The aqueous layer after dilution with water can be flushed down the drain. Sodium sulfate and Drierite, once free of ether, can be placed in the nonhazardous solid waste container. Ether recovered from distillation is placed in the organic solvents container.

2. *n*-Butylbarbituric Acid

To a perfectly dry 250-mL round-bottomed flask add 50 mL[1] of absolute ethanol. Absolute ethanol is 100% ethanol, specially dried. To the ethanol add 0.46 g (0.2 mole) of sodium metal. The sodium is most easily handled in the form of small spheres, which are stored in mineral spirits. Using a pair of tweezers rinse the spheres in dry ligroin, blot them dry, and weigh them in a beaker. Transfer the sodium to the flask containing the dry ethanol, and fit the flask with a reflux condenser (see Fig. 2 in Chapter 1). Add a calcium chloride tube (Fig. 1 in Chapter 31) to the top of the condenser to prevent moisture from diffusing into the flask. Warm the flask on the steam bath to dissolve the last traces of sodium. To the sodium ethoxide solution add 4.32 g (0.2 mole) of dry diethyl *n*-butylmalonate followed by a hot solution of 1.2 g (0.2 mole) of dry urea in 22 mL of absolute ethanol. Reflux the resulting mixture on the steam bath for 2 h. At first the reaction may bump. Clamp the apparatus firmly.

At the completion of the reaction acidify the solution with 20.0 mL of 10% hydrochloric acid (check with litmus paper) and then reduce the volume to one-half by removing the ethanol either by distillation or evaporation on a rotary evaporator. Cool the solution in an ice bath until no more product crystallizes, then collect it by vacuum filtration on the Büchner funnel. Empty the filter flask and then wash unreacted *n*-butyldiethyl malonate (detected by its odor) from the crystals with ligroin, press it dry, and then recrystallize the product from boiling water

---

1. Adjust quantities of all reagents in proportion to match the available quantity of diethyl *n*-butylmalonate.

(20 mL of water per gram of product). The acid crystallizes slowly. Cool the solution for at least 30 min before collecting the product by vacuum filtration. Dry the barbituric acid on filter paper until the next laboratory period, then weigh it, calculate the percent yield, and determine the mp. Pure *n*-butylbarbituric acid melts at 209–210°C.

Cleaning Up   Add scrap sodium metal to 1-propanol in a beaker. After all traces of sodium have disappeared, the alcohol can be diluted with water, neutralized with dilute hydrochloric acid, and flushed down the drain. Combine aqueous filtrates and mother liquors from crystallization, neutralize with sodium carbonate, and flush the solution down the drain. The ligroin wash goes in the organic solvents container.

Questions

1. The anion of diethyl malonate is usually made by reacting the diester with sodium methoxide. What weight of sodium would be required in the present experiment if that method were employed for the macroscale experiment?

2. Outline all steps in the synthesis of pentobarbital.

3. What problem might be encountered if $BrCH_2C(CH_3)_3$ were used as the halide in this synthesis?

44

Photochemistry: The Synthesis of Benzopinacol

Prelab Exercise: Draw a mechanism for the base-catalyzed cleavage of benzopinacol.

Photochemistry, as the name implies, is the chemistry of reactions initiated by light. A molecule can absorb light energy and then undergo isomerization, fragmentation, rearrangement, dimerization, or hydrogen atom abstraction. The last reaction is the subject of this experiment. On irradiation with sunlight, benzophenone abstracts a proton from the solvent 2-propanol and becomes reduced to benzopinacol.

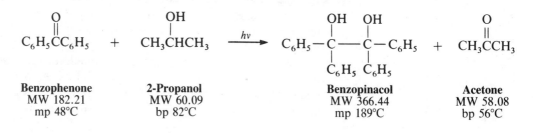

Benzophenone	2-Propanol	Benzopinacol	Acetone
MW 182.21	MW 60.09	MW 366.44	MW 58.08
mp 48°C	bp 82°C	mp 189°C	bp 56°C

The energy of light varies with its frequency where h is Planck's constant, ν is the frequency of the light, λ is its wavelength, and c is the speed of light.

$$\Delta E = h\nu = h\frac{c}{\lambda}$$

The fact that benzophenone is colorless means it does not absorb visible light, yet irradiation of benzophenone in alcohol results in a chemical change. Pyrex glass is not transparent to ultraviolet light with a wavelength shorter than 290 nm, so some wavelength between 290 nm and 400 nm (the edge of the visible region) must be responsible. The uv spectrum of benzophenone indicates an absorption band centered at about 355 nm. Light of that wavelength is absorbed by benzophenone.

What, in a general sense, occurs when a molecule absorbs light? A photon is absorbed only if its energy (wavelength) corresponds exactly to the difference between two electronic energy levels in the molecule. In benzophenone the electrons most loosely held and thus most easily excited are the two pairs of nonbonded electrons on the carbonyl oxygen, called the *n*-electrons.

$$\ce{\backslash C=\ddot{O}:} \quad \text{\textit{n}-electrons}$$

These electrons have paired spins $\ce{\backslash C=O}$ ↑↓, and only one is excited by the photon of light. The electron goes into the lowest unoccupied excited energy level, the π^*. Each electronic energy level of benzophenone has within it many vibrational and rotational energy levels. An electron can reside in many of these in the lower electronic energy level, the ground state, S_0, and it can be promoted to many vibrational and rotational energy levels in the first excited state (S_1). Hence the ultraviolet spectrum does not appear as a single sharp peak, but as a band made up of many peaks that arise from transitions between these many energy levels. See Fig. 1.

The spin of the electron cannot change in going from the ground state, S_0, to the excited singlet state, S_1 (conservation of angular momentum). Once in a higher vibrational or rotational state it can drop to S_1 by vibrational relaxation, losing some of its energy as heat. It then undergoes either fluorescence or intersystem crossing. As seen in Fig. 1, the rate for fluorescence, a light-emitting process in which the electron returns to the ground state, is 10^4 times slower than intersystem crossing; so benzophenone does not fluoresce. The electron flips its orientation during intersystem crossing so that it has the same orientation as the electron with

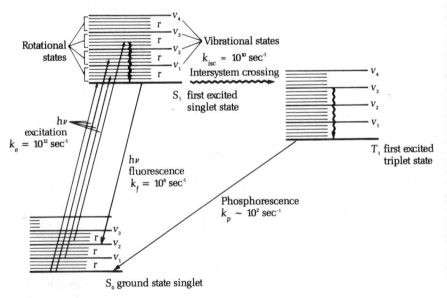

FIG. 1 Electronic energy levels and possible transitions.

which it was paired in the ground state. In this new state, called the *triplet*, it can lose energy as light; but this process, phosphorescence, is a slow one. The lifetime of the triplet state is long enough for chemical reactions to take place. The triplet can also lose energy as heat in a radiationless transition, but the probability of this happening is relatively low. This situation can be represented diagrammatically:

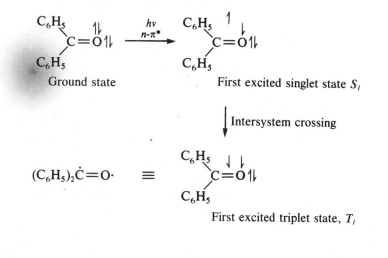

This T_1 state is a diradical and can abstract a methine proton from the solvent, then a hydroxyl proton to give acetone and diphenyl hydroxy radical, which then dimerizes to give the product.

$$(C_6H_5)_2\dot{C}=O\cdot \ + \ H-\underset{\underset{CH_3}{|}}{\overset{\overset{CH_3}{|}}{C}}-O-H \ \longrightarrow \ (C_6H_5)_2\dot{C}-OH \ + \ \cdot\underset{\underset{CH_3}{|}}{\overset{\overset{CH_3}{|}}{C}}-O-H$$

2-Propanol

$$(C_6H_5)_2\dot{C}=O\cdot \ + \ H-O-\underset{\underset{CH_3}{|}}{\overset{\overset{CH_3}{|}}{C}}\cdot \ \longrightarrow \ (C_6H_5)_2\dot{C}-OH \ + \ O=C\overset{CH_3}{\underset{CH_3}{}}$$

Acetone

$$2\ (C_6H_5)_2\dot{C}-OH \ \longrightarrow \ C_6H_5-\underset{\underset{C_6H_5}{|}}{\overset{\overset{OH}{|}}{C}}-\underset{\underset{C_6H_5}{|}}{\overset{\overset{OH}{|}}{C}}-C_6H_5$$

Benzopinacol

Experiments

1. Benzopinacol

The experiment should be done when there is good prospect for long hours of bright sunshine for several days. The benzopinacol is cleaved by alkali to benzhydrol and benzophenone and it is rearranged in acid to benzopinacolone.

In a 100-mL round-bottomed flask dissolve 10 g of benzophenone in 60–70 mL of isopropyl alcohol by warming on the steam bath, fill the flask to the neck with more of this alcohol, and add one drop of glacial acetic acid. (If the acid is omitted enough alkali may be derived from the glass of the flask to destroy the reaction product by the alkaline cleavage described in Experiment 2.) Stopper the flask with a well-rolled, tight-fitting cork, which is then wired in place. Invert the flask in a 100-mL beaker placed where the mixture will be most exposed to direct sunlight for some time. Since benzopinacol is but sparingly soluble in alcohol, its formation can be followed by the separation from around the walls of the flask of small, colorless crystals (benzophenone forms large, thick prisms). If the reaction mixture is exposed to direct sunlight throughout the daylight hours, the first crystals separate in about 5 h and the reaction is practically complete (95% yield) in four days. In winter the reaction may take as long as two weeks, and any benzophenone that crystallizes must be brought into solution by warming on the steam bath. When the reaction appears to be over, chill the flask if necessary and collect the product. The material should be pure, mp 188–189°C. If the yield is low, more material can be obtained by further exposure of the mother liquor to sunlight.

Reaction time: four days to two weeks

Cleaning Up   Dilute the isopropyl alcohol filtrate with water and flush the solution down the drain. Should any unreacted benzophenone precipitate, collect it by vacuum filtration, discard the filtrate down the drain, and place the recovered solid in the nonhazardous solid waste container.

2. Alkaline Cleavage

Suspend a small test sample of benzopinacol in alcohol and heat to boiling on a steam bath, making sure that the amount of solvent is not sufficient to dissolve the solid. Add a drop of sodium hydroxide solution, heat for a minute or two, and observe the result. The solution contains equal parts of benzhydrol and benzophenone, formed by the following reaction:

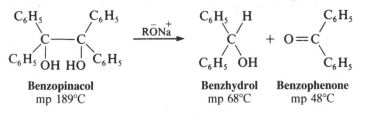

Benzopinacol   Benzhydrol   Benzophenone
mp 189°C   mp 68°C   mp 48°C

The low-melting products resulting from the cleavage are much more soluble than the starting material. Analyze the mixture by TLC.

Benzophenone can be converted into benzhydrol in nearly quantitative yield by following the procedure outlined above for the preparation of benzopinacol, modified by addition of a *very* small piece of sodium (5 mg) instead of the acetic acid. The reaction is complete when, after exposure to sunlight, the greenish-blue color disappears. To obtain the benzhydrol the solution is diluted with water, acidified, and evaporated. Benzopinacol is produced as before by photochemical reduction, but it is at once cleaved by the sodium alkoxide. The benzophenone formed by cleavage is converted into more benzopinacol, cleaved, and eventually consumed.

3. Pinacolone Rearrangement

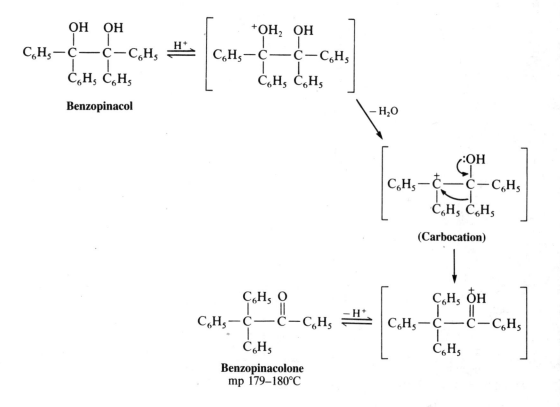

Benzopinacol

(Carbocation)

Benzopinacolone
mp 179–180°C

In a 100-mL round-bottomed flask place 5 g of benzopinacol, 25 mL of acetic acid, and two or three very small crystals of iodine (0.05 g). Heat to the boiling point for a minute or two under a reflux condenser until the crystals are dissolved, and then reflux the red solution for 5 min. On cooling, the pinacolone separates as a stiff paste. Thin the paste with

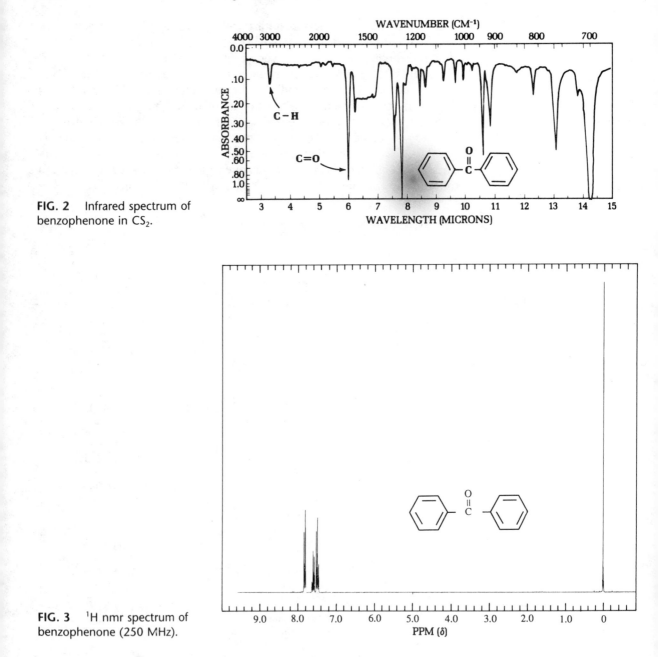

FIG. 2 Infrared spectrum of benzophenone in CS$_2$.

FIG. 3 ^{1}H nmr spectrum of benzophenone (250 MHz).

alcohol, collect the product, and wash it free from iodine with alcohol. The material should be pure; yield is 95%.

Cleaning Up   Dilute the filtrate with water, neutralize with sodium carbonate, and flush down the drain.

Question

Would the desired reaction occur if ethanol or *t*-butyl alcohol were used instead of isopropyl alcohol in the attempted photochemical dimerization of benzophenone?

45

Luminol: Synthesis of a Chemiluminescent Substance

Prelab Exercise: Write a balanced equation for the reduction of nitrophthalhydrazide to aminophthalhydrazide using sodium hydrosulfite, which is oxidized to bisulfite.

The oxidation of 3-aminophthalhydrazide, **1**, commonly known as luminol, is attended with a striking emission of blue-green light. Most exothermic chemical reactions produce energy in the form of heat, but a few produce light and release little or no heat. This phenomenon, chemiluminescence, is usually an oxidation reaction. In the case of luminol an alkaline solution of the compound is allowed to react with a mixture of hydrogen peroxide and potassium ferricyanide. The dianion **2** is oxidized to the triplet excited state (two unpaired electrons of like spin) of the amino phthalate ion, **3**. This slowly undergoes intersystem crossing to the singlet excited state (two unpaired electrons of opposite spin), **4**, which decays to the ground state ion, **5**, with the emission of one quantum of light (a photon) per molecule. Very few molecules are more efficient in chemiluminescence than luminol.

Luminol, **1**, is made by reduction of the nitro derivative, **8**, formed on thermal dehydration of a mixture of 3-nitrophthalic acid, **6**, and hydrazine, **7**. An earlier procedure for effecting the first step called for addition of hydrazine sulfate to an alkaline solution of the acid, evaporation to dryness, and baking the resulting mixture of the hydrazine salt and sodium sulfate at 165°C, and it required a total of 4.5 h for completion. This working time can be drastically reduced by adding high-boiling (bp 290°C) triethylene glycol to an aqueous solution of the hydrazine salt, distilling the excess water, and raising the temperature to a point where dehydration to **8** is complete within a few minutes. Nitrophthalhydrazide, **8**, is insoluble in dilute acid but soluble in alkali, by virtue of enolization; and it is conveniently reduced to luminol, **1**, by sodium hydrosulfite (sodium dithionite) in alkaline solution. In dilute, weakly acidic, or neutral solution luminol exists largely as the dipolar ion **9**, which exhibits beautiful blue fluorescence.[1]

An alkaline solution contains the doubly enolized anion **10** and displays particularly marked chemiluminescence when oxidized with a combination of hydrogen peroxide and potassium ferricyanide.

---

1. Several methods of demonstrating the chemiluminescence of luminol are described by E. H. Huntress, L. N. Stanley, and A. S. Parker. *J. Chem. Ed.*, **11**, 142 (1934). The mechanism of the reaction has been investigated by Emil H. White and co-workers (*J. Am. Chem. Soc.*, **86**, 940 and 942 (1964)).

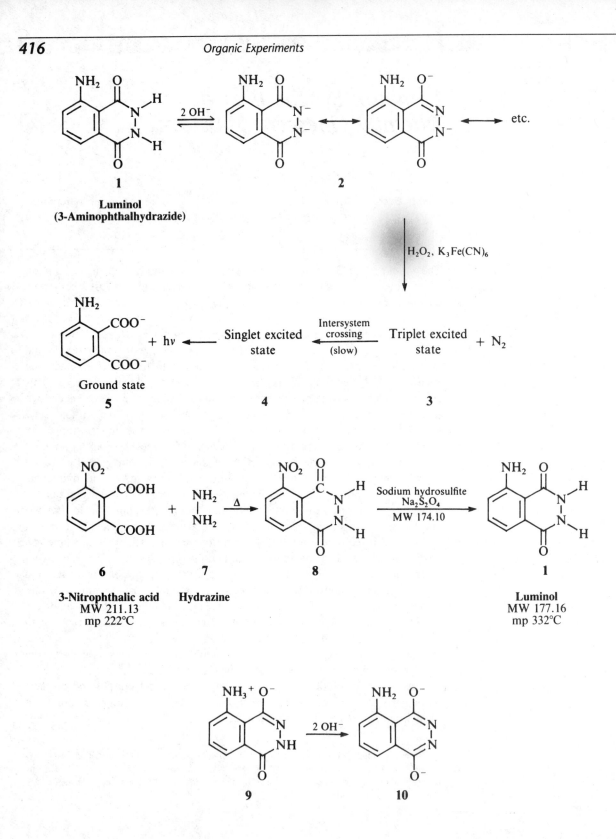

1

Luminol
(3-Aminophthalhydrazide)

2

etc.

H_2O_2, $K_3Fe(CN)_6$

Ground state
5

+ hν

Singlet excited
state
4

Intersystem
crossing
(slow)

Triplet excited
state
3

+ N_2

6

3-Nitrophthalic acid
MW 211.13
mp 222°C

7

Hydrazine

Δ

8

Sodium hydrosulfite
$Na_2S_2O_4$
MW 174.10

1

Luminol
MW 177.16
mp 332°C

9

2 OH⁻

10

Experiments

1. Synthesis of Luminol

First heat a flask containing 15 mL of water on the steam bath. Then heat a mixture of 1 g of 3-nitrophthalic acid and 2 mL of an 8% aqueous solution of hydrazine[2] (*caution*) in a 20 × 150-mm test tube over a Thermowell until the solid is dissolved, add 3 mL of triethylene glycol, and clamp the tube in a vertical position in a hot sand bath. Insert a thermometer, a boiling chip, and an aspirator tube connected to an aspirator, and boil the solution vigorously to distill the excess water (110–130°C). Let the temperature rise rapidly until it reaches 215°C (3–4 min). Remove the burner, note the time, and by intermittent gentle heating maintain a temperature of 214–220°C for 2 min. Remove the tube, cool to about 100°C (crystals of the product often appear), add the 15 mL of hot water, cool under the tap, and collect the light yellow granular nitro compound (**8**). Dry weight, 0.7 g.[3]

CAUTION: Hydrazine is a carcinogen. Handle with care. Wear gloves and carry out this experiment in the hood.

The nitro compound need not be dried and can be transferred at once, for reduction, to the uncleaned test tube in which it was prepared. Add 5 mL of 10% sodium hydroxide solution, stir with a rod, and to the resulting deep brown-red solution add 3 g of sodium hydrosulfite dihydrate. Wash the solid down the walls with a little water. Heat to the boiling point, stir, and keep the mixture hot for 5 min, during which time some of the reduction product may separate. Then add 2 mL of acetic acid, cool under the tap, and stir; collect the resulting precipitate of light yellow luminol (**1**). The filtrate on standing overnight usually deposits a further crop of luminol (0.1–0.2 g).

The two-step synthesis of a chemiluminescent substance can be completed in 25 min

$$Na_2S_2O_4 \cdot 2H_2O$$
Sodium hydrosulfite dihydrate
MW 210.15

CAUTION: Contact of solid sodium hydrosulfite with moist combustible material may cause fire.

Cleaning Up   Combine the filtrate from the first and second reactions, dilute with a few milliliters of water, neutralize with sodium carbonate, add 40 mL of household bleach (5.25% sodium hypochlorite solution), and heat the mixture to 50°C for 1 hr. This will oxidize any unreacted hydrazine and hydrosulfite. Dilute the mixture and flush it down the drain.

2. The Light-Producing Reaction

This reaction can be run on a scale five times larger. Dissolve the first crop of moist luminol (dry weight about 40–60 mg) in 2 mL of 10% sodium hydroxide solution and 18 mL of water; this is stock solution A. Prepare a second stock solution, B, by mixing 4 mL of 3% aqueous potassium ferricyanide, 4 mL of 3% hydrogen peroxide, and 32 mL of water. Now dilute 5 mL of solution A with 35 mL of water and, in a dark place, pour this solution and solution B simultaneously into an Erlenmeyer flask. Swirl the

---

2. Dilute 3.12 g of the commercial 64% hydrazine solution to a volume of 25 mL.

3. The reason for adding hot water and then cooling rather than adding cold water is that the solid is then obtained in more easily filterable form.

flask and, to increase the brilliance, gradually add further small quantities of alkali and ferricyanide crystals.

Ultrasonic sound can also be used to promote this reaction. Prepare stock solutions A and B again but omit the hydrogen peroxide. Place the combined solutions in an ultrasonic cleaning bath or immerse an ultrasonic probe into the reaction mixture. Spots of light are seen where the ultrasonic vibrations produce hydroxyl radicals.

And the sanguinary-minded can mix solutions A and B, omitting the ferricyanide from solution B. Light can be generated by adding blood dropwise to the reaction mixture.

Cleaning Up   Add 2 mL of 10% hydrochloric acid, dilute the solution with water, and flush the mixture down the drain.

Tetraphenylcyclopentadienone

Prelab Exercise: Write a detailed mechanism for the formation of tetraphenylcyclopentadienone from benzil and 1,3-diphenylacetone. To which general class of reactions does this condensation belong?

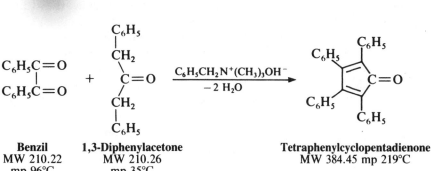

Benzil	**1,3-Diphenylacetone**	**Tetraphenylcyclopentadienone**
MW 210.22	MW 210.26	MW 384.45 mp 219°C
mp 96°C	mp 35°C	

Cyclopentadienone

Cyclopentadienone is an elusive compound that has been sought for many years but with little success. Molecular orbital calculations predict that it should be highly reactive, and so it is; it exists only as the dimer. The tetraphenyl derivative of this compound is to be synthesized in this experiment. This derivative is stable, and reacts readily with dienophiles. It is used not only for the synthesis of highly aromatic, highly arylated compounds, but also for examination of the mechanism of the Diels-Alder reaction itself. Tetraphenylcyclopentadienone has been carefully studied by means of molecular orbital methods in attempts to understand its unusual reactivity, color, and dipole moment. In Chapter 48 this highly reactive molecule is used to trap the fleeting benzyne to form tetraphenylnaphthalene. Indeed, this reaction constitutes evidence that benzyne does exist.

The literature procedure for condensation of benzil with 1,3-diphenylacetone in ethanol with potassium hydroxide as basic catalyst suffers from the low boiling point of the alcohol and the limited solubility of both potassium hydroxide and the reaction product in this solvent. Triethylene glycol is a better solvent and permits operation at a higher temperature. In the procedure that follows, the glycol is used with benzyltrimethylammonium hydroxide, a strong base readily soluble in organic solvents, which serves as catalyst.

Louis Fieser introduced the idea of using very high-boiling solvents to speed up this and many other reactions

Experiment

1. Tetraphenylcyclopentadienone

Short reaction period

Measure into a 25 × 150-mm test tube 2.1 g of benzil, 2.1 g of 1,3-diphenyl-acetone, and 10 mL of triethylene glycol, using the solvent to wash the walls of the test tube. Support the test tube in a hot sand bath, stir the mixture with a thermometer, and heat until the benzil is dissolved, then remove it from the sand. Measure 1 mL of a commercially available 40% solution of benzyltrimethylammonium hydroxide (Triton B) in methanol into a 10 × 75-mm test tube, adjust the temperature of the solution to exactly 100°C, remove from heat, add the catalyst, and stir once to mix. Crystallization usually starts in 10–20 s. Let the temperature drop to about 80°C and then cool under the tap, add 10 mL of methanol, stir to a thin crystal slurry, collect the product, and wash it with methanol until the filtrate is purple-pink, not brown. The yield of deep purple crystals is 3.3–3.7 g. If either the crystals are not well formed or the melting point is low, place 1 g of material and 10 mL of triethylene glycol in a vertically supported test tube, stir with a thermometer, raise the temperature to 220°C to bring the solid into solution, and let stand for crystallization (if initially pure material is recrystallized, the recovery is about 90%).

Caution: Triton B is toxic and corrosive.

Cleaning Up   Since the filtrate and washings from the reaction contain Triton B, they should be placed in the hazardous waste container. Crystallization solvent should be diluted with water and flushed down the drain.

Question

Draw the structure of the dimer of cyclopentadienone. Why doesn't tetraphenylcyclopentadienone undergo dimerization?

47

Hexaphenylbenzene and Dimethyl Tetraphenylphthalate

Prelab Exercise: Explain the driving force behind the loss of carbon monoxide from the intermediates formed in these two reactions.

This experiment illustrates two examples of the Diels-Alder reaction, which synthesizes molecules that would be extremely difficult to synthesize in any other way. Both reactions employ as the diene the tetraphenylcyclopentadienone prepared in the previous chapter. Although the Diels-Alder reaction is reversible, the intermediate in each of these reactions spontaneously loses carbon monoxide (why?) to form the products.

Hexaphenylbenzene melts at 465°C without decomposition. Few completely covalent organic molecules have higher melting points. Lead melts at 327.5°C. As is often the case, high melting point also means limited solubility. The solvent used to recrystallize hexaphenylbenzene, diphenyl ether, is very high-boiling (bp 259°C) and has superior solvent power.

In the first experiment the dienone is condensed with dimethyl acetylenedicarboxylate using as the solvent 1,2-dichlorobenzene. This solvent is chosen for its solvent properties as well as its high boiling point, which guarantees the reaction is complete in a minute or two.

Experiment

1. Dimethyl Tetraphenylphthalate

CAUTION: Work in a hood. Dimethyl acetylenedicarboxylate is corrosive and a lachrymator. 1,2-Dichlorobenzene is a toxic irritant.

Measure into a 25 × 150-mm test tube 2 g of tetraphenylcyclopentadienone, 10 mL of 1,2-dichlorobenzene, and 1 mL (1.1 g) of dimethyl acetylenedicarboxylate. Clamp the test tube in a hot sand bath, insert a thermometer, and raise the temperature to the boiling point (180–185°). Boil gently until there is no further color change and let the rim of condensate rise just high enough to wash the walls of the tube. The pure adduct is colorless, and if the starting ketone is adequately pure the color changes from purple to pale tan. A 5-min boiling period should be sufficient. Cool to 100°, slowly stir in 15 mL of 95% ethanol, and let crystallization proceed. Cool under the tap, collect the product, and rinse the tube with methanol.

Reaction time about 5 min Yield of colorless crystals should be in the range of 2.1–2.2 g.

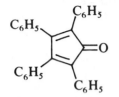

Tetraphenylcyclopentadienone
MW 384.45, mp 219°C

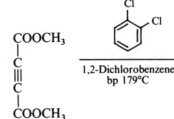

**Dimethyl
acetylenedicarboxylate**
MW 142.11, bp ~300°C

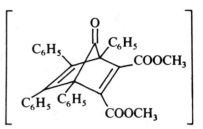

$-CO$

Dimethyl tetraphenylphthalate
MW 474.53, mp 258°C

Cleaning Up   Since the filtrate from this reaction contains 1,2-di-chlorobenzene, it should be placed in the halogenated organic solvents container.

2. Hexaphenylbenzene

Diphenylacetylene is a less reactive dienophile than dimethylacetylene-dicarboxylate; but when heated with tetraphenylcyclopentadienone without solvent a temperature (*ca.* 380–400°C) suitable for reaction can be attained. In the following procedure the dienophile is taken in large excess to serve as solvent. Since refluxing diphenylacetylene (bp about 300°C) keeps the temperature below the melting point of the product, removal of the diphenylacetylene lets the reaction mixture melt, which ensures completion of the reaction.

Procedure

Place 0.5 g each of tetraphenylcyclopentadienone and diphenylacetylene in a 25 × 150-mm test tube supported in a clamp and heat the mixture strongly with the free flame of a microburner held in the hand (do not insert a

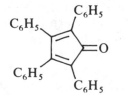

Tetraphenylcyclopentadienone
MW 384.45, mp 219°C

Diphenylacetylene
MW 178.22, mp 61°C

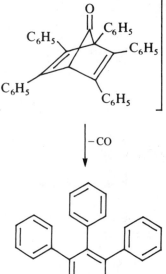

Hexaphenylbenzene
MW 534.66°C
mp 465°C

A hydrocarbon of very high mp. Although the small amount of carbon monoxide produced probably presents no hazard, work in a hood.

thermometer into the test tube; the temperature will be too high). Soon after the reactants have melted with strong bubbling, white masses of the product become visible. Let the diphenylacetylene reflux briefly on the walls of the tube and then remove some of the diphenylacetylene by letting it condense for a minute or two on a cold finger filled with water, but without fresh water running through it. Remove the flame, withdraw the cold finger, and wipe it with a towel. Repeat the operation until you are able, by strong heating, to melt the mixture completely. Then let the melt cool and solidify. Add 10 mL of diphenyl ether, using it to rinse the walls. Heat *carefully* over a free flame to dissolve the solid and then let the product crystallize. When cold, add 10 mL of toluene to thin out the mixture, collect the product, and wash with toluene. The yield of colorless plates, mp 465°, is 0.6–0.7 g.

Notes:

1. The melting point of the product can be determined with a Mel-Temp apparatus and a 500° thermometer. To avoid oxidation, seal the sample in an evacuated capillary tube.

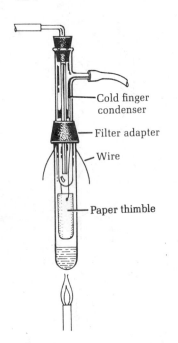

Cold finger condenser

Filter adapter

Wire

Paper thimble

FIG. 1   Soxhlet-type extractor.

2. In case the hexaphenylbenzene is contaminated with insoluble material, crystallization from a filtered solution can be accomplished as follows: Place 10 mL of diphenyl ether in a 25×150-mm test tube and pack the sample of hexaphenylbenzene into a 10-mm extraction thimble and suspend this in the test tube with two nichrome wires, as shown in Fig. 1. Insert a cold finger condenser supported by a filter adapter and adjust the condenser and the wires so that condensing liquid will drop into the thimble. Let the diphenyl ether reflux until the hexaphenylbenzene in the thimble is dissolved, and then let the product crystallize, add toluene, collect the product, and wash with toluene as described previously.

Cleaning Up   Place all filtrates in the organic solvents container.

Questions

1. What two factors probably contribute to the very high melting points of these two hexasubstituted benzenes?

2. What volume of carbon monoxide, measured at STP, is produced by the decomposition of 38 mg of tetraphenylcyclopentadienone?

FIG. 2   Research-quality Soxhlet extractor. Solvent vapors from the flask rise through A and up into the condenser. As they condense the liquid just formed returns to B, where sample to be extracted is placed. (Bottom of B is sealed at C.) Liquid rises in B to level D, at which time the automatic siphon, E, starts. Extracted material accumulates in the flask as more pure liquid vaporizes to repeat the process.

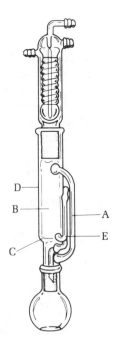

D

B

C

A

E

1,2,3,4-Tetraphenylnaphthalene via Benzyne

Prelab Exercise: Speculate on the reasons for the stability of the iodonium ion. What type of strain exists in the benzyne molecule?

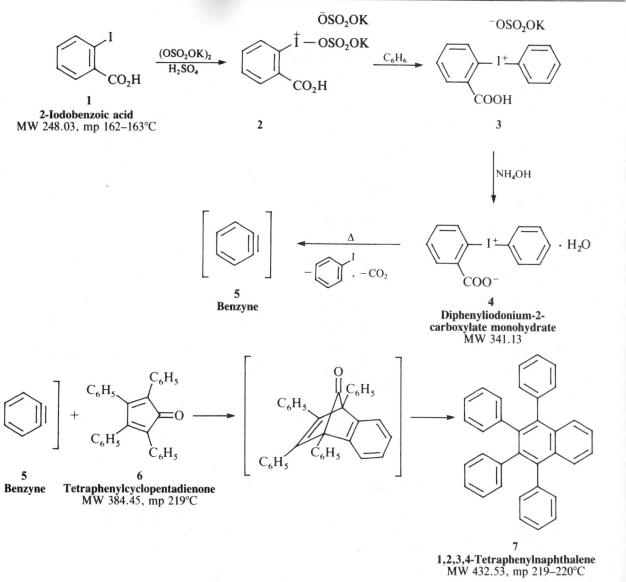

1
2-Iodobenzoic acid
MW 248.03, mp 162–163°C

2

3

4
Diphenyliodonium-2-carboxylate monohydrate
MW 341.13

5
Benzyne

5
Benzyne

6
Tetraphenylcyclopentadienone
MW 384.45, mp 219°C

7
1,2,3,4-Tetraphenylnaphthalene
MW 432.53, mp 219–220°C

Synthetic use of an intermediate not known as such

This synthesis of 1,2,3,4-tetraphenylnaphthalene (**7**) demonstrates the transient existence of benzyne (**5**), a hydrocarbon that has not been isolated as such. The precursor, diphenyliodonium-2-carboxylate (**4**), is heated in an inert solvent to a temperature at which it decomposes to benzyne, iodobenzene, and carbon dioxide in the presence of tetraphenylcyclopentadienone (**6**) as trapping agent. The preparation of the precursor (**4**) illustrates oxidation of a derivative of iodobenzene to an iodonium salt (**2**) and the Friedel-Crafts-like reaction of the substance with benzene to form the diphenyliodonium salt (**3**). Neutralization with ammonium hydroxide then liberates the precursor, inner salt (**4**), which, when obtained by crystallization from water, is the monohydrate.

Experiments

1. 2-Iodobenzoic Acid

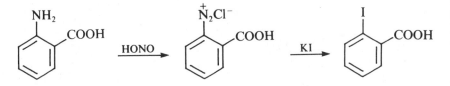

A 500-mL round-bottomed flask containing 13.7 g of anthranilic acid, 100 mL of water, and 25 mL of concentrated hydrochloric acid is heated until the solid is dissolved. The mixture is then cooled in ice while bubbling in nitrogen to displace the air. When the temperature reaches 0–5°C a solution of 7.1 g of sodium nitrite is added slowly. After 5 min a solution of 17 g of potassium iodide in 25 mL of water is added, when a brown complex partially separates. The mixture is let stand without cooling for 5 min (under nitrogen) and then warmed to 40°C, at which point a vigorous reaction ensues (gas evolution, separation of a tan solid). After reacting for 10 min the mixture is heated on the steam bath for 10 min and then cooled in ice. A pinch of sodium bisulfite is added to destroy any iodine present and the granular tan product collected and washed with water. The still moist product is dissolved in 70 mL of 95% ethanol, 35 mL of hot water is added, and the brown solution is treated with decolorizing charcoal, filtered, diluted at the boiling point with 35–40 mL of water, and let stand. 2-Iodobenzoic acid separates in large, slightly yellow needles of satisfactory purity (mp 164°C) for the experiment; yield is approximately 17 g (71%).

Cleaning Up   The reaction mixture, filtrate and mother liquor from the crystallization are combined, neutralized with sodium carbonate, and flushed down the drain with a large excess of water. Norit is placed in the nonhazardous solid waste container.

2. Diphenyliodonium-2-carboxylate Monohydrate (4)

K₂S₂O₈

Potassium persulfate
(Potassium peroxydisulfate)
MW 270.33

Strong oxidant!

1st step: about 20 min

CAUTION: *Benzene is a carcinogen. Carry out this experiment in the hood.*

2nd step: about 30 min

Measure 8 mL of concentrated sulfuric acid into a 25-mL Erlenmeyer flask and place the flask in an ice bath to cool. Place 2.0 g of *o*-iodobenzoic acid and 2.6 g of potassium persulfate[1] in a 125-mL Erlenmeyer flask, swirl the flask containing sulfuric acid vigorously in the ice bath for 2–3 min, and then remove it and wipe it dry. Place the larger flask in the ice bath and pour the chilled acid down the walls to dislodge any particles of solid. Swirl the flask in the ice bath for 4–5 min to produce an even suspension and then remove it and note the time. The reaction mixture foams somewhat and acquires a succession of colors. After it has stood at room temperature for 20 min, swirl the flask vigorously in an ice bath for 3–4 min, add 2 mL of benzene (*caution!*), and swirl and cool until the benzene freezes. Then remove and wipe the flask and note the time at which the benzene melts. Warm the flask in the palm of the hand and swirl frequently at room temperature for 20 min to promote completion of reaction in the two-phase mixture.

While the reaction is going to completion, place three 50-mL Erlenmeyer flasks in an ice bath to chill: one containing 19 mL of distilled water, another 23 mL of 29% ammonium hydroxide solution, and another 40 mL of dichloromethane (bp 40.8°C). At the end of the 20-min reaction period, chill the reaction mixture thoroughly in an ice bath, mount a separatory funnel to deliver into the flask containing the reaction mixture in benzene, and place in the funnel the chilled 19 mL of water. Swirl the reaction flask vigorously while running in the water slowly. The solid that separates is **3**, the potassium bisulfate salt of diphenyliodonium-2-carboxylic acid. Pour the chilled ammonia solution into the funnel and pour the chilled dichloromethane into the reaction flask so that it will be available for efficient extraction of the reaction product **4** as it is liberated from **3** on neutralization. While swirling the flask vigorously in the ice bath, run in the chilled ammonia solution during the course of about 10 min. The mixture must be alkaline (pH 9). If not, add more ammonia solution. Pour the mixture into a separatory funnel and rinse the flask with a little fresh dichloromethane. Let the layers separate and run the lower layer into a tared 125-mL Erlenmeyer flask, through a funnel fitted with a paper containing anhydrous sodium sulfate. Extract the aqueous solution with two 10-mL portions of dichloromethane and run the extracts through the drying agent into the tared flask. Evaporate the dried extracts to dryness on the steam bath (use an aspirator tube) and remove the solvent from the residual cake of solid by connecting the flask, with a rubber stopper, to the aspirator and heating the flask on the steam bath until the weight is constant; yield is about 2.4 g.

Note for the instructor

---

1. The fine granular material supplied by Fisher Scientific Co. is satisfactory. Persulfate in the form of large prisms should be finely ground prior to use.

For crystallization, dislodge the bulk of the solid with a spatula and transfer it onto weighing paper and then into a 50-mL Erlenmeyer flask. Measure 28 mL of distilled water into a flask and use part of the water to dissolve the residual material in the tared 125-mL flask by heating the mixture to the boiling point over a free flame. Pour this solution into the 50-mL flask. Add the remainder of the 28 mL of water to the 50-mL flask and bring the solid into solution at the boiling point. Add a small portion of charcoal for decolorization of the pale tan solution, swirl, and filter at the boiling point through a funnel, preheated on the steam bath and fitted with moistened filter paper. Diphenyliodonium-2-carboxylate monohydrate (**4**), the benzyne precursor, separates in colorless, rectangular prisms, mp 219–220°C, decomposes; yield is about 2.1 g.

Cleaning Up   Make the aqueous layer neutral with dilute hydrochloric acid. It should then be diluted with water and flushed down the drain. Allow the dichloromethane to evaporate from the drying agent and then place it in the nonhazardous solid waste container along with the Norit.

3. Preparation of 1,2,3,4-Tetraphenylnaphthalene (7)

Diels-Alder reaction

Use hood; CO is evolved

Place 1.0 g of the diphenyliodonium-2-carboxylate monohydrate just prepared and 1.0 g of tetraphenylcyclopentadienone in a 25 × 150-mm test tube. Add 6 mL of triethylene glycol dimethyl ether (bp 222°C) in a way that the solvent will rinse the walls of the tube. Support the test tube vertically, insert a thermometer, and heat with a Thermowell. When the temperature reaches 200°C remove from the heat and note the time. Then keep the mixture at 200–205°C by intermittent heating until the purple color is discharged, the evolution of gas (CO_2 + CO) subsides, and a pale yellow solution is obtained. In case a purple or red color persists after 3 min at 200–205°C, add additional small amounts of the benzyne precursor and continue to heat until all the solid is dissolved. Let the yellow solution cool to 90°C while heating 6 mL of 95% ethanol to the boiling point on the steam bath. Pour the yellow solution into a 25-mL Erlenmeyer flask and use portions of the hot ethanol, drawn into a capillary dropping tube, to rinse the test tube. Add the remainder of the ethanol to the yellow solution and heat at the boiling point. If shiny prisms do not separate at once, add a few drops of water by drops at the boiling point of the ethanol until prisms begin to separate. Let crystallization proceed at room temperature and then at 0°C. Collect the product and wash it with methanol. The yield of colorless prisms is 0.8–0.9 g. The pure hydrocarbon, **7**, exists in two crystalline forms (allotropes) and has a double melting point, the first of which is in the range of 196–199°C. Let the bath cool to about 195°C, remove the thermometer, and let the sample solidify. Then determine the second melting point, which, for the pure hydrocarbon, is 203–204°C.

Cleaning Up   The filtrate contains ethanol, iodobenzene, triglyme, and probably small amounts of the reactants and the product. Place the mixture in the halogenated organic solvents container.

Questions

1. To what general class of compounds does potassium persulfate belong?

2. Calculate the volume of carbon monoxide, at standard temperature and pressure, released during the reaction.

Triptycene via Benzyne

Prelab Exercise: Outline the preparation of triptycene, writing balanced equations for each reaction, including the reactions that are used to remove excess anthracene.

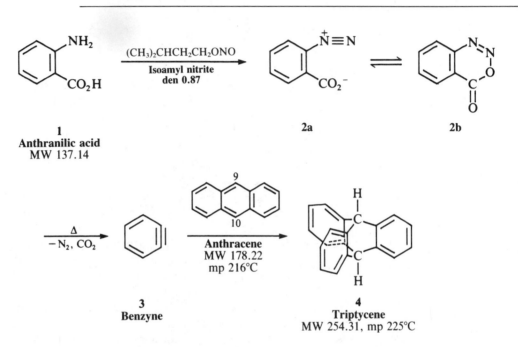

1
Anthranilic acid
MW 137.14

2a

2b

3
Benzyne

Anthracene
MW 178.22
mp 216°C

4
Triptycene
MW 254.31, mp 225°C

Aprotic means "no protons"; HOH and ROH are protic

This interesting cage-ring hydrocarbon results from 9,10-addition of benzyne to anthracene. In one procedure presented in the literature benzyne is generated under nitrogen from *o*-fluorobromobenzene and magnesium in the presence of anthracene, but the work-up is tedious and the yield only 24%. Diazotization of anthranilic acid to benzenediazonium-2-carboxylate (**2a**) or to the covalent form (**2b**) gives another benzyne precursor, but the isolated substance can be kept only at a low temperature and is sometimes explosive. However, isolation of the precursor is not necessary. On slow addition of anthranilic acid to a solution of anthracene and isoamyl nitrite in an aprotic solvent, the precursor **2** reacts with anthracene as fast as the precursor is formed. If the anthranilic acid is all present at the start, a side reaction of this substance with benzyne drastically reduces the yield. A low-boiling solvent (CH$_2$Cl$_2$, bp 41°C) is used, in which the desired reaction goes slowly, and a solution of anthranilic acid is added dropwise over a period of 4 h. To bring the reaction time into the limits of a laboratory period, the higher-boiling solvent 1,2-dimethoxy-ethane (bp 83°C, water soluble) is specified in this procedure and a large excess of anthranilic acid and isoamyl nitrite is used. Treatment of the dark reaction mixture with alkali removes acidic by-products and most of the color, but the crude product inevitably contains anthracene. However, brief heating with maleic anhydride at a suitable temperature leaves the triptycene untouched and converts the anthracene into its maleic anhydride adduct. Treatment of the reaction mixture with alkali converts the adduct into a water-soluble salt and affords colorless, pure triptycene.

Removal of anthracene

Experiments_____

1. Tryptycene

Test dimethoxyethane for peroxides. Carry out reaction in hood.

Place 2 g of anthracene,[1] 2 mL of isoamyl nitrite, and 20 mL of 1,2-dime-thoxyethane in a 125-mL round-bottomed flask mounted in a hot sand bath and fitted with a short reflux condenser. Insert a filter paper into a 55-mm short stem funnel, moisten the paper with 1,2-dimethoxyethane, and rest the funnel in the top of the condenser. Weigh 2.6 g of anthranilic acid on a folded paper, scrape the acid into the funnel with a spatula, and pack it down. Bring the mixture in the reaction flask to a gentle boil and note that the anthracene does not all dissolve. Measure 20 mL of 1,2-dimethoxyeth-ane into a graduate and use a capillary dropping tube to add small portions of the solvent to the anthranilic acid in the funnel, to slowly leach the acid into the reaction flask. If you make sure that the condenser is exactly vertical and the top of the funnel is centered, it should be possible to arrange for each drop to fall free into the flask and not touch the condenser wall. Once dripping from the funnel has started, add fresh batches of solvent to the acid but only 2–3 drops at a time. Plan to complete leaching the first charge of anthranilic acid in a period of not less than 20 min, using about 10 mL of solvent. Then add a second 2.6-g portion of anthranilic acid to the funnel, remove from heat, and by lifting the funnel up a little run in 2 mL of isoamyl nitrite through the condenser. Replace the funnel, resume heating, and leach the anthranilic acid as before in about 20 min time. Reflux for 10 min more and then add 10 mL of 95% ethanol and a solution of 3 g of sodium hydroxide in 40 mL of water to produce a suspension of solid in a brown alkaline liquor. Cool thoroughly in ice, and also cool a 4:1 methanol-water mixture for rinsing.

Technique for slow addition of a solid

Reaction time: 1 hour

Collect the solid on a small Büchner funnel and wash it with the chilled solvent to remove brown mother liquor. Transfer the moist, nearly color-less solid to a tared 125-mL round-bottomed flask and evacuate on the steam bath until the weight is constant; the anthracene-triptycene mixture (mp about 190–230°C) weighs 2.1 g. Add 1 g of maleic anhydride and 20 mL of triethylene glycol dimethyl ether ("triglyme," bp 222°C), heat the

CH₃OCH₂CH₂OCH₂
        |
CH₃OCH₂CH₂OCH₂
Triethylene glycol
dimethyl ether (triglyme)
Miscible with water

_____

1. Practical grade anthracene and anthranilic acid are satisfactory.

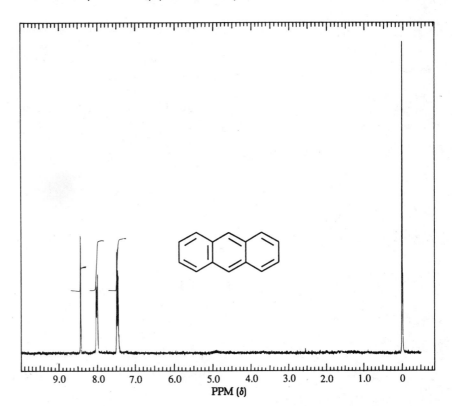

FIG. 1 ¹H nmr spectrum of anthracene (250 MHz).

mixture to the bp under reflux (see Fig. 15.1 for apparatus) and reflux for 5 min. Cool to about 100°C, add 10 mL of 95% ethanol and a solution of 3 g of sodium hydroxide in 40 mL of water; then cool in ice, along with 25 mL of 4:1 methanol-water for rinsing. Triptycene separates as nearly white crystals from the slightly brown alkaline liquor. The washed and dried product weighs 1.5 g and melts at 255°C. It will appear to be colorless, but it contains a trace of black insoluble material.

Dissolve the product in methylcyclohexane (23 mL/g), decant from the specks of black material, and allow the solution to cool slowly. The product crystallizes as flat, rectangular, laminated prisms.

Cleaning Up   Dilute the alkaline filtrate from the reaction with water and flush it down the drain. Methylcyclohexane mother liquor from the crystallization goes in the organic solvents container.

Optional Projects

In Chapter 48 diphenyliodonium-*o*-carboxylate is utilized as a benzyne precursor in the synthesis of 1,2,3,4-tetraphenylnaphthalene. For reasons unknown, the reaction of anthracene with benzyne generated in this way proceeds very poorly and gives only a trace of triptycene. What about the converse proposition? Would benzyne generated by diazotization of an-thranilic acid in the presence of tetraphenylcyclopentadienone afford 1,2,3,4-tetraphenylnaphthalene in satisfactory yield? In case you are interested in exploring this possibility, plan and execute a procedure and see what you can discover.

If benzyne is generated from anthranilic acid in the absence of anthracene, as in this experiment, then dibenzocyclobutadiene is formed.[2]

Devise a procedure for the isolation of this interesting hydrocarbon.

Questions

1. Write equations showing how benzyne might be generated from *o*-fluorobromobenzene and magnesium.

2. Study the mechanism of diazonium salt formation in Chapter 42 and then propose a mechanism for diazotization using isoamyl nitrite.

2. (a) F. M. Logullo, A. H. Seitz, and L. Friedman, *Org. Syn.* **48,** 12 (1968). (b) L. Friedman and F. M. Logullo, *J. Org. Chem.,* **34,** 3089 (1969).

Oxidative Coupling of Alkynes: 2,7-Dimethyl-3,5-octadiyn-2,7-diol

Prelab Exercise: Write a balanced equation for the coupling of 2-methyl-3-butyn-2-ol. What role does cuprous chloride play in this reaction?

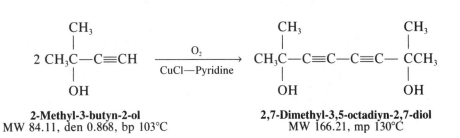

2-Methyl-3-butyn-2-ol
MW 84.11, den 0.868, bp 103°C

2,7-Dimethyl-3,5-octadiyn-2,7-diol
MW 166.21, mp 130°C

The starting material, 2-methyl-3-butyn-2-ol, is made commercially from acetone and acetylene and is convertible into isoprene. This experiment illustrates the oxidative coupling of a terminal acetylene to produce a diacetylene, the Glaser reaction.

Although Glaser reported the coupling of acetylenes using cuprous ion in 1869, the details of the mechanism of this reaction are still obscure. It appears to involve both Cu^+ and Cu^{2+} ions; as noted by Glaser, the cuprous acetylide is oxidized by oxygen,

$$C_6H_5C{\equiv}CH \xrightarrow[NH_4OH]{CuCl} C_6H_5C{\equiv}CCu \xrightarrow{air} C_6H_5C{\equiv}C-C{\equiv}CC_6H_5$$

but Cu^{2+} is the actual oxidizing agent. The amine seems to be needed to keep the acetylide in solution.

The reaction is very useful synthetically in the synthesis of polyenes, vitamins, fatty acids, and the annulenes. Baeyer used the reaction in his historic synthesis of indigo as long ago as 1882. The reaction allowed the unequivocal establishment of the carbon skeleton of this dye:

435

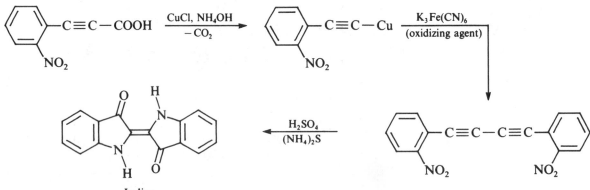

Indigo

FIG. 1 Balloon technique of oxygenation. About 10 lb oxygen pressure are needed to inflate the pipette bulbs.

Measure pyridine in hood

FIG. 2 Appearance of pipette bulb after oxygenation.

In the present experiment the reaction time is shortened by use of excess catalyst and by supplying oxygen under the pressure of a balloon; the state of the balloon provides an index of the course of the reaction.

Experiment

1. 2,7-Dimethyl-3,5-octadiyn-2,7-diol

The reaction vessel is a 125-mL filter flask or Erlenmeyer flask with side tube with a rubber bulb secured to the side arm with copper wire, evident from Figs. 1 or 2. Add 10 mL of 2-methyl-3-butyn-2-ol, 10 mL of methanol, 3 mL of pyridine, and 0.5 g of copper(I) chloride. Before going to the oxygen station,[1] practice capping the flask with a large serum stopper until you can do this quickly. You are to flush out the flask with oxygen and cap it before air can diffuse in; the reaction is about twice as fast in an atmosphere of oxygen as in air. Insert the oxygen delivery syringe into the flask with the needle under the surface of the liquid, open the valve, and let oxygen bubble through the solution in a brisk stream for 2 min. Close the valve and quickly cap the flask and wire the cap. Next thrust the needle of the delivery syringe through the center of the serum stopper, open the valve, and run in oxygen until you have produced a sizeable inflated bulb (see Fig. 1). Close the valve and withdraw the needle (the hole is self-sealing), note the time, and start swirling the reaction mixture. The rate of oxygen uptake depends on the efficiency of mixing of the liquid and gas phases. By vigorous and continuous swirling it is possible to effect deflation of the balloon to a 5-cm sphere (see Fig. 2) in 20–25 min. The reaction mixture warms up and becomes deep green. Introduce a second charge of oxygen of the same size as the first, note the time, and swirl. In 25–30 min the balloon reaches a constant

1. The valves of a cylinder of oxygen should be set to deliver gas at a pressure of 10 lb/sq in. when the terminal valve is opened. The barrel of a 2.5-mL plastic syringe is cut off and thrust into the end of a $\frac{1}{4} \times \frac{3}{16}$-in. rubber delivery tube. Read about the handling of compressed gas cylinders in Chapter 2. Be sure the cylinder is secured to a bench or wall.

volume (*e.g.*, a 5-cm sphere), and the reaction is complete. A pair of calipers is helpful in recognizing the constant size of the balloon and thus the end of the reaction.[2]

Open the reaction flask, cool if warm, add 5 mL of concentrated hydrochloric acid to neutralize the pyridine and keep copper compounds in solution, and cool again; the color changes from green to yellow. Use a spatula to dislodge copper salts adhering to the walls, leaving it in the flask. Then add 25 mL of saturated sodium chloride solution to precipitate the diol, and stir and cool the thick paste that results. Scrape out the paste onto a small Büchner funnel with the spatula; press down the material to an even cake. Rinse out the flask with water and wash the filter cake with enough more water to remove all the color from the cake and leave a colorless product. Since the moist product dries slowly, drying is accomplished in ether solution, and the operation is combined with recovery of diol retained in the mother liquor. Transfer the moist product to a 125-mL Erlenmeyer flask, extract the mother liquor in the filter flask with one portion of ether, wash the extract once with water, and run the ethereal solution into the flask containing the solid product. Add enough more ether to dissolve the product, transfer the solution to a separatory funnel, drain off the water layer, wash with saturated sodium chloride solution, and filter through anhydrous sodium sulfate into a tared flask. The solvent is evaporated on the steam bath and the solid residue is heated and evacuated until the weight is constant. Crystallization from toluene then gives colorless needles of the diol, mp 129–130°C. The yield of crude product is 7–8 g. In the crystallization, the recovery is practically quantitative.

Cleaning Up   Combine all aqueous and organic layers from the reaction, add sodium carbonate to make the aqueous layer neutral, and shake the mixture gently to extract pyridine into the organic layer. Place the organic layer in the organic solvents container. Remove any solid (copper salts) from the aqueous layer by filtration. The solid can be placed in the nonhazardous solid waste container and the aqueous layer diluted with water and flushed down the drain.

Questions

1. Write a balanced equation for this reaction.

2. What volume of oxygen at standard temperature and pressure is consumed in this reaction?

---

2. A magnetic stirrer does not materially shorten the reaction time.

CHAPTER

51

Sugars

Prelab Exercise: Write a balanced equation for the hydrolysis of the disaccharide sucrose, taking care to draw the stereochemistry of the products. Will the anomeric forms of glucose give different phenylosazones? Will they give different methyl glucosides?

The term sugar applies to mono-, di-, and oligosaccharides, which are all soluble in water and thereby distinguished from polysaccharides. Many natural sugars are sweet, but data of Table 1 show that sweetness varies greatly with stereochemical configuration and is exhibited by compounds of widely differing structural type.

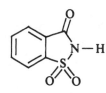

Saccharin

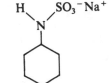

Sodium cyclamate

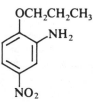

2-Amino-4-nitro-*n*-propoxybenzene

$$COOH$$

$$CH_2 \quad O \quad H \quad CH_2$$

$$H_2NCH-C-N-CH-COOCH_3$$

Aspartame

L-**Aspartyl-**L-**phenylalanine methyl ester**

Sugars are neutral and combustible and these properties distinguish them from other water-soluble compounds. Some polycarboxylic acids and some lower amines are soluble in water, but the solutions are acidic or basic. Water-soluble amine salts react with alkali with liberation of the amine, and sodium salts of acids are noncombustible.

One gram of sucrose dissolves in 0.5 mL of water at 25°C and in 0.2 mL at the boiling point, but the substance has marked, atypical crystallizing properties. In spite of the high solubility it can be obtained in beautiful, large crystals (rock candy). More typical sugars are obtainable in crystalline form only with difficulty, particularly in the presence of a trace of impurity, and even then give small and not well-formed crystals. Alcohol is often added to a water solution to decrease solubility and thus to induce crystallization. The amounts of 95% ethanol required to dissolve 1-g samples at 25° are sucrose, 170 mL; glucose, 60 mL; fructose, 15 mL. Some sugars have never been obtained in crystalline condition and are known only as viscous

Osazones

syrups. With phenylhydrazine many sugars form beautiful crystalline derivatives called osazones. Osazones are much less soluble in water than the parent sugars, because the molecular weight is increased by 178 units and the number of hydroxyl groups reduced by one. It is easier to isolate an osazone than to isolate the sugar, and sugars that are syrups often give crystalline osazones. Osazones of the more highly hydroxylic disaccharides are notably more soluble than those of monosaccharides.

TABLE 1   Relative Sweetness of Sugars and Sugar Substitutes

	Sweetness	
Compound	*To humans*	*To bees*
Monosaccharides		
D-Fructose	1.5	+
D-Glucose	0.55	+
D-Mannose	Sweet, then bitter	−
D-Galactose	0.55	−
D-Arabinose	0.70	−
Disaccharides		
Sucrose (glucose, fructose)	1	+
Maltose (2 glucose)	0.3	+
α-Lactose (glucose, galactose)	0.2	−
Cellobiose (2 glucose)	Indifferent	−
Gentiobiose (2 glucose)	Bitter	−
Synthetic sugar substitutes		
Aspartame (NutraSweet®)	180	
Saccharin	550	
2-Amino-4-nitro-1-*n*-propoxybenzene	4000	

Reducing sugars have an aldehyde form

Some disaccharides do not form osazones, but a test for formation or nonformation of the osazone is ambiguous, because the glycosidic linkage may suffer hydrolysis in a boiling solution of phenylhydrazine and acetic acid, with formation of an osazone derived from a component sugar and not from the disaccharide. If a sugar has reducing properties it is also capable of osazone formation; hence an unknown sugar is tested for reducing properties before preparation of an osazone is attempted. Three tests for differentiation between reducing and nonreducing sugars are described below; two are classical and the third modern.

Experiments

1. Fehling's Solution[1]

$$Cu^{2+} \rightarrow Cu_2O$$

Blue   Red

The reagent is made just prior to use by mixing equal volumes of Fehling's solution I, containing copper(II) sulfate, with solution II, containing tartaric acid and alkali. The copper, present as a deep blue complex anion, if reduced by a sugar from the copper(II) to the copper(I) state, precipitates as red copper(I) oxide. If the initial step in the reaction involved oxidation of the aldehydic group of the aldose to a carboxyl group, a ketose should not reduce Fehling's solution, or at least should react less rapidly than an aldose, but the comparative experiment that follows will show that this supposition is not the case. Hence, attack by an alkaline oxidizing agent must attack the α-ketol grouping common to aldoses and hexoses, and perhaps proceeds through an enediol, the formation of which is favored by alkali. A new α-ketol grouping is produced, and thus oxidation proceeds down the carbon chain.

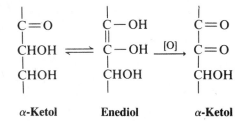

One milliliter of mixed solution will react with 5 mg of glucose; the empirically determined ratio is the basis for quantitative determination of the sugar. The Fehling's test is not specific to reducing sugars, because ordinary aldehydes reduce the reagent although by a different mechanism and at a different rate.

The following sugars are to be tested: 0.1 M solutions of glucose, fructose, lactose, maltose, sucrose (cane sugar).[2]

The scale of this experiment can be doubled.

Procedure

Fehling's test on five sugars

Introduce 5 drops of the 0.1 M solutions to be tested into each of five test tubes carrying some form of serial numbers resistant to heat and water (rubber bands), and prepare a beaker of hot water in which all the tubes can be heated at once. Measure 2.5 mL of Fehling's solution I into a 10-mL flask

Note for the instructor

1. *Solution I:* 34.64 g of $CuSO_4 \cdot 5H_2O$ dissolved in water and diluted to 500 mL. *Solution II:* 173 g of sodium potassium tartrate (Rochelle salt) and 65 g of sodium hydroxide dissolved in water and diluted to 500 mL.

2. To prepare 0.1 M test solutions dissolve the following amounts of substance in 100 mL of water each: D-glucose monohydrate, 1.98 g; fructose, 1.80 g; α-lactose monohydrate, 3.60 g; maltose monohydrate, 3.60 g; sucrose, 3.42 g; 1-butanol, 0.72 g.

and wash the pipette before using it to measure 2.5 mL of solution II into the same flask. Mix until all precipitate dissolves, measure 1 mL of mixed solution into each of the five test tubes, shake, put the tubes in the heating bath, note the time, and record any reaction as a function of time. Empty and wash the tubes with water, and then with dilute acid (leave the markers in place and continue heating the beaker of water for the next experiment).

Cleaning Up   Remove any precipitated copper by filtration. Place it in the nonhazardous solid waste container. The filtrate should be diluted with water and flushed down the drain.

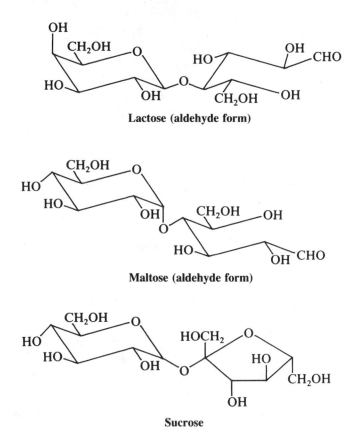

Lactose (aldehyde form)

Maltose (aldehyde form)

Sucrose

2. Tollens' Reagent

Tollens' reagent is a solution of a silver ammonium hydroxide complex that is reduced by aldoses and ketoses as well as by simple aldehydes. See Chapter 30 for the preparation of this reagent and the procedure for carrying out the test. The test is more sensitive than the Fehling's test and better able to reveal small differences in reactivity, but it is less reliable in distinguishing between reducing and nonreducing sugars.

Prepare the five test tubes according to Chapter 30. Into each tube put one drop of 0.1 M solution of glucose, fructose, lactose, maltose, and 1-butanol. (See Experiment 1, footnote 2.) Add 1 mL of Tollens' reagent to each tube and let the reaction proceed at room temperature. Watch closely and try to determine the order of reactivity as measured not by the color of solution but by the time of the first appearance of metal. At the completion of the reaction destroy all Tollens' reagent by adding nitric acid until the solution is clear.

Cleaning Up See Chapter 30 for the proper treatment of Tollens' reagent before disposal and also for cleaning silver mirrors from the test tubes.

3. Red Tetrazolium[3]

A sensitive test for reducing sugars

The reagent (RT) is a nearly colorless, water-soluble substance that oxidizes aldoses and ketoses, as well as other α-ketols, and is thereby reduced. The reduced form is a water-insoluble, intensely colored pigment, a diformazan.

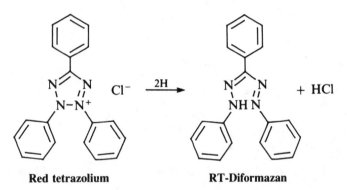

Red tetrazolium   **RT-Diformazan**

Red tetrazolium affords a highly sensitive test for reducing sugars and distinguishes between α-ketols and simple aldehydes more sharply than Fehling's and Tollens' tests.

Put one drop of each of the five 0.1 M test solutions of section 1 in the cleaned, marked test tubes, and to each tube add 1 mL of a 0.5% aqueous solution of red tetrazolium and one drop of 10% sodium hydroxide solution. Put the tubes in the beaker of hot water, note the time, and note the time of development of color for each tube.

For estimation of the sensitivity of the test, use the substance that you regard as the most reactive of the five studied. Dilute 1 mL of the 0.1 M solution with water to a volume of 100 mL, and run a test with RT on 0.2 mL of the diluted solution.

Note for the instructor

3. 2,3,5-Triphenyl-2H-tetrazolium chloride. Available from Aldrich and Eastman. Light-sensitive. Freshly prepared aqueous solutions should be used in tests. Any unused solution should be acidified and discarded.

Cleaning Up   The test solutions should be diluted with water and flushed down the drain.

4. Phenylosazones

Conduct procedure in hood. Avoid skin contact with phenylhydrazine.

This experiment can be carried out on a scale two or three times larger than that specified.

Put 1-mL portions of phenylhydrazine reagent[4] into each of four cleaned, numbered test tubes. Add 3.3-mL portions of 0.1 M solutions of glucose, fructose, lactose, and maltose and heat the tubes in the beaker of hot water for 20 min. Shake the tubes occasionally to relieve supersaturation and note the times at which osazones separate. If after 20 min no product has separated, cool and scratch the test tube to induce crystallization.

Collect and save the products for possible later use in identification of unknowns. Since osazones melt with decomposition, the bath in a mp determination should be heated at a standard rate (0.5°C per second).

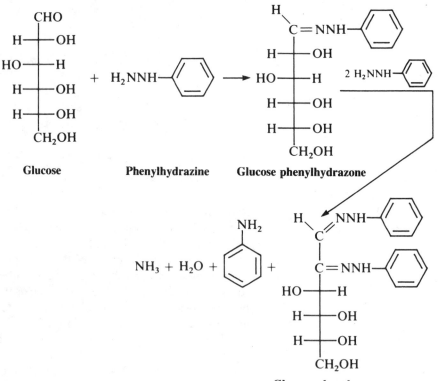

4. Phenylhydrazine (3 mL) is neutralized with 9 mL of acetic acid in a 50-mL Erlenmeyer flask, 15 mL of water is added and the mixture transferred to a graduated cylinder and the volume made up to 30 mL.

Cleaning Up The filtrate can be diluted with water and flushed down the drain. If it is necessary to destroy excess phenylhydrazine, neutralize the solution and add 2 mL of laundry bleach (5.25% sodium hypochlorite) for 1 mL of the reagent. Heat the mixture to 45–50°C for 2 hrs to oxidize the amine, cool the mixture, and flush down the drain.

5. Hydrolysis of a Disaccharide

Sucrose is heated with dil HCl

The object of this experiment is to determine conditions suitable for hydrolysis of a typical disaccharide. Put 1-mL portions of a 0.1 M solution of sucrose in each of five numbered test tubes and add five drops of concentrated hydrochloric acid to tubes 2–5. Let tube 2 stand at room temperature and heat the other four tubes in the hot water bath for the following periods of time: tube 3, 2.5 min; tube 4, 5 min; tubes 1 and 5, 15 min. As each tube is removed from the bath, it is cooled to room temperature, and if it contains acid, adjusted to approximate neutrality by addition of 15 drops of 10% sodium hydroxide. Measure one drop of each neutral solution into a new numbered test tube, add 1 mL of red tetrazolium solution and a drop of 10% sodium hydroxide, and heat the five tubes together for 2 min and watch them closely.

In which of the tubes was hydrolysis negligible, incomplete, and extensive? Does the comparison indicate the minimum heating period required for complete hydrolysis? If not, return the stored solutions to the numbered test tubes, treat each with 1 mL of stock phenylhydrazine reagent, and heat the tubes together for 5 min. On the basis of your results, decide upon a hydrolysis procedure to use in studying unknowns; the same method is applicable to the hydrolysis of methyl glycosides.

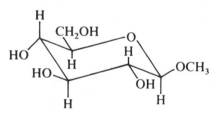

Methyl β-D-glucoside

6. Evaporation Test

Few solid derivatives suitable for identification of sugars are available. Osazones are not suitable since the same osazone can form from more than one sugar. Acetylation, in the case of a reducing sugar, is complicated by the possibility of formation of the α- or the β-anomeric form, or a mixture of both. (See Chapter 52.)

On thorough evaporation of all the water from a solution of glucose, fructose, mannose, or galactose, the sugar is left as a syrup that appears as a

Test for lactose, maltose

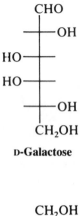

D-Galactose

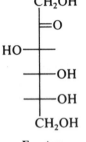

D-Fructose

glassy film on the walls of the container. Evaporation of solutions of lactose or maltose gives white solid products, which are distinguishable because the temperature ranges at which they decompose differ by about 100°C.

This experiment can, if necessary, be carried out on a scale two to four times larger than that given here.

Measure 2 mL of a 0.1 M solution of either lactose or maltose into a test tube, and add a boiling chip and an equal volume of cyclohexane (to hasten the evaporation). Connect the tube through a filter trap to the aspirator. Then turn the water running through the aspirator on at full force and rest the tube horizontally in the steam bath with all but the largest ring removed, so that the whole tube will be heated strongly. If evaporation does not occur rapidly, check the connections and trap to see what is wrong. If a water layer persists for a long time, disconnect and add 1–2 mL of cyclohexane to hasten evaporation. When evaporation appears to be complete, disconnect, rinse the walls of the tube with 1–2 mL of methanol, and evaporate again, when a solid should separate on the walls. Rinse this down with methanol and evaporate again to produce a thoroughly anhydrous product. Then scrape out the solid and determine the melting point (actually, the temperature range of decomposition).

Anhydrous α-maltose decomposes at about 100–120°C, and anhydrous α-lactose at 200–220°C. Note that in the case of an unknown a temperature of decomposition in one range or the other is valid as an index of identity only if the substance has been characterized as a reducing sugar. Before applying the test to an unknown, perform a comparable evaporation of a 0.1 M solution of glucose, fructose, galactose, or mannose.

Cleaning Up   The test solutions should be diluted with water and flushed down the drain.

7. Unknowns

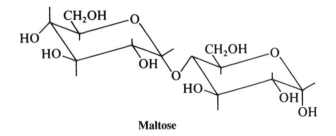

Maltose

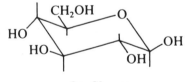

β-D-Glucose

The unknown, supplied as a 0.1 M solution, may be any one of the following substances:

D-Glucose   Maltose
D-Fructose   Sucrose
D-Galactose   Methyl β-D-glucoside
Lactose

You are to devise your own procedure of identification.

Questions

1. What do you conclude is the order of relative reactivity in the RT test of the compounds studied?

2. Which test do you regard as the most reliable for distinguishing reducing from nonreducing sugars, and which for differentiating an α-ketol from a simple aldehyde?

3. Write a mechanism for the acid-catalyzed hydrolysis of a disaccharide.

52

Synthesis of Vitamin K₁: Quinones

Prelab Exercise: Write balanced equations for the reduction of Orange II to aminonaphthol with sodium hydrosulfite and for the oxidation of aminonaphthol to 1,2-naphthoquinone using ferric chloride.

In the first experiment of this chapter the dye Orange II is reduced in aqueous solution with sodium hydrosulfite to water-soluble sodium sulfanilate and 1-amino-2-naphthol (**2**), which precipitates. This intermediate is purified as the hydrochloride and oxidized to 1,2-naphthoquinone (**3**).

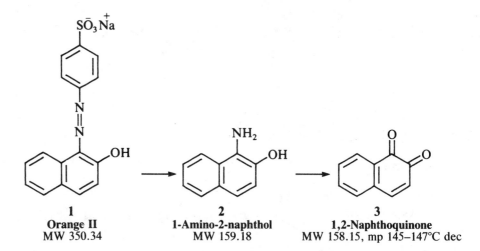

1
Orange II
MW 350.34

2
1-Amino-2-naphthol
MW 159.18

3
1,2-Naphthoquinone
MW 158.15, mp 145–147°C dec

The experiments of this chapter are a sequence of steps for the synthesis of the antihemorrhagic vitamin K₁ (or an analog) starting with a coal-tar hydrocarbon.

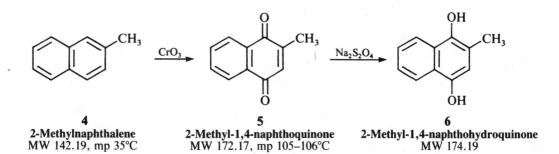

4
2-Methylnaphthalene
MW 142.19, mp 35°C

5
2-Methyl-1,4-naphthoquinone
MW 172.17, mp 105–106°C

6
2-Methyl-1,4-naphthohydroquinone
MW 174.19

2-Methylnaphthalene (**1**) is oxidized with chromic acid to 2-methyl-1,4-naphthoquinone (**5**); the yellow quinone is purified and reduced to its hydroquinone (**6**) by shaking an ethereal solution of the substance with aqueous hydrosulfite solution; the colorless hydroquinone is condensed with phytol; and the substituted hydroquinone (**7**) is oxidized to vitamin K₁ (**9**).

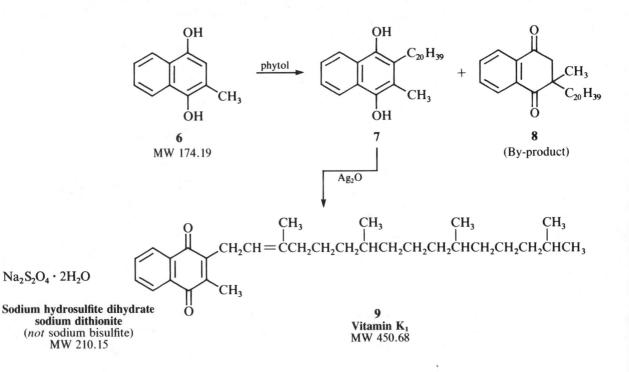

Phytol
MW 296.52, bp 145°C at 0.03–0.04 torr

6
MW 174.19

7

8
(By-product)

Ag₂O

Na₂S₂O₄ · 2H₂O

Sodium hydrosulfite dihydrate
sodium dithionite
(*not* sodium bisulfite)
MW 210.15

9
Vitamin K₁
MW 450.68

CAUTION: Contact of solid sodium hydrosulfite with moist combustible material may cause a fire.

An additional or alternative experiment is conversion of 2-methyl-1,4-naphthoquinone (**5**) through the oxide into the 3-hydroxy compound, phthiocol, which has been isolated from human tubercle bacilli after saponification, probably as a product of cleavage of vitamin K₁ (see Experiment 5).

Experiments

1. 1,2-Naphthoquinone (3)

Two short-time reactions

In a 125-mL Erlenmeyer flask dissolve 3.9 g of the dye Orange II in 50 mL of water and warm the solution to 40–50°C. Add 4.5 g of sodium hydrosulfite dihydrate and swirl until the red color is discharged and a cream-colored or pink precipitate of 1-amino-2-naphthol separates. To coagulate the product, heat the mixture nearly to boiling until it begins to froth, then cool in an ice bath, collect the product on a suction filter, and wash the residue with water. Prepare a solution of 1 mL of concentrated hydrochloric acid, 20 mL of water, and an estimated 50 mg of tin(II) chloride (antioxidant); transfer the precipitate of aminonaphthol to this solution and wash in material adhering to the funnel. Swirl, warm gently, and when all but a little fluffy material has dissolved, filter the solution by suction through a thin layer of decolorizing charcoal. Transfer the filtered solution to a clean flask, add 4 mL of concentrated hydrochloric acid, heat over a hot plate until the precipitated aminonaphthol hydrochloride has been brought into solution, and then cool thoroughly in an ice bath. Collect the crystalline, colorless hydrochloride and wash it with a mixture of 1 mL of concentrated hydrochloric acid and 4 mL of water. Leave the air-sensitive crystalline product in the funnel while preparing a solution for its oxidation. Dissolve 5.5 g of iron(III) chloride crystals ($FeCl_3 \cdot 6H_2O$) in 2 mL of concentrated hydrochloric acid and 10 mL of water by heating, cool to room temperature, and filter by suction.

Wash the crystalline aminonaphthol hydrochloride into a beaker, stir, add more water, and warm to about 35°C until the salt is dissolved. Filter the solution quickly by suction from a trace of residue and stir in the iron(III) chloride solution. 1,2-Naphthoquinone separates at once as a voluminous precipitate and is collected on a suction filter and washed thoroughly to remove all traces of acid. The yield from pure, salt-free Orange II is usually about 75%.

1,2-Naphthoquinone, highly sensitive and reactive, does not have a well-defined melting point but decomposes at about 145–147°C. Suspend a sample in hot water and add concentrated hydrochloric acid. Dissolve a small sample in cold methanol and add a drop of aniline; the red product is 4-anilino-1,2-naphthoquinone.

Cleaning Up   All aqueous solutions can be diluted with water and flushed down the drain since tin and iron are considered of low toxicity. If desired, they can be precipitated as the hydroxides by adding sodium carbonate, collected by vacuum filtration, and placed in the nonhazardous solid waste container.

2. 2-Methyl-1,4-naphthoquinone (5)

CAUTION: CrO₃ is highly toxic and a cancer suspect agent. Work in a hood with this material and avoid skin contact. Wear gloves.

In the hood, clamp a separatory funnel in place to deliver into a 600-mL beaker, which can be cooled in an ice bath when required. The oxidizing solution to be placed in the funnel is prepared by dissolving 50 g of chromium(VI) oxide (CrO_3, chromic anhydride) in 35 mL of water and diluting the dark red solution with 35 mL of acetic acid.[1] In the beaker prepare a mixture of 14.2 g of 2-methylnaphthalene and 150 mL of acetic acid, and without cooling run in small portions of the oxidizing solution. Stir with a thermometer until the temperature rises to 60°C. At this point ice cooling will be required to prevent a further rise in temperature. By alternate addition of reagent and cooling, the temperature is maintained close to 60°C throughout the addition, which can be completed in about 10 min. When the temperature begins to drop spontaneously the solution is heated gently on the steam bath (85–90°C) for 1 h to complete the oxidation.

Reaction time: 1.25 h

Dilute the dark green solution with water nearly to the top of the beaker, stir well for a few minutes to coagulate the yellow quinone, collect the product on a Büchner funnel, and wash it thoroughly with water to remove chromium(III) acetate. The crude material can be crystallized from methanol (40 mL) while still moist, and gives 6.5–7.3 g of satisfactory 2-methyl-1,4-naphthoquinone, mp 105–106°C. The product is to be saved for the preparation of **6**, **9**, and **11**. This quinone must be kept away from light, which converts it into a pale yellow, sparingly soluble polymer.

Cleaning Up   Add 3 M sulfuric acid until the pH is 1. Complete the reduction of any remaining chromic ion by adding solid sodium thiosulfate until the solution becomes cloudy and blue colored. Neutralize with sodium carbonate and filter the flocculent precipitate of $Cr(OH)_3$ through Celite in a Büchner funnel. The filtrate can be diluted with water and flushed down the drain, while the precipitate and Celite should be placed in the heavy metals hazardous waste container.

3. 2-Methyl-1,4-naphthohydroquinone (6)

Short-time reaction

In an Erlenmeyer flask dissolve 2 g of 2-methyl-1,4-naphthoquinone (**5**) in 35 mL of ether by warming on a steam bath, pour the solution into a separatory funnel, and shake with a fresh solution of 4 g of sodium hydrosulfite in 30 mL of water. After passing through a brown phase (quinhydrone) the solution should become colorless or pale yellow in a few minutes; if not, add more hydrosulfite solution. After removing the aqueous layer, shake the ethereal solution with 25 mL of saturated sodium chloride solution and 1–2 mL of saturated hydrosulfite solution to remove the bulk of the water. Filter the ethereal layer by gravity through a filter paper

Note for the instructor

---

1. The chromium(VI) oxide is hygroscopic; weigh it quickly in the hood and do not leave the bottle unstoppered. The substance dissolves very slowly in acetic acid–water mixtures, and solutions are prepared by adding the acetic acid only after the substance has been completely dissolved in water.

Aspirator tube

moistened with ether and about one-third filled with anhydrous sodium sulfate. Evaporate the filtrate on the steam bath until nearly all the solvent has been removed, cool, and add petroleum ether. The hydroquinone separating as a white or grayish powder is collected, washed with petroleum ether, and dried; the yield is about 1.9 g (the substance has no sharp mp).

Cleaning Up   The aqueous layer and saturated sodium chloride solution should be diluted with water and flushed down the drain. After the solvent is allowed to evaporate from the drying agent in the hood, it is placed in the nonhazardous solid waste container. The petroleum ether from the crystallization should be placed in the organic solvents container.

4. Vitamin K₁ (2-Methyl-3-phytyl-1,4-naphthoquinone, (9))

CAUTION: Dioxane is a carcinogen. Carry out this experiment in a hood.

Place 1.5 g of phytol[2] and 10 mL of dioxane (see margin note) in a 50-mL Erlenmeyer flask and warm to 50°C on a hot plate. Prepare a solution of 1.5 g of 2-methyl-1,4-naphthohydroquinone (**6**) and 1.5 mL of boron trifluoride etherate in 10 mL of dioxane, and add this in portions with a Pasteur pipette in the course of 15 min with constant swirling and while maintaining a temperature of 50°C (do not overheat). Continue in the same way for 20 min longer. Cool to 25°C, wash the solution into a separatory funnel with 40 mL of ether, and wash the orange-colored ethereal solution with two 40-mL portions of water to remove boron trifluoride and dioxane. Extract the unchanged hydroquinone with a freshly prepared solution of 2 g of sodium hydrosulfite in 40 mL of 2% aqueous sodium hydroxide and 10 mL of a saturated sodium chloride solution (which helps break in the resulting emulsion). Shake vigorously for a few minutes, during which time any red color should disappear and the alkaline layer should acquire a bright yellow color. After releasing the pressure through the stopcock, allow the layers to separate, keeping the funnel stoppered as a precaution against oxidation. Draw off the yellow liquor and repeat the extraction a second and a third time, or until the alkaline layer remains practically colorless. Separate the faintly colored ethereal solution, dry it over anhydrous sodium sulfate, filter into a tared flask, and evaporate the filtrate on the steam bath—eventually with evacuation at the aspirator. The total oil, which becomes waxy on cooling, amounts to 1.7–1.9 g.

Add 10 mL of petroleum ether (bp 35–60°C) and boil and manipulate with a spatula until the brown mass has changed to a white paste. Wash the paste into small centrifuge tubes with 10–20 mL of fresh petroleum ether, make up the volume of paired tubes to the same point, cool well in ice, and centrifuge. Decant the brown supernatant liquor into the original tared

Note for the instructor

2. Supplier: Aldrich Chemical Co. Geraniol can be used instead of expensive phytol; it reacts with 2-methyl-1,4-naphthohydroquinone to give a product similar in chemical and physical properties to the natural vitamin and with pronounced antihemorrhagic activity.

flask, fill the tubes with fresh solvent, and stir the white sludge to an even suspension. Then cool, centrifuge, and decant as before. Evaporation of the decanted liquor and washings gives 1.1–1.3 g of residual oil, which can be discarded. Dissolve the portions of washed white sludge of vitamin K_1 hydroquinone in a total of 10–15 mL of absolute ether, and add a little decolorizing charcoal for clarification, if the solution is pink or dark. Add 1 g of silver oxide and 1 g of anhydrous sodium sulfate. Shake for 20 min, filter into a tared flask, and evaporate the clear yellow solution on the steam bath, removing traces of solvent at the water pump. Undue exposure to light should be avoided when the material is in the quinone form. The residue is a light yellow, rather mobile oil consisting of pure vitamin K_1; the yield is about 0.6–0.9 g. A sample for preservation is transferred with a Pasteur pipette to a small specimen vial wrapped in metal foil or black paper to exclude light.

To observe a characteristic color reaction, transfer a small bit of vitamin on the end of a stirring rod to a test tube, stir with 0.1 mL of alcohol, and add 0.1 mL of 10% alcoholic potassium hydroxide solution; the end pigment responsible for the red color is phthiocol.

Cleaning Up   The aqueous layer from the first part of the reaction should be neutralized with sodium carbonate, diluted with water, and flushed down the drain. The aqueous layer containing hydrosulfite, saturated sodium chloride, and sodium hydroxide are treated with household bleach (5.25% sodium hypochlorite solution) until no further reaction is evident. After neutralizing the solution with dilute hydrochloric acid, the solution is diluted with water and flushed down the drain. Shake the mixture of silver oxide, silver, and sodium sulfate with water to dissolve the sodium sulfate, and then recover the mixture of silver and silver oxide by vacuum filtration.

5. Phthiocol (2-Methyl-3-hydroxy-1,4-naphthoquinone, (11))

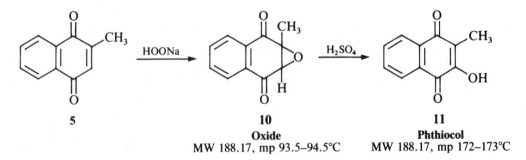

5

10
Oxide
MW 188.17, mp 93.5–94.5°C

11
Phthiocol
MW 188.17, mp 172–173°C

Dissolve 1 g of 2-methyl-1,4-naphthoquinone (**5**) in 10 mL of alcohol by heating, and let the solution stand while the second reagent is prepared by dissolving 0.2 g of anhydrous sodium carbonate in 5 mL of water and

Two short-time reactions

CAUTION: *Avoid contamination of hydrogen peroxide with metal salts, organic compounds, or metal spatulas: explosive.*

adding (cold) 1 mL of 30% hydrogen peroxide solution. Cool the quinone solution under the tap until crystallization begins, add the peroxide solution all at once, and cool the mixture. The yellow color of the quinone should be discharged immediately. Add about 100 mL of water, cool in ice, and collect the colorless, crystalline epoxide (**10**); yield 0.97 g, mp 93.5–94.5°C (pure: 95.5–96.5°C).

To 1 g of **10** in a 25-mL Erlenmeyer flask add 5 mL of concentrated sulfuric acid; stir if necessary to produce a homogeneous deep red solution, and after 10 min cool this in ice and slowly add 20 mL of water. The precipitated phthiocol can be collected, washed, and crystallized by dissolving in methanol (25 mL), adding a few drops of hydrochloric acid to give a pure yellow color, treating with decolorizing charcoal, concentrating the filtered solution, and diluting to the saturation point. Alternatively, the yellow suspension is washed into a separatory funnel and the product extracted with a mixture of 25 mL each of toluene and ether. The organic layer is dried over anhydrous sodium sulfate and evaporated to a volume of about 10 mL for crystallization. The total yield of pure phthiocol (**11**), mp 172–173°C, is about 0.84–0.88 g.

Cleaning Up   Combine all aqueous solutions, neutralize with sodium carbonate, and flush down the drain. Organic filtrates go in the organic solvents container.

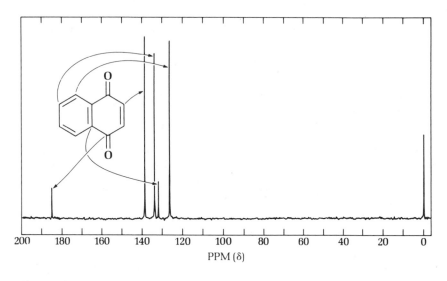

FIG. 1   ¹³C nmr spectrum of 1,4-naphthoquinone (22.6 MHz).

Questions

1. To what general class of compounds does phytol belong?

2. Write a mechanism for the reaction of 2-methyl-1,4-naphthoquinone epoxide (**10**) with sulfuric acid to form phthiocol.

The Friedel-Crafts Reaction: Anthraquinone and Anthracene

Prelab Exercise: Draw the mechanism for the cyclization of 2-benzoylbenzoic acid to anthraquinone using concentrated sulfuric acid.

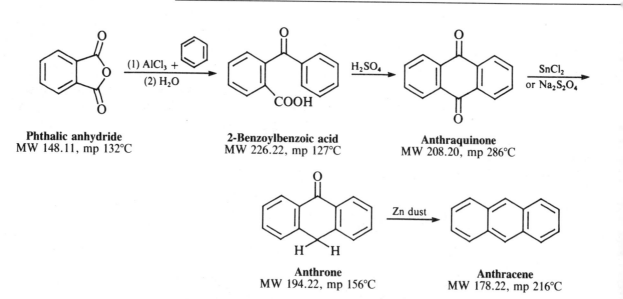

Phthalic anhydride
MW 148.11, mp 132°C

2-Benzoylbenzoic acid
MW 226.22, mp 127°C

Anthraquinone
MW 208.20, mp 286°C

Anthrone
MW 194.22, mp 156°C

Anthracene
MW 178.22, mp 216°C

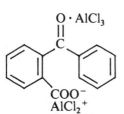

Complex salt

The Friedel-Crafts reaction of phthalic anhydride with excess benzene as solvent and two equivalents of aluminum chloride proceeds rapidly and gives a complex salt of 2-benzoylbenzoic acid in which one mole of aluminum chloride has reacted with the acid function to form the salt $RCO_2^-AlCl_2^+$ and a second mole is bound to the carbonyl group. On addition of ice and hydrochloric acid the complex is decomposed and basic aluminum salts are brought into solution.

Treatment of 2-benzoylbenzoic acid with concentrated sulfuric acid effects cyclodehydration to anthraquinone, a pale-yellow, high-melting compound of great stability. Because anthraquinone can be sulfonated only under forcing conditions, a high temperature can be used to shorten the reaction time without loss in yield of product; the conditions are so adjusted that anthraquinone separates from the hot solution in crystalline form favoring rapid drying.

Reduction of anthraquinone to anthrone can be accomplished rapidly on a small scale with tin(II) chloride in acetic acid solution. A second

$Na_2S_2O_4$
Sodium Hydrosulfite

method, which involves refluxing anthraquinone with an aqueous solution of sodium hydroxide and sodium hydrosulfite, is interesting to observe because of the sequence of color changes: anthraquinone is reduced first to a deep red liquid containing anthrahydroquinone diradical dianion; the red color then gives place to a yellow color characteristic of anthranol radical anion; as the alkali is neutralized by the conversion of $Na_2S_2O_4$ to $2\ NaHSO_3$, anthranol ketonizes to the more stable anthrone. The second method is preferred in industry, because sodium hydrosulfite costs less than half as much as tin(II) chloride and because water is cheaper than acetic acid and no solvent recovery problem is involved.

Reduction of anthrone to anthracene is accomplished by refluxing in aqueous sodium hydroxide solution with activated zinc dust. The method has the merit of affording pure, beautifully fluorescent anthracene.

Fieser's Solution

A solution of 2 g of sodium anthraquinone-2-sulfonate and 15 g of sodium hydrosulfite in 100 mL of a 20% aqueous solution of potassium hydroxide affords a blood-red solution of the diradical dianion:

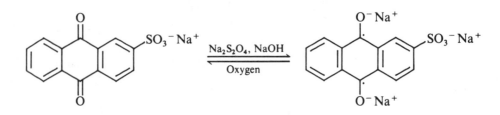

This solution has a remarkable affinity for oxygen. It is used to remove traces of oxygen from gases such as nitrogen or argon when it is desirable to render them absolutely oxygen-free. This solution has a capacity of about 800 mL of oxygen. The color fades and the solution turns brown when it is exhausted.

Experiments

1. 2-Benzoylbenzoic Acid

This Friedel-Crafts reaction is conducted in a 500-mL round-bottomed flask equipped with a short condenser. A trap for collecting liberated hydrogen chloride is connected to the top of the condenser by rubber tubing of sufficient length to make it possible either to heat the flask on the steam bath or to plunge it into an ice bath. The trap is a suction flask half filled with water and fitted with a delivery tube inserted to within 1 cm of the surface of the water (see Fig. 1 in Chapter 13).

CAUTION: *Benzene is a carcinogen. Carry out this experiment in a hood. Avoid contact of the benzene with the skin or wear gloves. Do not carry out this reaction if adequate hood facilities are unavailable.*

Short reaction period, high yield

Fifteen grams of phthalic anhydride and 75 mL of reagent grade benzene (see margin note) are placed in the flask and this solution is cooled in an ice bath until the benzene begins to crystallize. Ice cooling serves to moderate the vigorous reaction, which otherwise might be difficult to control. Thirty grams of anhydrous aluminum chloride[1] is added, the condenser and trap are connected, and the flask is shaken well and warmed for a few minutes by the heat of the hand. If the reaction does not start, the flask is warmed *very gently* by holding it for a few seconds over the steam bath. At the first sign of vigorous boiling, or evolution of hydrogen chloride, the flask is held over the ice bath in readiness to cool it if the reaction becomes too vigorous. This gentle, cautious heating is continued until the reaction is proceeding smoothly enough to be refluxed on the steam bath. This point is reached in about 5 min. Continue the heating on the steam bath, swirl the mixture, and watch it carefully for sudden separation of the addition compound, since the heat of crystallization is such that it may be necessary to plunge the flask into the ice bath to moderate the process. Once the addition compound has separated as a thick paste, heat the mixture for 10 min more on the steam bath, remove the condenser, and swirl the flask in an ice bath until cold. (Should no complex separate, heat for 10 min more and then proceed as directed.) Take the flask and ice bath to the hood, weigh out 100 g of ice, add a few small pieces of ice to the mixture, swirl and cool as necessary, and wait until the ice has reacted before adding more. After the 100 g of ice have been added and the reaction of decomposition has subsided, add 20 mL of concentrated hydrochloric acid, 100 mL of water, swirl vigorously, and make sure that the mixture is at room temperature. Then add 50 mL of water, swirl vigorously, and again make sure the mixture is at room temperature. Add 50 mL of ether and, with a flattened stirring rod, dislodge solid from the neck and walls of the flask and break up lumps at the bottom. To further promote hydrolysis of the addition compound, extraction of the organic product, and solution of basic aluminum halides, stopper the flask with a cork and shake vigorously for several minutes.

When most of the solid has disappeared, pour the mixture through a funnel into a separatory funnel until the separatory funnel is nearly filled. Discard the lower aqueous layer. Pour the rest of the mixture into the separatory funnel, rinse the reaction flask with fresh ether, and again drain off the aqueous layer. To reduce the fluffy, dirty precipitate that appears at the interface, add 10 mL of concentrated hydrochloric acid and 25 mL of water, shake vigorously for 2–3 min, and drain off the aqueous layer. If some interfacial dirty emulsion still persists, decant the benzene-ether solution through the mouth of the funnel into a filter paper for gravity

---

1. This is best weighed in a stoppered test tube. The chloride should be from a freshly opened bottle and should be weighed and transferred in the hood. It is very hygroscopic; work rapidly to avoid exposure of the compound to air. It releases hydrogen chloride on exposure to water. The quality of the aluminum chloride determines the success of this experiment.

filtration and use fresh ether to rinse the funnel. Clean the funnel and pour in the filtered benzene-ether solution. Shake the solution with a portion of dilute hydrochloric acid, and then isolate the reaction product by either of the following procedures:

Alternative procedures

1. Add 50 mL of 3M sodium hydroxide solution, shake thoroughly, and separate the aqueous layer.[2] Extract with a further 25-mL portion of aqueous alkali and combine the extracts. Wash with 10 mL of water, and add this aqueous solution to the 75 mL of aqueous extract already collected. Discard the benzene-ether solution. Acidify the 85 mL of combined alkaline extract with concentrated hydrochloric acid to pH 1–2 and, if the *o*-benzoylbenzoic acid separates as an oil, cool in ice and rub the walls of the flask with a stirring rod to induce crystallization of the hydrate; collect the product and wash it well with water. This material is the monohydrate $C_6H_5COC_6H_4CO_2H \cdot H_2O$. To convert it into anhydrous *o*-benzoylbenzoic acid, put it in a tared, 250-mL round-bottomed flask, evacuate the flask at the full force of the aspirator, and heat it in the open rings of a steam bath covering the flask with a towel. Check the weight of the flask and contents for constancy after 45 min, 1 h, and 1.25 h. The yield is usually 19–21 g, mp 126–127°C.

2. Filter the benzene-ether solution through anhydrous sodium sulfate for superficial drying, put it into a 250-mL round-bottomed flask, and distill over the steam bath through a condenser into an ice-cooled receiver until the volume in the distilling flask is reduced to about 55 mL. Add ligroin slowly until the solution is slightly turbid and let the product crystallize at 25°C and then at 5°C. The yield of anhydrous, colorless, well-formed crystals, mp 127–128°C, is about 18–20 g.

Discard benzene-ether filtrates in the container provided

Drying time: about 1 h

Extinguish flames

2. Anthraquinone

Handle hot sulfuric acid with care: highly corrosive

Reaction time: 10 min

Place 5.0 g of 2-benzoylbenzoic acid[3] (anhydrous) in a 125-mL round-bottomed flask, add 25 mL of concentrated sulfuric acid, and heat on the steam bath with swirling until the solid is dissolved. Then clamp the flask over a microburner, insert a thermometer, raise the temperature to 150°C, and heat to maintain a temperature of 150–155°C for 5 min. Let the solution cool to 100°C, remove the thermometer after letting it drain, and, with a Pasteur pipette, add 5 mL of water by drops with swirling to keep the precipitated material dissolved as long as possible so that it will separate as small, easily filtered crystals. Let the mixture cool further, dilute with water until the flask is full, again let cool, collect the product by suction filtration, and wash well with water. Then remove the filtrate and wash the filter flask, return the funnel to the filter flask without applying suction, and test the

2. The nature of a yellow pigment that appears in the first alkaline extract is unknown; the impurity is apparently transient, for the final product dissolves in alkali to give a colorless solution.

3. Available from the Aldrich Chemical Co.

filter cake for unreacted starting material as follows: Dilute 10 mL of concentrated ammonia solution with 50 mL of water, pour the solution onto the filter, and loosen the cake so that it is well leached. Then apply suction, wash the cake with water, and acidify a few milliliters of the filtrate. If there is no precipitate upon acidifying, the yield of anthraquinone should be close to the theoretical because it is insoluble in water. Dry the product to constant weight but do not take the melting point since it is so high (mp 286°C).

Cleaning Up   The aqueous filtrate, after neutralization with sodium carbonate, is diluted with water and flushed down the drain.

3. Anthrone

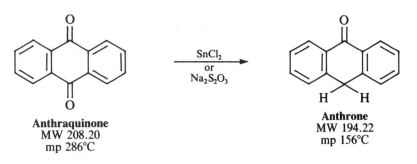

Anthraquinone
MW 208.20
mp 286°C

Anthrone
MW 194.22
mp 156°C

Tin(II) Chloride Reduction

Procedure 1

$SnCl_2 \cdot 2H_2O$
Tin(II) chloride dihydrate

In a 100-mL round-bottomed flask provided with a reflux condenser put 5.0 g of anthraquinone, 40 mL of acetic acid, and a solution made by warming 13 g of tin(II) chloride dihydrate with 13 mL of concentrated hydrochloric acid. Add a boiling stone to the reaction mixture, note the time, and reflux gently until crystals of anthraquinone have completely disappeared (8–10 min); then reflux 15 min longer and record the total time. Disconnect the flask, heat it on the steam bath, and add water (about 12 mL) in 1-mL portions until the solution is saturated. Let the solution stand for crystallization. Collect and dry the product and take the melting point (156°C given). Dispose of the filtrate in the container provided. The yield of pale-yellow crystals is 4.3 g.

Cleaning Up   Dilute the filtrate with a large volume of water and flush the solution down the drain.

Hydrosulfite Reduction

Procedure 2

In a 500-mL round-bottomed flask, which can be heated under reflux, put 5.0 g of anthraquinone, 6 g of sodium hydroxide, 15 g of sodium hydrosulfite, and 130 mL of water. Heat over a free flame and swirl for a few minutes to convert the anthraquinone into the deep red anthra-hydroquinone anion. Note that particles of different appearance begin to separate even before the anthraquinone has all dissolved. Arrange for refluxing and reflux for 45 min; cool, filter the product, and wash it well and let dry. Note the weight of the product and melting point of the crude material and then crystallize it from 95% ethanol; the solution may require filtration to remove insoluble impurities. Record the approximate volume of solvent used and, if the first crop of crystals recovered is not satisfactory, concentrate the mother liquor and secure a second crop.

Cleaning Up   The aqueous filtrate is treated with household bleach (sodium hypochlorite) until reaction is complete. Remove the solid by filtration. It may be placed in the nonhazardous solid waste container. The filtrate can be diluted with water and flushed down the drain.

4. Comparison of Results

Compare your results with those obtained by neighbors using the alternative procedure with respect to yield, quality of product, and working time. Which is the better laboratory procedure? Then consider the cost of the three solvents concerned, the cost of the two reducing agents (current prices can be looked up in a catalog), the relative ease of recovery of the organic solvents, and the prudent disposal of by-products, and decide which method would be preferred as a manufacturing process.

5. Fluorescent Anthracene

Reflux time: 30 min

Put 10 g of zinc dust into a 500-mL round-bottomed flask and activate the dust by adding 60 mL of water and 1 mL of copper(II) sulfate solution (Fehling's solution I) and swirling for a minute or two. Add 4.0 g of anthrone, 10 g of sodium hydroxide, and 100 mL of water; attach a reflux condenser, heat to boiling, note the time, and start refluxing the mixture. Anthrone at first dissolves as the yellow anion of anthranol, but anthracene soon begins to separate as a white precipitate. In about 15 min the yellow color initially observed on the walls disappears, but refluxing should be continued for a full 30 min. Then stop the heating, use a water wash bottle to rinse down anthracene that has lodged in the condenser, and filter the still hot mixture on a large Büchner funnel. It usually is possible to decant from, and so remove, a mass of zinc. After liberal washing with water, blow or shake out the gray cake into a 400-mL beaker and rinse funnel and paper with water. To remove most of the zinc metal and zinc oxide, add 20 mL of concentrated hydrochloric acid, heat on the steam bath with stirring for 20–25 min when initial frothing due to liberated hydrogen should have ceased. Collect the now nearly white precipitate on a large Büchner funnel and, after liberal washing with water, release the suction, rinse the walls of the funnel with methanol, use enough methanol to cover the cake, and then

Extinguish flames. Hydrogen liberated.

Methanol removes water without dissolving much anthracene

apply suction. Wash again with enough methanol to cover the cake and then remove solvent thoroughly by suction.

The product need not be dried before crystallization from toluene. Transfer the methanol-moist material to a 125-mL Erlenmeyer and add 60 mL of toluene; a liberal excess of solvent is used to avoid crystallization in the funnel. Make sure there are no flames nearby and heat the mixture on the hot plate to bring the anthracene into solution with a small residue of zinc remaining. Filter by gravity by the usual technique to remove zinc. From the filtrate anthracene is obtained as thin, colorless, beautifully fluorescent plates; yield about 2.8 g. In washing the equipment with acetone, you should be able to observe the striking fluorescence of very dilute solutions. The fluorescence is quenched by a bare trace of impurity.

Caution: Highly flammable solvent

Cleaning Up   The aqueous solution is diluted with water, neutralized with sodium carbonate, filtered under vacuum to remove zinc hydroxide, and the filtrate diluted with water and flushed down the drain. The solid waste can be placed in the nonhazardous solid waste container. Toluene and acetone are placed in the organic solvents container.

Questions

1. Calculate the number of moles of hydrogen chloride liberated in the microscale synthesis of 2-(4-benzoyl)benzoic acid. If this gaseous acid were dissolved in water, hydrochloric acid would be formed. How many milliliters of concentrated hydrochloric acid would be formed in this reaction? The concentrated acid is 12 molar in HCl.

2. Write a mechanism for the formation of anthraquinone from 2-benzoylbenzoic acid, indicating clearly the role of sulfuric acid. What is the name commonly given to this type of reaction?

Derivatives of 1,2-Diphenylethane—
A Multistep Synthesis[1]

Procedures are given in the next seven chapters for rapid preparation of small samples of twelve related compounds starting with benzaldehyde and phenylacetic acid. The quantities of reagents specified in the procedures are often such as to provide somewhat more of each intermediate than is required for completion of subsequent steps in the sequence of reactions. If the experiments are dovetailed, the entire series of preparations can be completed in very short working time. For example, one can start the preparation of benzoin (record the time of starting and do not rely on memory), and during the reaction period start the preparation of α-phenylcinnamic acid; this requires refluxing for 35 min, and while it is proceeding the benzoin preparation can be stopped when the time is up and the product let crystallize. The α-phenylcinnamic acid mixture can be let stand (and cooled) until one is ready to work it up. Also, while a crystallization is proceeding one may want to observe the crystals occasionally but should utilize most of the time for other operations.

Points of interest concerning stereochemistry and reaction mechanisms are discussed in the introductions to the individual chapters. Since several of the compounds have characteristic ultraviolet or infrared absorption spectra, pertinent spectroscopic constants are recorded and brief interpretations of the data are presented.

Question. Starting with 150 mg of benzaldehyde and assuming an 80% yield on each step, what yield of diphenylacetylene, in grams, might you expect? From the information given and assuming an 80% yield on the last two reactions, what yield of stilbene dibromide would you expect employing 150 mg of benzaldehyde?

Note for the instructor

1. If the work is well organized and proceeds without setbacks, the experiments can be completed in about four laboratory periods. The instructor may elect to name a certain number of periods in which the student is to make as many of the compounds as possible; the instructor may also decide to require submission only of the end products in each series.

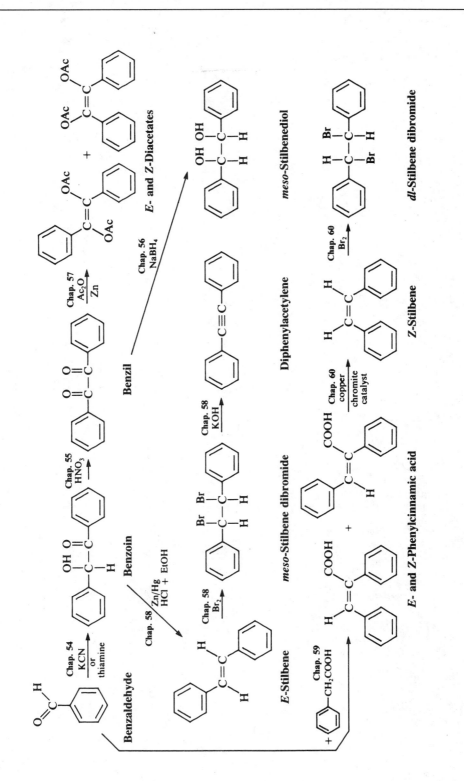

Benzaldehyde

Chap. 54 KCN or thiamine

Benzoin

Chap. 55 HNO₃

Benzil

Chap. 57 Ac₂O Zn

E- and *Z-*Diacetates

Chap. 56 NaBH₄

*meso-*Stilbenediol

Chap. 60 Br₂

*dl-*Stilbene dibromide

Chap. 58 Zn/Hg HCl + EtOH

Chap. 58 Br₂

*meso-*Stilbene dibromide

Chap. 58 KOH

Diphenylacetylene

*E-*Stilbene

Chap. 59 —CH₂COOH

E- and *Z-*Phenylcinnamic acid

Chap. 60 copper chromite catalyst

*Z-*Stilbene

54

The Benzoin Condensation: Cyanide Ion and Thiamine Catalyzed

Prelab Exercise: What purpose does the sodium hydroxide serve in the thiamine-catalyzed benzoin condensation?

The reaction of two moles of benzaldehyde to form a new carbon-carbon bond is known as the benzoin condensation. It is catalyzed by two rather different catalysts—cyanide ion and the vitamin thiamine—which, on close examination, are seen to function in exactly the same way.

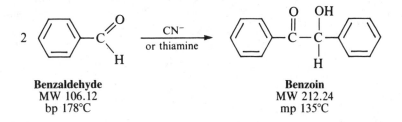

Benzaldehyde
MW 106.12
bp 178°C

Benzoin
MW 212.24
mp 135°C

Consider first the cyanide ion-catalyzed reaction. The cyanide ion attacks the carbonyl oxygen to form a stable cyanohydrin, mandelonitrile, a liquid of bp 170°C that under the basic conditions of the reaction loses a proton to give a resonance-stabilized carbanion, **A**. The carbanion attacks another molecule of benzaldehyde to give **B**, which undergoes a proton transfer and loses cyanide to give benzoin. Evidence for this mechanism lies in the failure of 4-nitrobenzaldehyde to undergo the reaction, because the nitro group reduces the nucleophilicity of the anion in **A**. On the other hand a strong electron-donating group in the 4-position of the phenyl ring makes the loss

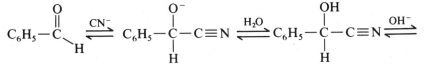

Benzaldehyde

Mandelonitrile
bp 170°C

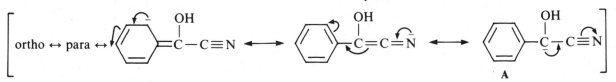

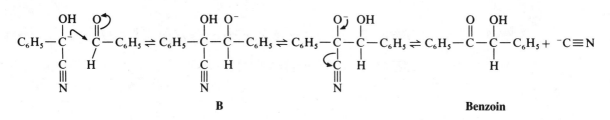

B **Benzoin**

of the proton from the cyanohydrin very difficult, and thus 4-dimethyl-aminobenzaldehyde also does not undergo the benzoin condensation with itself.

A number of biochemical reactions bear a close resemblance to the benzoin condensation but are not, obviously, catalyzed by the highly toxic cyanide ion. Some 30 years ago Breslow proposed that vitamin B_1, thiamine hydrochloride, in the form of the coenzyme thiamine pyrophosphate, can function in a manner completely analogous to cyanide ion in promoting reactions like the benzoin condensation. The resonance-stabilized conjugate base of the thiazolium ion, thiamine, and the resonance-stabilized carbanion, **C**, which it forms, are again the keys to the reaction. Like the cyanide ion, the thiazolium ion has just the right balance of nucleophilicity, ability to stabilize the intermediate anion, and good leaving group qualities.

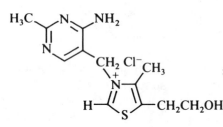

Cyanide ion binds irreversibly to hemoglobin, rendering it useless as a carrier of oxygen

Thiamine hydrochloride **Thiamine**

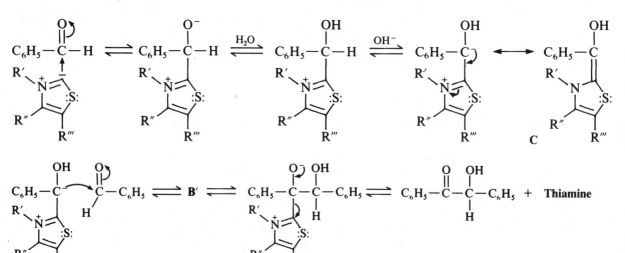

The importance of thiamine is evident in that it is a vitamin, an essential substance that must be provided in the diet to prevent beriberi, a nervous system disease.

In the reactions that follow, cyanide ion functions as a fast and efficient catalyst, although in large quantities it is highly toxic. The thiamine-catalyzed reaction is much slower but the catalyst is edible.

Experiments

1. Cyanide Ion-Catalyzed Benzoin Condensation

CAUTION: Potassium cyanide is poisonous. Do not handle if you have open cuts on your hands.

Never acidify a cyanide solution (HCN gas is evolved). Wash hands after handling cyanide.

Place 1.5 g of potassium cyanide (see margin note) in a 125-mL round-bottomed flask, dissolve it in 15 mL of water, add 30 mL of 95% ethanol and 15 mL of pure benzaldehyde,[1] introduce a boiling stone, attach a short condenser, and reflux the solution gently on the flask heater for 30 min (Fig. 1). Remove the flask, cool it in an ice bath, and, if no crystals appear within a few minutes, withdraw a drop on the stirring rod and rub it against the neck of the flask to induce crystallization. When crystallization is complete, collect the product and wash it free of yellow mother liquor with a 1:1 mixture of 95% ethanol and water. Usually this first-crop material is colorless and of satisfactory melting point (134–135°C); usual yield 10–12 g.[2]

Cleaning Up   Add the aqueous filtrate to 10 mL of a 1% sodium hydroxide solution. Add 50 mL of household bleach (5.25% sodium hypochlorite) to oxidize the cyanide ion. The resulting solution can be tested for cyanide using the Prussian blue test described in Chapter 70, Part 3a. When cyanide is not present, the solution can be diluted with water and flushed down the drain. Ethanol used in crystallization should be placed in the organic solvents container.

2. Thiamine-Catalyzed Benzoin Condensation

Reaction time: 30 min

Place 2.6 g of thiamine hydrochloride in a 125-mL Erlenmeyer flask, dissolve it in 8 mL of water, add 30 mL of 95% ethanol, and cool the solution in an ice bath. Add 5 mL of 3 M sodium hydroxide dropwise with swirling such that the temperature of the solution does not rise above 20°C.

Note for the instructor

---

1. Commercial benzaldehyde inhibited against autoxidation with 0.1% hydroquinone is usually satisfactory. If the material available is yellow or contains benzoic acid crystals it should be shaken with equal volumes of 5% sodium carbonate solution until carbon dioxide is no longer evolved and the upper layer dried over calcium chloride and distilled (bp 178–180°C), with avoidance of exposure of the hot liquid to air. The distillation step can be omitted if the benzaldehyde is colorless.

2. Concentration of the mother liquor to a volume of 20 mL gives a second crop (1.8 g, mp 133–134.5°C); best total yield 13.7 g (87%). Recrystallization can be accomplished with either methanol (11 mL/g) or 95% ethanol (7 mL/g) with 90% recovery in the first crop.

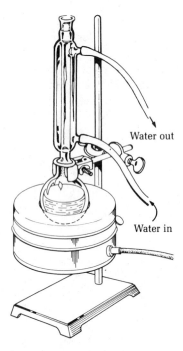

FIG. 1 Reflux apparatus. A steam bath or electric flask heater is preferred to a burner.

To the yellow solution add 15 mL of pure benzaldehyde[3] and heat the mixture at 60°C for 1–1.5 h. The progress of the reaction can be followed by thin-layer chromatography. Alternatively, the reaction mixture can be stored at room temperature for at least 24 h. (The rate of most organic reactions doubles for each 10°C rise in temperature.)

Cool the reaction mixture in an ice bath. If crystallization does not occur, withdraw a drop of solution on a stirring rod and rub it against the inside surface of the flask to induce crystallization. Collect the product by suction filtration and wash it free of yellow mother liquor with a 1:1 mixture of 95% ethanol and water. The product should be colorless and of sufficient purity (mp 134–135°C) to use in subsequent reactions; usual yield is 10–12 g. If desired, the moist product can be recrystallized from 95% ethanol (8 mL per g).

Cleaning Up   The aqueous filtrate, after neutralization with dilute hydrochloric acid, is diluted with water and flushed down the drain. Ethanol used in crystallization should be placed in the organic solvents container.

Questions

1. Speculate on the structure of the compound formed when 4-dimethylaminobenzaldehyde is condensed with 4-chlorobenzaldehyde.

2. Why might the presence of benzoic acid be deleterious to the benzoin condensation?

3. How many π-electrons are in the thiazoline ring of thiamine hydrochloride? of thiamine?

---

3. See footnote 1.

55

Nitric Acid Oxidation. Preparation of Benzil from Benzoin. Synthesis of a Heterocycle: Diphenylquinoxaline

Prelab Exercise: Write a detailed mechanism for the formation of 2,3-dimethylquinoxaline.

Benzoin can be oxidized to the α-diketone, benzil, very efficiently by nitric acid or by copper(II) sulfate in pyridine. On oxidation with sodium dichromate in acetic acid the yield is lower because some material is converted into benzaldehyde by cleavage of the bond between two oxidized carbon atoms and activated by both phenyl groups (a). Similarly, hydrobenzoin on oxidation with dichromate or permanganate yields chiefly benzaldehyde and only a trace of benzil (b).

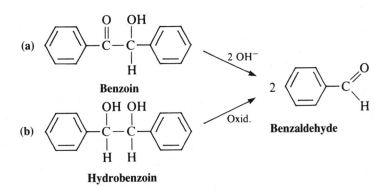

Test for the Presence of Unoxidized Benzoin

Dissolve about 0.5 mg of crude or purified benzil in 0.5 mL of 95% ethanol or methanol and add one drop of 10% sodium hydroxide. If benzoin is present the solution soon acquires a purplish color owing to a complex of benzil with a product of autoxidation of benzoin. If no color develops in 2–3 min, an indication that the sample is free from benzoin, add a small amount of benzoin, observe the color that develops, and note that if the test tube is stoppered and shaken vigorously the color momentarily disappears; when the solution is allowed to stand, the color reappears.

FIG. 1   The ultraviolet spectrum of benzoin. λ_{max}^{EtOH} 247 nm (ε = 13,200). Concentration: 12.56 mg/L = 5.92×10^{-5} mole/L. See Chapter 21 (Ultraviolet Spectroscopy) for the relationship between the extinction coefficient, ε, absorbance, A, and concentration, C. The absorption band at 247 nm is attributable to the presence of the phenyl ketone group,

$$\begin{matrix} & O \\ & \| \\ C_6H_5 — & C — \end{matrix},$$

in which the carbonyl group is conjugated with the benzene ring. Aliphatic α,β-unsaturated ketones, R—CH=CH—C=O, show selective absorption of ultraviolet light of comparable wavelength.

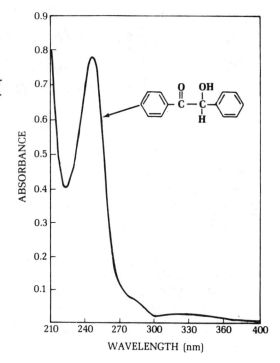

FIG. 2   Infrared spectrum of benzoin in $CHCl_3$.

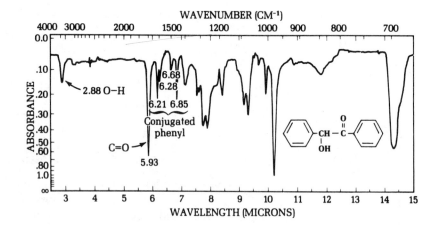

Experiments

1. Nitric Acid Oxidation of Benzoin

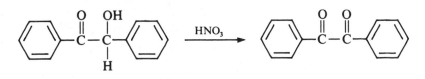

Benzoin	Benzil
MW 212.24, mp 135°C	MW 210.23, mp 94–95°C

Reaction time: 11 min

Heat a mixture of 4 g of benzoin and 14 mL of concentrated nitric acid on the steam bath for 11 min. Carry out the reaction under a hood or use an aspirator tube near the top of the flask to remove nitrogen oxides. Add 75 mL of water to the reaction mixture, cool to room temperature, and swirl for a minute or two to coagulate the precipitated product; collect and wash the yellow solid on a Hirsch funnel, pressing the solid well on the filter to squeeze out the water. This crude product (dry weight 3.7–3.9 g) need not be dried but can be crystallized at once from ethanol. Dissolve the product in 10 mL of hot ethanol, add water dropwise to the cloud point, and set aside to crystallize. Record the yield, crystalline form, color, and mp of the purified benzil.

Test for the Presence of Unoxidized Benzoin. Dissolve about 0.5 mg of crude or purified benzil in 0.5 mL of 95% ethanol or methanol and add one drop of 10% sodium hydroxide. If benzoin is present the solution soon acquires a purplish color owing to a complex of benzil with a product of autoxidation of benzoin. If no color develops in 2–3 min, an indication that the sample is free from benzoin, add a small amount of benzoin, observe the observe the color that develops, and note that if the test tube is stoppered and shaken vigorously the color momentarily disappears; when the solution is then let stand, the color reappears.

Cleaning Up   The aqueous filtrate should be neutralized with sodium carbonate, diluted with water, and flushed down the drain. Ethanol used in crystallization should be placed in the organic solvents container.

2. Benzil Quinoxaline Preparation

Handle o-phenylenediamine with care. Similar compounds (hair dyes) are mild carcinogens. Carry out reaction in hood.

Commercial *o*-phenylenediamine is usually badly discolored (air oxidation) and gives a poor result unless purified as follows. Place 200 mg of material in a 20 × 150-mm test tube, evacuate the tube at full aspirator suction, clamp it in a horizontal position, and heat the bottom of the tube with a hot sand bath to sublime colorless *o*-phenylenediamine from the dark residue

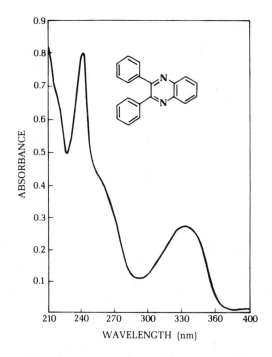

FIG. 3   Ultraviolet spectrum of quinoxaline derivative. λ_{max}^{EtOH} 244 nm (ε = 37,400), 345 nm (ε = 12,700). Spectrum recorded on Cary Model 17 spectrometer.

into the upper half of the tube. Let the tube cool in position until the melt has solidified, and scrape out the white solid.

Weigh 0.20 g of benzil (theory = 210 mg) and 0.10 g of your purified *o*-phenylenediamine (theory = 108 mg) into a 20 × 150-mm test tube and heat in a steam bath for 10 min, which changes the initially molten mixture to a light tan solid. Dissolve the solid in hot methanol (about 5 mL) and let the solution stand undisturbed. If crystallization does not occur within 10 min, reheat the solution and dilute it with a little water to the point of saturation. The crystals should be filtered as soon as formed, for brown oxidation products accumulate on standing. The quinoxaline forms colorless needles, mp 125–126°C; yield is 185 mg.

Reaction time: 10 min

Cleaning Up   The residues from the distillation of 1,2-phenylenediamine and the solvent from the reaction should all be placed in the aromatic amines hazardous waste container since the diamine may be a carcinogen.

56

Borohydride Reduction of a Ketone: Hydrobenzoin from Benzil

Prelab Exercise: Compare the reductive abilities of lithium aluminum hydride with those of sodium borohydride.

Sodium borohydride was discovered in 1943 by H. I. Schlesinger and H. C. Brown. Brown devoted his entire scientific career to this reagent, making it and other hydrides the most useful and versatile of reducing reagents. He received a Nobel prize for his work.

Considering the extreme reactivity of most hydrides (such as sodium hydride and lithium aluminum hydride) toward water, sodium borohydride is somewhat surprisingly sold as a stabilized aqueous solution 14 molar in sodium hydroxide containing 12% sodium borohydride. Unlike lithium aluminum hydride, sodium borohydride is insoluble in ether and soluble in methanol and ethanol.

Sodium borohydride is a mild and selective reducing reagent. In ethanol solution it reduces aldehydes and ketones rapidly at 25°C, esters very slowly, and is inert toward functional groups that are readily reduced by lithium aluminum hydride: carboxylic acids, epoxides, lactones, nitro groups, nitriles, azides, amides, and acid chlorides.

The present experiment is a typical sodium borohydride reduction. These same conditions and isolation procedures could be applied to hundreds of other ketones and aldehydes.

Experiment

Sodium Borohydride Reduction of Benzil

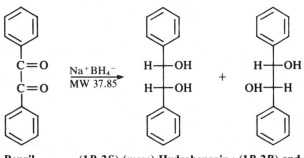

Benzil	(1*R*,2*S*)-(*meso*)-Hydrobenzoin	(1*R*,2*R*) and (1*S*, 2*S*)-Hydrobenzoin
MW 210.22	mp 137°C, MW 214.25	mp 120°C

Addition of two atoms of hydrogen to benzoin or of four atoms of hydrogen to benzil gives a mixture of stereoisomeric diols, of which the predominant isomer is the nonresolvable (2*R*,3*S*)-hydrobenzoin, the *meso* isomer, accompanied by the enantiomeric (2*R*,3*R*) and (2*S*,3*S*) compounds. The reaction proceeds rapidly at room temperature; the intermediate borate ester is hydrolyzed with water to give the product alcohol.

$$4 \ R_2C{=}O \ + \ Na^+BH_4^- \longrightarrow (R_2CHO)_4B^-Na^+$$

$$(R_2CHO)_4B^-Na^+ \ + \ 2 \ H_2O \longrightarrow 4 \ R_2CHOH \ + \ Na^+BO_2^-$$

The procedure that follows specifies use of benzil rather than benzoin because you can then follow the progress of the reduction by the discharge of the yellow color of the benzil.

Procedure

In a 50-mL Erlenmeyer flask, dissolve 0.5 g of benzil in 5 mL of 95% ethanol and cool the solution under the tap to produce a fine suspension. Then add 0.1 g of sodium borohydride (large excess). The benzil dissolves, the mixture warms up, and the yellow color disappears in 2–3 min. After a total of 10 min, add 5 mL of water, heat to the boiling point, filter in case the solution is not clear, dilute to the point of saturation with more water (10 mL), and set the solution aside to crystallize, *meso*-Hydrobenzoin separates in lustrous thin plates, mp 136–137°C; yield is about 0.35 g.

Cleaning Up   The aqueous filtrate should be diluted with water and neutralized with acetic acid (to destroy borohydride) before flushing the mixture down the drain.

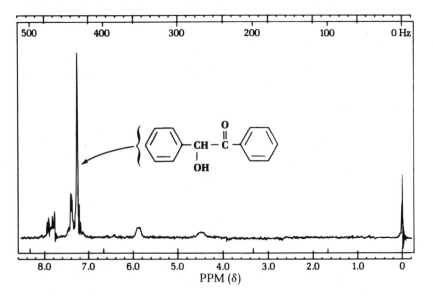

FIG. 1   ¹H nmr spectrum of benzoin (60 MHz).

Questions

1. Draw and name, using the *R,S* system of nomenclature, all of the isomers of hydrobenzoin.

2. Calculate the theoretical weight of sodium borohydride needed to reduce 50 mg of benzil.

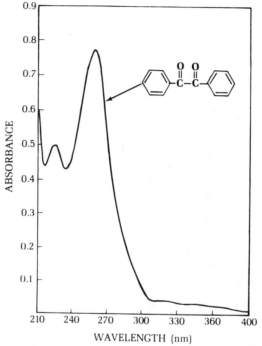

FIG. 2   Ultraviolet spectrum of benzil. λ_{max}^{EtOH} 260 nm (ε = 19,800). One-cm cells and 95% ethanol have been employed for all the uv spectra in this chapter.

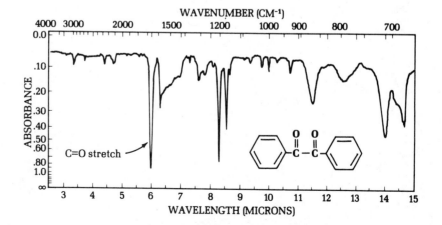

FIG. 3   Infrared spectrum of benzil in $CHCl_3$.

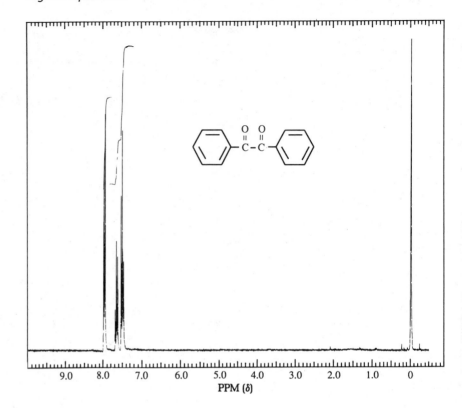

FIG. 4 ¹H nmr spectrum of
benzil (250 MHz).

CHAPTER
57

1,4-Addition: Reductive Acetylation of Benzil

Prelab Exercise: Write the complete mechanism for the reaction of acetic anhydride with an alcohol.

In one of the first demonstrations of the phenomenon of 1,4-addition, Johannes Thiele (1899) established that reduction of benzil with zinc dust in a mixture of acetic anhydride–sulfuric acid involves 1,4-addition of hydrogen to the α-diketone grouping and acetylation of the resulting enediol before it can undergo ketonization to benzoin. The process of reductive acetylation results in a mixture of the *E*- and *Z*-isomers **1** and **2**. Thiele and subsequent investigators isolated the more soluble, lower-melting *Z*-stilbenediol diacetate (**2**) in only impure form, mp 110°C. Separation of the two isomers by chromatography is not feasible because they are equally adsorbable on alumina. However, separation is possible by fractional crystallization (described in the following procedure) and both isomers can be isolated in pure condition. In the method prescribed here for the preparation of the isomer mixture, hydrochloric acid is substituted for sulfuric acid because the latter acid gives rise to colored impurities and is reduced to sulfur and to hydrogen sulfide.[1]

Ac = *acetyl group*

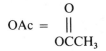

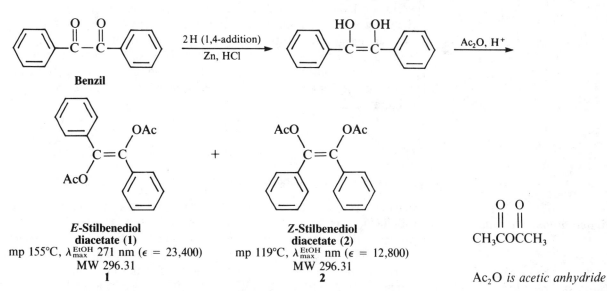

E-Stilbenediol diacetate (**1**)
mp 155°C, λ_{max}^{EtOH} 271 nm ($\epsilon = 23{,}400$)
MW 296.31
1

Z-Stilbenediol diacetate (**2**)
mp 119°C, λ_{max}^{EtOH} nm ($\epsilon = 12{,}800$)
MW 296.31
2

$$CH_3COCCH_3$$

Ac$_2$O *is acetic anhydride*

---

1. If acetyl chloride (2 mL) is substituted for the hydrochloric acid-acetic anhydride mixture in the procedure, the *Z*-isomer is the sole product.

FIG. 1   Ultraviolet spectra of *Z*-
and *E*-stilbene diacetate. *Z*: λ_{max}^{EtOH}
223 nm ($\varepsilon = 20{,}500$), 269 nm
($\varepsilon = 10{,}800$); *E*: λ_{max}^{EtOH} 272 nm
($\varepsilon = 20{,}800$). In the *Z*-diacetate
the two phenyl rings cannot be
coplanar, which prevents
overlap of the *p*-orbitals of the
phenyl groups with those of the
central double bond. This steric
inhibition of resonance accounts
for the diminished intensity of
the *Z*-isomer relative to the *E*.
Both spectra were run at the
same concentration.

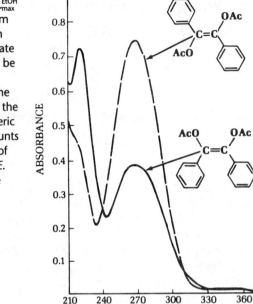

Assignment of configuration

The configurations of this pair of geometrical isomers remained
unestablished for over 50 years, but the tentative inference that the higher-
melting isomer has the more symmetrical *E* configuration eventually was
found to be correct. Evidence of infrared spectroscopy is of no avail; the
spectra are nearly identical in the interpretable region (2–8 μ) characterizing the acetoxyl groups but differ in the fingerprint region (8–12 μ).
However, the isomers differ markedly in ultraviolet absorption (Fig. 1);
and, in analogy to *E*- and *Z*-stilbene the conclusion is justified that the
higher-melting isomer, because it has an absorption band at longer wavelength and higher intensity than its isomer, does indeed have the
configuration **1**.

Experiment

1. Reductive Acetylation of Benzil

Reaction time: 10 min

Carry out reaction in hood

Place one test tube (20 × 150-mm) containing 7 mL of acetic anhydride and
another (13 × 100-mm) containing 1 mL of concentrated hydrochloric acid
in an ice bath. When both are thoroughly chilled, transfer the acid to the
anhydride dropwise in not less than one minute by means of a capillary
dropping tube. Wipe the test tube dry, pour the chilled solution into a 50-mL
Erlenmeyer flask containing 1 g of pure benzil and 1 g of zinc dust, and swirl
for 2–3 min in an ice bath. Remove the flask and hold it in the palm of the

Handle acetic anhydride and hydrochloric acid with care

hand; if it begins to warm up, cool further in ice. When there is no further exothermic effect, let the mixture stand for 5 min and then add 25 mL of water. Swirl, break up any lumps of product, and allow a few minutes for hydrolysis of excess acetic anhydride. Then collect the mixture of product and zinc dust by vacuum filtration, wash with water, and press and apply suction to the cake until there is no further drip. Digest the solid (drying is not necessary) with 70 mL of ether to dissolve the organic material, add about 4 g of anhydrous sodium sulfate, and swirl briefly; filter the solution into a 125-mL Erlenmeyer flask, concentrate the filtrate (steam bath, boiling stone, water aspirator) to a volume of approximately 15 mL,[2] and let the flask stand, corked and undisturbed.

The *E*-diacetate (**1**) soon begins to separate in prismatic needles, and after 20–25 min crystallization appears to stop. Remove the crystals by filtration on the Hirsch funnel. Wash them once with ether and then evaporate the ether to dryness. Dissolve the residue in 10 mL of methanol and transfer the hot solution to a 25-mL Erlenmeyer flask.

Let the solution stand undisturbed for about 10 min, and drop in one tiny crystal of the *E*-diacetate (**1**). This should give rise, in 20–30 min, to a second crop of the *E*-diacetate. Then concentrate the mother liquor and washings to a volume of 7–8 mL, let cool to room temperature as before, and again seed with a crystal of *E*-diacetate; this usually affords a third crop of the *E*-diacetate. The three crops might be: 300 mg, mp 154–156°C; 50 mg, mp 153–156°C; and 50 mg, mp 153–155°C.

At this point the mother liquor should be rich enough in the more soluble *Z*-diacetate (**2**) for its isolation. Concentrate the methanol mother liquor and washings from the third crop of **1** to a volume of 4–5 mL, stopper the flask, and let the solution stand undisturbed overnight. The *Z*-diacetate (**2**) sometimes separates spontaneously in large rectangular prisms of great beauty. If the solution remains supersaturated, addition of a seed crystal of **2** causes prompt separation of the *Z*-diacetate in a paste of small crystals (e.g., 215 mg, mp 118–119°C; then: 70 mg, mp 116–117°C).

Cleaning Up   Spread the zinc out on a watch glass for about 20 min before wetting it and placing it in the nonhazardous solid waste container. Sometimes zinc dust from a reaction like this is pyrophoric (spontaneously flammable in air) because it is so finely divided and has such a large, clean surface area able to react with air. The aqueous filtrate should be neutralized with sodium carbonate, zinc salts removed by vacuum filtration, and the filtrate diluted with water and flushed down the drain. The zinc salts are placed in the nonhazardous solid waste container. The filtrate from the fractional crystallization should all be placed in the organic solvents container. Allow the solvent to evaporate from the sodium sulfate in the hood and then place it in the nonhazardous solid waste container.

---

2. Measure 15 mL of a solvent into a second flask of the same size and compare the levels in the two flasks.

58

Synthesis of an Alkyne from an Alkene. Bromination and Dehydrobromination: Stilbene and Diphenylacetylene

Prelab Exercise: Calculate the theoretical quantities of thionyl chloride and of sodium borohydride needed to convert benzoin to *E*-stilbene.

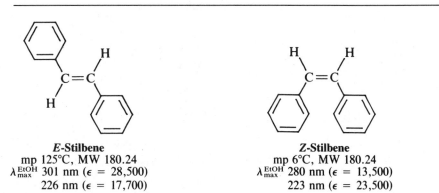

E-Stilbene
mp 125°C, MW 180.24
λ_{max}^{EtOH} 301 nm ($\epsilon = 28,500$)
226 nm ($\epsilon = 17,700$)

Heat of hydrogenation, −20.1 kcal/mole

Z-Stilbene
mp 6°C, MW 180.24
λ_{max}^{EtOH} 280 nm ($\epsilon = 13,500$)
223 nm ($\epsilon = 23,500$)

Heat of hydrogenation, −25.8 kcal/mole

In this experiment, benzoin, prepared in Chapter 54, is converted to the alkene *trans*-stilbene (*E*-stilbene), which is in turn brominated and dehydrobrominated to form the alkyne, diphenylacetylene.

One method of preparing *E*-stilbene is reduction of benzoin with zinc amalgam in ethanol-hydrochloric acid, presumably through an intermediate:

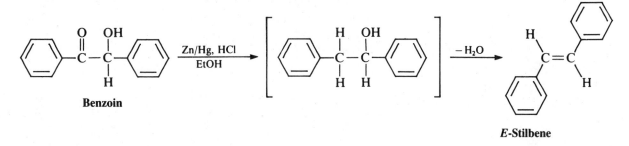

Benzoin

E-Stilbene

The procedure that follows is quick and affords very pure hydrocarbon. It involves three steps: (1) replacement of the hydroxyl group of benzoin by chlorine to form desyl chloride, (2) reduction of the keto group with sodium borohydride to give what appears to be a mixture of the two diastereoisomeric chlorohydrins, and (3) elimination of the elements of hypochlorous acid with zinc and acetic acid. The last step is analogous to the debromination of an olefin dibromide.

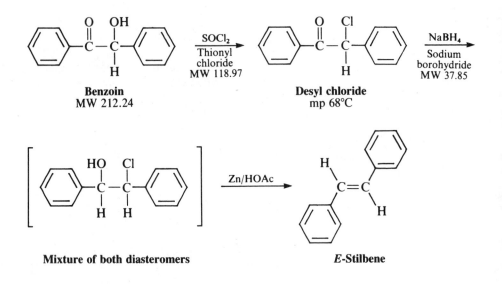

Experiments

1. Stilbene

Carry out procedure in hood

Place 4 g of benzoin (crushed to a powder) in a 100-mL round-bottomed flask, cover it with 4 mL of thionyl chloride,[1] warm gently on the steam bath (hood) until the solid has all dissolved, and then more strongly for 5 min.

Caution: If the mixture of benzoin and thionyl chloride is left standing at room temperature for an appreciable time before being heated, an undesired reaction intervenes[2] and the synthesis of *E*-stilbene is spoiled.

To remove excess thionyl chloride (bp 77°C), evacuate at the aspirator for a few minutes, add 10 mL of petroleum ether (bp 30–60°C), boil it off (hood), and evacuate again. Desyl chloride is thus obtained as a viscous, pale yellow oil (it will solidify if let stand). Dissolve this in 40 mL of 95%

---

1. The reagent can be dispensed from a burette or measured by pipette; in the latter case the liquid should be drawn into the pipette with a pipetter, **not by mouth.**

2. $C_6H_5C=O$ $\xrightarrow{SOCl_2}$ C_6H_5C-O $\xrightarrow{NaBH_4}$ C_6H_5CO
    | ‖ SO |
C_6H_5CHOH C_6H_5C-O $C_6H_5CH_2$

Desoxybenzoin

ethanol, cool under the tap, and add 360 mg of sodium borohydride (an excess is harmful). Stir, break up any lumps of the borohydride, and after 10 min add to the solution of chlorohydrins 2 g of zinc dust and 4 mL of acetic acid and reflux for 1 h. Then cool under the tap. When white crystals separate add 50 mL of ether and decant the solution from the bulk of the zinc into a separatory funnel. Wash the solution twice with an equal volume of water containing 1–2 mL of concentrated hydrochloric acid (to dissolve basic zinc salts) and then, in turn, with 5% sodium carbonate solution and saturated sodium chloride solution. Dry the ether over anhydrous sodium sulfate (4 g), filter to remove the drying agent, evaporate the filtrate to dryness, dissolve the residue in the minimum amount of hot 95% ethanol (30–40 mL), and let the product crystallize. *E*-Stilbene separates in diamond-shaped iridescent plates, mp 124–125°C; yield is 1.8–2.2 g.

Cleaning Up   Combine washings from the cotton in the trap and all aqueous layers, neutralize with sodium carbonate, remove zinc salts by vacuum filtration, and flush the filtrate down the drain with excess water. The zinc salts are placed in the nonhazardous solid waste container. Allow the ether to evaporate from the sodium sulfate in the hood and then place it in the nonhazardous solid waste container. Ethanol mother liquor goes in the organic solvents container. Any zinc isolated should be spread out on a watch glass to dry and air oxidize. It should then be wetted with water and placed in the nonhazardous solid waste container.

2. *meso*-Stilbene Dibromide

E-Stilbene reacts with bromine predominantly by the usual process of *trans*-addition and affords the optically inactive, nonresolvable *meso*-dibromide; the much lower melting *dl*-dibromide (Chapter 60) is a very minor product of the reaction.

Procedure

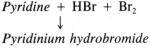

Pyridine + HBr + Br$_2$

↓

Pyridinium hydrobromide perbromide

CAUTION: *Corrosive, lachrymator.*

Total time required: 10 min
Carry out procedure in hood

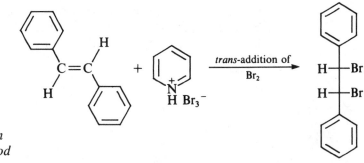

E-Stilbene	Pyridinium hydro-	(1*R*,2*S*)-Stilbene dibromide (*meso* isomer)
MW 180.24	bromide perbromide	mp 238°C, MW 340.07
	MW 319.86	

In a 125-mL Erlenmeyer flask dissolve 2 g of *E*-stilbene in 40 mL of acetic acid, by heating on the steam bath, and then add 4 g of pyridinium hydrobromide perbromide.[3] Mix by swirling; if necessary, rinse crystals of reagent down the walls of the flask with a little acetic acid; and continue the heating for 1–2 min longer. The dibromide separates almost at once in small plates. Cool the mixture under the tap, collect the product, and wash it with methanol; yield of colorless crystals, mp 236–237°C, about 3.2 g. Use 0.5 g of this material for the preparation of diphenylacetylene and turn in the remainder.

Cleaning Up　　To the filtrate add sodium bisulfite (until a negative test with starch-iodide paper is observed) to destroy any remaining perbromide, neutralize with sodium carbonate, extract the pyridine released with ether, which goes in the organic solvents container. The aqueous layer can then be diluted with water and flushed down the drain.

Diphenylacetylene

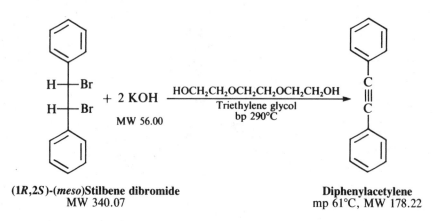

(1R,2S)-(*meso*)Stilbene dibromide
MW 340.07

Diphenylacetylene
mp 61°C, MW 178.22

One method for the preparation of diphenylacetylene involves oxidation of benzil dihydrazone with mercuric oxide; the intermediate diazo compound loses nitrogen as formed to give the hydrocarbon:

---

3. Can be purchased from Aldrich or see Chapter 13 for synthesis.

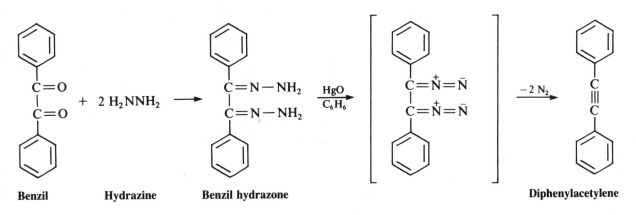

Benzil   **Hydrazine**   **Benzil hydrazone**   **Diphenylacetylene**

The method used in this procedure involves dehydrohalogenation of *meso*-stilbene dibromide. An earlier procedure called for refluxing the dibromide with 43% ethanolic potassium hydroxide in an oil bath at 140°C for 24 h. In the following procedure the reaction time is reduced to a few minutes by use of the high-boiling triethylene glycol as solvent to permit operation at a higher reaction temperature.

3. Synthesis of Diphenylacetylene

Reaction time: 5 min

Caution: Potassium hydroxide is corrosive to skin.

In a 20 × 150-mm test tube place 0.5 g of *meso*-stilbene dibromide, 3 pellets of potassium hydroxide[4] (250 mg), and 2 mL of triethylene glycol. Insert a thermometer into a 10 × 75-mm test tube containing enough triethylene glycol to cover the bulb, and slip this assembly into the larger tube. Clamp the tube in a vertical position in a hot sand bath, and heat the mixture to a temperature of 160°C, when potassium bromide begins to separate. By intermittent heating, keep the mixture at 160–170°C for 5 min more, then cool to room temperature, remove the thermometer and small tube, and add 10 mL of water. The diphenylacetylene that separates as a nearly colorless, granular solid is collected by suction filtration. The crude product need not be dried but can be crystallized directly from 95% ethanol. Let the solution stand undisturbed in order to observe the formation of beautiful, very large spars of colorless crystals. After a first crop has been collected, the mother liquor on concentration affords a second crop of pure product; total yield, about 0.23 g; mp 60–61°C.

Cleaning Up   Combine the crystallization mother liquor with the filtrate from the reaction, dilute with water, neutralize with 10% hydrochloric acid, and flush down the drain.

---

4. Potassium hydroxide pellets are 85% KOH and 15% water.

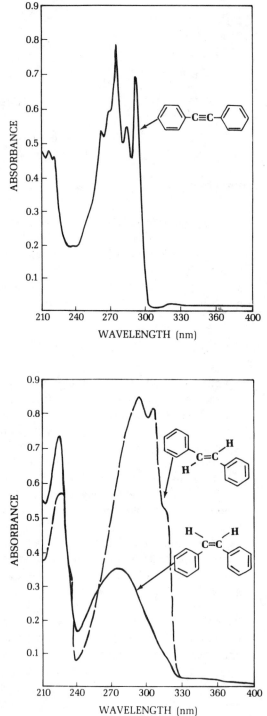

FIG. 1   Ultraviolet spectrum of diphenylacetylene. λ_{max}^{EtOH} 279 nm ($\varepsilon = 31{,}400$). This spectrum is characterized by considerable fine structure (multiplicity of bands) and a high extinction coefficient.

FIG. 2   Ultraviolet spectra of Z- and E-stilbene. Z: λ_{max}^{EtOH} 224 nm ($\varepsilon = 23{,}300$), 279 nm ($\varepsilon = 11{,}100$); E: λ_{max}^{EtOH} 226 nm ($\varepsilon = 18{,}300$), 295 nm ($\varepsilon = 27{,}500$). Like the diacetates, steric hindrance and lack of coplanarity in these hydrocarbons cause the long wavelength absorption of the Z-isomer to be of diminished intensity relative to the E-isomer.

59

The Perkin Reaction: Synthesis of α-Phenylcinnamic Acid

Prelab Exercise: At the end of this reaction the products are present as mixed anhydrides. What are these, how are they formed, and how are they converted to product?

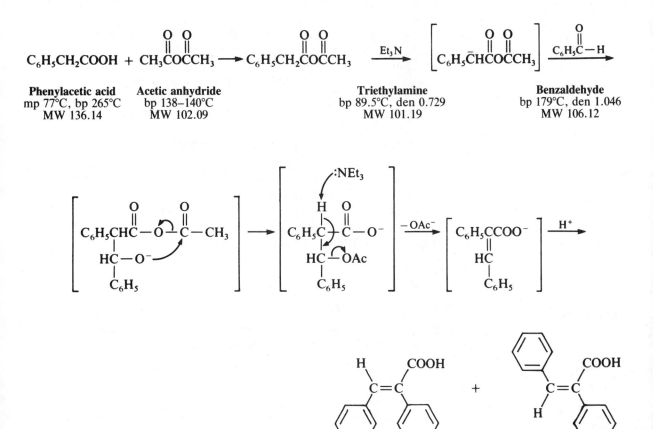

Phenylacetic acid	Acetic anhydride		Triethylamine	Benzaldehyde
mp 77°C, bp 265°C	bp 138–140°C		bp 89.5°C, den 0.729	bp 179°C, den 1.046
MW 136.14	MW 102.09		MW 101.19	MW 106.12

E-α-Phenylcinnamic acid[1]
mp 174°C, pK_a 6.1
MW 224.25

+

Z-α-Phenylcinnamic acid
mp 138°C, pK_a 4.8
MW 224.25

1. pK_a measured in 60% ethanol.

The reaction of benzaldehyde with phenylacetic acid to produce a mixture of the α-carboxylic acid derivatives of Z- and E-stilbene,[2] a form of aldol condensation known as the Perkin reaction, is effected by heating a mixture of the components with acetic anhydride and triethylamine. In the course of the reaction the phenylacetic acid is probably present both as anion and as the mixed anhydride resulting from equilibration with acetic anhydride. A reflux period of 5 h specified in an early procedure has been shortened by a factor of 10 by restriction of the amount of the volatile acetic anhydride, use of an excess of the less expensive, high-boiling aldehyde component, and use of a condenser that permits some evaporation and consequent elevation of the reflux temperature.

E-Stilbene is a by-product of the condensation, but experiment has shown that neither the E- nor Z-acid undergoes decarboxylation under the conditions of the experiment.

At the end of the reaction the α-phenylcinnamic acids are present in part as the neutral mixed anhydrides, but these can be hydrolyzed by addition of excess hydrochloric acid. The organic material is taken up in ether and the acids extracted with alkali. Neutralization with acetic acid (pK_a 4.76) then causes precipitation of only the less acidic E-acid (see pK_a values under the formulas); the Z-acid separates on addition of hydrochloric acid.

Whereas Z-stilbene is less stable and lower-melting than E-stilbene, the reverse is true of the α-carboxylic acids, and in this preparation the more stable, higher-melting E-acid is the predominant product. Evidently the steric interference between the carboxyl and phenyl groups in the Z-acid is greater than that between the two phenyl groups in the E-acid. Steric hindrance is also evident from the fact that the Z-acid is not subject to Fischer esterification (ethanol and an acid catalyst) whereas the E-acid is.

Synthesis of α-Phenylcinnamic Acid

Reflux time: 35 min

Measure into a 25-mL round-bottomed flask 2.5 g of phenylacetic acid, 3 mL of benzaldehyde, 2 mL of triethylamine, and 2 mL of acetic anhydride. Add a water-cooled condenser and a boiling stone and reflux the

2. Stereochemistry of alkenes can be designated by the E,Z-system of nomenclature [see J. L. Blackwood, C. L. Gladys, K. L. Loening, A. E. Petrarca, and J. E. Rush, *J. Am. Chem. Soc.*, **90**, 509 (1968)] in which the groups attached to the double bond are given an order of priority according to the Cahn, Ingold, and Prelog system [see R. S. Cahn, C. K. Ingold, and V. Prelog, *Experientia*, **12**, 81 (1956)]. The atom of highest atomic number attached directly to the alkene is given highest priority. If two atoms attached to the alkene are the same, one goes to the second or third atom, etc., away from the alkene carbons. When the two groups of highest priority are on adjacent sides of the double bond, the stereochemistry is designated as Z (German *zusammen,* together). When the two groups are on opposite sides of the double bond, the stereochemistry is designated E (German *entgegen,* opposed).

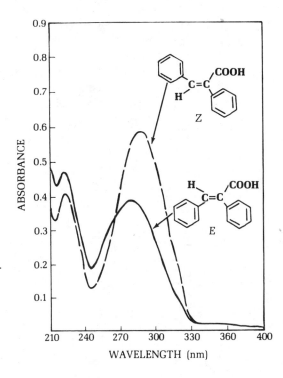

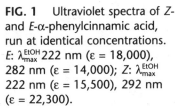

FIG. 1 Ultraviolet spectra of Z- and E-α-phenylcinnamic acid, run at identical concentrations. E: λ_{max}^{EtOH} 222 nm (ε = 18,000), 282 nm (ε = 14,000); Z: λ_{max}^{EtOH} 222 nm (ε = 15,500), 292 nm (ε = 22,300).

mixture for 35 min. Cool the yellow melt, add 4 mL of concentrated hydrochloric acid, and swirl, whereupon the mixture sets to a stiff paste. Add ether, warm to dissolve the bulk of the solid, and transfer to a separatory funnel with use of more ether. Wash the ethereal solution twice with water and then extract it with a mixture of 25 mL of water and 5 mL of 10% sodium hydroxide solution.[3] Repeat the extraction twice more and discard the dark-colored ethereal solution.[4] Acidify the combined, colorless, alkaline extract to pH 6 by adding 5 mL of acetic acid, collect the E-acid that precipitates, and save the filtrate and washings. The yield of E-acid, mp 163–166°C, is usually about 2.9 g. Crystallize 0.3 g of material by dissolving it in 8 mL of ether, adding 8 mL of petroleum ether (bp 30–60°C), heating briefly to the boiling point, and letting the solution stand. Silken needles form, mp 173–174°C.

Addition of 5 mL of concentrated hydrochloric acid to the aqueous filtrate from precipitation of the E-acid produces a cloudy emulsion, which,

---

3. If stronger alkali is used the sodium salt may separate.

4. For isolation of stilbene, wash this ethereal solution with saturated sodium bisulfite solution for removal of benzaldehyde, dry, evaporate, and crystallize the residue from a little methanol. Large, slightly yellow spars, mp 122–124°C, separate (9 mg).

on standing for about one-half hour, coagulates to crystals of Z-acid; yield, about 0.3 g, mp 136–137°C.[5]

Cleaning Up   The dark-colored ether solution and the mother liquor from crystallization are placed in the organic solvents container. After the Z-acid has been removed from the reaction mixture, the acidic solution should be diluted with water and flushed down the drain.

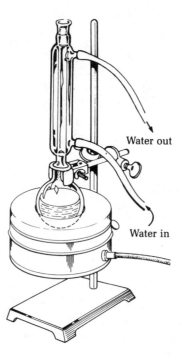

FIG. 2   Cold finger reflux condenser.

5. The *E*-acid can be recrystallized by dissolving 0.3 g in 5 mL of ether, filtering if necessary from a trace of sodium chloride, adding 10 mL of petroleum ether (bp 30–60°C), and evaporating to a volume of 5 mL; the acid separates as a hard crust of prisms, mp 138–139°C.

Decarboxylation: Synthesis of cis-Stilbene

Prelab Exercise: Calculate the volume (measured at STP) of carbon dioxide evolved in this reaction.

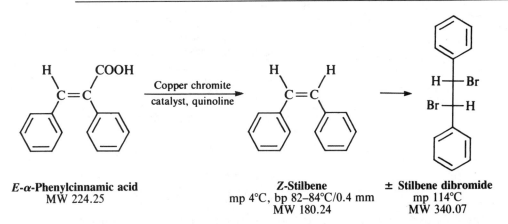

E-α-Phenylcinnamic acid
MW 224.25

Z-Stilbene
mp 4°C, bp 82–84°C/0.4 mm
MW 180.24

± **Stilbene dibromide**
mp 114°C
MW 340.07

The catalyst for this reaction is copper chromite, $2\,CuO \cdot Cr_2O_3$, a relatively inexpensive commercially available catalyst used for both hydrogenation and dehydrogenation as well as decarboxylation.

Decarboxylation of *E-α*-phenylcinnamic acid is effected by refluxing the acid in quinoline in the presence of a trace of copper chromite catalyst; both the basic properties and boiling point (237°C) of quinoline make it a particularly favorable solvent. *Z*-Stilbene, a liquid at room temperature, can be characterized by *trans* addition of bromine to give the crystalline ±-dibromide. A little *meso*-dibromide derived from *E*-stilbene in the crude hydrocarbon starting material is easily separated by virtue of its sparing solubility.

Although free rotation is possible around the single bond connecting the chiral carbon atoms of the stilbene dibromides and hydrobenzoins, evidence from dipole-moment measurements indicates that the molecules tend to exist predominantly in the specific shape or conformation in which the two phenyl groups repel each other and occupy positions as far apart as possible. The optimal conformations of the + or − dibromide and the *meso*-dibromide are represented in Fig. 1 by Newman projection formulas, in which the molecules are viewed along the axis of the bond connecting the two chiral carbon atoms. In the *meso*-dibromide the two repelling phenyl groups are on opposite sides of the molecule, and so are the two large bromine atoms.

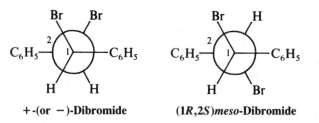

FIG. 1 Favored conformations of stilbene dibromide.

Hence, the structure is much more symmetrical than that of the + (or −) dibromide. X-ray diffraction measurements of the dibromides in the solid state confirm the conformations indicated in Fig. 1. The Br-Br distances found are *meso*-dibromide, 4.50 Å; ±-dibromide, 3.85 Å. The difference in symmetry of the two optically inactive isomers accounts for the marked contrast in properties:

	Mp (°C)	Solubility in ether (18°C)
±-Dibromide	114	1 part in 3.7 parts
meso-Dibromide	237	1 part in 1025 parts

Stilbene Dibromide

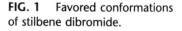

Copper chromite is toxic; weigh it in hood

2CuO · Cr₂O₃
Copper chromite

Reaction time in first step: 10 min

Quinoline, bp 237°C

Because a trace of moisture causes troublesome spattering, the reactants and catalyst are dried prior to decarboxylation. Stuff 2.5 g of crude, "dry" *E*-α-phenylcinnamic acid and 0.2 g of copper chromite catalyst[1] into a 20 × 150-mm test tube, add 3 mL of quinoline[2] (bp 237°C), and let it wash down the solids. Make connection with a rubber stopper to the aspirator and turn it on full force. Make sure that you have a good vacuum (pressure gauge) and heat the tube strongly on the steam bath with most of the rings removed. Heat and evacuate for 5–10 min to remove all traces of moisture. Then wipe the outside walls of the test tube dry, insert a thermometer, clamp the tube over a microburner, raise the temperature to 230°C, and note the time. Then maintain a temperature close to 230°C for 10 min. Cool the yellow solution containing suspended catalyst to 25°C, add 30 mL of ether, and filter the solution by gravity (use more ether for rinsing). Transfer the solution to a separatory funnel and remove the quinoline by extraction twice with about 15 mL of water containing 3–4 mL of concentrated hydrochloric acid. Then shake the ethereal solution well with water containing a little sodium hydroxide solution, draw off the alkaline liquor, and acidify it. A substantial precipitate will show that decarboxylation was incomplete, in which case

---

1. The preparations described in *J. Am. Chem. Soc.*, **54**, 1138 (1932) and *ibid.*, **72**, 2626 (1950) are both satisfactory.

2. Material that has darkened in storage should be redistilled over a little zinc dust.

the starting material can be recovered and the reaction repeated. If there is only a trace of precipitate, shake the ethereal solution with saturated sodium chloride solution for preliminary drying, dry the ethereal solution over sodium sulfate, remove the drying agent by filtration, and evaporate the ether. The residual brownish oil (1.3–1.8 g) is crude *Z*-stilbene containing a little *E*-isomer formed by rearrangement during heating.

Second step requires about one-half hour

Carry out procedure in hood

Dissolve the crude *Z*-stilbene (e.g., 1.5 g) in 10 mL of acetic acid and, in subdued light, add double the weight of pyridinium hydrobromide perbromide (e.g., 3.0 g). Warm on the steam bath until the reagent is dissolved, and then cool under the tap and scratch to effect separation of a small crop of plates of the *meso*-dibromide (10–20 mg). Filter the solution by suction, dilute extensively with water, and extract with ether. Wash the solution twice with water and then with 5% sodium bicarbonate solution until neutral; shake with saturated sodium chloride solution, dry over sodium sulfate, and evaporate to a volume of about 10 mL. If a little more of the sparingly soluble *meso*-dibromide separates, remove it by gravity filtration and then evaporate the remainder of the solvent. The residual *dl*-dibromide is obtained as a dark oil that readily solidifies when rubbed with a rod. Dissolve it in a small amount of methanol and let the solution stand to crystallize. The *dl*-dibromide separates as colorless prismatic plates, mp 113–114°C; yield, about 0.6 g.

Cleaning Up   The catalyst removed by filtration should be placed in the nonhazardous solid waste container. The combined aqueous layers from all parts of the experiment containing quinoline and pyridine should be treated with a small quantity of bisulfite to destroy any bromine, neutralized with sodium carbonate, the quinoline and pyridine released extracted into ligroin, and the ligroin solution placed in the organic solvents container. The aqueous layer should be diluted with water and flushed down the drain. After ether is allowed to evaporate from the sodium sulfate in the hood, it can be placed in the nonhazardous solid waste container. Methanol from the crystallization goes in the organic solvents container.

Questions

1. Draw the mechanism that shows how the bromination of *E*-stilbene produces *meso*-stilbene dibromide.

2. Can ^{1}H nmr spectroscopy be used to distinguish between *meso*- and ±-stilbene dibromide?

61

Martius Yellow

Prelab Exercise: Prepare a time-line for this experiment, indicating clearly which experiments can be carried out simultaneously.

This experiment was introduced by Louis Fieser of Harvard University over half a century ago. It has long been the basis for an interesting laboratory competition. The present author was a winner a third of a century ago.

Starting with 5 g of 1-naphthol, a skilled operator familiar with the procedures can prepare pure samples of the seven compounds in 3–4 h. In a first trial of the experiment, a particularly competent student, who plans his or her work in advance, can complete the program in two laboratory periods (6 h).

The first compound of the series, Martius Yellow, a mothproofing dye for wool (1 g of Martius Yellow dyes 200 g of wool) discovered in 1868 by Karl Alexander von Martius, is the ammonium salt of 2,4-dinitro-1-naphthol (**1**), shown below. Compound **1** in the series of reactions is obtained by sulfonation of 1-naphthol with sulfuric acid and treatment of the resulting disulfonic acid with nitric acid in aqueous medium. The exchange of groups occurs with remarkable ease, and it is not necessary to isolate the disulfonic acid. The advantage of introducing the nitro groups in this indirect way is that 1-naphthol is very sensitive to oxidation and would be partially destroyed on direct nitration. Martius Yellow is prepared by reaction of the acidic phenolic group of **1** with ammonia to form the ammonium salt. A small portion of this salt (Martius Yellow) is converted by acidification and crystallization into pure 2,4-dinitro-1-naphthol (**1**), a sample of which is saved. The rest is suspended in water and reduced to diaminonaphthol with sodium hydrosulfite according to the equation:

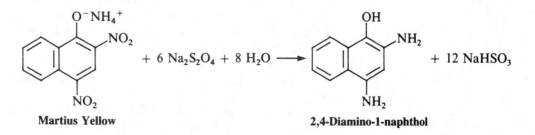

Martius Yellow ... 2,4-Diamino-1-naphthol

$$+\ 6\ Na_2S_2O_4\ +\ 8\ H_2O\ \longrightarrow\ +\ 12\ NaHSO_3$$

The diaminonaphthol separates in the free condition, rather than as an ammonium salt, because the diamine, unlike the dinitro compound, is a very weakly acidic substance.

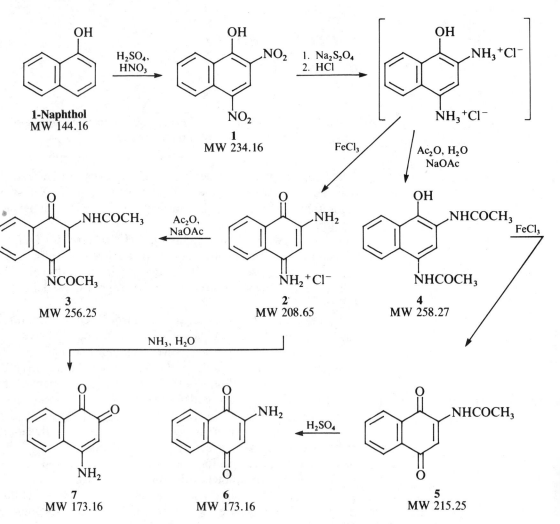

1-Naphthol
MW 144.16

1
MW 234.16

3
MW 256.25

2
MW 208.65

4
MW 258.27

7
MW 173.16

6
MW 173.16

5
MW 215.25

Since 2,4-diamino-1-naphthol is exceedingly sensitive to air oxidation as the free base, it is at once dissolved in dilute hydrochloric acid. The solution of diaminonaphthol dihydrochloride is clarified with decolorizing charcoal and divided into equal parts. One part on oxidation with iron(III) chloride affords the fiery red 2-amino-1,4-naphthoquinonimine hydrochloride (**2**). Since this substance, like many other salts, has no melting point, it is converted for identification to the yellow diacetate, **3**. Compound **2** is remarkable in that it is stable enough to be isolated. On hydrolysis it affords the orange 4-amino-1,2-naphthoquinone (**7**).

The other part of the diaminonaphthol dihydrochloride solution is treated with acetic anhydride and then sodium acetate; the reaction in aqueous solution effects selective acetylation of the amino groups and affords 2,4-diacetylamino-1-naphthol (**4**). Oxidation of **4** by Fe^{3+} and oxygen from the air is attended with cleavage of the acetylamino group at the 4-

1-Naphthol

position and the product is 2-acetylamino-1,4-naphthoquinone (**5**). This yellow substance is hydrolyzed by sulfuric acid to the red 2-amino-1,4-naphthoquinone (**6**), the last member of the series. The reaction periods are brief and the yields high; however, remember to scale down quantities of reagents and solvents if the quantity of starting material is less than that called for.[1]

CAUTION: Toxic irritant

Experiments

1. Preparation of 2,4-Dinitro-1-naphthol(1)

Use care when working with hot concentrated sulfuric and nitric acids

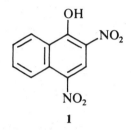

1

Avoid contact of the yellow product and its orange NH_4^+ salt with the skin

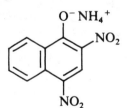

Martius Yellow

Place 5 g of pure 1-naphthol[2] in a 125-mL Erlenmeyer flask, add 10 mL of concentrated sulfuric acid, and heat the mixture with swirling on the steam bath for 5 min, when the solid should have dissolved and an initial red color should be discharged. Cool in an ice bath, add 25 mL of water, and cool the solution rapidly to 15°C. Measure 6 mL of concentrated nitric acid into a test tube and transfer it with a Pasteur pipette in small portions (0.5 mL) to the chilled aqueous solution while keeping the temperature in the range 15–20°C by swirling the flask vigorously in the ice bath. When the addition is complete and the exothermic reaction has subsided (1–2 min), warm the mixture gently to 50°C (1 min), when the nitration product should separate as a stiff yellow paste. Apply the full heat of the steam bath for 1 min more, fill the flask with water, break up the lumps and stir to an even paste, collect the product **1** on a Büchner funnel, wash it well with water, and then wash it into a 600-mL beaker with water (100 mL). Add 150 mL of hot water and 5 mL of concentrated ammonia solution (den 0.90), heat to the boiling point, and stir to dissolve the solid. Filter the hot solution by suction if it is dirty, add 10 g of ammonium chloride to the filtrate to salt out the ammonium salt (Martius Yellow), cool in an ice bath, collect the orange salt, and wash it with water containing 1–2% of ammonium chloride. The salt does not have to be dried (dry weight 7.7 g, 88.5%).

---

1. This series of reactions lends itself to a laboratory competition, the rules for which might be as follows: (1) No practice or advance preparation is allowable except collection of reagents not available at the contestant's bench (ammonium chloride, sodium hydrosulfite, iron(III) chloride solution, acetic anhydride). (2) The time scored is the actual working time, including that required for bottling the samples and cleaning the apparatus and bench; labels can be prepared out of the working period. (3) Time is not charged during an interim period (overnight) when solutions are let stand to crystallize or solids are let dry, on condition that during this period no adjustments are made and no cleaning or other work is done. (4) Melting point and color test characterizations are omitted. (5) Successful completion of the contest requires preparation of authentic and macroscopically crystalline samples of all seven compounds. (6) Judgment of the winners among the successful contestants is based upon quality and quantity of samples, technique and neatness, and working time. (Superior performance 3–4 h.)

2. If the 1-naphthol is dark it can be purified by distillation at atmospheric pressure. The colorless distillate is most easily pulverized before it has completely cooled and hardened.

Set aside an estimated 0.3 g of the moist ammonium salt. This sample is to be dissolved in hot water, the solution acidified (HCl), and the free 2,4-dinitro-1-naphthol (**1**) crystallized from methanol or ethanol (use decolorizing charcoal if necessary); it forms yellow needles, mp 138°C.

Cleaning Up Combine aqueous filtrates, dilute with water, neutralize with sodium carbonate, and flush the solution down the drain. Recrystallization solvents go in the organic solvents container.

Preparation of Unstable 2,4-Diamino-1-naphthol

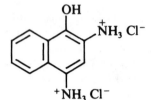

2,4-Diamino-1-naphthol dihydrochloride

$Na_2S_2O_4$

Sodium hydrosulfite (sodium dithionite, not sodium bisulfite)

Wash the rest of the ammonium salt into a beaker with a total of about 200 mL of water, add 40 g of sodium hydrosulfite, stir until the original orange color has disappeared and a crystalline tan precipitate has formed (5–10 min), and cool in ice. Make ready a solution of 1–2 g of sodium hydrosulfite in 100 mL of water for use in washing and a 400-mL beaker containing 6 mL of concentrated hydrochloric acid and 25 mL of water. In collecting the precipitate by suction filtration, use the hydrosulfite solution for rinsing and washing, avoid even briefly sucking air through the cake after the reducing agent has been drained away, and wash the solid at once into the beaker containing the dilute hydrochloric acid and stir to convert all the diamine to the dihydrochloride.

The acid solution, often containing suspended sulfur and filter paper, is clarified by filtration by suction through a moist charcoal bed made by shaking 2 g of the decolorizing carbon with 25 mL of water in a stoppered flask to produce a slurry and pouring this on the paper of an 85-mm Büchner funnel. Pour the water out of the filter flask and then filter the solution of dihydrochloride. Divide the pink or colorless filtrate into approximately two equal parts and at once add the reagents for conversion of one part to **2** and the other part to **4**.

Cleaning Up Neutralize the filtrate with sodium carbonate, dilute with water, and then, in the hood, cautiously add household bleach (5.25% sodium hypochlorite solution) to the mixture until a test with 5% silver nitrate shows no more hydrosulfite is present (absence of a black precipitate). Neutralize the solution and filter through Celite to remove suspended solids. The filtrate should be diluted with water and flushed down the drain. The solid residue goes in the nonhazardous solid waste container.

2. Preparation of 2-Amino-1,4-naphthoquinonimine Hydrochloride (2)

To one-half of the diamine dihydrochloride solution add 25 mL of 1.3 M iron(III) chloride solution,[3] cool in ice, and, if necessary, initiate crys-

Note for the instructor

3. Dissolve 90 g of $FeCl_3 \cdot 6H_2O$ (MW 270.32) in 100 mL of water and 100 mL of concentrated hydrochloric acid by warming, cool, and filter (248 mL of solution).

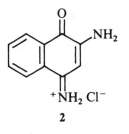

2

tallization by scratching. Rub the liquid film with a glass stirring rod at a single spot slightly above the surface of the liquid. If efforts to induce crystallization are unsuccessful, add more hydrochloric acid. Collect the red product and wash with dilute HCl. Dry weight is 2.4–2.7 g.

Divide the moist product into three equal parts and spread out one part to dry for conversion to **3**. The other two parts can be used while still moist for conversion to **7** and for recrystallization. Dissolve the other two-thirds by gentle warming in a little water containing 2–3 drops of hydrochloric acid, shake for a minute or two with decolorizing charcoal, filter, and add concentrated hydrochloric acid to decrease the solubility.

Cleaning Up   Neutralize the filtrate with sodium carbonate, collect the iron hydroxide by vacuum filtration through Celite on a Büchner funnel. The solid goes into the nonhazardous solid waste container while the filtrate is diluted with water and flushed down the drain.

3. Preparation of 2-Amino-1,4-naphthoquinonimine Diacetate (3)

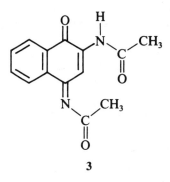

3

A mixture of 0.5 g of the dry quinonimine hydrochloride (**2**), 0.5 g of sodium acetate (anhydrous), and 3 mL of acetic anhydride is stirred in a test tube and warmed gently on a hot plate or steam bath. With thorough stirring the red salt should soon change into yellow crystals of the diacetate. The solution may appear red, but as soon as particles of red solid have disappeared the mixture can be poured into about 10 mL of water. Stir until the excess acetic anhydride has either dissolved or become hydrolyzed, collect and wash the product (dry weight 0.5 g), and (drying is unnecessary) crystallize it from ethanol or methanol; yellow needles, mp 189°C.

Cleaning Up   The filtrate should be diluted with water and flushed down the drain. Crystallization solvent goes in the organic solvents container.

4. Preparation of 2,4-Diacetylamino-1-naphthol (4)

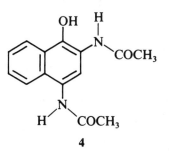

4

To one-half of the diaminonaphthol dihydrochloride solution saved from Section 1 add 3 mL of acetic anhydride, stir vigorously, and add a solution of 3 g of sodium acetate (anhydrous) and about 100 mg of sodium hydrosulfite in 20–30 mL of water. The diacetate may precipitate as a white powder or it may separate as an oil that solidifies when chilled in ice and rubbed with a rod. Collect the product and, to hydrolyze any triacetate present, dissolve it in 5 mL of 10% sodium hydroxide and 50 mL of water by stirring at room temperature. If the solution is colored, a pinch of sodium hydrosulfite may bleach it. Filter by suction and acidify by gradual addition of well-diluted hydrochloric acid (2 mL of concentrated acid). The diacetate tends to remain in supersaturated solution; hence, either to initiate crystallization or to ensure maximum separation, it is advisable to stir well, rub the walls with a rod, and cool in ice. Collect the product, wash it with water, and divide it into thirds (dry weight 2.1–2.6 g).

Two-thirds of the material can be converted without drying into **5** and the other third used for preparation of a crystalline sample. Dissolve the third reserved for crystallization (moist or dry) in enough hot acetic acid to bring about solution, add a solution of a small crystal of tin(II) chloride in a few drops of dilute hydrochloric acid to inhibit oxidation, and dilute gradually with 5–6 volumes of water at the boiling point. Crystallization may be slow, and cooling and scratching may be necessary. The pure diacetate forms colorless prisms, mp 224°C, dec.

Cleaning Up   Combine all filtrates including the acetic acid used to crystallize the product, dilute with water, neutralize with sodium carbonate, and flush the solution down the drain.

5. Preparation of 2-Acetylamino-1,4-naphthoquinone (5)

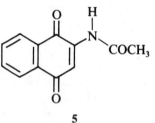

5

Dissolve 1.5 g of the moist diacetylaminonaphthol (**4**) in 10 mL of acetic acid (hot), dilute with 20 mL of hot water, and add 10 mL of 0.13 M iron(III) chloride solution. The product separates promptly in flat, yellow needles, which are collected (after cooling) and washed with a little alcohol;

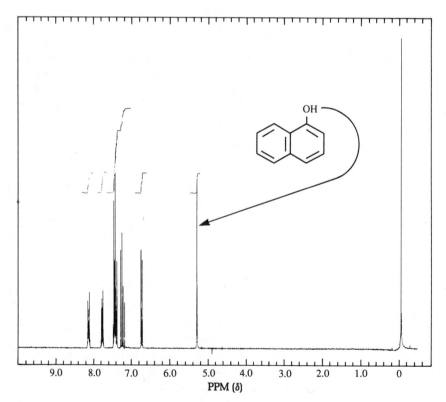

FIG. 1   ¹H nmr spectrum of 1-naphthol (250 MHz).

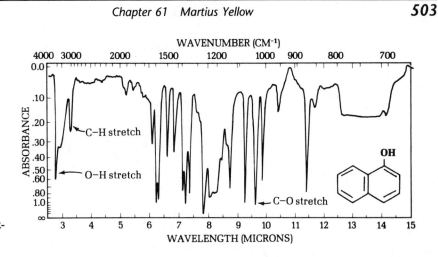

FIG. 2 Infrared spectrum of α-naphthol.

the yield is usually 1.2 g. Dry one-half of the product for conversion to **6** and crystallize the rest from 95% ethanol; mp 204°C.

Cleaning Up   Dilute the filtrate with water, neutralize it with sodium carbonate, and flush it down the drain. A negligible quantity of iron is disposed of in this way.

6. Preparation of 2-Amino-1,4-naphthoquinone (6)

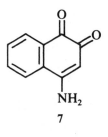

6

To 0.5 g of dry 2-acetylamino-1,4-naphthoquinone (**5**) contained in a 25-mL Erlenmeyer flask add 2 mL of concentrated sulfuric acid and heat the mixture on the steam bath with swirling to promote rapid solution (1–2 min). After 5 min cool the deep red solution, dilute extensively with water, and collect the precipitated product; wash it with water and crystallize the moist sample (dry weight about 0.4 g) from alcohol or alcohol-water; red needles, mp 206°C.

Cleaning Up   The filtrate is neutralized with sodium carbonate, diluted with water, and flushed down the drain.

7. Preparation of 4-Amino-1,2-naphthoquinone (7)

Moist 2 is satisfactory

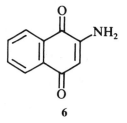

7

Dissolve 1 g of the aminonaphthoquinonimine hydrochloride (**2**) reserved from Experiment 2 in 25 mL of water, add 2 ml of concentrated ammonia solution (den 0.90), and boil the mixture for 5 min. The free quinonimine initially precipitated is hydrolyzed to a mixture of the aminoquinone **7** and the isomer **6**. Cool, collect the precipitate and suspend it in about 50 mL of water, and add 25 mL of 10% sodium hydroxide solution. Stir well, remove the small amount of residual 2-amino-1,4-naphthoquinone (**6**) by filtration, and acidify the filtrate with acetic acid. The orange precipitate of **7** is collected, washed, and crystallized while still wet from 500–600 mL of hot water (the separation is slow). The yield of orange needles, which decomposes about 270°C, is about 0.4 g.

Cleaning Up   The filtrate is neutralized with sodium carbonate, diluted with water, and flushed down the drain.

Question_____

Write a balanced equation for the preparation of 2-amino-1,4-naphtho-quinonimine hydrochloride (2).

62

Catalytic Hydrogenation

Prelab Exercise: Calculate the volume of hydrogen gas generated when 3 mL of 1 M sodium borohydride reacts with concentrated hydrochloric acid. Write a balanced equation for the reaction of sodium borohydride with platinum chloride. Calculate the volume of hydrogen that can be liberated by reacting 1 g of zinc with acid.

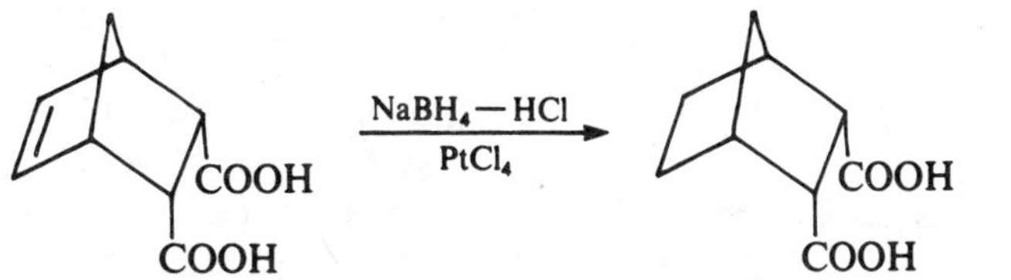

cis-**Norbornene-5,6-*endo*-**
dicarboxylic acid
mp 180–190°C, dec
MW 182.17

cis-**Norbornane-5,6-*endo*-**
dicarboxylic acid
mp 170–175°C, dec
MW 184.19

Hydrogen generated in situ

This experiment presents two methods for carrying out catalytic hydrogenation on a small scale. Conventional procedures for hydrogenation in the presence of a platinum catalyst employ hydrogen drawn from a cylinder of compressed gas and require elaborate equipment. H. C. Brown and C. A. Brown introduced the simple procedure of generating hydrogen *in situ* from sodium borohydride and hydrochloric acid and prepared a highly active supported catalyst by reduction of platinum chloride and sodium borohydride in the presence of decolorizing carbon. The special apparatus described by these authors is here dispensed with in favor of a balloon technique also employed for catalytic oxygenation (Chapter 50).

Experiment

Brown[1] Hydrogenation

The reaction vessel is a 125-mL filter flask with a white rubber pipette bulb wired onto the side arm (Fig. 1). Introduce 10 mL of water, 1 mL of platinum(IV) chloride solution,[2] and 0.5 g of decolorizing charcoal and swirl during addition of 3 mL of stabilized 1 M sodium borohydride solu-

---

1. H. C. Brown and C. A. Brown, *J. Am. Chem. Soc.*, **84**, 1495 (1962).

2. A solution of 200 mg of PtCl₄ in 4 mL of water.

Handle platinum chloride carefully to prevent any waste

Reaction time: about 15 min

FIG. 1   Apparatus for Brown hydrogenation.

Common ion effect

tion.[3] While allowing 5 min for formation of the catalyst, dissolve 1 g of *cis*-norbornene-5,6-*endo*-dicarboxylic acid in 10 mL of hot water. Pour 4 mL of concentrated hydrochloric acid into the reaction flask, followed by the hot solution of the unsaturated acid. Cap the flask with a large serum stopper and wire it on. Draw 1.5 mL of the stabilized sodium borohydride solution into the barrel of a plastic syringe, thrust the needle through the center of the stopper, and add the solution dropwise with swirling. The initial uptake of hydrogen is so rapid that the balloon may not inflate until you start injecting a second 1.5 mL of borohydride solution through the stopper. When the addition is complete and the reaction appears to be reaching an end point (about 5 min), heat the flask on the steam bath with swirling and try to estimate the time at which the balloon is deflated to a constant size (about 5 min). When balloon size is constant, heat and swirl for 5 min more and then release the pressure by injecting the needle of an open syringe through the stopper.

Filter the hot solution by suction and place the catalyst in a jar marked "Catalyst Recovery." Cool the filtrate and extract it with three 15-mL portions of ether. The combined extracts are to be washed with saturated sodium chloride solution and dried over anhydrous sodium sulfate. Evaporation of the ether gives about 0.8–0.9 g of white solid.

The only solvent of promise for crystallization of the saturated *cis*-diacid is water, and the diacid is very soluble in water and crystallizes extremely slowly and with poor recovery. However, the situation is materially improved by addition of a little hydrochloric acid to decrease the solubility of the diacid.

Scrape out the bulk of the solid product and transfer it to a 25-mL Erlenmeyer flask. Add 1–2 mL of water to the 125-mL flask, heat to boiling to dissolve residual solid, and pour the solution into the 25-mL flask. Bring the material into solution at the boiling point with a total of not more than 3 mL of water (as a guide, measure 3 mL of water into a second 25-mL Erlenmeyer). With a Pasteur pipette add 3 drops of concentrated hydrochloric acid and let the solution stand for crystallization. Clusters of heavy prismatic needles soon separate; the recovery is about 90%. The product should give a negative test for unsaturation with acidified permanganate solution.

Observe what happens when a sample of the product is heated in a melting point capillary to about 170°C. Account for the result. You may be able to confirm your inference by letting the oil bath cool until the sample solidifies and then noting the mp temperature and behavior on remelting.

---

3. Dissolve 1.6 g of sodium borohydride and 0.30 g of sodium hydroxide (stabilizer) in 40 mL of water. When not in use, the solution should be stored in a refrigerator. If left for some time at room temperature in a tightly stoppered container, gas pressure may develop sufficient to break the vessel.

Cleaning Up   The catalyst removed by filtration may be pyrophoric (spontaneously flammable in air). Immediately remove it from the Hirsch funnel, wet it with water, and place it either in the hazardous waste container or in the catalyst recovery container. It should be kept wet with water at all times. The combined aqueous filtrates, after neutralization, are diluted with water and flushed down the drain. After the ether is allowed to evaporate from the sodium sulfate in the hood, it can be placed in the nonhazardous solid waste container.

Questions

1. How would you convert the reduced bicyclic diacid product to the corresponding bicyclic anhydride?

2. Why does the addition of hydrochloric acid cause the solubility of the bicyclic diacid in water to decrease?

3. Why is decolorizing charcoal added to the hydrogenation reaction mixture?

63

Dichlorocarbene

Prelab Exercise: Propose a synthesis of ⬡—O—CH₂CH₂CH₂CH₃
using the principles of phase transfer catalysis. Why is phase transfer catalysis particularly suited to this reaction?

Dichlorocarbene is a highly reactive intermediate of bivalent carbon with only six valence electrons around the carbon. It is electrically neutral and a powerful electrophile. As such it reacts with alkenes, forming cyclopropane derivatives by *cis*-addition to the double bond.

There are a dozen or so ways by which dichlorocarbene may be generated. In this experiment thermal decomposition of anhydrous sodium trichloroacetate in an aprotic solvent in the presence of *cis,cis*-1,5-cyclooctadiene generates dichlorocarbene to give 5,5,10,10-tetrachlorotricyclo[7.1.0.0$^{4.6}$]decane (**1**).

Thermal Decomposition of Sodium Trichloroacetate; Reaction of Dichlorocarbene with 1,5-Cyclooctadiene

CH₃OCH₂CH₂OCH₂CH₂OCH₃

Diglyme
MW 134.17, bp 161°C
miscible with water

The thermal decomposition of sodium trichloroacetate initially gives the trichloromethyl anion (Eq. 1). In the presence of a proton-donating solvent (or moisture) this anion gives chloroform; in the absence of these reagents the anion decomposes by loss of chloride ion to give dichlorocarbene (Eq. 2).

The conventional method for carrying out the reaction is to add the salt portionwise, during 1–2 h, to a magnetically stirred solution of the olefin in diethylene glycol dimethyl ether (''diglyme'') at a temperature maintained in a bath at 120°C. Under these conditions the reaction mixture becomes almost black, isolation of a pure product is tedious, and the yield is low.

Test the diglyme for peroxides

Tetrachloroethylene, a nonflammable solvent widely used in the dry cleaning industry, boils at 121°C and is relatively inert toward electrophilic dichlorocarbene. On generation of dichlorocarbene from either chloroform or sodium trichloroacetate in the presence of tetrachloroethylene, the yield of hexachlorocyclopropane (mp 104°C) is only 0.2–10% (W. R. Moore, 1963; E. K. Fields, 1963). The first idea for simplifying the procedure[1] was

---

1. L. F. Fieser and David H. Sachs, *J. Org. Chem.*, **29**, 1113 (1964).

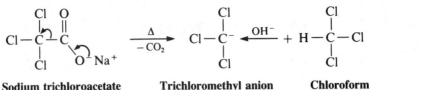

Sodium trichloroacetate   **Trichloromethyl anion**   **Chloroform**
MW 185.39   MW 119.39, bp 61°C

(1)

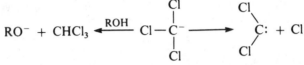

Dichlorocarbene

(2)

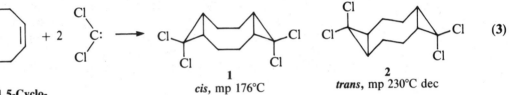

(3)

cis,cis-1,5-Cyclo-   **1**   **2**
octadiene   *cis*, mp 176°C   *trans*, mp 230°C dec
MW 108.14
bp 149–150°C   **5,5,10,10-Tetrachlorotricyclo[7.1.0.0$^{4.6}$]decane**
MW 274.03

Rationale of the procedure

X-ray analysis

to use tetrachloroethylene to control the temperature to the desired range, but sodium trichloroacetate is insoluble in this solvent and no reaction occurs. Diglyme, or an equivalent, is required to provide some solubility. The reaction proceeds better in a 7:10 mixture of diglyme to tetrachloroethylene than in diglyme alone, but the salt dissolves rapidly in this mixture and has to be added in several small portions and the reaction mixture becomes very dark. The situation is vastly improved by the simple expedient of decreasing the amount of diglyme to a 2.5:10 ratio. The salt is so sparingly soluble in this mixture that it can be added at the start of the experiment and it dissolves slowly as the reaction proceeds. The boiling and evolution of carbon dioxide provide adequate stirring, the mixture can be left unattended, and what little color develops is eliminated by washing the crude product with methanol.

The main reaction product crystallizes from ethyl acetate in beautiful prismatic needles, mp 174–175°C, and this was the sole product encountered in runs made in the solvent mixture recommended. In earlier runs made in diglyme with manual control of temperature, the ethyl acetate mother liquor material on repeated crystallization from toluene afforded small amounts of a second isomer, mp 230°C, dec. Analyses checked for a pair of *cis-trans* isomers and both gave negative permanganate tests. To distinguish between them, the junior author of the paper undertook an X-ray analysis, which showed that the lower-melting isomer is *cis* and the higher-melting isomer is *trans*.

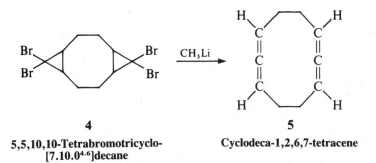

4

**5,5,10,10-Tetrabromotricyclo-
[7.10.0⁴·⁶]decane**

5

Cyclodeca-1,2,6,7-tetracene

A striking reaction of a bis dihalocarbene adduct

The bis adduct (**4**) of *cis,cis*-1,5-cyclooctadiene with dibromocarbene is described as melting at 174–180°C and may be a *cis-trans* mixture. Treatment of the substance with methyl lithium at −40°C gave a small amount of the ring-expanded bisallene (**5**).

1. Sodium Trichloroacetate

Use caution in handling the corrosive trichloroacetic acid. Carry out procedure in hood.

Use commercial sodium trichloroacetate (dry) or prepare it as follows. Place 12.8 g of trichloroacetic acid in a 125-mL Erlenmeyer flask, dissolve 3.2 g of sodium hydroxide pellets in 12 mL of water in a 50-mL Erlenmeyer flask, cool the solution thoroughly in an ice bath, and swirl the flask containing the acid in the ice bath while slowly dropping in about nine-tenths of the alkali solution. Then add a drop of 0.04% Bromocresol Green solution to produce a faint yellow color, visible when the flask is dried and placed on white paper. With a Pasteur pipette, titrate the solution to an end point where a single drop produces a change from yellow to blue. If the end point is overshot, add a few crystals of acid and titrate more carefully. Close the flask with a rubber stopper, connect to an aspirator, and place the flask within the rings of the steam bath and wrap a towel around both for maximum heat. Turn on the water at full force for maximum suction. The evaporation requires no further attention and should be complete in 15–20 min. When you have an apparently dry white solid, scrape it out with a spatula and break up the large lumps. If you see any evidence of moisture, or in case the weight exceeds the theory, place the solid in a 25 × 150-mm test tube and rest this on its side on a drying tray mounted 5 cm above the base of a 70-watt hot plate and let the drier operate overnight.

Make sure that the salt is completely dry

2. Dichlorocarbene Reaction

Place 14.2 g of *dry* sodium trichloroacetate in a 250-mL round-bottomed flask mounted over a flask heater and add 20 mL of tetrachloroethylene, 5 mL of diglyme, and 5 mL (4.4 g) of *cis,cis*-1,5-cyclooctadiene.[2] Attach a

Note for the instructor

2. Store diglyme, tetrachloroethylene, and 1,5-cyclooctadiene over Linde 5A molecular sieves or anhydrous magnesium sulfate to guarantee that the reagents are dry. The reaction will not work if any reagent is wet.

Reflux time: about 75 min

reflux condenser and in the top opening of the condenser insert a rubber stopper carrying a glass tube connected by rubber tubing to a short section of glass tube, which can be inserted below the surface of 2–3 mL of tetrachloroethylene in a 20 × 150-mm test tube mounted at a suitable level. This bubbler will show when the evolution of carbon dioxide ceases; the solvent should be the same as that of the reaction mixture, should there be a suckback. Heat to boiling, note the time, and reflux gently until the reaction is complete. You will notice foaming, due to liberated carbon dioxide, and separation of finely divided sodium chloride. Inspection of the bottom of the flask will show lumps of undissolved sodium trichloroacetate, which gradually disappear. A large flask is specified because it will serve later for removal of tetrachloroethylene by steam distillation. Make advance preparation for this operation.

Easy workup

When the reaction is complete add 75 mL of water to the hot mixture, heat over a hot electric flask heater and steam distill until the tetrachloroethylene is eliminated and the product separates as an oil or semisolid. Cool the flask to room temperature, decant the supernatant liquid into a separatory funnel, and extract with dichloromethane. Run the extract into the reaction flask and use enough more dichloromethane to dissolve the product; use a Pasteur pipette to rinse down material adhering to the adapter. Run the dichloromethane solution of the product into an Erlenmeyer flask through a cone of anhydrous sodium sulfate in a funnel and evaporate the solvent (bp 41°C). The residue is a tan or brown solid (8 g). Cover it with methanol, break up the cake with a flattened stirring rod, and crush the lumps. Cool in ice, collect, and wash the product with methanol. The yield of colorless, or nearly colorless, 5,5,10,10-tetrachlorotricyclo[7.1.0.0^{4,6}]-decane is 3.3 g. This material, mp 174–175°C, consists almost entirely of the *cis* isomer. Dissolve it in ethyl acetate (15–20 mL) and let the solution stand undisturbed for crystallization at room temperature. The pure *cis* isomer separates in large, prismatic needles, mp 175–176°C.

Cleaning Up   The aqueous layer can be diluted with water and flushed down the drain. Organic layers, which contain halogenated material, must be placed in the halogenated organic solvents container. After the solvent has been allowed to evaporate from the drying agent in the hood, it can be placed in the nonhazardous solid waste container. Methanol and ethyl acetate used in crystallization can be placed in the organic solvents container.

$$CCl_3^- + H_2O \longrightarrow CHCl_3 + OH^-$$
$$\mathbf{3}$$

$$:CCl_2 + 2\,H_2O$$
$$\mathbf{4} \downarrow$$
$$CH_2Cl_2 + 2\,OH^-$$

3. Phase Transfer Catalysis—Reaction of Dichlorocarbene with Cyclohexene

Ordinarily water must be scrupulously excluded from carbene-generating reactions. Both the intermediate trichloromethyl anion (**3**) and dichlorocarbene (**4**) react with water. But in the presence of a phase transfer catalyst (a quaternary ammonium salt such as benzyltriethylammonium chloride) it is possible to carry out the reaction in aqueous medium. Chloroform, 50% aqueous sodium hydroxide, and the olefin, in the presence of a catalytic amount of the quaternary ammonium salt, are stirred for a few minutes to produce an emulsion; and after the exothermic reaction is complete (about 30 min) the product is isolated.

Both benzyltriethylammonium chloride (**1**) and benzyltriethylammonium hydroxide (**2**) partition between the aqueous and organic phases. In the aqueous phase, the quaternary ammonium chloride (**1**) reacts with concentrated hydroxide to give the quaternary ammonium hydroxide (**2**). In the chloroform phase, **2** reacts with chloroform to give the trichloromethyl anion (**3**), which eliminates chloride ion to give dichlorocarbene, :CCl$_2$, (**4**), and **1**. The carbene (**4**) reacts with the olefin to give product (**5**).

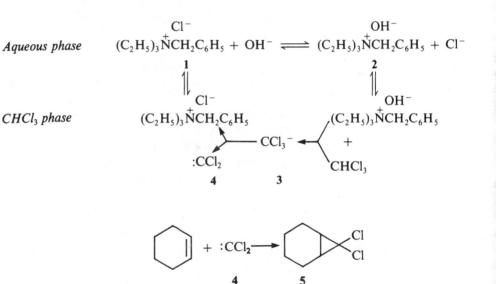

Procedure

CAUTION: *Chloroform is a carcinogen and should be handled in the hood. Avoid all contact with the skin.*

To a mixture of 8.2 g of cyclohexene, 12.0 g of chloroform (*caution!*), and 20 mL of 50% aqueous sodium hydroxide in a 125-mL Erlenmeyer flask containing a thermometer, add 0.2 g of benzyltriethylammonium chloride.[3] Swirl the mixture to produce a thick emulsion. The temperature of the reaction will rise gradually at first and then markedly accelerate. As it approaches 60°C prepare to immerse the flask in an ice bath. With the flask alternately in and out of the ice bath, stir the thick paste and maintain the temperature between 50 and 60°C. After the exothermic reaction is com-

Note for the instructor

3. Many other quaternary ammonium salts, commercially available, work equally well (e.g., cetyltrimethylammonium bromide).

plete (about 10 min) allow the mixture to cool spontaneously to 35°C and then dilute with 50 mL of water. Separate the layers (test to determine which is the organic layer), extract the aqueous layer once with 10 mL of ether, and wash the combined organic layer and ether extract once with 20 mL of water. Dry the cloudy organic layer by shaking it with anhydrous sodium sulfate until the liquid is clear. Decant into a 50-mL round-bottomed flask and distill the mixture (boiling chip), first from a steam bath to remove ether and some unreacted cyclohexene and chloroform, then over a free flame. Collect 7,7-dichlorobicyclo[4.1.0]heptane over the range 195–200°C. Yield, about 9 g. A purer product will result if the distillation is carried out under reduced pressure (Chapter 7).

Cleaning Up   Organic layers go in the organic solvents container. The aqueous layer should be diluted with water and flushed down the drain. After the ether evaporates from the sodium sulfate in the hood, it can be placed in the nonhazardous solid waste container. The pot residue from the distillation goes in the halogenated organic solvents container.

The success of this reaction is critically dependent on forming a thick emulsion, for only then will the surface between the two phases be large enough for complete reaction to occur. James Wilbur at Southwest Missouri State University found that when freshly distilled, pure reagents were used, no exothermic reaction occurred, and very little or no dichloronorcarane was formed. We have subsequently found that old, impure cyclohexene gives very good yields of product. Cyclohexene is notorious

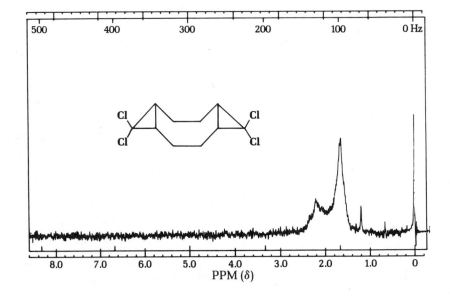

FIG. 1   ¹H nmr spectrum of 5,5,10,10-tetrachlorotricyclo-[7,1,0,0⁴,⁶]decane (60 MHz).

for forming peroxides at the allylic position. Under the reaction conditions this peroxide will be converted to the ketone:

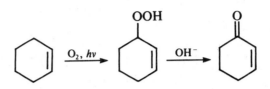

Apparently, small amounts of this ketone are responsible for the formation of stable emulsions. This appears to be one of the rare cases when impure reagents lead to successful completion of a reaction. If only pure reagents are available, then the reaction mixture must be stirred vigorously over-night, using a magnetic stirrer. Alternatively, a small quantity of cyclohex-ene hydroperoxide can be prepared by pulling air through the alkene for a few hours, employing an aspirator, and using this impure reagent, or—as Wilbur found—by adding 1 mL of cyclohexanol to the reaction mixture. Detergents such as sodium lauryl sulfate do not increase the rate of reaction or emulsion formation.

Questions

1. If water is not rigorously excluded from sodium trichloroacetate dichlorocarbene synthesis, what side reaction occurs?

2. Why is vigorous stirring or emulsion formation necessary in a phase transfer reaction?

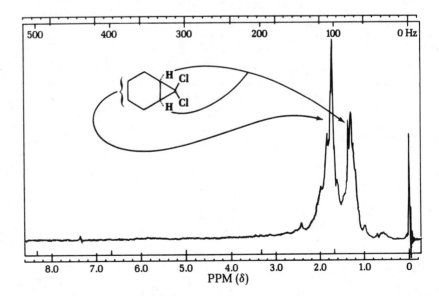

FIG. 2   ¹H nmr spectrum of 7,7-dichlorobicyclo[4.1.0]-heptane (60 MHz).

Enzymatic Reduction: A Chiral Alcohol from a Ketone

Prelab Exercise: Study the biochemistry of the conversion of glucose to ethanol (Chapter 18). Which enzyme might be responsible for the reduction of ethyl acetoacetate?

Reduction of an achiral ketone with the usual laboratory reducing agents such as sodium borohydride or lithium aluminum hydride will not give a chiral alcohol because the chances for attack on two sides of the planar carbonyl group are equal. However, if the reducing agent is chiral, there is the possibility of obtaining a chiral alcohol. Organic chemists in recent years have devised a number of such chiral reducing agents, but few of them are as efficient as those found in nature.

In this experiment we will use the enzymes found in baker's yeast to reduce ethyl acetoacetate to $S(+)$-ethyl 3-hydroxybutanoate. This compound is a very useful synthetic building block. At least eight chiral natural product syntheses are based on this hydroxyester.[1]

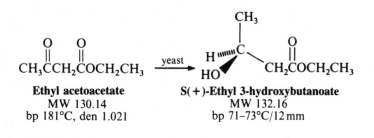

Ethyl acetoacetate
MW 130.14
bp 181°C, den 1.021

$S(+)$-Ethyl 3-hydroxybutanoate
MW 132.16
bp 71–73°C/12 mm

There are a large number of different enzymes present in yeast. The primary ones responsible for the conversion of glucose to ethanol are discussed in Chapter 18. While this fermentation reaction is taking place, certain ketones can be reduced to chiral alcohols.

Whenever a chiral product is produced from an achiral starting material, the chemical yield as well as the optical yield is important—in other words, how stereoselective the reaction has been. The usual method for recording this is to calculate the enantiomeric excess (ee). A sodium borohydride reduction will produce 50% R and 50% S alcohol with no enantiomer in excess. If 93% $S(+)$ and 7% $R(-)$ isomer are produced, then

---

1. R. Amstutz, E. Hungerbühler, and D. Seebach, *Helv. Chim. Acta,* **64,** 1796 (1981) and D. Seebach, *Tetrahedron Lett.* 159 (1982).

the enantiomeric excess is 86%. In the present yeast reduction of ethyl acetoacetate various authors have reported enantiomeric excesses ranging from 70% to 97%. This optical yield is distinct from the chemical yield, which depends on how much material is isolated from the reaction mixture.

The use of enzymes to carry out stereospecific chemical reactions is not new, but it is not always possible to predict if an enzymatic reaction (unlike a purely chemical reaction) will occur, or how stereospecific it will be if it does. Because this experiment is easily carried out, it might be an interesting research project to explore the range of possible ketones that yeast will reduce to chiral alcohols. For example, butyrophenone can be reduced to the corresponding chiral alcohol. For a review, see Sih and Rosazza.[2] The present experiment is based on the work of Seebach,[3] Mori,[4] and Ridley.[5] See also the work of Bucciarelli, et al.[6]

Experiments

1. Enzymatic Resolution

In a 500-mL flask (see Chapter 18, Biosynthesis of Ethanol) dissolve 80 g of sucrose and 0.5 g of disodium hydrogenphosphate in 300 mL of warm (35°C) tap water. Add two packets (16 g) of dry yeast and swirl to suspend the yeast throughout the solution. After 15 min, while fermentation is progressing vigorously, add 5 g of ethyl acetoacetate. Store the flask in a warm place—ideally at 30–35°C—for at least 48 h (a longer time will do no harm). At the end of this time add 20 g of Celite filtration aid and remove the yeast cells by filtration on a 10-cm Büchner funnel (see Chapter 18). Wash the cells with 50 mL of water; then saturate the filtrate with sodium chloride in order to decrease the solubility of the product. See a handbook for the solubility of sodium chloride in water to decide approximately how much to use. Extract the resulting solution five times with 50-mL portions of ether, taking care to shake the separatory funnel hard enough to mix the layers, but not so hard as to form an emulsion between the ether and water. Addition of a small amount of methanol may help to break up emulsions. Dry the ether layer by adding anhydrous sodium sulfate until it no longer clumps together. After approximately 15 min decant the ether solution into a tared distilling flask and remove the ether by distillation or by evaporation. The residue should weigh about 3.5 g. It should, unlike the starting material, give a negative iron(III) chloride test (see p. 579). It can be analyzed by TLC (use dichloromethane as the solvent) to determine

Keep the mixture warm; the reaction is slow at low temperatures

Extinguish all flames when working with ether

2. C. J. Sih and J. P. Rosazza, *Application of Biochemical Systems in Organic Chemistry,* Part 1 (J. B. Jones, C. J. Sih, and D. Perlman, Eds.), pp. 71–78, Wiley, New York, 1976.

3. D. Seebach, *Tetrahedron Lett.* 159 (1982).

4. K. Mori, *Tetrahedron,* **37**, 1341 (1981).

5. D. D. Ridley and M. Stralow, *Chem. Commun.,* **400** (1975).

6. M. Bucciarelli, A. Forni, I. Moretti, and G. Torre, *Synthesis,* **1983**, 897.

whether unreacted ethyl acetoacetate is present. Infrared spectroscopy should show the presence of the hydroxyl group and may show unreduced methyl ketone. The nmr spectrum of the product is easily distinguished from that of starting material.

Cleaning Up   The aqueous layer can be diluted with water and flushed down the drain. After the ether evaporates from the sodium sulfate in the hood, it can be placed in the nonhazardous solid waste container. Dichloromethane (the TLC solvent) goes in the halogenated organic solvents container. Ether distillate goes in the organic solvents container.

2. Determination of Optical Purity

The optical purity of the product can be determined by measuring the optical rotation in a polarimeter. The specific rotation, $[\alpha]_D^{25}$, of $S(+)$-ethyl 3-hydroxybutanoate has been reported to vary from $+31.3°$ to $+41.7°$ in chloroform. The specific rotation, $[\alpha]_D^{25}$, of $37.2°$ (chloroform, C 1.3) corresponds to an enantiomeric excess of 85%.

A better method for determining optical purity of a small quantity of material is to use an nmr chiral shift reagent.[7] The use of shift reagents is discussed extensively in Chapter 20. A shift reagent will complex with a basic center (the hydroxyl group in ethyl 3-hydroxybutanoate) and cause the nmr peaks to shift, usually downfield. Protons nearest the shift reagent/hydroxyl group shift more than those far away. A chiral shift reagent forms diastereomeric complexes so that peaks from one enantiomer shift downfield more than peaks from the other enantiomer. By comparing the areas of the peaks the enantiomeric excess can be calculated.

The nmr experiment is easily accomplished by adding 20-mg increments of tris[3-(heptafluoropropylhydroxymethylene)-(+)-campharato]-europium(III) shift reagent (Aldrich 16,474–7) to a solution of 30 mg of pure, dry ethyl 3-hydroxybutanoate in 0.3 mL of carbon tetrachloride.[8]

Cleaning Up   Place the contents of the nmr tube in the europium shift reagent hazardous waste container.

3. Preparation of 3,5-Dinitrobenzoate

One of the best ways of purifying the product is to convert it to a crystalline derivative, the 3,5-dinitrobenzoate. 3,5-Dinitrobenzoates are common alcohol derivatives (see Chapter 71); they are easily prepared, the two nitro groups add considerably to the molecular weight, and they are easily crystallized.

The 3,5-dinitrobenzoates of a racemic mixture of *R*- and *S*-ethyl 3-hydroxybutanoate, like almost all racemates, crystallize together to give

---

7. K. B. Lipkowitz and J. L. Mooney, *J. Chem. Ed.*, **64**, 985 (1987).

8. See footnote 7.

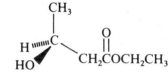

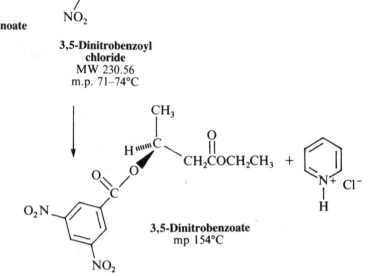

Pyridine
MW 79.10
bp 115°C
den 0.978

S(+)-Ethyl-3-hydroxybutanoate

**3,5-Dinitrobenzoyl
chloride**
MW 230.56
m.p. 71–74°C

3,5-Dinitrobenzoate
mp 154°C

crystals that in this case melt at 146°C. No amount of recrystallization causes one enantiomer to crystallize out while the other remains in solution since they are, after all, mirror images of each other. However, if one enantiomer is in large excess it is possible to effect a separation by crystallization. In the present case the *S* enantiomer predominates and it crystallizes out, leaving most of the *R* + *S* racemate in solution. Repeated crystallization increases the melting point and the purity to the point at which the optical purity of the crystalline product is almost 100%. At this point the 3,5-dinitrobenzoate has a melting point of 154°C. Treatment of the derivative with excess acidified ethanol regenerates 100% ee *S*(+)-ethyl 3-hydroxybutanoate.

Procedure

To 0.5 g of the *S*(+)-ethyl 3-hydroxybutanoate in a 25 mL round-bottomed flask fitted with condenser add 0.875 g of pure[9] 3,5-dinitrobenzoyl chloride and 5 mL of pyridine. Reflux the mixture for 15 min and then transfer it with a Pasteur pipette to 20 mL of water in an Erlenmeyer flask. Remove the solvent from the crystals and shake the crystals with 10 mL of 5% sodium carbonate solution to remove dinitrobenzoic acid. Remove the bicarbonate

---

9. Check the melting point of the 3,5-dinitrobenzoyl chloride. If it is below 70°C it should be recrystallized from dichloromethane.

solution and recrystallize the derivative from ethanol. Dry a portion and determine the melting point. If it is not near 154°C repeat the crystallization. Dry the pure derivative and calculate the percent yield.

Treatment of the derivative with excess acidified ethanol will regenerate 100% ee $S(+)$-ethyl 3-hydroxybutanoate. If several crops of crude product are pooled, distillation at reduced pressure (see Chapter 7) can give chemically pure ethyl 3-hydroxybutanoate (bp 71–73° at 12 mm). Both the R and S enantiomers have the same boiling points, so the optical purity will not be changed by distillation.

Cleaning Up   The pyridine solvent goes in the organic solvents container. The aqueous layer should be diluted with water and flushed down the drain. Ethanol from the crystallization goes in the organic solvents container, along with the pot residue if the final product is distilled.

Question

What can you say about the purity of the reduction product whose ¹H nmr spectrum is illustrated in Fig. 1? Assign the peaks of ethyl 3-hydroxybutanoate to specific protons in the molecule.

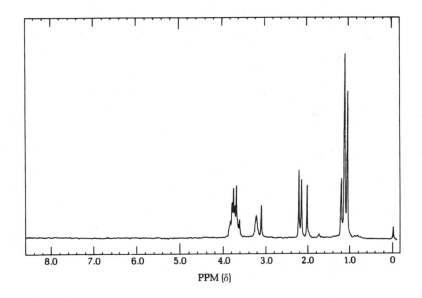

FIG. 1 ¹H nmr spectrum of yeast reduction reaction mixture (90 MHz).

65

Enzymatic Resolution of DL-Alanine

Prelab Exercise: Discuss at least three other methods for resolving DL-alanine into its enantiomers. Judging from its name, would you expect acylase to be equally effective on the acetyl derivatives of other amino acids? Explain why the enzyme acylase will hydrolyze only one of a pair of enantiomers and how a mixture of L-alanine and N-acetyl-D-alanine can be separated at the end of the reaction.

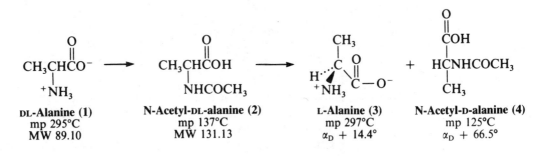

DL-Alanine (1)
mp 295°C
MW 89.10

N-Acetyl-DL-alanine (2)
mp 137°C
MW 131.13

L-Alanine (3)
mp 297°C
α_D + 14.4°

N-Acetyl-D-alanine (4)
mp 125°C
α_D + 66.5°

Resolution of DL-alanine (**1**) is accomplished by heating the N-acetyl derivative (**2**) in weakly alkaline solution with acylase, a proteinoid preparation from porcine kidney containing an enzyme that promotes rapid hydrolysis of N-acyl derivatives of natural L-amino acids but acts only immeasurably slowly on the unnatural D-isomers. N-Acetyl-DL-alanine (**2**) can thus be converted into a mixture of L-(+)-alanine (**3**) and N-acetyl-D-alanine (**4**). The mixture is easily separable into the components, because the free amino acid (**3**) is insoluble in ethanol and the N-acetyl derivative (**4**) is readily soluble in this solvent. Note that, in contrast to the weakly levorotatory D-(−)-alanine (α_D − 14.4°), its acetyl derivative is strongly dextrorotatory.

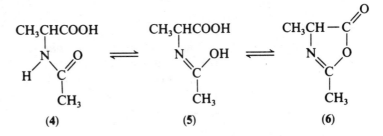

(4)   (5)   (6)

The acetylation of an α-amino acid presents the difficulty that, if the conditions are too drastic, the N-acetyl derivative (**4**) is converted in part through the enol (**5**) to the azlactone (**6**).[1] However, under critically controlled conditions of concentration, temperature, and reaction time, N-acetyl-DL-alanine can be prepared easily in high yield.

Experiment

1. Acetylation of DL-Alanine

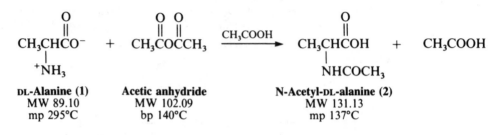

$$\underset{\substack{\text{DL-Alanine (1)}\\\text{MW 89.10}\\\text{mp 295°C}}}{\underset{\substack{|\\^+NH_3}}{CH_3CHCO^-}} + \underset{\substack{\text{Acetic anhydride}\\\text{MW 102.09}\\\text{bp 140°C}}}{CH_3COCCH_3} \xrightarrow{CH_3COOH} \underset{\substack{\text{N-Acetyl-DL-alanine (2)}\\\text{MW 131.13}\\\text{mp 137°C}}}{\underset{\substack{|\\NHCOCH_3}}{CH_3CHCOH}} + CH_3COOH$$

Check the pressure gauge

FIG. 1 Drying *N*-acetyl-DL-alanine under reduced pressure.

Place 2 g of DL-alanine and 5 mL of acetic acid in a 25 × 150-mm test tube, insert a thermometer, and clamp the tube in a vertical position. Measure 3 mL of acetic anhydride, which is to be added when the alanine/acetic acid mixture is at exactly 100°C. Heat the test tube cautiously with a small flame, with stirring, until the temperature of the suspension has risen a little above 100°C. Stir the suspension, let the temperature gradually fall, and when it reaches 100°C add the 3-mL portion of acetic anhydride and note the time. In the course of 1 min the temperature falls (91–95°C, cooled by added reagent), rises (100–103°C, the acetylation is exothermic), and begins to fall with the solid largely dissolving. Stir to facilitate reaction of a few remaining particles of solid, let the temperature drop to 80°C, pour the solution into a tared 100-mL round-bottomed flask, and rinse the thermometer and test tube with a little acetone. Add 10-mL of water to react with excess anhydride, connect the flask to the aspirator operating at full force, put the flask *inside* the rings of the steam bath, and wrap the flask with a towel. Evacuation and heating for about 5–10 min should remove most of the acetic acid and water and leave an oil or thick syrup. Add 10 mL of cyclohexane and evacuate and heat as before for 5–10 min. Traces of water in the syrup are removed as a cyclohexane/water azeotrope. If the product has not yet separated as a white solid or semisolid, determine the weight of the product, add 10 mL more cyclohexane, and repeat the process. When the weight becomes constant, the yield of acetyl DL-alanine should be close to the theoretical amount. The product has a pronounced tendency to remain in supersaturated solution and hence does not crystallize readily.

---

1. The azlactone of DL-alanine is known only as a partially purified liquid.

2. Enzymatic Resolution of N-Acetyl-DL-Alanine

Add 10 mL of distilled water[2] to the reaction flask, grasp this with a clamp, swirl the mixture over a free flame to dissolve all the product, and cool under the tap. Remove a drop of the solution on a stirring rod, add it to 0.5 mL of a 0.3% solution of ninhydrin in water, and heat to boiling. If any unacetylated DL-alanine is present a purple color will develop and should be noted. Pour the solution into a 20 × 150-mm test tube and rinse the flask with a little water. Add 1.5 mL of concentrated ammonia solution, stir to mix, check the pH with Hydrion paper, and, if necessary, adjust to pH 8 by addition of more ammonia with a Pasteur pipette. Add 10 mg of commercial acylase powder, or 2 mL of fresh acylase solution,[3] mix with a stirring rod, rinse the rod with distilled water, and make up the volume until the tube is about half full. Then stopper the tube, mark it for identification (Fig. 3), and let the mixture stand at room temperature overnight, or at 37°C[4] for 4 h.

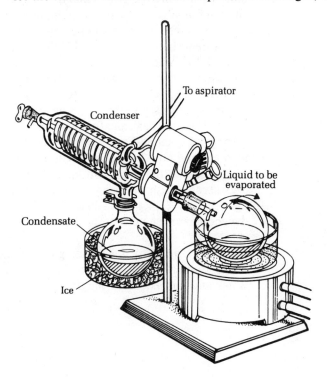

FIG. 2   Rotary evaporation.

Note for the instructor

2. Tap water may contain sufficient heavy metal ion to deactivate the enzyme.

3. Commercial porcine kidney acylase is available from Schwarz/Mann, Division of Becton, Dickinson and Co., Orangeburg, NY 10962, or from Sigma Chemical Co., P.O. Box 14503, St. Louis, MO 63172.

4. A reasonably constant heating device that will hold 15 tubes is made by filling a 1-L beaker with water, adjusting to 37°C, and maintaining this temperature by the heat of a 250-watt infrared drying lamp shining horizontally on the beaker from a distance of about 40 cm. The capacity can be tripled by placing other beakers on each side of the first one and a few cm closer to the lamp.

FIG. 3 Filter paper identification marker.

Work-up time $\frac{1}{2}-\frac{3}{4}$ *h*

At the end of the incubation period add 3 mL of acetic acid to denature the enzyme and if the solution is not as acidic as pH 5 add more acid. Rinse the cloudy solution into a 125-mL Erlenmeyer flask, add 100 mg of decolorizing carbon (0.5-cm column in a 13 × 100-mm test tube), heat and swirl over a free flame for a few moments to coagulate the protein, and filter the solution by suction. Transfer the solution to a 100-mL round-bottomed flask and evaporate on a rotary evaporator under vacuum, or add 20 mL of cyclohexane (to prevent frothing) and a boiling stone and evaporate on the steam bath under vacuum to remove water and acetic acid as completely as possible. Remove the last traces of water and acid by adding 15 mL of cyclohexane and evaporating again to remove water and acetic acid as azeotropes. The mixture of L-alanine and acetyl D-alanine separates as a white scum on the walls. Add 15 mL of 95% ethanol, digest on the steam bath, and dislodge some of the solid with a spatula. Cool well in ice for a few minutes, and then scrape as much of the crude L-alanine as possible onto a suction funnel, and wash it with ethanol. Save the ethanol mother liquor.[5] To recover the L-alanine retained by the flask, add 2 mL of water and warm on the steam bath until the solid is all dissolved, then transfer the solution to a 25-mL Erlenmeyer flask by means of a Pasteur pipette, rinse the flask with 2 mL more water, and transfer in the same way. Add the filtered L-alanine, dissolve by warming, and filter the solution by gravity into a 50-mL Erlenmeyer flask (use the dropping tube to effect the transfer of solution to filter). Rinse the flask and funnel with 1 mL of water and then with 5 mL of warm 95% ethanol. Then heat the filtrate on the steam bath and add more 95% ethanol (10–15 mL) in portions until crystals of L-alanine begin to separate from the hot solution. Let crystallization proceed. Collect the crystals and wash with ethanol. The yield of colorless needles of L-alanine, α_D + 13.7 to + 14.4°[6] (in 1 N hydrochloric acid) varies from 0.40 to 0.56 g, depending on the activity of the enzyme.

Cleaning Up The decolorizing carbon goes in the nonhazardous solid waste container. Ethanol used in the crystallization can be diluted with water and flushed down the drain.

---

5. In case the yield of L-alanine is low, evaporation of this mother liquor may reveal the reason. If the residue solidifies readily and crystallizes from acetone to give acetyl-DL-alanine, mp 130°C or higher, the acylase preparation is recognized as inadequate in activity or amount. Acetyl-D-alanine is much more soluble and slow to crystallize.

6. Determination of optical activity can be made in the student laboratory with a Zeiss Pocket Polarimeter, which requires no monochromatic light source and no light shield. For construction of a very inexpensive polarimeter, see W. H. R. Shaw, *J. Chem. Ed.,* **32,** 10 (1955).

Questions

1. The melting point of N-acetyl-DL-alanine is 137°C and that of N-acetyl-D-alanine is 125°C. What would you expect the melting point of N-acetyl-L-alanine to be, or is this impossible to predict?

2. Would the ^{13}C nmr spectrum of D-alanine differ from that of L-alanine (Fig. 4)?

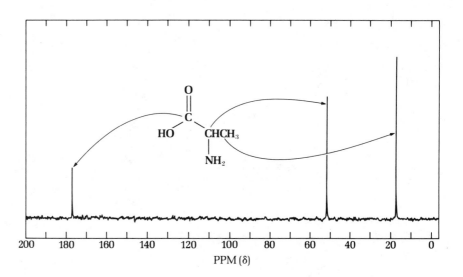

FIG. 4 ^{13}C nmr spectrum of L-alanine (22.6 MHz).

CHAPTER 66

Dyes and Dyeing[1]

Prelab Exercise: Operating on the simple hypothesis that the intensity of a dye on a fiber will depend on the number of strongly polar or ionic groups in the fiber molecule, predict the relative intensities produced by methyl orange when it is used to dye a variety of different fibers, such as are found in the Multifiber Fabric.

Since prehistoric times man has been dyeing cloth. The "wearing of the purple" has long been synonymous with royalty, attesting to the cost and rarity of Tyrian purple, a dye derived from the sea snail *Murex brandaris*. The organic chemical industry originated with William Henry Perkin's discovery of the first synthetic dye, Perkin's Mauve, in 1856.

In this experiment several dyes will be synthesized and these and other dyes will be used to dye a representative group of natural and man-made fibers. You will receive several pieces of Multifiber Fabric 43,[2] which has 13 strips of different fibers woven into it.

Below the black thread at the top, the fibers are acetate rayon (cellulose di- or triacetate, SEF (Monsanto's modacylic), Arnel (cellulose triacetate), cotton, Creslan (polyacrylonitrile), Dacron 54 and 64 (polyester without and with a brightener), nylon 6.6 (polyamide), Orlon 75 (polyacrylonitrile), silk (polyamide), polypropylene, viscose rayon (regenerated cellulose), and wool (polyamide).

Acetate rayon is cellulose (from any source) in which about two of the hydroxyl groups in each unit have been acetylated. This renders the polymer soluble in acetone from which it can be spun into fiber. The smaller number of hydroxyl groups in acetate rayon compared to cotton makes direct dyeing of rayon more difficult than of cotton.

Cotton is pure cellulose. Nylon is a polyamide and made by polymerizing adipic acid and hexamethylenediamine. The nylon polymer chain can be prepared with one acid and one amine group at the termini, or with both acids or both amines. Except for these terminal groups, there are no polar centers in nylon and consequently it is difficult to dye. Similarly Dacron, a polyester made by polymerizing ethylene glycol and terephthalic

1. For a detailed discussion of the chemistry of dyes and dyeing see *Topics in Organic Chemistry*, by Louis F. Fieser and Mary Fieser, Reinhold Publishing Corp., New York, 1963, pp. 357–417.

2. Obtained from Testfabrics, Inc., P.O. Box 240, 200 Blackford Ave., Middlesex, NJ 08846, (908) 469-6446. Cut into ¾" strips.

acid, has few polar centers within the polymer and consequently is difficult to dye. Even more difficult to dye is Orlon, a polymer of acrylonitrile. Wool and silk are polypeptides crosslinked with disulfide bridges. The acidic and basic amino acids (e.g., glutamic acid and lysine) provide many polar groups in wool and silk to which a dye can bind, making these fabrics easy to dye. In this experiment note the marked differences in shade produced by the same dye on different fibers.

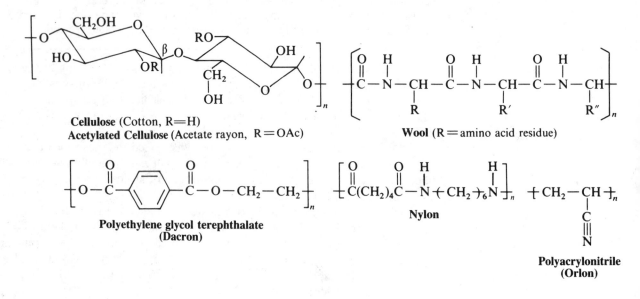

Cellulose (Cotton, R=H)
Acetylated Cellulose (Acetate rayon, R=OAc)

Wool (R=amino acid residue)

Polyethylene glycol terephthalate
(Dacron)

Nylon

Polyacrylonitrile
(Orlon)

Part 1. Dyes

Azo group

The most common dyes are the azo dyes, formed by coupling diazotized amines to phenols. The dye can be made in bulk, or, as we shall see, the dye molecule can be developed on and in the fiber by combining the reactants in the presence of the fiber.

One dye, Orange II, is made by coupling diazotized sulfanilic acid with 2-naphthol in alkaline solution; another, Methyl Orange, is prepared by coupling the same diazonium salt with N,N-dimethylaniline in a weakly acidic solution. Methyl Orange is used as an indicator as it changes color at pH 3.2–4.4. The change in color is due to transition from one chromophore (azo group) to another (quinonoid system).

You are to prepare one of these two dyes and then exchange samples with a neighbor and do the tests with both dyes. Both substances dye wool, silk, and skin, and you must work carefully to avoid getting them on your hands or clothes. The dye will eventually wear off your hands or they can be cleaned by soaking them in warm, slightly acidic (H_2SO_4) permanganate solution until heavily stained with manganese dioxide and then removing the stain in a bath of warm, dilute bisulfite solution.

Experiments

1. Diazotization of Sulfanilic Acid

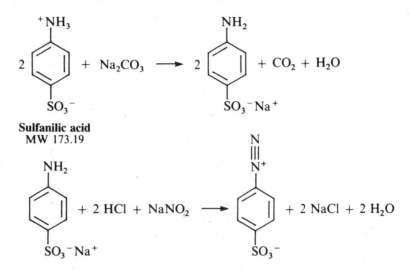

Sulfanilic acid
MW 173.19

Avoid skin contact with diazonium salts. Some diazonium salts are explosive when dry. Always use in solution.

In a 125-mL Erlenmeyer flask dissolve, by boiling, 4.8 g of sulfanilic acid monohydrate in 50 mL of 2.5% sodium carbonate solution (or use 1.33 g of anhydrous sodium carbonate and 50 mL of water). Cool the solution under the tap, add 1.9 g of sodium nitrite, and stir until it is dissolved. Pour the solution into a flask containing about 25 g of ice and 5 mL of concentrated hydrochloric acid. In a minute or two a powdery white precipitate of the diazonium salt should separate and the material is then ready for use. The product is not collected but is used in the preparation of the dye Orange II and/or Methyl Orange while in suspension. It is more stable than most diazonium salts and will keep for a few hours.

Choice of 2 or 3

2. Orange II (1-*p*-Sulfobenzeneazo-2-Naphthol Sodium Salt)

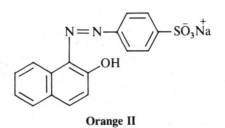

Orange II

In a 400-mL beaker dissolve 3.6 g of 2-naphthol in 20 mL of cold 10% sodium hydroxide solution and pour into this solution, with stirring, the suspension of diazotized sulfanilic acid from Section 2. Rinse the Erlenmeyer flask with a small amount of water and add it to the beaker. Coupling occurs very rapidly and the dye, being a sodium salt, separates easily from the solution because a considerable excess of sodium ion from the carbonate, the nitrite, and the alkali is present. Stir the crystalline paste thoroughly to effect good mixing and, after 5–10 min, heat the mixture until the solid dissolves. Add 10 g of sodium chloride to further decrease the solubility of the product, bring this all into solution by heating and stirring, set the beaker in a pan of ice and water, and let the solution cool undisturbed. When near room temperature, cool further by stirring and collect the product on a Büchner funnel. Use saturated sodium chloride solution rather than water for rinsing the material out of the beaker and for washing the filter cake free of the dark-colored mother liquor. The filtration is somewhat slow.[3]

The product dries slowly and it contains about 20% of sodium chloride. The crude yield is thus not significant, and the material need not be dried before being purified. This particular azo dye is too soluble to be crystallized from water; it can be obtained in a fairly satisfactory form by adding saturated sodium chloride solution to a hot, filtered solution in water and cooling, but the best crystals are obtained from aqueous ethanol. Transfer the filter cake to a beaker, wash the material from the filter paper and funnel with water, and bring the cake into solution at the boiling point. Avoid a large excess of water, but use enough to prevent separation of solid during filtration (use about 50 mL). Filter by suction through a Büchner funnel that has been preheated on the steam bath. Pour the filtrate into an Erlenmeyer flask, rinse the filter flask with a small quantity of water, add it to the flask, estimate the volume, and if greater than 60 mL evaporate by boiling. Cool to 80°C, add 100–125 mL of ethanol, and allow crystallization to proceed. Cool the solution well before collecting the product. Rinse the beaker with mother liquor and wash finally with a little ethanol. The yield of pure, crystalline material is 6.8 g. Orange II separates from aqueous alcohol with two molecules of water of crystallization and allowance for this should be made in calculating the yield. If the water of hydration is eliminated by drying at 120°C the material becomes fiery red.

Cleaning Up   The filtrate from the reaction, although highly colored, contains little dye, but is very soluble in water. It can be diluted with a large quantity of water and flushed down the drain or, with the volume kept as small as possible, it can be placed in the aromatic amines hazardous waste container or it can be reduced with tin(II) chloride (see Experiment 4). The crystallization filtrate should go into the organic solvents container.

---

3. If the filtration must be interrupted, fill the funnel, close the rubber suction tubing (while the aspirator is still running) with a screw pinchclamp placed close to the filter flask, and then disconnect the tubing from the trap. Set the unit aside; thus, suction will be maintained, and filtration will continue.

3. Methyl Orange (*p*-Sulfobenzeneazo-4-Dimethylaniline Sodium Salt)

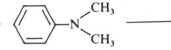 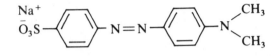

**Diazotized
sulfanilic acid**
4-Diazobenzenesulfonic acid

N,N-Dimethylaniline
MW 121.18
bp 194°C

Methyl orange
4-[4-(Dimethylamino)phenylazo]
benzenesulfonic acid, sodium salt
MW 327.34
λ_{max} 507 nm

In a test tube, thoroughly mix 3.2 mL of dimethylaniline and 2.5 mL of glacial acetic acid. To the suspension of diazotized sulfanilic acid from Experiment 1 contained in a 400-mL beaker add, with stirring, the solution of dimethylaniline acetate. Rinse the test tube with a small quantity of water and add it to the beaker. Stir and mix thoroughly and within a few minutes the red, acid-stable form of the dye should separate. A stiff paste should result in 5–10 min and 35 mL of 10% sodium hydroxide solution is then added to produce the orange sodium salt. Stir well and heat the mixture to the boiling point, when a large part of the dye should dissolve. Place the beaker in a pan of ice and water and allow the solution to cool undisturbed. When cooled thoroughly, collect the product on a Büchner funnel, using saturated sodium chloride solution rather than water to rinse the flask and to wash the dark mother liquor from the filter cake.

The crude product need not be dried but can be crystallized from water after making preliminary solubility tests to determine the proper conditions. The yield is 5–6 g.

Methyl Orange is an acid-base indicator

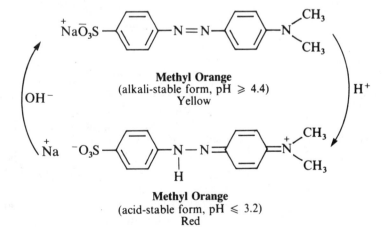

Methyl Orange
(alkali-stable form, pH ≥ 4.4)
Yellow

Methyl Orange
(acid-stable form, pH ≤ 3.2)
Red

Cleaning Up   The highly colored filtrates from the reaction and crystallization are very water soluble. After dilution with a large quantity of water, they can be flushed down the drain since the amount of solid is small. Alternatively, the combined filtrates should be placed in the hazardous waste container or the mixture can be reduced with tin(II) chloride (see Experiment 4).

4. Tests

Solubility and Color

Compare the solubility in water of Orange II and Methyl Orange and account for the difference in terms of structure. Treat the first solution with alkali and note the change in shade due to salt formation; to the other solution alternately add acid and alkali.

Reduction

Characteristic of an azo compound is the ease with which the molecule is cleaved at the double bond by reducing agents to give two amines. Since amines are colorless, the reaction is easily followed by the color change. The reaction is of use in preparation of hydroxyamino and similar compounds, in analysis of azo dyes by titration with a reducing agent, and in identification of azo compounds from an examination of the cleavage products.

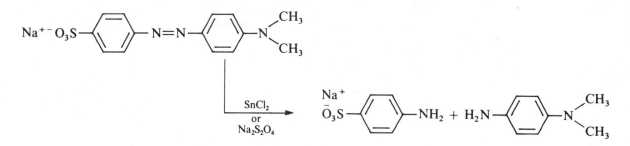

This reaction can, if necessary, be run on five times the indicated scale. Dissolve about 0.1 g of tin(II) chloride in 0.2 mL of concentrated hydrochloric acid, add a small quantity of the azo compound (20 mg), and heat. A colorless solution should result and no precipitate should form on adding water. The aminophenol or the diamine products are present as the soluble hydrochlorides; the other product of cleavage, sulfanilic acid, is sufficiently soluble to remain in solution.

Cleaning Up   Dilute the reaction mixture with water, neutralize with sodium carbonate, and remove the solids by vacuum filtration. The solids go in the aromatic amines hazardous waste container and the filtrate can be flushed down the drain.

Part 2. Dyeing

Experiments

1. Direct Dyes

With good laboratory technique your hands will not be dyed; use care

The sulfonate groups on the Methyl Orange and Orange II molecules are polar and thus enable these dyes to combine with polar sites in the fibers. Wool and silk have many polar sites on their polypeptide chains and hence bind strongly to a dye of this type. Martius Yellow, picric acid, and eosin are also highly polar dyes and thus dye directly to wool and silk.

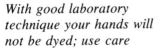

Picric Acid

Orange II or Methyl Orange

The dye bath is prepared from 50 mg of Orange II or Methyl Orange, 0.5 mL of 10% sodium sulfate solution, 15 mL of water, and 5 drops of 10% sulfuric acid in a 30-mL beaker. Place a piece of test fabric, a strip 3/4-in. wide, in the bath for 5 min at a temperature near the boiling point. Remove the fabric from the dye bath, allow it to cool, and then wash it thoroughly with soap under running water before drying it.

Dye untreated test fabric and one or more of the pieces of test fabric that have been treated with a mordant following this same procedure. See Experiment 3 for application of mordants.

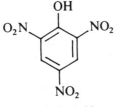

Martius Yellow

Picric Acid or Martius Yellow

Dissolve 50 mg of one of these acidic dyes in 15 mL of hot water to which a few drops of dilute sulfuric acid have been added. Heat a piece of test fabric in this bath for one minute, then remove it with a stirring rod, rinse well, scrub with soap and water, and dry. Describe the results.

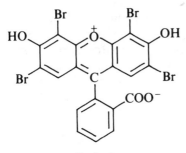

Eosin A
(λ_{max} 516,483 nm)

Eosin

Dissolve 10 mg of sodium eosin in 20 mL of water and dye a piece of test fabric by heating it with the solution for about 10 min. Eosin is the dye used in red ink. Also dye pieces of mordanted cloth in eosin (see Experiment 3). Wash and rinse the dyed cloth in the usual way.

Cleaning Up   In each case dilute the dye bath with a large quantity of water and flush it down the drain.

2. Substantive Dyes

Cotton and the rayons do not have the anionic and cationic carboxyl and amine groups of wool and silk and hence do not dye well with direct dyes, but they can be dyed with substances of rather high molecular weight showing colloidal properties. Such dyes probably become fixed to the fiber by hydrogen bonding. Such a dye is Congo Red, a substantive dye.

Congo Red, a Benzidine Dye

Dissolve 10 mg of Congo Red in 40 mL of water, add about 0.1 mL each of 10% solutions of sodium carbonate and sodium sulfate, heat to a temperature just below the boiling point, and introduce a piece of test fabric. At the end of 10 min remove the fabric and wash in warm water as long as the dye is removed. Place pieces of the dyed material in very dilute hydrochloric acid solution and observe the result. Rinse and wash the material with soap.

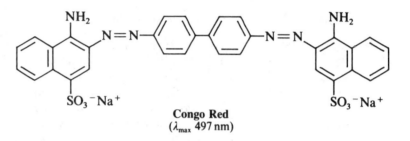

Congo Red
(λ_{max} 497 nm)

Cleaning Up   The dye bath should be diluted with water and flushed down the drain.

3. Mordant Dyes

Chromium functioning as a mordant

One of the oldest known methods of producing wash-fast colors involves the use of metallic hydroxides, which form a link, or mordant (L. *mordere*, to bite), between the fabric and the dye. Other substances, such as tannic acid, also function as mordants. The color of the final product depends on both the dye used and the mordant. For instance, the dye Turkey Red (alizarin) is red with an aluminum mordant, violet with an iron mordant, and brownish-red with a chromium mordant. Some important mordant dyes possess a structure based on triphenylmethane, as do Crystal Violet and Malachite Green.

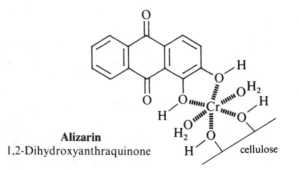

Alizarin
1,2-Dihydroxyanthraquinone

Applying Mordants—Tannic acid, Fe, Sn, Cr, Cu, Al

Mordant pieces of test fabric by allowing them to stand in a hot (nearly boiling) solution of 0.1 g of tannic acid in 50 mL of water for 30 min. The tannic acid mordant must now be fixed to the cloth; otherwise it would wash out. For this purpose, transfer the cloth to a hot bath made from 20 mg of potassium antimonyl tartrate (tartar emetic) in 20 mL of water. After 5 min, wring the cloth and dry it as much as possible over a warm hot plate.

Mordant 1/2-in. strips of test cloth in the following mordants, which are 0.1 M solutions of the indicated salts. Immerse pieces of cloth in the solutions, which are kept near the boiling point, for about 15 to 20 min or longer. The mordants are ferrous sulfate, stannous chloride, potassium dichromate, copper sulfate, and potassium aluminum sulfate (alum). The alum and dichromate solutions should also contain 0.05 M oxalic acid. These mordanted pieces of cloth can then be dyed with alizarin (1,2-dihydroxyanthraquinone) and either Methyl Orange or Orange II in the usual way.

Cleaning Up   Mix the mordant baths. The Fe^{2+} and Sn^{2+} will reduce the Cr^{6+} to Cr^{3+}. The mixture can then be diluted with water and flushed down the drain since the quantity of metal ions is extremely small. Alternatively, precipitate them as the hydroxides, collect by vacuum filtration, and place the solid in the hazardous waste container.

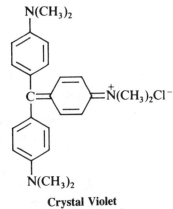

Crystal Violet
(λ_{max} 591,540 nm)

Dyeing with a Triphenylmethane Dye—Crystal Violet or Malachite Green

A dye bath is prepared by dissolving 10 mg of either Crystal Violet or Malachite Green in 20 mL of boiling water. Dye the mordanted cloth in this bath for 5–10 min at a temperature just below the boiling point. Dye another piece of cloth that has not been mordanted and compare the two. In each case allow as much of the dye to drain back into the beaker as possible and then, using glass rods, wash the dyed cloth under running water with soap, blot, and dry.

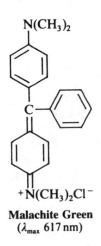

Malachite Green
(λ_{max} 617 nm)

Cleaning Up   The stains on glass produced by triphenylmethane dyes can be removed with a few drops of concentrated hydrochloric acid and washing with water, as HCl forms a di- or trihydrochloride more soluble in water than the original monosalt.

The dye bath and acid washings are diluted with water and flushed down the drain since the quantity of dye is extremely small.

4. Developed Dyes

A superior method of applying azo dyes to cotton, patented in England in 1880, is that in which cotton is soaked in an alkaline solution of a phenol and then in an ice-cold solution of a diazonium salt; the azo dye is developed directly on the fiber. The reverse process (ingrain dyeing) of impregnating cotton with an amine, which is then diazotized and developed by immersion in a solution of the phenol, was introduced in 1887. The first ingrain dye was Primuline Red, obtained by coupling the sulfur dye Primuline, after application to the cloth and diazotization, with 2-naphthol. Primuline (substantive to cotton) is a complex thiazole, prepared by heating *p*-toluidine with sulfur and then introducing a solubilizing sulfonic acid group.

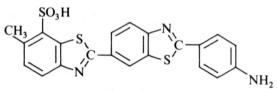

Primuline

Primuline Red

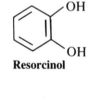

Resorcinol

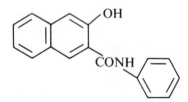

Naphthol AS

Dye three pieces of cotton cloth in a solution of 20 mg of Primuline and 0.5 mL of sodium carbonate solution in 50 mL of water, at a temperature just below the boiling point for 15 min. Wash the cloth twice in about 50 mL of water. Prepare a diazotizing bath by dissolving 20 mg of sodium nitrite in 50 mL of water containing a little ice and, just before using the bath, add 0.5 mL of concentrated hydrochloric acid. Allow the cloth dyed with Primuline to stay in this diazotizing bath for about 5 min. Now prepare three baths for the coupling reaction. Dissolve 10 mg of 2-naphthol in 0.2 mL of 5% sodium hydroxide solution and dilute with 10 mL of water; prepare similar baths from phenol, resorcinol, Naphthol AS, or other phenolic substances.

Transfer the cloth from the diazotizing bath to a beaker containing about 50 mL of water and stir. Put one piece of cloth in each of the developing baths and allow them to stay for 5 min. Primuline coupled to 2-naphthol gives the dye called Primuline Red. Draw the structure of the dye.

Para Red, an Ingrain Color

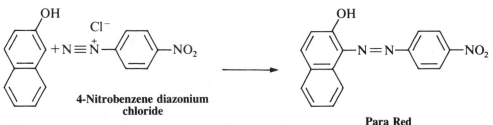

2-Naphthol

4-Nitrobenzene diazonium chloride

Para Red

A solution is prepared by suspending 300 mg of 2-naphthol in 10 mL of water, stirring well, and adding 10% sodium hydroxide solution, a drop at a time, until the naphthol just dissolves. Do not add excess alkali. The material to be dyed is first soaked in or painted with this solution and then dried, preferably in an oven.

CAUTION: 4-Nitroaniline is toxic. Handle with care.

Prepare a solution of 4-nitrobenzenediazonium chloride as follows: dissolve 140 mg of 4-nitroaniline in a mixture of 3 mL of water and 0.6 mL of 10% hydrochloric acid by heating. Cool the solution in ice (the hydrochloride of the amine may crystallize), add all at once a solution of 80 mg of sodium nitrite in about 0.5 mL of water, and stir. In about 10 min a clear solution of the diazonium salt will be obtained. Just before developing the dye on the cloth add a solution of 80 mg of sodium acetate in 0.5 mL of cold water. Stir in the acetate well, add 30 mL of water, and immediately add the cloth. The diazonium chloride solution may also be painted onto the cloth.

Good results can be obtained by substituting Naphthol-AS for 2-naphthol; in this case it is necessary to warm the Naphthol-AS with alkali and to break the lumps with a flattened stirring rod in order to bring the naphthol into solution.

Cleaning Up   The dye baths should be diluted with water and flushed down the drain since the quantity of dye is extremely small.

5. Vat Dyes

Vat dyeing depends upon the reduction of some dyes (e.g., indigo) to a colorless, or leuco, derivative, which is soluble in dilute alkali. If fabric is immersed in this alkaline solution, the leuco compound is adsorbed by hydrogen bonding. On exposure to air the leuco compound is oxidized to the dye, which remains fixed to the cloth. Vat dyes are all quinonoid substances that are readily reduced to colorless hydroquinonoid compounds that are reoxidizable by oxygen in the air.

The indigo so formed is very insoluble in all solvents. However, it is not covalently bound to the cotton, only adhering to the surface of the fiber. Hence, it is subject to removal by abrasion. This explains why the knees and other parts of blue jeans (dyed exclusively with indigo) subject to wear will gradually turn white. It also explains the unique appearance of stone-washed denim.

Dyeing with Indigo—The Denim Dye

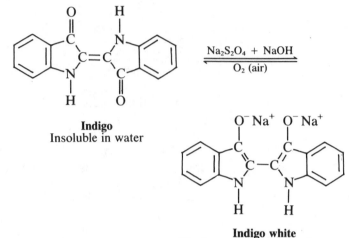

Indigo
Insoluble in water

Indigo white
The leuco form, soluble in water

Use 100 mg of indigo and a drop of detergent or a pinch of soap powder. Boil the dye with 50 mL of water, 2.5 mL of 10% sodium hydroxide solution, and about 0.5 g of sodium hydrosulfite until the dye is reduced. At this point a clear solution will be seen through the side of the beaker. Add more sodium hydrosulfite if necessary. Introduce a piece of cloth and boil the solution gently for 10 min. Rinse the cloth well in water and allow it to dry. To increase the intensity of the dye, repeat the process several times with no drying. Describe what happens during the drying process.

    Other dyes that can be used in this procedure are Indanthrene Brilliant Violet and Indanthrene Yellow.

Cleaning Up   Add household bleach (5.25% sodium hypochlorite solution) to the dye bath to oxidize it to the starting material. The mixture can be diluted with water and flushed down the drain or the small amount of solid removed by filtration and placed in the aromatic amines hazardous waste container.

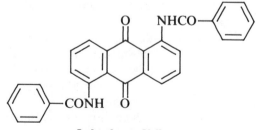

Indanthrene Yellow

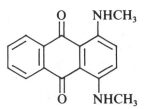

Celliton Fast Blue B

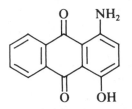

Celliton Fast Pink B

6. Disperse Dyes

Fibers such as Dacron, acetate rayon, nylon, and polypropylene are difficult to dye with conventional dyes because they contain so few polar groups. These fibers are dyed with substances that are insoluble in water but that, at elevated temperatures (pressure vessels), are soluble in the fiber as true solutions. They are applied to the fiber in the form of a dispersion of finely divided dye (hence the name). The Cellitons are typical disperse dyes.

In this experiment Disperse Red, a brilliant red dye used commercially, is synthesized.

Diazotization of 2-Amino-6-methoxybenzothiazole

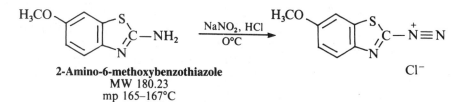

2-Amino-6-methoxybenzothiazole
MW 180.23
mp 165–167°C

To 135 mg (0.75 mmole) of 2-amino-6-methoxybenzothiazole in 1.5 mL of water in a test tube add 0.175 mL of concentrated hydrochloric acid, then cool the solution to 0–5°C. To this mixture add, dropwise, an *ice-cold* solution of 55 mg of sodium nitrite that has been dissolved in 0.75 mL of water. The reaction mixture changes color and some of the diazonium salt crystallizes out, but it should not foam. Foaming, caused by the evolution of nitrogen, is an indication the mixture is too warm. Keep the mixture ice-cold until used in the coupling reaction.

Disperse Red

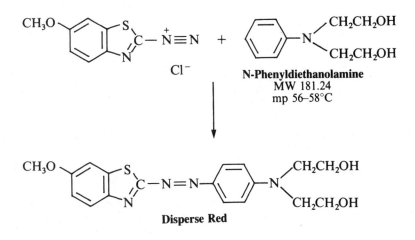

N-Phenyldiethanolamine
MW 181.24
mp 56–58°C

Disperse Red

To 135 mg (0.75 mmole) of N-phenyldiethanolamine in 0.75 mL of hot water add just enough 10% hydrochloric acid to bring the amine into solution. This amount is less than 0.5 mL. Cool the resulting solution to 0°C in ice and add to it, dropwise and with *very thorough mixing,* the diazonium chloride solution. Mix the solution well by drawing into the Pasteur pipette and expelling it into the cold reaction tube. Allow the mixture to come to room temperature over a period of 10 min, then add 225 mg of sodium chloride, and heat the mixture to boiling. The sodium chloride decreases the solubility of the product in water. Allowing the hot solution to cool slowly to room temperature should afford easily filterable crystals. Collect the dye on the Hirsch funnel, wash it with a few drops of saturated sodium chloride solution, and press it dry on the funnel.

Often the reaction gives a noncrystalline product that looks like purple tar. This is the dye and it can be used to dye the multifiber test cloth, so don't discard the reaction mixture.

Save the filtrate. Add the material on the filter paper to 50 mL of boiling water and dye a piece of test cloth for 5 min. Do the same with the filtrate. Wash the cloth with soap and water, rinse and dry it, and compare the results.

Cleaning Up   The dye baths should be diluted with water and flushed down the drain since the quantity of dye is extremely small.

7. Fiber Reactive Dyes

Among the newest of the dyes are the fiber reactive compounds, which form a covalent link to the hydroxyl groups of cellulose. The reaction involves an amazing and little understood nucleophilic displacement of a chloride ion from the triazine part of the molecule by the hydroxyl groups of cellulose; yet the reaction occurs in aqueous solution.

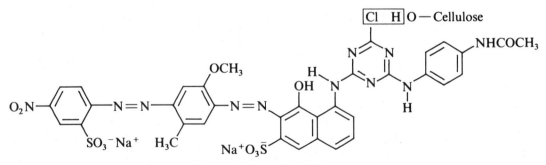

Chlorantin Light Blue 8G

8. Optical Brighteners—Fluorescent White Dyes

Most modern detergents contain a blue-white fluorescent dye that is adsorbed on the cloth during the washing process. These dyes fluoresce, that is, absorb ultraviolet light and reemit light in the visible blue region of the spectrum. This blue color counteracts the pale yellow color of white goods, which develops because of a buildup of insoluble lipids. The modern-day use of optical brighteners has replaced a past custom of using bluing (ferriferrocyanide).

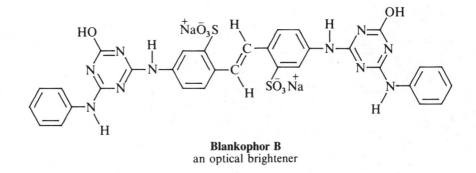

Blankophor B
an optical brightener

Dyeing with Detergents

Immerse a piece of test fabric in a hot solution (0.5 g of detergent, 200 mL of water) of a commercial laundry detergent that you suspect may contain an optical brightener (e.g., Tide and Cheer) for 15 min. Rinse the fabric thoroughly, dry, and compare with an untreated fabric sample under an ultraviolet lamp.

Cleaning Up   The solution should be diluted with water and flushed down the drain.

Questions _____

1. Write reactions showing how nylon can be synthesized such that it will react with (a) basic dyes and (b) acidic dyes.

2. Draw the resonance form of dimethylaniline that is most prone to react with diazotized sulfanilic acid.

3. Draw a resonance form of indigo that would be present in base.

4. Draw a resonance form of indigo that has been reduced and is therefore colorless.

Prelab Exercise: In the preparation of nylon by interfacial polymerization sebacoyl chloride is synthesized from decanedioic acid and thionyl chloride. What volume of hydrogen chloride is produced in this reaction? What volume of sulfur dioxide is produced?

Polymers are ubiquitous. Natural polymers such as proteins (polyamino acids), DNA (polynucleotides), and cellulose (polyglucose) are the basic building blocks of plant and animal life. Synthetic organic polymers, or plastics, are now among our most common structural materials. In the United States we make and use more synthetic polymers than we do steel, aluminum, and copper combined—in 1984, 46 billion pounds, worth $18 billion dollars.

We use more polymers than steel, aluminum, and copper combined

Polymers, from the Greek meaning "many parts," are high-molecular-weight molecules made up of repeating units of smaller molecules. Most polymers consist of long chains held together by hydrogen bonds, van der Waals forces, and the tangling of the long chains. When heated, the covalent bonds of some polymers, which are thermoplastic, do not break, but the chains slide over one another to adopt new shapes. These shapes can be films, sheets, extrusions, or molded parts in a myriad of forms.

Nitrocellulose

The first man-made plastic was nitrocellulose, made in 1862 by nitrating the natural polymer, cellulose. Nitrocellulose, when mixed with a plasticizer such as camphor to make it more workable, was originally used as a replacement for ivory in billiard balls and piano keys and to make Celluloid collars. This material, from which the first movie film was made, is notoriously flammable.

Cellulose acetate

Cellulose acetate, made by treating cellulose with acetic acid and acetic anhydride, was originally used as a waterproof varnish to coat the fabric of airplanes during World War I. It later became important as a photographic film base and as acetate rayon.

Bakelite

The first completely synthetic organic polymer was Bakelite, named for its discoverer Leo Baekeland. He was a Belgian chemistry professor who invented the first successful photographic paper, Velox. He came to America at the age of 35 and sold his invention to George Eastman for $1,000,000 in 1899. He then turned his attention to finding a replacement for shellac, which comes from the Asian lac beetle. At the time, shellac was coming into great demand in the fledgling electrical industry as an insulator. The polymer Baekeland produced is still used for electrical plugs, distributor caps in automobiles, switches, and the black handles and knobs on pots and pans. It has superior electrical insulating properties and very high heat resistance. It is made by the base-catalyzed reaction of excess formaldehyde with phenol. In a low-molecular-weight form it is used to glue together the plies of plywood or mixed with a filler such as sawdust. When it is heated to a high temperature, crosslinking occurs as the polymer "cures."

Thermoplastic polymers

Most polymers are amorphous, linear macromolecules that are thermoplastic and soften at high temperature. In Bakelite the polymer crosslinks to form a three-dimensional network and the polymer becomes a dark, insoluble, infusible substance. Such polymers are said to be thermosetting. Natural rubber is thermoplastic. It becomes a thermosetting polymer when heated with sulfur, as Charles Goodyear discovered. With 2% sulfur the rubber becomes crosslinked but is still elastic; at 30% sulfur it can be made into bowling balls. Some other important thermosetting polymers are ureaformaldehyde resins and melamine-formaldehyde resins. The latter are among the hardest of polymers and take on a high-gloss finish. Melamine is used extensively to manufacture plastic dinnerware.

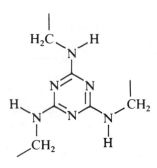

**Melamine,
a thermosetting polymer**

Even though vinyl chloride was discovered in 1835, polyvinyl chloride was not produced until 1912. It is now one of our most common polymers; production in 1984 was over 6 billion pounds. The monomer is made by the pyrolysis of 1,2-dichloroethane, formed by chlorination of ethylene. Free radical polymerization follows Markovnikov's rule to give the head-to-tail polymer with high specificity:

Vinyl chloride

$$n\text{CH}_2\!=\!\text{CHCl} \longrightarrow -(\text{CH}_2\text{CHCl})_n-$$

Vinyl chloride   **Polyvinyl chloride**

Plasticizers

Pure polyvinyl chloride (PVC) is an extremely hard polymer. It, along with some other polymers, can be modified by the addition of plasticizers; the greater the ratio of plasticizer to polymer, the greater the flexibility of the polymer. PVC pipe is rigid and contains little plasticizer, while shower curtains contain a large percentage. The most common plasticizer is di(2-ethylhexyl) phthalate, which can be added in concentrations of up to 50%.

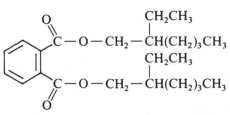

Di(2-ethylhexyl)phthalate (Dioctyl phthalate)

PVC is used for raincoats, house siding, and artificial leather for handbags, briefcases, and inexpensive shoes. It is found in garden hose, floor covering, swimming pool liners, and automobile upholstery. When vinyl upholstery is exposed to high temperatures, as in the interior of an automobile, the plasticizer distills out. The result is an opaque, difficult-to-remove film on the insides of the windows, and upholstery that is hard and brittle.

Polymerization methods

Monomers can be polymerized in the gas phase, in bulk, as suspensions, and as emulsions. The most common method of making PVC is by emulsifying the monomer, vinyl chloride, in water with surfactants (soaps), water-soluble catalysts, and heat. The monomer is polymerized to solid particles, which are suspended in the aqueous phase. This product can be centrifuged and dried or used as such. Chemists can control the average molecular weight, which can become very high in emulsion polymerization. A high molecular weight means a more rigid and stronger polymer, but also one that is more difficult to work with. An emulsion of polyvinyl acetate is sold as latex paint. When the vehicle, water, evaporates, the polymer is left as a hard film. A thicker emulsion of polyvinyl acetate is an excellent adhesive, the familiar white glue. When vinyl acetate and vinyl chloride are polymerized together, a copolymer results that has properties all of its own. This copolymer is particularly good for detailed moldings and is used to make phonograph records.

Polyvinylidene chloride is primarily extruded as a film that has low permeability to water vapor and air and is therefore used as the familiar clinging plastic food wrap, Saran Wrap.

Teflon

Polytetrafluoroethylene, Teflon, another of the halogenated polymers, has a number of unique properties. It has a very high melting point, 327°C, it does not dissolve in any solvent, and nothing sticks to it. It is also an extremely good electrical insulator. A product of the DuPont company, Teflon is one of the densest of the polymers and also one of the most expensive. The surface of the polymer must be etched with metallic sodium to form free radicals, to which glue can adhere. The polymer has an extremely low coefficient of friction, which makes it useful for bearings. Its chemical inertness makes it an ideal liner for chemical reagent bottle caps, and its no-stick property is ideal for the coating on the inside of frying pans. At 380°C Teflon is still so viscous that it cannot be injection-molded. Instead it is molded by pressing the powdered polymer at high temperature and pressure, a process called sintering.

$$-(CH_2CHCl)_n- \qquad -(CH_2CCl_2)_n- \qquad -(CF_2CF_2)_n- \qquad -(CH_2CH_2)_n-$$

Polyvinyl chloride   Polyvinylidene chloride   Polytetrafluoroethylene   Polyethylene

The polymer produced in highest volume is polyethylene. Invented by the British, who call it polythene, and put into production in 1939, it could for a long time only be produced by the oxygen-catalyzed polymerization of ethylene at pressures near 40,000 lb/in.^{2}. Such pressures are expensive and dangerous to maintain on an industrial scale. The polyethylene produced has a low density and is used primarily to make film for bags of all types—from sandwich bags to trash can liners. The opaque appearance of polyethylene is due to crystallites, regions of order in the polymer that resemble crystals. In the 1950s Karl Ziegler and Giulio Natta developed catalysts composed of $TiCl_4$, alkyl aluminum, and transition metal halides with which ethylene can be polymerized at pressures of just 450 lb/in.^{2}. The resulting product has a higher density and a 20°C higher softening temperature than the low-density material. The catalysts which Ziegler and Natta developed, and for which they received the 1963 Nobel prize, cause stereoregular polymerization and thus a crystalline product. High-density polyethylene is as rigid as polystyrene and yet has high impact resistance. It is used to mold very large articles such as luggage, the cases for domestic appliances, trash cans, and soft drink crates.

High-density polyethylene

Polystyrene is a brilliantly clear, high-refractive-index polymer familiar in the form of disposable drinking glasses. It is brittle and produces sharp, jagged edges when fractured. It softens in boiling water and it burns readily with a very smoky flame. But it foams readily and makes a very good insulator; witness the disposable, white, hot-drink cup. It is used extensively for insulation when properly protected from ignition. The addition of a small quantity of butadiene to the styrene makes a polymer that is no longer transparent but that has high impact resistance. Blends of acrylonitrile, butadiene, and styrene (ABS) have excellent molding properties and are used to make car bodies. One formulation can be chrome-plated for automobile grills and bumpers.

Polystyrene

$$CH_2{=}CHCN \quad C_6H_5CH{=}CH_2 \quad CH_2{=}\;CH{-}CH{=}CH_2$$

Acrylonitrile   **Styrene**   **1,3-Butadiene**

Rubber

Joseph Priestley, the discoverer of oxygen, named rubber for its ability to remove lead pencil marks. Rubber is an elastomer, defined as a substance that can be stretched to at least twice its length and return to its original size. The Germans, cut off from a supply of natural rubber, began manufacturing synthetic rubber during World War I. Called "buna" for *bu*tadiene and sodium, *Na,* the polymerization catalyst, it was not the ideal substitute. Cars with buna tires had to be jacked up when not in use because their tires would develop flat spots. The addition of about 25% styrene greatly improved the qualities of the product; styrene-butadiene synthetic rubber now dominates the market, a principal outlet being automobile tires. Addition of 30% acrylonitrile to butadiene produces nitrile rubber, which is used to make conveyor belts, tank liners, rubber hose, and gaskets.

Synthetic rubber, a styrene-butadiene copolymer

The chemist classifies polymers in several ways. There are thermosetting plastics such as Bakelite and melamine and the much larger category of thermoplastic materials, which can be molded, blown, and formed after polymerization. There are the arbitrary distinctions made among plastics, elastomers, and fibers. And there are the two broad categories formed by the polymerization reaction itself: (1) addition polymers (e.g., vinyl polymerizations), in which a double bond of a monomer is transformed into a single bond between monomers, (2) condensation poly-

Addition polymers
Condensation polymers

mers (e.g., Bakelite), in which a small molecule, such as water or alcohol, is split out as the polymerization reaction occurs.

Nylon

One of the most important condensation polymers is nylon, a name so ingrained into our language that it has lost trademark status. It was developed by Wallace Carothers, director of organic chemicals research at DuPont, and was the outgrowth of his fundamental research into polymer chemistry. Introduced in 1938, it was the first totally synthetic fiber. The most common form of nylon is the polyamide formed by the condensation of hexamethylene diamine and adipic acid:

$$n\text{H}_2\text{N(CH}_2)_6\text{NH}_2 + n\text{HOOC(CH}_2)_4\text{COOH}$$

| **Hexamethylene** **diamine** | **Adipic acid** |

$$n\text{H}_3\overset{+}{\text{N}}\text{(CH}_2)_6\overset{+}{\text{N}}\text{H}_3 + n\overset{\text{O}}{\underset{\|}{\text{OC}}}\ \text{(CH}_2)_4\overset{\text{O}}{\underset{\|}{\text{CO}}}$$

Pressure, 280°C

$$\text{H}\!+\!\text{NH(CH}_2)_6\text{NH}\overset{\text{O}}{\underset{\|}{\text{C}}}\text{(CH}_2)_4\overset{\text{O}}{\underset{\|}{\text{C}}}\!+\!_n\text{OH} + (2n - 1)\text{H}_2\text{O}$$

Nylon 6.6

The reactants are mixed together to form a salt that melts at 180°C. This is converted into the polyamide by heating to 280°C under pressure, which eliminates water. Nylon 6.6 is used to make textiles, while nylon 6.10, from the 10-carbon diacid, is used for bristles and high-impact sports equipment. Nylon can also be made by interfacial and by ring-opening polymerization, both of which are used in the following experiments.

The condensation polymer made by reacting ethylene glycol with 1,4-benzene dicarboxylic acid (terephthalic acid) produces a polymer that is almost exclusively converted into the fiber Dacron. The polymerization is run as an ester interchange reaction using the methyl ester of terephthalic acid:

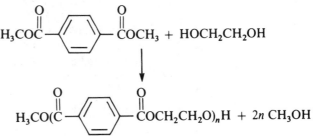

Polyethylene glycol terephthalate (Dacron)

The structure of a polymer is not simple. For example, polymerization of styrene produces a chiral carbon at each benzyl position. We can ask whether the phenyl rings are all on the same side of the long carbon chain, whether they alternate positions, or whether they adopt some random configuration. In the case of copolymers we can ask whether the two components alternate: ABABABABAB . . . or whether they adopt a random configuration: AABBBABAAB . . . or whether they polymerize as short chains of one and then the other: AAAABBBBBAAABBBB. . . . These questions are important because the physical properties of the resulting polymer depend on them. The polymer chemist is concerned with finding the answers and discovering catalysts and reaction conditions that can control these parameters.

In the experiments that follow, you will find that the preparation of nylon by interfacial polymerization is a spectacular and reliable experiment easily carried out in one afternoon. The synthesis of Bakelite works well; it requires overnight heating in an oven to complete the polymerization. Nylon by ring-opening polymerization requires skill and care because of the high temperatures involved. The polymerization of styrene also requires care, but is somewhat easier to carry out.

Experiments

1. Nylon by Interfacial Polymerization[1]

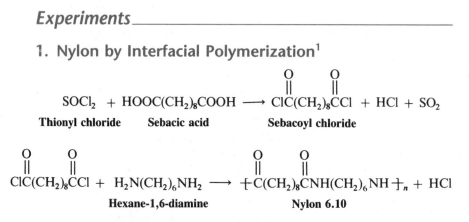

In this experiment a diamine dissolved in water is carefully floated on top of a solution of a diacid chloride dissolved in an organic solvent. Where the two solutions come in contact (the interface), an S_N2 reaction occurs to form a film of a polyamide. The reaction stops there unless the polyamide is removed. In the case of nylon 6.10, the product of this reaction, the film is so strong that it can be picked up with a wire hook and continuously removed in the form of a rope.

This reaction works because the diamine is soluble in both water and carbon tetrachloride, the organic solvent used. As the diamine diffuses into

1. P. W. Morgan and S. L. Kwolek, *J. Chem. Ed.*, **36**, 182 (1959).

the organic layer, reaction occurs immediately to give the insoluble polymer. The HCl produced reacts with the sodium hydroxide in the aqueous layer. The acid chloride does not hydrolyze before reacting with the amine because it is not very soluble in water. The acid chloride is conveniently prepared using thionyl chloride.

Procedure [2]

In a 4-in. test tube fitted with a gas trap like the one in Fig. 1 in Chapter 13, place 1.0 g of sebacic acid (1,8-octane dicarboxylic acid, decanedioic acid), 1.0 mL of thionyl chloride, and 0.05 mL of N,N-dimethylformamide. Heat the tube to 60–70°C in a water bath in the hood. This is best accomplished by putting very hot water in a beaker and placing the beaker on a steam bath to keep it at 60–70°C. As the reaction proceeds the product forms a liquid layer on the bottom of the tube. Use this liquid to wash down unreacted acid as the reaction proceeds. When the acid has all reacted and gas evolution has ceased (about 10–15 min), transfer the product, which should be a clear liquid at this point, to a 250-mL beaker using 50 mL of dichloromethane. Carefully pour onto the top of this dichloromethane solution 1.0 g of hexane-1,6-diamine (hexamethylenediamine) that has been dissolved in 25 mL of water containing 0.5 g of sodium hydroxide. Pick up the polymer film at the center with a copper wire and lead it over the outer surface of a bottle, beaker, or graduated cylinder as it is removed. Remove as much of the polymer as possible, wash it thoroughly in water, and press it as dry as possible. After the polymer has dried determine its weight and calculate the yield. Try dissolving the polymer in two or three solvents. Attach a piece of the polymer to your laboratory report.

Cleaning Up   Add the cotton from the trap to the used reaction mixture and stir the mixture vigorously to cause nylon to precipitate. Decant the water and dichloromethane and then squeeze the solid as dry as possible; place it in the nonhazardous solid waste container. The dichloromethane should be placed in the halogenated organic solvents container, and the aqueous layer, after neutralization, should be diluted with water and flushed down the drain.

2. The Condensation Polymerization of Phenol and Formaldehyde: Bakelite

Condensation of phenol with formaldehyde is a base-catalyzed process in which one resonance form of the phenoxide ion attacks formaldehyde. The resulting trimethylol phenol is then crosslinked by heat, presumably by dehydration with the intermediate formation of benzylcarbocations. The resulting polymer is Bakelite. Since the cost of phenol is relatively high and

2. G. C. East and S. Hassell, *J. Chem. Ed.*, **60**, 69 (1983).

the polymer is somewhat brittle, it is common practice to add an extender such as sawdust to the material before crosslinking. The mixture is placed in molds and heated to form the polymer. The resulting polymer, like other thermosetting polymers, is not soluble in any solvent and does not soften when heated.

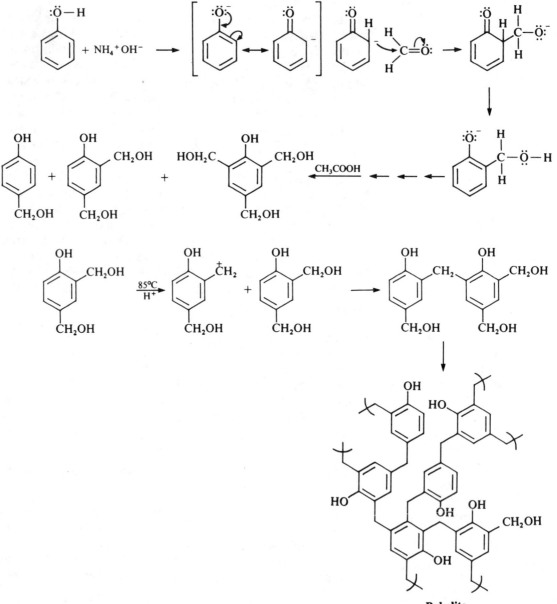

Bakelite

Procedure

In a 25-mL round-bottomed flask place 3.0 g of phenol and 10 mL of 37% by weight aqueous formaldehyde solution. The formaldehyde solution contains 10–15% methanol, which has been added as a stabilizer to prevent the formaldehyde from polymerizing. Add 1.5 mL of concentrated ammonium hydroxide to the solution and reflux it for 5 min beyond the point at which the solution turns cloudy, a total reflux time of about 10 min. In the hood pour the warm solution into a test tube and draw off the upper layer. Immediately clean the flask with a small amount of acetone. Warm the viscous milky lower layer on the steam bath and add acetic acid dropwise with thorough mixing until the layer is clear, even when the polymer is cooled to room temperature. Heat the tube on a water bath at 60–65°C for 30 min. Then, after placing a wood stick in the polymer to use as a handle, leave the tube, with your name attached, in an 85°C oven overnight or until the next laboratory period. To free the polymer the tube may need to be broken (see margin note). Attach a piece of the polymer to your lab report.

3. Nylon by Ring-Opening Polymerization[3]

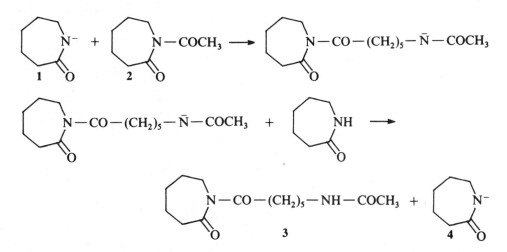

While interfacial polymerization in the manner described above is not a commercial process, the ring-opening of caprolactam is. The nylon produced, nylon 6, is used extensively in automobile tire cord and for gears and bearings in small mechanical devices.

The catalyst used in this reaction is sodium hydride; therefore, this is referred to as an anionic polymerization. The sodium hydride removes the acidic lactam proton to form an anion (**1**) that attacks the coinitiator, acetylcaprolactam (**2**), which has an electron-attracting acetyl group at-

3. L. J. Mathias, R. A. Vaidya, and J. B. Canterbury, *J. Chem. Ed.,* **61**, 805 (1984).

tached to the nitrogen. The ring of the acetylcaprolactam is attacked by the anion and the acetylcaprolactam ring opens, forming a substituted caprolactam (**3**) that still has an electron-attracting group attached to nitrogen. A proton transfer reaction occurs, generating a new caprolactam anion (**4**), and **4** then attacks **3**, and so on.

Ordinarily anionic polymerizations must be run in the absence of oxygen, but the addition of polyethylene glycol serves to complex with the sodium ion just as 18-crown-6 does and enhances the catalytic activity of the sodium hydride.

Procedure

Set the heater to its maximum setting on your sand bath. The sand must be quite hot before beginning this experiment. Heating can also be done over a small Bunsen burner flame.

Into a disposable 4-in. test tube place 4 g of caprolactam, 0.25 g of polyethylene glycol, and 2 drops of N-acetylcaprolactam. Heat the mixture and as soon as it has melted remove it from the heat and add 50 mg of gray (not white) sodium hydride (50% dispersion in mineral oil). Mix the catalyst with the reactants by stirring with a Pasteur pipette and heat the mixture *rapidly* to boiling (200–230°C). This should take place over a 2-min period. Polymerization takes place rapidly, as indicated by an increase in viscosity. If polymerization has not occurred within 3 min remove the tube from the heat, cool it somewhat, and add another 50 mg of sodium hydride. When the solution is so viscous that it will barely flow, insert a wood stick and with help from a neighbor draw fibers from the melt. After it cools the nylon-6 can usually be removed from the tube as one cylindrical piece. Try dissolving a piece of the nylon or the fibers in various solvents. Test the physical properties of the fibers by stretching them to the breaking point. Describe your observations and attach a piece of fiber to your report. Don't forget to turn off the heater.

Cleaning Up   Place the used tube, if coated with polymer, in the nonhazardous solid waste container. Should it be necessary to destroy sodium hydride, add it to excess 1-butanol (38 mL/g of hydride). After the reaction has ceased, cautiously add water, then dilute with more water and flush the solution down the drain.

4. Polystyrene by Free-Radical Polymerization

Polystyrene, the familiar crystal-clear brittle plastic used to make disposable drinking glasses and, when foamed, the lightweight white cups for hot drinks, is usually made by free-radical polymerization. Commercially an initiator is not used because polymerization begins spontaneously at elevated temperatures. At lower temperatures a variety of initiators could be used (e.g., 2,2′-azobis-(2-methylpropionitrile) which was used in the free-radical chlorination of 1-chlorobutane). In this experiment we use benzoyl peroxide as the initiator. On mild heating it splits into two benzoyloxy radicals

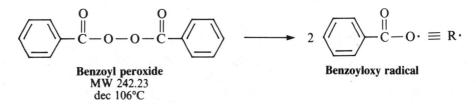

Benzoyl peroxide
MW 242.23
dec 106°C

Benzoyloxy radical

which react with styrene through initiation, propagation, and termination steps to form polystyrene:

Initiation:

$$R \cdot + CH_2{=}CH \longrightarrow RCH_2CH \cdot$$
$$\qquad\quad\; C_6H_5 \qquad\qquad C_6H_5$$

Propagation:

$$RCH_2CH \cdot + CH_2{=}CH \longrightarrow RCH_2CH{-}CH_2CH \cdot + CH_2{=}CH \longrightarrow RCH_2CH{-}CH_2CH{-}CH_2CH \cdot, \text{ etc.}$$
$$\;\; C_6H_5 \qquad\quad C_6H_5 \qquad\qquad C_6H_5 \quad\; C_6H_5 \qquad\quad C_6H_5 \qquad\qquad C_6H_5 \quad\; C_6H_5 \quad\; C_6H_5$$

Termination:

$$2\,R \cdot \longrightarrow R{-}R, \quad RCH_2CH \cdot + R \cdot \longrightarrow RCH_2CH{-}R, \quad 2\,RCH_2CH \cdot \longrightarrow RCH_2CH{-}CHCH_2R$$
$$\qquad\qquad\qquad\qquad\; C_6H_5 \qquad\qquad\qquad\quad C_6H_5 \qquad\qquad C_6H_5 \qquad\qquad\; C_6H_5 \; C_6H_5$$

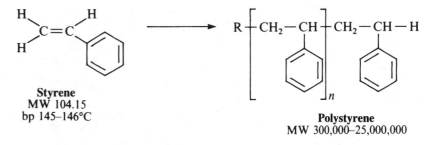

Styrene
MW 104.15
bp 145–146°C

Polystyrene
MW 300,000–25,000,000

The final polymer has about 3000 monomer units in a single chain, but it can be made with up to 240,000 monomers per chain.

To prevent styrene from polymerizing in the bottle in which it is sold, the manufacturer adds 10 to 15 parts per million of 4-*tert*-butylcatechol, a radical inhibitor (a particularly good chain terminator). This must be removed by passing the styrene through a column of alumina before the styrene can be polymerized.

Procedure

Styrene is flammable, an irritant, and has a bad odor. Work with it in the hood.

CAUTION: Benzoyl peroxide is flammable and may explode on heating or on impact. There is no need for more than a gram or two in the laboratory at any one time. It should be stored in and dispensed from waxed paper containers and not in metal or glass containers with screw-cap lids.

In a Pasteur pipette loosely place a very small piece of cotton followed by 2.5 g of alumina. Add to the top of the pipette 1.5 mL of styrene and collect 1 mL in a disposable 10 × 75-mm test tube. Add to the tube 50 mg of benzoyl peroxide and a thermometer and heat the tube over a hot sand bath. When the temperature reaches about 135°C polymerization begins and, since it is an exothermic process, the temperature rises. Keep the reaction under control by cautious heating. The temperature rises, perhaps to 180°C, well above the boiling point of styrene (145°C); the viscosity also increases. Pull the thermometer from the melt from time to time to form fibers; when a cool fiber is found to be brittle remove the thermometer. A boiling stick can be added to the tube and the polymer allowed to cool. It can then be removed from the tube or the tube can be broken from the polymer. Should the polymer be sticky the polymerization can be completed in an oven overnight at a temperature of about 85°C.

Cleaning Up   Shake the alumina out of the pipette and place it in the styrene/alumina hazardous waste container. Clean up spills of benzoyl peroxide immediately. It can be destroyed by reacting each gram with 1.4 g of sodium iodide in 28 mL of acetic acid. After 30 min the brown solution should be neutralized with sodium carbonate, diluted with water, and flushed down the drain.

Questions

1. What might the products be from an explosion of smokeless gunpowder? How many moles of CO_2 and H_2O would come from one mole of trinitroglucose? Does the molecule contain enough oxygen for the production of these two substances?

2. Write a balanced equation for the reaction of sebacoyl chloride with water.

3. In the final step in the synthesis of Bakelite the partially polymerized material is heated at 85°C for several hours. What other product is produced in this reaction?

4. Give the detailed mechanism of the first step of the ring-opening polymerization of caprolactam.

Epoxidation of Cholesterol

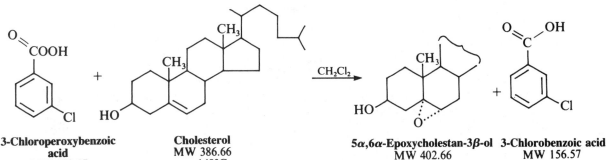

3-Chloroperoxybenzoic acid
MW 172.57
mp 92–94°C (dec)

Cholesterol
MW 386.66
mp 149°C

5α,6α-Epoxycholestan-3β-ol
MW 402.66

3-Chlorobenzoic acid
MW 156.57
mp 157°C

This experiment carries out an epoxidation reaction on cholesterol, which is a representative of a very important group of molecules, the steroids. The rigid cholesterol molecule gives products of well-defined stereochemistry. The epoxidation reaction is stereospecific and the product can be used to carry out further stereospecific reactions.

Cholesterol itself is the principal constituent of gallstones and can be readily isolated from them (see Chapter 22). The average person contains about 200 g of cholesterol, primarily in brain and nerve tissue. The closing of arteries by cholesterol leads to the disease arteriosclerosis (hardening of the arteries).

Certain naturally occurring and synthetic steroids have powerful physiological effects. Progesterone and estrone are the female sex hormones, and testosterone is the male sex hormone; they are responsible for the development of secondary sex characteristics. The closely related synthetic steroid, norethisterone, is an oral contraceptive, and addition of four hydrogen atoms (reduction of the ethynyl group to the ethyl group) and a methyl group gives an anabolic steroid, ethyltestosterone. This muscle-building steroid is now outlawed for use by Olympic athletes. Fluorocortisone is used to treat inflammations such as arthritis, and ergosterol on irradiation with ultraviolet light is converted to vitamin D_2.

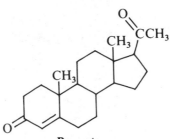

Progesterone

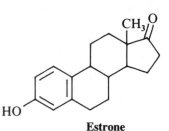

Estrone

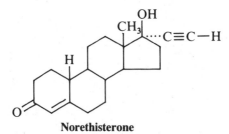

Norethisterone

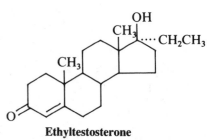

Ethyltestosterone

Much of our present knowledge about the stereochemistry of reactions was developed from steroid chemistry. In this experiment the double bond of cholesterol is stereospecifically converted to the 5α,6α epoxide. The α designation indicates the epoxide is on the backside of the molecule. A substituent on the topside is designated β. Study of molecular models reveals that the angular methyl group prevents topside attack on the double bond by the perbenzoic acid; hence the epoxide forms exclusively on the back, or α side of the molecule.

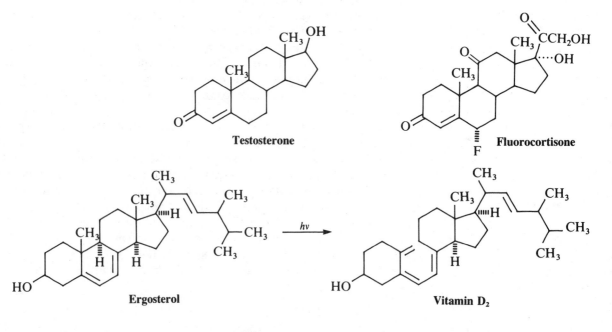

Testosterone

Fluorocortisone

Ergosterol

Vitamin D₂

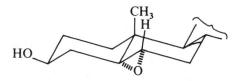

Epoxides are most commonly formed by reaction of a peroxycarboxylic acid with an olefin at room temperature. It is a one-step cycloaddition reaction:

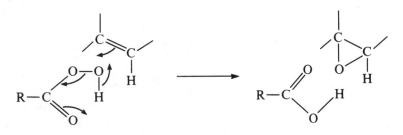

Some peroxyacids are explosive; the reagent used in the present experiment is a particularly stable and convenient peroxycarboxylic acid.

The reaction is carried out in an inert solvent, dichloromethane, and the product is isolated by chromatography. No great care is required in the chromatography to collect fractions because the 3-chlorobenzoic acid, being polar, is adsorbed strongly onto the alumina while the relatively nonpolar product is eluted easily by ether. After removal of ether the product is easily recrystallized from a mixture of acetone and water.

Experiment

Cholesterol Epoxide

To 1 g of cholesterol dissolved by gentle warming in 4 mL of dichloromethane in a 25-mL Erlenmeyer flask is added a solution obtained by gently warming 0.6 g of 3-chloroperoxybenzoic acid in 4 mL of dichloromethane. The two solutions must be cool before mixing because the reaction is exothermic. Clamp the flask in a beaker of water at 40°C for 30 min to complete the reaction. The progress of the reaction can be followed by thin-layer chromatography on silica gel plates using ether as the eluent.

The reaction mixture is pipetted onto a chromatography column prepared from 15 g of alumina following the procedure described in Chapter 10, except that ether is used to fill the column and to prepare the alumina slurry. The 3-chlorobenzoic acid will be adsorbed strongly by the alumina. The product is eluted with 150 mL of ether collected in a tared (previously weighed) 250-mL round-bottomed flask.

Most of the ether is removed on the rotary evaporator and the last traces are removed using the apparatus depicted in Fig. 5 in Chapter 10 for drying a solid under reduced pressure. The residue should weigh more than 0.75 g. If it does not, pass more ether through the column and collect the product as before. Dissolve the product in 7.5 mL of warm acetone and, using a Pasteur pipette, transfer it to a 25-mL Erlenmeyer flask. Add 1 mL of water to the solution, warm the mixture to bring the solid into solution, and then let the flask and contents cool slowly to room temperature. Cool the mixture in ice and collect the product on a Hirsch funnel. Press the solid down on the filter to squeeze solvent from the crystals and then wash the product with 1 mL of ice-cold 90% acetone. Spread the product out on a watch glass to dry. Determine the weight and mp of the product and calculate the percent yield.

Cleaning Up   Place dichloromethane solutions in the halogenated organic solvents container and organic solvents in the organic solvents container. The alumina should be placed in the alumina hazardous waste container. If it should be necessary to destroy 3-chloroperoxybenzoic acid, add it to an excess of an ice-cold solution of saturated sodium bisulfite in the hood. A peracid will give a positive starch/iodide test (blue-purple color).

Questions

1. What are the numbers of moles of the two reactants used in this experiment? Assume the 3-chloroperoxybenzene acid is 80% pure.

2. What simple test could you perform to show that 3-chlorobenzoic acid is not eluted from the chromatography column?

Glass Blowing

Prelab Exercise: Grasp a pencil in the left hand, palm down, and grasp another in the right hand, palm up. Touch the erasers together and rotate both pencils at the same rate. This technique is needed to heat a glass tube, without twisting it, prior to bending or blowing a bulb in the tube.

A knowledge of elementary glass-blowing techniques is useful to every chemist and scientist, and glass blowing can be an enjoyable pastime as well.

In the laboratory you will encounter primarily Kimax or Pyrex glass, a borosilicate glass with a low coefficient of thermal expansion, which gives it remarkable resistance to thermal shock. It cannot easily be worked in a Bunsen burner flame because it has a working temperature of 820°C. Consequently, many laboratories stock soft or lime glass tubing and rod, which, with a softening point of about 650°C, can be worked in a Bunsen burner. However, except for very simple bends and joints, soft glass is not used for laboratory glass blowing because of the ease with which it breaks on being subjected to a thermal gradient. Since the two types of glass are not compatible it is important to be able to distinguish between them. This is easily done by immersing the glass object in a solution having exactly the same refractive index as the glass. In such a solution the glass will seem to disappear, whereas a glass of different refractive index will be plainly visible. A solution of 14 parts (by volume) of methanol and 86 parts of toluene has a refractive index of 1.474, which is the same as that of Pyrex 7740 glass, the most common of the various Pyrexes. In this solution Pyrex will not be visible. Store the solution in a wide-mouth jar with a close-fitting cap. It will keep indefinitely.

Glass Tubing

To cut a glass tube, first make a fine, straight scratch, extending about a quarter of the way around the tube, with a glass scorer. This is done by applying firm pressure on the scorer and rotating the glass tube slightly (Fig. 1). Only one scratch should be made; in no case should you try to saw a groove in the tube. The tube is then grasped with the scratch away from the body and the thumbs pressed together at the near side of the tube just opposite the scratch, with the arms pressed tightly against the body (Fig. 2). A straight, clean break will result when *slight* pressure is exerted with the thumbs and a strong force applied to pull the tube apart. It is a matter of 90% pull and 10% bend.

Break by pulling

Fire Polishing

The sharp edges that result from breaking a glass rod or tube will cut you as well as the corks, rubber stoppers, and tubing being fitted over them. Remove these sharp edges by holding the end of the rod (or tube) in a Bunsen burner flame and rotating the rod until the sharp edges melt and disappear. This fire-polishing process can be done even for Pyrex glass if the flame is hot enough. Open the air inlet at the bottom of the burner barrel to its maximum; the hottest part of the flame is about 7 mm above the inner blue cone. A stirring rod with a flattened head, useful for crushing lumps of solid against the bottom of a flask, is made by heating a glass rod until a short section at the end is soft, and quickly pressing the end onto a smooth metal surface.

Bends

Rotate constantly

The secret to successful glass working is to have the glass thoroughly and uniformly heated before an operation. Since Pyrex glass softens at 820°C, and soft glass softens at 650°C, the best way to work Pyrex is with a gas-oxygen torch; but with patience it can be satisfactorily heated over an ordinary Bunsen burner with a wing top attached (Fig. 3). Stopper the tube at the left-hand end; then grasp in the left hand with palm down and in the right hand with palm up so you can swing the open end of the tube into position for blowing without interruption of the synchronous rotation of the two ends. Adjust the air intake of the burner for the maximum amount of air possible (too much will blow out the flame) and rotate the tube constantly, holding it about 7 mm above the inner blue cone. A bit of coordination is needed to rotate both ends at the same speed once the glass begins to soften; when the flame is thoroughly tinged with yellow (from sodium ions escaping from the hot glass) and the tube begins to sag, remove the tube from the flame and bend it in the vertical plane with the ends upward and the bend at the bottom. Should the tube become constricted at the bend, blow into the open end immediately upon completion of the bend to expand the glass to its full size.

The Gas-Oxygen Torch

The following operations are best carried out using Pyrex tubing and a gas-oxygen torch. To light the torch turn on the gas first to give a large luminous flame. *Gradually* turn on the oxygen until a long thin blue flame with a clearly defined inner blue cone is formed. The hottest part of the flame is at the tip of the inner blue cone. To turn off the flame always turn off the oxygen first. Wear glass blower's didymium goggles to protect the eyes from the blinding glare of hot Pyrex.

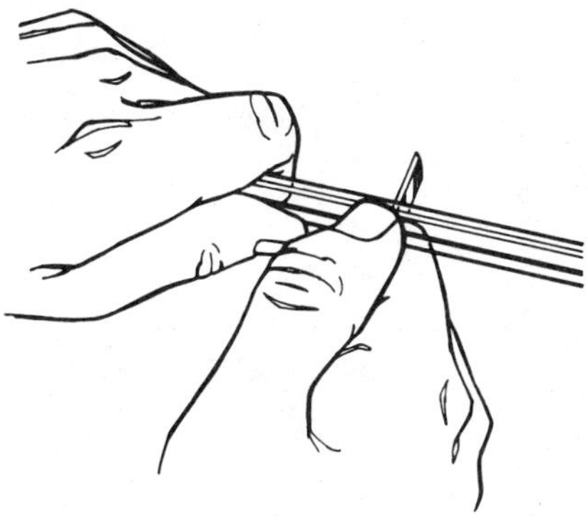

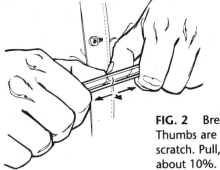

FIG. 1   Scratching glass tubing with glass scorer prior to breaking. The scratch is about one-fourth the tube's circumference in length.

FIG. 2   Breaking glass tubing. Thumbs are opposite the scratch. Pull, about 90%; bend, about 10%.

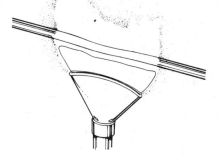

FIG. 3   Heating glass tubing prior to bending. The wing top produces a broad flame that heats enough of the tubing to allow a good bend to be made. The tubing is held about 7 mm above the inner blue flame.

Flaring

It will often be necessary to flare the end of a tube in order to make a joint. Heat the end of the tube until the glass begins to sag, then remove the glass from the fire. While rotating the tube, insert a tool such as the tine of a file or a carbon rod and press it sufficiently to form the flare (Fig. 4).

Test Tube Ends

To close a tube heat it strongly at some convenient point while rotating both ends simultaneously. When the glass is soft remove the tube from the flame and pull the ends rapidly for a few inches while maintaining the rotation. Allow the glass to cool slightly, then heat the tube at A (Fig. 5) and pull the tube into two pieces. Heat the point B with a sharp flame to collapse it and blow it out slightly as in C. Heat the whole end of the tube until it shrinks as in D and finally blow it to a uniform hemisphere as in E. Maintain uniform rotation while carrying out all of these operations.

FIG. 4   Flaring a glass tube.

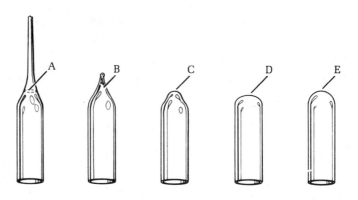

FIG. 5   Blowing a test tube end.

Straight Seal

To join two tubes end-on, cork one and hold it with the palm down in the left hand. Hold the open tube in the right hand, palm up. Rotate both tubes simultaneously in the flame so that the glass becomes soft just at the end and forms a rounded edge, but it does not constrict. Move the tubes to a cool part of the flame, press the ends lightly and evenly together on the same axis, and then with no hesitation pull the tubes apart slightly to reduce the thickness of the glass (Fig. 6). While rotating the tubing with both hands, heat the joint and cause it to shrink to about half its original diameter. Remove the tubing from the flame and blow gently to expand the joint slightly larger than the tubing diameter. Heat the joint once more and pull the tubing if necessary to restore the glass thickness to that of the original tubing. The joint, when finished, should closely resemble ordinary tubing.

T-Seal

Cork the end of a tube and heat a small spot on the side of the tube with a sharp flame (Fig. 7). Blow a small bulge on the tube. Reheat this bulge carefully at its tip and blow sharply to form a very thin walled bulb, which is broken off. Cork the other end of the tube, heat uniformly the edges of the opening as well as the end of the side tube, which has previously been slightly flared. After the edges of the openings are fairly soft remove the tubes from the flame, press them together, and then pull slightly as soon as complete contact has been made. Blow slightly to remove any irregularities. If necessary reheat, shrink, and blow until all irregularities are removed. Finally, heat the whole joint to obtain the correct angles between the tubes.

Ring Seals

Ring seals can be made in two different ways. In one method (Fig. 8) a flared tube of the appropriate length is dropped inside a test tube and centered with a smaller diameter tube through a cork (Fig. 8A). Heat the bottom of

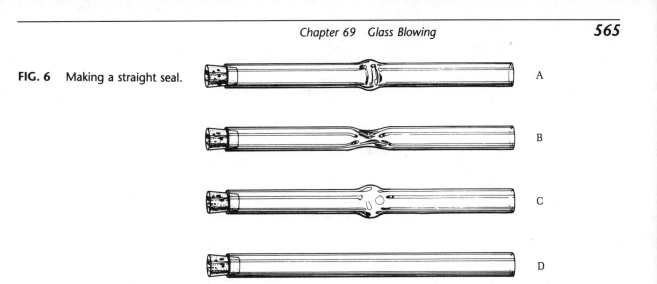

FIG. 6   Making a straight seal.

A

B

C

D

FIG. 7   Making a T-seal.

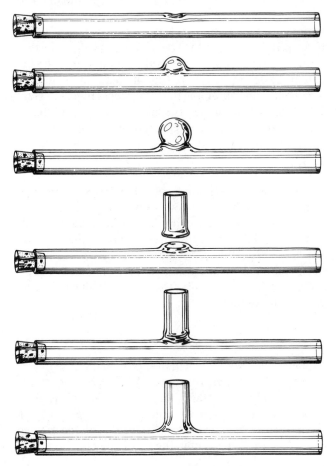

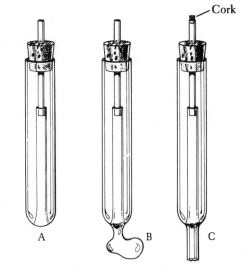

FIG. 8   Making a ring seal, first method.

FIG. 9 Making a ring seal, second method.

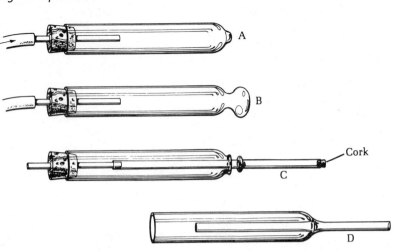

the test tube until the inner tube forms a seal, then blow out a bulb as in Fig. 8B. Cork the guide tube and attach another flared tube to give the finished seal as seen in Fig. 8C.

In the second method for making ring seals a test tube is heated at the bottom and a small bulge is blown (Fig. 9A). This is heated again and a thin bulb blown out (Fig. 9B) and broken off. The resulting hole should be the same size as the tube to be sealed in. At an appropriate place in the tube to be sealed in, blow a small bulge and assemble the two pieces (Fig. 9C). Heat the joint with a small flame until it is completely sealed then blow and shrink the glass alternately until irregularities are removed (Fig. 9D).

Blowing a Bulb in a Tube

Cork one end of a 6-mm tube, grasp it left hand palm down, right hand palm up, and heat it uniformly over a length of about 25 mm while maintaining constant rotation. Surface tension will cause the glass to thicken as the ends of the tube are brought closer. When the tube has the appearance of Fig. 10A, carefully blow a bulb (Fig. 10B). The success of this procedure is governed by the first operation. It is necessary to have enough hot glass gathered in the hot section such that the bulb, when blown, will have a wall thickness equal to that of the tubing.

Annealing

The rapid cooling of hot glass will put strains in the glass. In the case of soft glass these strains will often cause the glass to crack. With Pyrex the strains will make the glass mechanically weak where they occur, even if the glass does not crack. These strains are relieved by cooling the glass slowly from the molten state. On a small scale this can be done by turning off the oxygen of the burner and holding the hot glass joint in the relatively cool luminous

FIG. 10   Blowing a bulb.

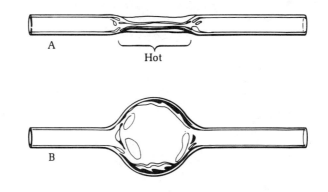

A

Hot

B

12 Mesh
iron screen

FIG. 11   Squaring the end of a
jagged tube.

flame until the joint is coated with an even layer of soot. During the minute or two this requires, many of the strains will be relieved. Large and complex pieces are annealed in an oven with a controlled temperature drop.

Squaring a Jagged Break

It is often desired to square the end of a tube that has a jagged end prior to carrying out the fire-polishing operation. This can be accomplished by stroking the glass with a 13-cm square of wire screen over a waste container (Fig. 11). The glass is removed as very small chips and dust. Wear safety glasses when doing this. Much potentially hazardous apparatus can be repaired in this way, followed by fire polishing of the resulting opening.

70

Qualitative Organic Analysis

Prelab Exercise: In the identification of an unknown organic compound certain procedures are more valuable than others. For example, much more information is obtained from an infrared spectrum than from a refractive index measurement. Outline, in order of priority, the steps you will employ in identifying your unknown.

Identification and characterization of the structures of unknown substances are an important part of organic chemistry. It is often, of necessity, a micro process, e.g., in drug analyses. It is sometimes possible to establish the structure of a compound on the basis of spectra alone (ir, uv, and nmr), but these spectra must usually be supplemented with other information about the unknown: physical state, elementary analysis, solubility, and confirmatory tests for functional groups. Conversion of the unknown to a solid derivative of known melting point will often provide final confirmation of structure.

Procedures

All experiments in this chapter can, if necessary, be run on 2 to 3 times the indicated quantities of material.

Physical State

Check for Sample Purity

Distill or recrystallize as necessary. Constant bp and sharp mp are indicators, but beware of azeotropes and eutectics. Check homogeneity by TLC, gas, HPLC, or paper chromatography.

Note the Color

Common colored compounds include nitro and nitroso compounds (yellow), α-diketones (yellow), quinones (yellow to red), azo compounds (yellow to red), and polyconjugated olefins and ketones (yellow to red). Phenols and amines are often brown to dark-purple because of traces of air oxidation products.

Note the Odor

CAUTION: Do not taste an unknown compound. To note the odor cautiously smell the cap of the container and do it only once.

Some liquid and solid amines are recognizable by their fishy odors; esters are often pleasantly fragrant. Alcohols, ketones, aromatic hydrocarbons, and aliphatic olefins have characteristic odors. On the unpleasant side are thiols, isonitriles, and low-molecular-weight carboxylic acids.

Make an Ignition Test

Heat a small sample on a spatula; first hold the sample near the side of a microburner to see if it melts normally and then burns. Heat it in the flame. If a large ashy residue is left after ignition, the unknown is probably a metal salt. Aromatic compounds often burn with a smoky flame.

Spectra

Obtain infrared and nuclear magnetic resonance spectra following the procedures of Chapters 19 and 20. If these spectra indicate the presence of conjugated double bonds, aromatic rings, or conjugated carbonyl compounds obtain the ultraviolet spectrum following the procedures of Chapter 21. Interpret the spectra as fully as possible by reference to the sources cited at the end of the various spectroscopy chapters.

Elementary Analysis, Sodium Fusion

Explanation

This method for detection of nitrogen, sulfur, and halogen in organic compounds depends on the fact that fusion of substances containing these elements with sodium yields NaCN, Na_2S, and NaX (X = Cl, Br, I). These products can, in turn, be readily identified. The method has the advantage that the most usual elements other than C, H, and O present in organic compounds can all be detected following a single fusion, although the presence of sulfur sometimes interferes with the test for nitrogen. Unfortunately, even in the absence of sulfur, the test for nitrogen is sometimes unsatisfactory (nitro compounds in particular). Practicing organic chemists rarely perform this test. Either they know what elements their unknowns contain, or they have access to a mass spectrometer or atomic absorption instrument.

Rarely performed by professional chemists

Place a 3-mm cube of sodium[1] (30 mg, no more)[2] in a 10×75 mm Pyrex test tube and support the tube in a vertical position (Fig. 1). Have a microburner with small flame ready to move under the tube, place an estimated 20 mg of solid on a spatula or knife blade, put the burner in place, and heat until the sodium first melts and then vapor rises 1.5–2.0 cm in the tube. Remove the burner and at once drop the sample onto the hot sodium. If the substance is a liquid add 2 drops of it. If there is a flash or small explosion the fusion is complete; if not, heat briefly to produce a flash or a charring. Then let the tube cool to room temperature, be sure it is cold, add a drop of methanol, and let it react. Repeat until 10 drops have been added.

CAUTION: Manipulate sodium with a knife and forceps; never touch it with the fingers. Wipe it free of kerosene with a dry towel or filter paper; return scraps to the bottle or destroy scraps with methyl or ethyl alcohol, never with water. Safety glasses! Hood!

Notes for the instructor

---

1. Sodium spheres $\frac{1}{16}''$ to $\frac{1}{4}''$ are convenient.

2. A dummy 3-mm cube of rubber can be attached to the sodium bottle to indicate the correct amount.

FIG. 1   Sodium fusion, just
prior to addition of sample.

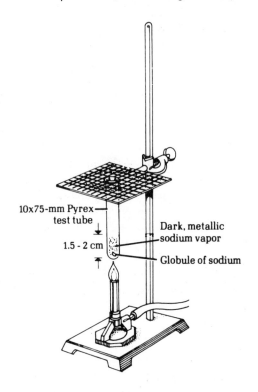

10x75-mm Pyrex
test tube

1.5 - 2 cm

Dark, metallic
sodium vapor

Globule of sodium

With a stirring rod break up the char to uncover sodium. When you are sure
that all sodium has reacted, empty the tube into a 13×100 mm test tube,
hold the small tube pointing away from you or a neighbor, and pipette into it
1 mL of water. Boil and stir the mixture and pour the water into the larger
tube; repeat with 1 mL more water. Then transfer the solution with a
Pasteur pipette to a 2.5-cm funnel (fitted with a fluted filter paper) resting in
a second 13×100 mm test tube. Portions of the alkaline filtrate are used for
the tests that follow:

Do not use $CHCl_3$ *or* CCl_4 *as
samples in sodium fusion.
They react extremely
violently.*

(a) Nitrogen

The test is done by boiling a portion of the alkaline solution from the
solution fusion with iron(II) sulfate and then acidifying. Sodium cyanide
reacts with iron(II) sulfate to produce ferrocyanide, which combines with
iron(III) salts, inevitably formed by air oxidation in the alkaline solution, to
give insoluble Prussian Blue, $NaFe[Fe(CN)_6]$. Iron(II) and iron(III) hydroxide precipitate along with the blue pigment but dissolve on
acidification.

*Run each test on a known
and an unknown*

   Place 50 mg of powdered iron(II) sulfate (this is a large excess) in a
10×75 mm test tube, add 0.5 mL of the alkaline solution from the fusion,
heat the mixture gently with shaking to the boiling point, and then—without
cooling—acidify with dilute sulfuric acid (hydrochloric acid is unsatisfac-

tory). A deep blue precipitate indicates the presence of nitrogen. If the coloration is dubious, filter through a 2.5-cm funnel and see if the paper shows blue pigment.

Cleaning Up   The test solution should be diluted with water and flushed down the drain.

(b) Sulfur

$$Na_2(NO)Fe(CN)_6 \cdot 2H_2O$$
Sodium nitroprusside

(1) Dilute one drop of the alkaline solution with 1 mL of water and add a drop of sodium nitroprusside; a purple coloration indicates the presence of sulfur. (2) Prepare a fresh solution of sodium plumbite by adding 10% sodium hydroxide solution to 0.2 mL of 0.1 M lead acetate solution until the precipitate just dissolves, and add 0.5 mL of the alkaline test solution. A black precipitate or a colloidal brown suspension indicates the presence of sulfur.

Cleaning Up   The test solution should be diluted with water and flushed down the drain.

(c) Halogen

Differentiation of the halogens

Do not waste silver nitrate

Acidify 0.5 mL of the alkaline solution from the fusion with dilute nitric acid (indicator paper) and, if nitrogen or sulfur has been found present, boil the solution (hood) to expel HCN or H_2S. On addition of a few drops of silver nitrate solution, halide ion is precipitated as silver halide. Filter with minimum exposure to light on a 2.5-cm funnel, wash with water, and then with 1 mL of concentrated ammonia solution. If the precipitate is white and readily soluble in ammonium hydroxide solution it is AgCl; if it is pale yellow and not readily soluble it is AgBr; if it is yellow and insoluble it is AgI. Fluorine is not detected in this test since silver fluoride is soluble in water.

Cleaning Up   The test solution should be diluted with water and flushed down the drain.

> **Run tests on knowns in parallel with unknowns for all qualitative organic reactions. In this way, interpretations of positive reactions are clarified and defective test reagents can be identified and replaced.**

Beilstein Test for Halogens

A fast, easy, reliable test

Heat the tip of a copper wire in a burner flame until no further coloration of the flame is noticed. Allow the wire to cool slightly, then dip it into the unknown (solid or liquid) and again heat it in the flame. A green flash is indicative of chlorine, bromine, and iodine; fluorine is not detected since copper fluoride is not volatile. The Beilstein test is very sensitive; halogen-containing impurities may give misleading results. Run the test on a compound known to contain halogen for comparison with your unknown.

Solubility Tests

Weigh and measure carefully

Like dissolves like; a substance is most soluble in that solvent to which it is most closely related in structure. This statement serves as a useful classification scheme for all organic molecules. The solubility measurements are done at room temperature with 1 drop of a liquid, or 5 mg of a solid (finely crushed), and 0.2 mL of solvent. The mixture should be rubbed with a rounded stirring rod and shaken vigorously. Lower members of a homologous series are easily classified; higher members become more like the hydrocarbons from which they are derived.

If a very small amount of the sample fails to dissolve when added to some of the solvent, it can be considered insoluble; and, conversely, if several portions dissolve readily in a small amount of the solvent, the substance is obviously soluble.

If an unknown seems to be more soluble in dilute acid or base than in water, the observation can be confirmed by neutralization of the solution; the original material will precipitate if it is less soluble in a neutral medium.

If both acidic and basic groups are present, the substance may be amphoteric and therefore soluble in both acid and base. Aromatic aminocarboxylic acids are amphoteric, like aliphatic ones, but they do not exist as zwitterions. They are soluble in both dilute hydrochloric acid and sodium hydroxide, but not in bicarbonate solution. Aminosulfonic acids exist as zwitterions; they are soluble in alkali but not in acid.

The solubility tests are not infallible and many borderline cases are known.

Carry out the tests according to the scheme of Fig. 2 and the following Notes to Solubility Tests and tentatively assign the unknown to one of the groups I–X.

Cleaning Up   Since the quantities of material used in these tests are extremely small and since no hazardous substances are handed out as unknowns, it is possible to dilute the material with a large quantity of water and flush it down the drain.

Classification Tests

After the unknown is assigned to one of the solubility groups (Fig. 2) on the basis of solubility tests, the possible type should be further narrowed by application of classification tests, e.g., for alcohols, or methyl ketones, or esters.

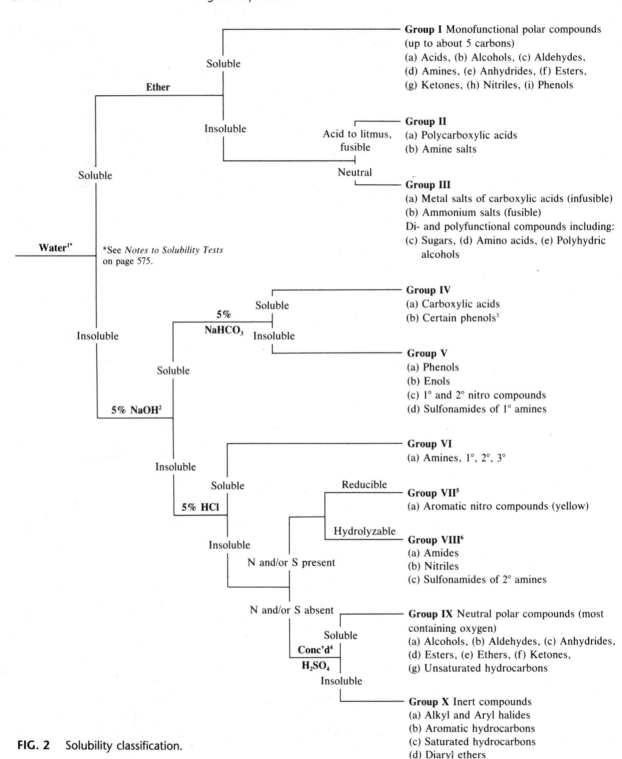

FIG. 2   Solubility classification.

Notes to Solubility Tests

See chart on page 574

1. Groups I, II, III (soluble in water). Test the solution with pH paper. If the compound is not easily soluble in cold water, treat it as water-insoluble but test with indicator paper.

2. If the substance is insoluble in water but dissolves partially in 5% sodium hydroxide, add more water; the sodium salts of some phenols are less soluble in alkali than in water. If the unknown is colored, be careful to distinguish between the *dissolving* and the *reacting* of the sample. Some quinones (colored) *react* with alkali and give highly colored solutions. Some phenols (colorless) *dissolve and then* become oxidized to give colored solutions. Some compounds (e.g., benzamide) are hydrolyzed with such ease that careful observation is required to distinguish them from acidic substances.

3. Nitrophenols (yellow), aldehydophenols, and polyhalophenols are sufficiently strongly acidic to react with sodium bicarbonate.

4. Oxygen- and nitrogen-containing compounds form oxonium and ammonium ions in concentrated sulfuric acid and dissolve.

5. On reduction in the presence of hydrochloric acid these compounds form water-soluble amine hydrochlorides. Dissolve 250 mg of tin(II) chloride in 0.5 mL of concentrated hydrochloric acid, add 50 mg of the unknown, and warm. The material should dissolve with the disappearance of the color and give a clear solution when diluted with water.

6. Most amides can be hydrolyzed by short boiling with 10% sodium hydroxide solution; the acid dissolves with evolution of ammonia. Reflux 100 mg of the sample and 10% sodium hydroxide solution for 15–20 min under a cold finger condenser. Test for the evolution of ammonia, which confirms the elementary analysis for nitrogen and establishes the presence of a nitrile or amide.

Complete Identification—Preparation of Derivatives

Once the unknown has been classified by functional group, the physical properties should be compared with those of representative members of the group (see tables at the end of this chapter). Usually, several possibilities present themselves, and the choice can be narrowed by preparation of derivatives. Select derivatives that distinguish most clearly among the possibilities.

Classification Tests

Group I. Monofunctional Polar Compounds (up to ca. 5 carbons)

(a) Acids

(Table 1; Derivatives, page 591)
No classification test is necessary. Carboxylic and sulfonic acids are

detected by testing aqueous solutions with litmus. Acyl halides may hydrolyze during the solubility test.

(b) Alcohols

(Table 2; Derivatives, page 592)

Jones Oxidation. Dissolve 5 mg of the unknown in 0.5 mL of pure acetone in a test tube and add to this solution 1 small drop of Jones reagent (chromic acid in sulfuric acid). A positive test is formation of a green color within 5 s upon addition of the orange-yellow reagent to a primary or secondary alcohol. Aldehydes also give a positive test, but tertiary alcohols do not.

CAUTION: Cr^{+6} dust is toxic

    Reagent: Dissolve/suspend 13.4 g of chromium trioxide in 11.5 mL of concentrated sulfuric acid and add this carefully with stirring to enough water to bring the volume to 50 mL.

Cleaning Up   Place the test solution in the hazardous waste container.

Handle dioxane with care. It is a cancer suspect agent.

Cerium(IV) Nitrate Test (Ammonium Hexanitratocerium(IV) Test). Dissolve 15 mg of the unknown in a few drops of water or dioxane in a reaction tube. Add to this solution 0.25 mL of the reagent and mix thoroughly. Alcohols cause the reagent to change from yellow to red.

    Reagent: Dissolve 22.5 g of ammonium hexanitratocerium(IV), $Ce(NH_4)_2(NO_3)_6$, in 56 mL of 2 N nitric acid.

Cleaning Up   The solution should be diluted with water and flushed down the drain.

(c) Aldehydes

(Table 3; Derivatives, page 594)

2,4-Dinitrophenylhydrazones. All aldehydes and ketones readily form bright-yellow to dark-red 2,4-dinitrophenylhydrazones. Yellow derivatives are formed from isolated carbonyl groups and orange-red to red derivatives from aldehydes or ketones conjugated with double bonds or aromatic rings.

    Dissolve 10 mg of the unknown in 0.5 mL of ethanol and then add 0.75 mL of 2,4-dinitrophenylhydrazine reagent. Mix thoroughly and let sit for a few minutes. A yellow to red precipitate is a positive test.

    Reagent: Dissolve 1.5 g of 2,4-dinitrophenylhydrazine in 7.5 mL of concentrated sulfuric acid. Add this solution, with stirring, to a mixture of 10 mL of water and 35 mL of ethanol.

Cleaning Up   Place the test solution in the hazardous waste container.

Schiff Test. Add 1 drop (30 mg) of the unknown to 1 mL of Schiff's reagent. A magenta color will appear within 10 min with aldehydes. Compare the color of your unknown with that of a known aldehyde.

Reagent: Prepare 50 mL of a 0.1 percent aqueous solution of *p*-rosaniline hydrochloride (fuchsin). Add 2 mL of a saturated aqueous solution of sodium bisulfite. After 1 h add 1 mL of concentrated hydrochloric acid.

Bisulfite Test. Follow the procedure on page 313, Chapter 30. Nearly all aldehydes and most methyl ketones form solid, water-soluble bisulfite addition products.

Tollens' Test. Follow the procedure on page 311, Chapter 30. A positive test, deposition of a silver mirror, is given by most aldehydes, but not by ketones.

Destroy used Tollens' reagent promptly with nitric acid. It can form explosive fulminates.

(d) Amines

(Tables 5 and 6; Derivatives, page 596, 597)

Hinsberg Test. Follow the procedure on page 373, Chapter 40, using benzenesulfonyl chloride to distinguish between primary, secondary, and tertiary amines.

(e) Anhydrides and Acid Halides

(Table 7; Derivatives, pages 597–598) Anhydrides and acid halides will react with water to give acidic solutions, detectable with litmus paper. They easily form benzamides and acetamides.

Acidic Iron(III) Hydroxamate Test. With iron(III) chloride alone a number of substances give a color which can interfere with this test. Dissolve 2 drops (or about 30 mg) of the unknown in 1 mL of ethanol and add 1 mL of 1 N hydrochloric acid followed by 1 drop of 10% aqueous iron(III) chloride solution. If any color except yellow appears you will find it difficult to interpret the results from the following test.

Add 2 drops (or about 30 mg) of the unknown to 0.5 mL of a 1 N solution of hydroxylamine hydrochloride in alcohol. Add 2 drops of 6 M hydrochloric acid to the mixture, warm it slightly for 2 min, and boil it for a few seconds. Cool the solution and add 1 drop of 10% ferric chloride solution. A red-blue color is a positive test.

Cleaning Up   Neutralize the reaction mixture with sodium carbonate, dilute with water, and flush down the drain.

(f) Esters

(Table 8, page 598. Derivatives are prepared from component acid and alcohol obtained on hydrolysis.)

Esters, unlike anhydrides and acid halides, do not react with water to give acidic solutions and do not react with acidic hydroxylamine hydrochloride. They do, however, react with alkaline hydroxylamine.

Alkaline Iron(III) Hydroxamate Test. First test the unknown with iron(III) chloride alone. (See under Group I(e), Acidic Iron(III) Hydroxamate Test.)

To a solution of one drop (30 mg) of the unknown in 0.5 mL of 0.5 N hydroxylamine hydrochloride in ethanol add two drops of 20% sodium hydroxide solution. Heat the solution to boiling, cool slightly, and add 1 mL of 1 N hydrochloric acid. If cloudiness develops add up to 1 mL of ethanol. Add 10% iron(III) chloride solution dropwise with thorough mixing. A red-blue color is a positive test. Compare your unknown with a known ester.

Cleaning Up   Neutralize the solutions with sodium carbonate, dilute with water, and flush down the drain.

(g) Ketones

(Table 14; Derivatives, page 602)

2,4-Dinitrophenylhydrazone. See under Group I(c), Aldehydes. All ketones react with 2,4-dinitrophenylhydrazine reagent.

Iodoform Test for Methyl Ketones. Follow the procedure in Chapter 30. A positive iodoform test is given by substances containing the $CH_3\overset{\displaystyle O}{\overset{\|}{C}}$ — group or by compounds easily oxidized to this group, e.g., CH_3COR, CH_3CHOHR, CH_3CH_2OH, CH_3CHO, $RCOCH_2COR$. The test is negative for compounds of the structure CH_3COOR, CH_3CONHR, and other compounds of similar structure that give acetic acid on hydrolysis. It is also negative for $CH_3COCH_2CO_2R$, CH_3COCH_2CN, $CH_3COCH_2NO_2$.

Bisulfite Test. Follow the procedure in Chapter 30. Aliphatic methyl ketones and unhindered cyclic ketones form bisulfite addition products. Methyl aryl ketones, such as acetophenone, $C_6H_5COCH_3$, fail to react.

(h) Nitriles

(Table 15, page 603. Derivatives prepared from the carboxylic acid obtained by hydrolysis.)

At high temperature nitriles (and amides) are converted to hydroxamic acids by hydroxylamine:

$$RCN + 2\ H_2NOH \longrightarrow RCONHOH + NH_3$$

The hydroxamic acid forms a red-blue complex with iron(III) ion. The unknown must first give a negative test with hydroxylamine at lower temperature (Group I(f), Alkaline Hydroxamic Acid Test) before trying this test.

Hydroxamic Acid Test for Nitriles (and Amides). To 1 mL of a 1 M hydroxylamine hydrochloride solution in propylene glycol add 15 mg of the unknown dissolved in the minimum amount of propylene glycol. Then add

0.5 mL of 1 N potassium hydroxide in propylene glycol and boil the mixture for 2 min. Cool the mixture and add 0.1 to 0.25 mL of 10% iron(III) chloride solution. A red-blue color is a positive test for almost all nitrile and amide groups, although benzanilide fails to give a positive test.

Cleaning Up   Since the quantity of material is extremely small, the test solution can be diluted with water and flushed down the drain.

(i) Phenols

(Table 17, page 604)
Iron(III) Chloride Test. Dissolve 15 mg of the unknown compound in 0.5 mL of water or water-alcohol mixture and add one or two drops of 1% iron(III) chloride solution. A red, blue, green, or purple color is a positive test.

Cleaning Up   Since the quantity of material is extremely small, the test solution can be diluted with water and flushed down the drain.

A more sensitive test for phenols consists of dissolving or suspending 15 mg of the unknown in 0.5 mL of chloroform and adding 1 drop of a solution made by dissolving 0.1 g of iron(III) chloride in 10 mL of chloroform. (*Caution!* $CHCl_3$ is a carcinogen.) Addition of a drop of pyridine, with stirring, will produce a color if phenols or enols are present.

Group II. Water-soluble Acidic Salts, Insoluble in Ether

Amine Salts

(Table 5 (1° and 2° amines); Table 6 (3° amines))
The free amine can be liberated by addition of base and extraction into ether. Following evaporation of the ether the Hinsberg test, Group I(d), can be applied to determine if the compound is a primary, secondary, or tertiary amine.

The acid iron(III) hydroxamate test, Group I(d), can be applied directly to the amine salt.

Group III. Water-soluble Neutral Compounds, Insoluble in Ether

(a) Metal Salts of Carboxylic Acids

(Table 1, carboxylic acids; Derivatives, page 586)
The free acid can be liberated by addition of acid and extraction into an appropriate solvent, after which the carboxylic acid can be characterized by mp or bp before proceeding to prepare a derivative.

(b) Ammonium Salts

(Table 1, carboxylic acids; Derivatives, page 586)

Ammonium salts on treatment with alkali liberate ammonia, which can be detected by its odor and the fact that it will turn red litmus, blue. A more sensitive test utilizes the copper(II) ion, which is blue in the presence of ammonia [see Group VIII a(i)]. Ammonium salts will not give a positive hydroxamic acid test (Ih) as given by amides.

(c) Sugars

See page 311, Chapter 30, for Tollens' test and page 444, Chapter 51, for phenylosazone formation.

(d) Amino Acids

Add 2 mg of the suspected amino acid to 1 mL of ninhydrin reagent, boil for 20 s, and note the color. A blue color is a positive test.

Reagent: Dissolve 0.2 g of ninhydrin in 50 mL of water.

Cleaning Up   Since the quantity of material is extremely small, the test solution can be diluted with water and flushed down the drain.

(e) Polyhydric Alcohols

(Table 2; Derivatives, page 586)

Periodic Acid Test for vic-Glycols.[3] Vicinal glycols (hydroxyl groups on adjacent carbon atoms) can be detected by reaction with periodic acid. In addition to 1,2-glycols, a positive test is given by α-hydroxy aldehydes, α-hydroxy ketones, α-hydroxy acids, and α-amino alcohols, as well as 1,2-diketones.

To 2 mL of periodic acid reagent add one drop (no more) of concentrated nitric acid and shake. Then add one drop or a small crystal of the unknown. Shake for 15 s and add 1 or 2 drops of 5% aqueous silver nitrate solution. Instantaneous formation of a white precipitate is a positive test.

Reagent: Dissolve 0.25 g of paraperiodic acid (H_5IO_6) in 50 mL of water.

Cleaning Up   Since the quantity of material is extremely small, the test solution can be diluted with water and flushed down the drain.

---

3. R. L. Shriner, R. C. Fuson, D. Y. Curtin, and T. C. Morill, *The Systematic Identification of Organic Compounds,* 6th ed., John Wiley & Sons, Inc., New York, 1980.

Group IV. Certain Carboxylic Acids, Certain Phenols, and Sulfonamides of 1° Amines

(a) Carboxylic Acids

Solubility in both 5% sodium hydroxide and sodium bicarbonate is usually sufficient to characterize this class of compounds. Addition of mineral acid should regenerate the carboxylic acid. The neutralization equivalent can be obtained by titrating a known quantity of the acid (ca. 50 mg) dissolved in water-ethanol with 0.1 N sodium hydroxide to a phenolphthalein end point.

(b) Phenols

Negatively substituted phenols such as nitrophenols, aldehydophenols, and polyhalophenols are sufficiently acidic to dissolve in 5% sodium bicarbonate. See Group I(i), page 579, for the iron(III) chloride test for phenols; however, this test is not completely reliable for these acidic phenols.

Group V. Acidic Compounds, Insoluble in Bicarbonate

(a) Phenols

See Group I(i).

(b) Enols

See Group I(i).

(c) 1° and 2° Nitro Compounds

(Table 16, Derivatives, page 603)

Iron (II) Hydroxide Test. To a small vial (capacity 1–2 mL) add 5 mg of the unknown to 0.5 mL of freshly prepared ferrous sulfate solution. Add 0.4 mL of a 2 N solution of potassium hydroxide in methanol, cap the vial, and shake it. The appearance of a red-brown precipitate of iron(III) hydroxide within 1 min is a positive test. Almost all nitro compounds give a positive test within 30 s.

   Reagents: Dissolve 2.5 g of ferrous ammonium sulfate in 50 mL of deoxygenated (by boiling) water. Add 0.2 mL of concentrated sulfuric acid and a piece of iron to prevent oxidation of the ferrous ion. Keep the bottle tightly stoppered. The potassium hydroxide solution is prepared by dissolving 5.6 g of potassium hydroxide in 50 mL of methanol.

Cleaning Up   Since the quantity of material is extremely small, the test solution can be diluted with water and flushed down the drain after neutralization with dilute hydrochloric acid.

(d) Sulfonamides of 1° Amines

An extremely sensitive test for sulfonamides (Feigl, *Spot Tests in Organic Analysis*) consists of placing a drop of a suspension or solution of the unknown on sulfonamide test paper followed by a drop of 0.5% hydrochloric acid. A red color is a positive test for sulfonamides.

The test paper is prepared by dipping filter paper into a mixture of equal volumes of a 1% aqueous solution of sodium nitrite and a 1% methanolic solution of N,N-dimethyl-1-naphthylamine. Allow the filter paper to dry in the dark.

CAUTION: Handle N,N-dimethyl-1-naphthylamine with care. As a class aromatic amines are quite toxic and many are carcinogenic. Handle them all with care—in a hood if possible.

Cleaning Up   Place the test paper in the solid hazardous waste container.

Group VI. Basic Compounds, Insoluble in Water, Soluble in Acid

Amines

See Group I(d).

Group VII. Reducible, Neutral N- and S-Containing Compounds

Aromatic Nitro Compounds

See Group V(c).

Group VIII. Hydrolyzable, Neutral N- and S-Containing Compounds (Identified through the acid and amine obtained on hydrolysis)

(a) Amides

Unsubstituted amides are detected by the hydroxamic acid test, Group I(h).
(1) Unsubstituted Amides. Upon hydrolysis, unsubstituted amides liberate ammonia, which can be detected by reaction with cupric ion (Group III(b)).

To 1 mL of 20% sodium hydroxide solution add 25 mg of the unknown. Cover the mouth of the reaction tube with a piece of filter paper moistened with a few drops of 10% copper(II) sulfate solution. Boil for 1 min. A blue color on the filter paper is a positive test for ammonia.

Cleaning Up   Neutralize the test solution with 10% hydrochloric acid, dilute with water, and flush down the drain.

(2) Substituted Amides. The identification of substituted amides is not easy. There are no completely general tests for the substituted amide groups and hydrolysis is often difficult.

Hot sodium hydroxide solution is corrosive; use care

Hydrolyze the amide by refluxing 250 mg with 2.5 mL of 20% sodium hydroxide for 20 min. Isolate the primary or secondary amine produced, by extraction into ether, and identify as described under Group I(d). Liberate the acid by acidification of the residue, isolate by filtration or extraction, and characterize by bp or mp and the mp of an appropriate derivative.

Cleaning Up   The test solution can be diluted with water and flushed down the drain.

Use care in shaking concentrated sulfuric acid

Dichromate dust is carcinogenic, when inhaled. Cr^{+6} is not a carcinogen when applied to the skin or ingested.

(3) Anilides. Add 50 mg of the unknown to 1.5 mL of concentrated sulfuric acid. Carefully stopper the reaction tube with a rubber stopper and shake vigorously. **Caution.** Add 25 mg of finely powdered potassium dichromate. A blue-pink color is a positive test for an anilide that does not have substituents on the ring (e.g., acetanilide).

Cleaning Up   Carefully add the solution to water, neutralize with sodium carbonate, and flush down the drain.

(b) Nitriles

See Group I(h).

(c) Sulfonamides

See Group V(d), page 582.

Group IX. Neutral Polar Compounds, Insoluble in Dilute Hydrochloric Acid, Soluble in Concentrated Sulfuric Acid (most compounds containing oxygen)

(a) Alcohols

See Group I(b).

(b) Aldehydes

See Group I(c).

(c) Anhydrides

See Group I(e).

(d) Esters

See Group I(f).

(e) Ethers

(Table 9, page 599)

Ethers are very unreactive. Care must be used to distinguish ethers from those hydrocarbons that are soluble in concentrated sulfuric acid.

Cleaning Up   Place the mixture in the hazardous waste container.

$Fe[Fe(SCN)_6]$
**Iron(III) hexathiocyanato-
ferrate(III)**

Ferrox Test. In a dry test tube grind together, with a stirring rod, a crystal of iron(III) ammonium sulfate (or iron(III) chloride) and a crystal of potassium thiocyanate. Iron(III) hexathiocyanatoferrate(III) will adhere to the stirring rod. In a clean tube place 3 drops of a liquid unknown or a saturated toluene solution of a solid unknown and stir with the rod. The salt will dissolve if the unknown contains oxygen to give a red-purple color, but it will not dissolve in hydrocarbons or halocarbons. Diphenyl ether does not give a positive test.

Alkyl ethers are generally soluble in concentrated sulfuric acid; alkyl aryl and diaryl ethers are not soluble.

Cleaning Up   Place the mixture in the hazardous waste container.

(f) Ketones

See Group I(g).

(g) Unsaturated Hydrocarbons

(Table 12, page 600)

Use care in working with the bromine solution

Bromine in Carbon Tetrachloride. Dissolve 1 drop (20 mg) of the unknown in 0.5 mL of carbon tetrachloride. Add a 2% solution of bromine in carbon tetrachloride dropwise with shaking. If more than 2 drops of bromine solution are required to give a permanent red color, unsaturation is indicated. The bromine solution must be fresh.

Cleaning Up   Place the mixture in the halogenated solvents container.

Potassium Permanganate Solution. Dissolve 1 drop (20 mg) of the unknown in reagent grade acetone and add a 1% aqueous solution of potassium permanganate dropwise with shaking. If more than one drop of reagent is required to give a purple color to the solution, unsaturation or an easily oxidized functional group is present. Run parallel tests on pure acetone and, as usual, a compound known to be an alkene.

Cleaning Up   The solution should be diluted with water and flushed down the drain.

Group X. Inert Compounds. Insoluble in Concentrated Sulfuric Acid

(a) Alkyl and Aryl Halides

(Table 10, page 599)

Alcoholic Silver Nitrate. Add one drop of the unknown (or saturated solution of 10 mg of unknown in ethanol) to 0.2 mL of a saturated solution of silver nitrate. A precipitate which forms within 2 min is a positive test for an alkyl bromide, or iodide, or a tertiary alkyl chloride, as well as allyl halides.

Do not waste silver nitrate.

If no precipitate forms within 2 min, heat the solution to boiling. A precipitate of silver chloride will form from primary and secondary alkyl chlorides. Aryl halides and vinyl halides will not react.

Cleaning Up   The quantity of material is extremely small. It can be diluted with water and flushed down the drain.

(b) Aromatic Hydrocarbons

(Table 13; Derivatives, page 601)

Aromatic hydrocarbons are best identified and characterized by uv and nmr spectroscopy, but the Friedel-Crafts reaction produces a characteristic color with certain aromatic hydrocarbons.

Keep moisture away from aluminum chloride

Friedel-Crafts Test. Heat a test tube containing about 50 mg of anhydrous aluminum chloride in a hot flame to sublime the salt up onto the sides of the tube. Add a solution of about 10 mg of the unknown dissolved in a drop of chloroform to the cool tube in such a way that it comes into contact with the sublimed aluminum chloride. Note the color that appears.

CAUTION: *Chloroform is carcinogenic. Carry out this test in a hood.*

Nonaromatic compounds fail to give a color with aluminum chloride, benzene and its derivatives give orange or red colors, naphthalenes a blue or purple color, biphenyls a purple color, phenanthrene a purple color, and anthracene a green color.

Cleaning Up   Place the test mixture in the halogenated organic solvents container.

(c) Saturated Hydrocarbons

Saturated hydrocarbons are best characterized by nmr and ir spectroscopy, but they can be distinguished from aromatic hydrocarbons by the Friedel-Crafts test (Group X(b)).

(d) Diaryl Ethers

Because they are so inert, diaryl ethers are difficult to detect and may be mistaken for aromatic hydrocarbons. They do not give a positive Ferrox Test, Group IX(e), for ethers, and do not dissolve in concentrated sulfuric acid. Their infrared spectra, however, are characterized by an intense C—O single-bond, stretching vibration in the region 1270–1230 cm^{-1}.

Derivatives

1. Acids

(Table 1)

p-Toluidides and Anilides. Reflux a mixture of the acid (100 mg) and thionyl chloride (0.5 mL) in a reaction tube for 0.5 h. Cool the reaction mixture and add 0.25 g of aniline or p-toluidine in 3 mL of toluene. Warm the mixture on the steam bath for 2 min, and then wash with 1-mL portions of water, 5% hydrochloric acid, 5% sodium hydroxide and water. The toluene is dried briefly over anhydrous sodium sulfate and evaporated in the hood; the derivative is recrystallized from water or ethanol-water.

CAUTION: p-Toluidine is a highly toxic irritant.

Cleaning Up   The combined aqueous layers can be diluted with water and flushed down the drain. The drying agent should be placed in the hazardous waste container.

Amides. Reflux a mixture of the acid (100 mg) and thionyl chloride (0.5 mL) for 0.5 h. Transfer the cool reaction mixture into 1.4 mL of ice-cold concentrated ammonia. Stir until reaction is complete, collect the product by filtration, and recrystallize it from water or water-ethanol.

Thionyl chloride is an irritant. Use it in a hood.

Cleaning Up   Neutralize the aqueous filtrate with 10% hydrochloric acid, dilute with water, and flush down the drain.

2. Alcohols

(Table 2)

3,5-Dinitrobenzoates. Gently boil 100 mg of 3,5-dinitrobenzoyl chloride and 25 mg of the alcohol for 5 min. Cool the mixture, pulverize any solid that forms, and add 2 mL of 2% sodium carbonate solution. Continue to grind and stir the solid with the sodium carbonate solution (to remove 3,5-dinitrobenzoic acid) for about a minute, filter, and wash the crystals with water. Dissolve the product in about 2.5–3 mL of hot ethanol, add water to the cloud point, and allow crystallization to proceed. Wash the 3,5-dinitrobenzoate with water-alcohol and dry.

Note to instructor: Check to ascertain that the 3,5-dinitrobenzoyl chloride has not hydrolyzed. The mp should be > 65°C. Reported mp is 68–69°C.

Cleaning Up   The aqueous filtrate should be diluted with water and flushed down the drain.

Phenylurethanes. Mix 100 mg of anhydrous alcohol (or phenol) and 100 mg of phenyl isocyanate (or α-naphthylurethane) and heat on the steam bath for 5 min. (If the unknown is a phenol add a drop of pyridine to the reaction mixture.) Cool, add about 1 mL of ligroin, heat to dissolve the product,

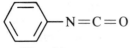

Phenyl isocyanate

filter hot to remove a small amount of diphenylurea which usually forms, and cool the filtrate in ice, with scratching, to induce crystallization.

Cleaning Up   The ligroin filtrate should be placed in the organic solvents container.

3. Aldehydes

(Table 3)
Semicarbazones. See page 310, Chapter 30. Use 0.5 mL of the stock solution and an estimated 1 millimole of the unknown aldehyde (or ketone).
2,4-Dinitrophenylhydrazones. See page 308, Chapter 30. Use 1 mL of the stock solution of 0.1 M 2,4-dinitrophenylhydrazine and an estimated 0.1 millimole of the unknown aldehyde (or ketone).

4. Primary and Secondary Amines

(Table 5)
Benzamides. Add about 0.25 g of benzoyl chloride in small portions with vigorous shaking and cooling to a suspension of 0.5 millimole of the unknown amine in 0.5 mL of 10% aqueous sodium hydroxide solution. After about 10 min of shaking the mixture is made pH 8 (pH paper) with dilute hydrochloric acid. The lumpy product is removed by filtration, washed thoroughly with water, and recrystallized from ethanol-water.

Cleaning Up   The filtrate should be diluted with water and flushed down the drain.

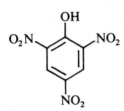

**Picric acid
(2,4,6-Trinitrophenol)**

Handle pure acid with care (explosive). It is sold as a moist solid. Do not allow to dry out.

Picrates. Add a solution of 30 mg of the unknown in 1 mL of ethanol (or 1 mL of a saturated solution of the unknown) to 1 mL of a saturated solution of picric acid (2,4,6-trinitrophenol, a strong acid) in ethanol, and heat the solution to boiling. Cool slowly, remove the picrate by filtration, and wash with a small amount of ethanol. Recrystallization is not usually necessary; in the case of hydrocarbon picrates the product is often too unstable to be recrystallized.

Cleaning Up   See page 375 for the treatment of solutions containing picric acid.

Acetamides. Reflux about 0.5 millimole of the unknown with 0.2 mL of acetic anhydride for 5 min, cool, and dilute the reaction mixture with 2.5 mL of water. Initiate crystallization by scratching, if necessary. Remove the crystals by filtration and wash thoroughly with dilute hydrochloric acid to remove unreacted amine. Recrystallize the derivative from alcohol-water. Amines of low basicity, e.g., *p*-nitroaniline, should be

Acetic anhydride is corrosive. Work with this in a hood.

refluxed for 30 to 60 min with 1 mL of pyridine as a solvent. The pyridine is removed by shaking the reaction mixture with 5 mL of 2% sulfuric acid solution; the product is isolated by filtration and recrystallized.

Cleaning Up The filtrate from the usual reaction should be neutralized with sodium carbonate. It can then be diluted with water and flushed down the drain. If pyridine is used as the solvent, the filtrate should be neutralized with sodium carbonate and extracted with ligroin. The ligroin/pyridine goes in the organic solvents container while the aqueous layer can be diluted with water and flushed down the drain.

$$R_3N + CH_3I$$
$$\downarrow$$
$$R_3\overset{+}{N}CH_3I^-$$

Methyl iodide is a cancer suspect agent.

5. Tertiary Amines

(Table 6)

Picrates. See under Primary and Secondary Amines.

Methiodides. Reflux 100 mg of the amine and 100 mg of methyl iodide for 5 min on the steam bath. Cool, scratch to induce crystallization, and recrystallize the product from ethyl alcohol or ethyl acetate.

Cleaning Up Since the filtrate may contain some methyl iodide, it should be placed in the halogenated solvents container.

6. Anhydrides and Acid Chlorides

(Table 7)

Acids. Reflux 40 mg of the acid chloride or anhydride with 1 mL of 5% sodium carbonate solution for 20 min or less. Extract unreacted starting material with 1 mL of ether, if necessary, and acidify the reaction mixture with dilute sulfuric acid to liberate the carboxylic acid.

Cleaning Up Ether goes in the organic solvents container and the aqueous layer should be diluted with water and flushed down the drain.

Amides. Since the acid chloride (or anhydride) is already present, simply mix the unknown (50 mg) and 0.7 mL of ice-cold concentrated ammonia until reaction is complete, collect the product by filtration, and recrystallize it from water or ethanol-water.

Cleaning Up Neutralize the filtrate with dilute hydrochloric acid and flush it down the drain.

Anilides. Reflux 40 mg of the acid halide or anhydride with 100 mg of aniline in 2 mL of toluene for 10 min. Wash the toluene solution with 5-mL portions each of water, 5% hydrochloric acid, 5% sodium hydroxide and water. The toluene solution is dried over anhydrous sodium sulfate and evaporated; the anilide is recrystallized from water or ethanol-water.

Cleaning Up The combined aqueous layers are diluted with water and flushed down the drain. The sodium sulfate should be placed in the aromatic amines hazardous waste container.

7. Aryl Halides

(Table 11)

Nitration. Add 0.4 mL of concentrated sulfuric acid to 100 mg of the aryl halide (or aromatic compound) and stir. Add 0.4 mL of concentrated nitric acid dropwise with stirring and shaking while cooling the reaction mixture in water. Then heat and shake the reaction mixture in a water bath at about 50°C for 15 min, pour into 2 mL of cold water, and collect the product by filtration. Recrystallize from methanol to constant melting point.

To nitrate unreactive compounds use fuming nitric acid in place of concentrated nitric acid.

Use great care when working with fuming nitric acid

Cleaning Up   Dilute the filtrate with water, neutralize with sodium carbonate, and flush the solution down the drain.

Sidechain Oxidation Products. Dissolve 0.2 g of sodium dichromate in 0.6 mL of water and add 0.4 mL of concentrated sulfuric acid. Add 50 mg of the unknown and boil for 30 min. Cool, add 0.4 to 0.6 mL of water, and then remove the carboxylic acid by filtration. Wash the crystals with water and recrystallize from methanol-water.

Cleaning Up   Place the filtrate from the reaction, after neutralization with sodium carbonate, in the hazardous waste container.

8. Hydrocarbons: Aromatic

(Table 13)

Nitration. See preceding, under Aryl Halides.
Picrates. See preceding, under Primary and Secondary Amines (page 377).

9. Ketones

(Table 14)

Semicarbazones and 2,4-dinitrophenylhydrazones. See preceding directions under Aldehydes.

10. Nitro Compounds

(Table 16)

Reduction to Amines. Place 100 mg of the unknown in a reaction tube, add 0.2 g of tin, and then—in portions—2 mL of 10% hydrochloric acid. Reflux for 30 min, add 1 mL of water, then add slowly, with good cooling, sufficient 40% sodium hydroxide solution to dissolve the tin hydroxide. Extract the reaction mixture with three 1-mL portions of ether, dry the ether extract over anhydrous sodium sulfate, wash the drying agent with ether, and evaporate the ether to leave the amine. Determine the boiling point or melting point of the amine and then convert it into a benzamide or acetamide as described under **4. Primary and Secondary Amines p. 587.**

Cleaning Up   Neutralize the aqueous layer with 10% hydrochloric acid, remove the tin hydroxide by filtration, and discard it in the nonhazardous solid waste container. The filtrate should be diluted with water and flushed down the drain. Sodium sulfate, after the ether evaporates from it, can be placed in the nonhazardous waste container.

11. Phenols

(Table 17)

α-**Naphthylurethane.** Follow the procedure for preparation of a phenylurethane under Alcohols, page 586.

Use great care when working with bromine. Should any touch the skin wash it off with copious quantities of water. Work in a hood and wear disposable gloves.

Bromo Derivative. In a reaction tube dissolve 160 mg of potassium bromide in 1 ml of water. *Carefully* add 100 mg of bromine. In a separate flask dissolve 20 mg of the phenol in 0.2 mL of methanol and add 0.2 mL of water. Add about 0.3 mL of the bromine solution with swirling (hood); continue the addition of bromine until the yellow color of unreacted bromine persists. Add 0.6 to 0.8 mL of water to the reaction mixture and shake vigorously. Remove the product by filtration and wash well with water. Recrystallize from methanol-water.

Cleaning Up   Any unreacted bromine should be destroyed by adding sodium bisulfite solution dropwise until the color dissipates. The solution is then diluted with water and flushed down the drain.

TABLE 1 Acids

Bp	Mp	Compound	Derivatives		
			p-Toluidide[a]	Anilide[b]	Amide[c]
			Mp	Mp	Mp
101		Formic acid	53	47	43
118		Acetic acid	126	106	79
139		Acrylic acid	141	104	85
141		Propionic acid	124	103	81
162		*n*-Butyric acid	72	95	115
163		Methacrylic acid		87	102
165		Pyruvic acid	109	104	124
185		Valeric acid	70	63	106
186		2-Methylvaleric acid	80	95	79
194		Dichloroacetic acid	153	118	98
202–203		Hexanoic acid	75	95	101
237		Octanoic acid	70	57	107
254		Nonanoic acid	84	57	99
	31–32	Decanoic acid	78	70	108
	43–45	Lauric acid	87	78	100
	47–49	Bromoacetic acid		131	91
	47–49	Hydrocinnamic acid	135	92	105
	54–55	Myristic acid	93	84	103
	54–58	Trichloroacetic acid	113	97	141
	61–62	Chloroacetic acid	162	137	121
	61–62.5	Palmitic acid	98	90	106
	67–69	Stearic acid	102	95	109
	68–69	3,3-Dimethylacrylic acid		126	107
	71–73	Crotonic acid	132	118	158
	77–78.5	Phenylacetic acid	136	118	156
	101–102	Oxalic acid dihydrate		257	400 (dec)
	98–102	Azelaic acid (nonanedioic)	164 (di)	107 (mono) 186 (di)	93 (mono) 175 (di)
	103–105	*o*-Toluic acid	144	125	142
	108–110	*m*-Toluic acid	118	126	94
	119–121	DL-Mandelic acid	172	151	133
	122–123	Benzoic acid	158	163	130
	127–128	2-Benzoylbenzoic acid		195	165
	129–130	2-Furoic acid	107	123	143
	131–133	DL-Malic acid	178 (mono) 207 (di)	155 (mono) 198 (di)	163 (di)
	131–134	Sebacic acid	201	122 (mono) 200 (di)	170 (mono) 210 (di)

a. For preparation, see page 586.
b. For preparation, see page 586.
c. For preparation, see page 586.

TABLE 1   continued

Bp	Mp	Compound	p-Toluidide[a] Mp	Anilide[b] Mp	Amide[c] Mp
	134–135	*E*-Cinnamic acid	168	153	147
	134–136	Maleic acid	142 (di)	198 (mono) 187 (di)	260 (di)
	135–137	Malonic acid	86 (mono) 253 (di)	132 (mono) 230 (di)	
	138–140	2-Chlorobenzoic acid	131	118	139
	140–142	3-Nitrobenzoic acid	162	155	143
	144–148	Anthranilic acid	151	131	109
	147–149	Diphenylacetic acid	172	180	167
	152–153	Adipic acid	239	151 (mono) 241 (di)	125 (mono) 220 (di)
	153–154	Citric acid	189 (tri)	199 (tri)	210 (tri)
	157–159	4-Chlorophenoxyacetic acid		125	133
	158–160	Salicylic acid	156	136	142
	163–164	Trimethylacetic acid		127	178
	164–166	5-Bromosalicylic acid		222	232
	166–167	Itaconic acid		190	191 (di)
	171–174	D-Tartaric acid		180 (mono) 264 (di)	171 (mono) 196 (di)
	179–182	3,4-Dimethoxybenzoic acid		154	164
	180–182	4-Toluic acid	160	145	160
	182–185	4-Anisic acid	186	169	167
	187–190	Succinic acid	180 (mono) 255 (di)	143 (mono) 230 (di)	157 (mono) 260 (di)
	201–203	3-Hydroxybenzoic acid	163	157	170
	203–206	3,5-Dinitrobenzoic acid		234	183
	210–211	Phthalic acid	150 (mono) 201 (di)	169 (mono) 253 (di)	149 (mono) 220 (di)
	214–215	4-Hydroxybenzoic acid	204	197	162
	225–227	2,4-Dihydroxybenzoic acid		126	228
	236–239	Nicotinic acid	150	132	128
	239–241	4-Nitrobenzoic acid	204	211	201
	299–300	Fumaric acid		233 (mono) 314 (di)	270 (mono) 266 (di)
	>300	Terephthalic acid		334	

a. For preparation, see page 586.
b. For preparation, see page 586.
c. For preparation, see page 586.

TABLE 2   Alcohols

			Derivatives	
			3,5-Dinitrobenzoate[a]	Phenylurethane[b]
Bp	Mp	Compound	Mp	Mp
65		Methanol	108	47
78		Ethanol	93	52
82		2-Propanol	123	88
83		*t*-Butyl alcohol	142	136
96–98		Allyl alcohol	49	70
97		1-Propanol	74	57
98		2-Butanol	76	65
102		2-Methyl-2-butanol	116	42
104		2-Methyl-3-butyn-2-ol	112	
108		2-Methyl-1-propanol	87	86
114–115		Propargyl alcohol		63
114–115		3-Pentanol	101	48
118		1-Butanol	64	61
118–119		2-Pentanol	62	oil
123		3-Methyl-3-pentanol	96(62)	43
129		2-Chloroethanol	95	51
130		2-Methyl-1-butanol	70	31
132		4-Methyl-2-pentanol	65	143
136–138		1-Pentanol	46	46
139–140		Cyclopentanol	115	132
140		2,4-Dimethyl-3-pentanol	75	95
146		2-Ethyl-1-butanol	51	
151		2,2,2-Trichloroethanol	142	87
157		1-Hexanol	58	42
160–161		Cyclohexanol	113	82
170		Furfuryl alcohol	80	45
176		1-Heptanol	47	60(68)
178		2-Octanol	32	oil
178		Tetrahydrofurfuryl alcohol	83	61
183–184		2,3-Butanediol		201 (di)
183–186		2-Ethyl-1-hexanol		34
187		1,2-Propanediol		153 (di)
194–197		Linaloöl		66
195		1-Octanol	61	74
196–198		Ethylene glycol	169	157 (di)
204		1,3-Butanediol		122
203–205		Benzyl alcohol	113	77
204		1-Phenylethanol	95	92
219–221		2-Phenylethanol	108	78
230		1,4-Butanediol		183 (di)
231		1-Decanol	57	59

a. For preparation, see page 586.
b. For preparation, see page 586.

TABLE 2 continued

Bp	Mp	Compound	Derivatives 3,5-Dinitrobenzoate[a] Mp	Phenylurethane[b] Mp
259		4-Methoxybenzyl alcohol		92
	33–35	Cinnamyl alcohol	121	90
	38–40	1-Tetradecanol	67	74
	48–50	1-Hexadecanol	66	73
	58–60	1-Octadecanol	77(66)	79
	66–67	Benzhydrol	141	139
	147	Cholesterol	195	168

a. For preparation, see page 586.
b. For preparation, see page 586.

TABLE 3 Aldehydes

Bp	Mp	Compound	Derivatives Semicarbazone[a] Mp	2,4-Dinitrophenylhydrazone[b] Mp
21		Acetaldehyde	162	168
46–50		Propionaldehyde	89(154)	148
63		Isobutyraldehyde	125(119)	187(183)
75		Butyraldehyde	95(106)	123
90–92		3-Methylbutanal	107	123
98		Chloral	90	131
104		Crotonaldehyde	199	190
117		2-Ethylbutanal	99	95(130)
153		Heptaldehyde	109	108
162		2-Furaldehyde	202	212(230)
163		2-Ethylhexanal	254	114(120)
179		Benzaldehyde	222	237
195		Phenylacetaldehyde	153	121(110)
197		Salicylaldehyde	231	248
204–205		4-Tolualdehyde	234(215)	232
209–215		2-Chlorobenzaldehyde	146(229)	213
247		2-Ethoxybenzaldehyde	219	
248		4-Anisaldehyde	210	253
250–252		*E*-Cinnamaldehyde	215	255
	33–34	1-Naphthaldehyde	221	254
	37–39	2-Anisaldehyde	215	254
	42–45	3,4-Dimethoxybenzaldehyde	177	261
	44–47	4-Chlorobenzaldehyde	230	254
	57–59	3-Nitrobenzaldehyde	246	293
	81–83	Vanillin	230	271

a. For preparation, see page 587.
b. For preparation, see page 587.

TABLE 4   Amides

Bp	Mp	Name of Compound	Mp	Name of Compound
153		N,N-Dimethylformamide	127–129	Isobutyramide
164–166		N,N-Dimethylacetamide	128–129	Benzamide
210		Formamide	130–133	Nicotinamide
243–244		N-Methylformanilide	177–179	4-Chloroacetanilide
	26–28	N-Methylacetamide		
	79–81	Acetamide		
	109–111	Methacrylamide		
	113–115	Acetanilide		
	116–118	2-Chloroacetamide		

TABLE 5   Primary and Secondary Amines

			Derivatives		
			Benzamide[a]	Picrate[b]	Acetamide[c]
Bp	Mp	Compound	Mp	Mp	Mp
33–34		Isopropylamine	71	165	
46		t-Butylamine	134	198	
48		n-Propylamine	84	135	
53		Allylamine		140	
55		Diethylamine	42	155	
63		s-Butylamine	76	139	
64–71		Isobutylamine	57	150	
78		n-Butylamine	42	151	
84		Di-isopropylamine		140	
87–88		Pyrrolidine	oil	112	
106		Piperidine	48	152	
111		Di-n-propylamine	oil	75	
118		Ethylenediamine	244 (di)	233	172 (di)
129		Morpholine	75	146	
137–139		Di-isobutylamine		121	86
145–146		Furfurylamine		150	
149		N-Methylcyclohexylamine	85	170	
159		Di-n-butylamine	oil	59	
182–185		Benzylamine	105	199	60
184		Aniline	163	198	114
196		N-Methylaniline	63	145	102
199–200		2-Toluidine	144	213	110
203–204		3-Toluidine	125	200	65
205		N-Ethylaniline	60	138(132)	54

a. For preparation, see page 588.
b. For preparation, see page 588.
c. For preparation, see page 588.

TABLE 5　continued

Bp	Mp	Compound	Derivatives		
			Benzamide[a]	Picrate[b]	Acetamide[c]
			Mp	Mp	Mp
208–210		2-Chloroaniline	99	134	87
210		2-Ethylaniline	147	194	111
216		2,6-Dimethylaniline	168	180	177
218		2,4-Dimethylaniline	192	209	133
218		2,5-Dimethylaniline	140	171	139
221		N-Ethyl-*m*-toluidine	72		
225		2-Anisidine	60(84)	200	85
230		3-Chloroaniline	120	177	72(78)
231–233		2-Phenetidine	104		79
241		4-Chloro-2-methylaniline	142		140
242		3-Chloro-4-methylaniline	122		105
250		4-Phenetidine	173	69	137
256		Dicyclohexylamine	153(57)	173	103
	35–38	N-Phenylbenzylamine	107	48	58
	41–44	4-Toluidine	158	182	147
	49–51	2,5-Dichloroaniline	120	86	132
	52–54	Diphenylamine	180	182	101
	57–60	4-Anisidine	154	170	130
	57–60	2-Aminopyridine	165 (di)	216(223)	
	60–62	N-Phenyl-1-naphthylamine	152		115
	62–65	2,4,5-Trimethylaniline	167		162
	64–66	1,3-Phenylenediamine	125 (mono) 240 (di)	184	87 (mono) 191 (di)
	66	4-Bromoaniline	204	180	168
	68–71	4-Chloroaniline	192	178	179(172)
	71–73	2-Nitroaniline	110(98)	73	92
	97–99	2,4-Diaminotoluene	224 (di)		224 (di)
	100–102	1,2-Phenylenediamine	301	208	185
	104–107	2-Methyl-5-nitroaniline	186		151
	107–109	2-Chloro-4-nitroaniline	161		139
	112–114	3-Nitroaniline	157(150)	143	155(76)
	115–116	4-Methyl-2-nitroaniline	148		99
	117–119	4-Chloro-2-nitroaniline			104
	120–122	2,4,6-Tribromoaniline	198(204)		232
	131–133	2-Methyl-4-nitroaniline			202
	138–140	2-Methoxy-4-nitroaniline	149		
	138–142	1,4-Phenylenediamine	128 (mono) 300 (di)		162 (mono) 304 (di)
	148–149	4-Nitroaniline	199	100	215
	162–164	4-Aminoacetanilide			304
	176–178	2,4-Dinitroaniline	202(220)		120

a. For preparation, see page 587.
b. For preparation, see page 587.
c. For preparation, see page 587.

TABLE 6   Tertiary Amines

		Derivatives	
		Picrate[a]	Methiodide[b]
Bp	Compound	Mp	Mp
85–91	Triethylamine	173	280
115	Pyridine	167	117
128–129	2-Picoline	169	230
143–145	2,6-Lutidine	168(161)	233
144	3-Picoline	150	92(36)
145	4-Picoline	167	149
155–158	Tri-*n*-propylamine	116	207
159	2,4-Lutidine	180	113
183–184	N,N-Dimethylbenzylamine	93	179
216	Tri-*n*-butylamine	105	186
217	N,N-Diethylaniline	142	102
237	Quinoline	203	133(72)

a. For preparation, see page 588.
b. For preparation, see page 588.

TABLE 7   Anhydrides and Acid Chlorides

			Derivatives		
			Acid[a]	Amide[b]	Anilide[c]
Bp	Mp	Compound	Bp Mp	Mp	Mp
52		Acetyl chloride	118	82	114
77–79		Propionyl chloride	141	81	106
102		Butyryl chloride	162	115	96
138–140		Acetic anhydride	118	82	114
167		Propionic anhydride	141	81	106
198–199		Butyric anhydride	162	115	96
198		Benzoyl chloride	122	130	163
225		3-Chlorobenzoyl chloride	158	134	122
238		2-Chlorobenzoyl chloride	142	142	118
	32–34	*cis*-1,2-Cyclohexanedicarboxylic anhydride	192		
	35–37	Cinnamoyl chloride	133	147	151
	39–40	Benzoic anhydride	122	130	163
	54–56	Maleic anhydride	130	181 (mono) 266 (di)	173 (mono) 187 (di)
	72–74	4-Nitrobenzoyl chloride	241	201	211

a. For preparation, see page 588.
b. For preparation, see page 588.
c. For preparation, see page 588.

TABLE 7   continued

			Derivatives				
			Acid[a]		Amide[b]	Anilide[c]
Bp	Mp	Compound	Bp	Mp	Mp	Mp
---	---	---	---	---	---	---
	119–120	Succinic anhydride		186	157 (mono)	148 (mono)
					260 (di)	230 (di)
	131–133	Phthalic anhydride		206	149 (mono)	170 (mono)
					220 (di)	253 (di)
	254–258	Tetrachlorophthalic anhydride		250		
	267–269	1,8-Naphthalic anhydride		274		250–282 (di)

a. For preparation, see page 588.
b. For preparation, see page 588.
c. For preparation, see page 588.

TABLE 8   Esters

Bp	Mp	Compound	Bp	Mp	Compound
34		Methyl formate	169–170		Methyl acetoacetate
52–54		Ethyl formate	180–181		Dimethyl malonate
72–73		Vinyl acetate	181		Ethyl acetoacetate
77		Ethyl acetate	185		Diethyl oxalate
79		Methyl propionate	198–199		Methyl benzoate
80		Methyl acrylate	206–208		Ethyl caprylate
85		Isopropyl acetate	208–210		Ethyl cyanoacetate
93		Ethyl chloroformate	212		Ethyl benzoate
94		Isopropenyl acetate	217		Diethyl succinate
98		Isobutyl formate	218		Methyl phenylacetate
98		t-Butyl acetate	218–219		Diethyl fumarate
99		Ethyl propionate	222		Methyl salicylate
99		Ethyl acrylate	225		Diethyl maleate
100		Methyl methacrylate	229		Ethyl phenylacetate
101		Methyl trimethylacetate	234		Ethyl salicylate
102		n-Propyl acetate	268		Dimethyl suberate
106–113		s-Butyl acetate	271		Ethyl cinnamate
120		Ethyl butyrate	282		Dimethyl phthalate
127		n-Butyl acetate	298–299		Diethyl phthalate
128		Methyl valerate	298–299		Phenyl benzoate
130		Methyl chloroacetate	340		Dibutyl phthalate
131–133		Ethyl isovalerate		56–58	Ethyl p-nitrobenzoate
142		n-Amyl acetate		88–90	Ethyl p-aminobenzoate
142		Isoamyl acetate		94–96	Methyl p-nitrobenzoate
143		Ethyl chloroacetate		95–98	n-Propyl p-hydroxybenzoate
154		Ethyl lactate		116–118	Ethyl p-hydroxybenzoate
168		Ethyl caproate (ethyl hexanoate)		126–128	Methyl p-hydroxybenzoate

TABLE 9 Ethers

Bp	Mp	Compound	Bp	Mp	Compound
32		Furan	215		4-Bromoanisole
33		Ethyl vinyl ether	234–237		Anethole
65–67		Tetrahydrofuran	259		Diphenyl ether
94		n-Butyl vinyl ether	273		2-Nitroanisole
154		Anisole	298		Dibenzyl ether
174		4-Methylanisole		50–52	4-Nitroanisole
175–176		3-Methylanisole		56–60	1,4-Dimethoxybenzene
198–203		4-Chloroanisole		73–75	2-Methoxynaphthalene
206–207		1,2-Dimethoxybenzene			

TABLE 10 Halides

Bp	Compound	Bp	Compound
34–36	2-Chloropropane	100–105	1-Bromobutane
40–41	Dichloromethane	105	Bromotrichloromethane
44–46	Allyl chloride	110–115	1,1,2-Trichloroethane
57	1,1-Dichloroethane	120–121	1-Bromo-3-methylbutane
59	2-Bromopropane	121	Tetrachloroethylene
68	Bromochloromethane	123	3,4-Dichloro-1-butene
68–70	2-Chlorobutane	125	1,3-Dichloro-2-butene
69–73	Iodoethane	131–132	1,2-Dibromoethane
70–71	Allyl bromide	140–142	1,2-Dibromopropane
71	1-Bromopropane	142–145	1-Bromo-3-chloropropane
72–74	2-Bromo-2-methylpropane	146–150	Bromoform
74–76	1,1,1-Trichloroethane	147	1,1,2,2-Tetrachloroethane
81–85	1,2-Dichloroethane	156	1,2,3-Trichloropropane
87	Trichloroethylene	161–163	1,4-Dichlorobutane
88–90	2-Iodopropane	167	1,3-Dibromopropane
90–92	1-Bromo-2-methylpropane	177–181	Benzyl chloride
91	2-Bromobutane	197	(2-Chloroethyl)benzene
94	2,3-Dichloro-1-propene	219–223	Benzotrichloride
95–96	1,2-Dichloropropane	238	1-Bromodecane
96–98	Dibromomethane		

TABLE 11   Aryl Halides

			Derivatives			
			Nitration Product[a]		Oxidation Product[b]	
Bp	Mp	Compound	Position	Mp	Name	Mp
132		Chlorobenzene	2, 4	52		
156		Bromobenzene	2, 4	70		
157–159		2-Chlorotoluene	3, 5	63	2-Chlorobenzoic acid	141
162		4-Chlorotoluene	2	38	4-Chlorobenzoic acid	240
172–173		1,3-Dichlorobenzene	4, 6	103		
178		1,2-Dichlorobenzene	4, 5	110		
196–203		2,4-Dichlorotoluene	3, 5	104	2,4-Dichlorobenzoic acid	164
201		3,4-Dichlorotoluene	6	63	3,4-Dichlorobenzoic acid	206
214		1,2,4-Trichlorobenzene	5	56		
279–281		1-Bromonaphthalene	4	85		
	51–53	1,2,3-Trichlorobenzene	4	56		
	54–56	1,4-Dichlorobenzene	2	54		
	66–68	1,4-Bromochlorobenzene	2	72		
	87–89	1,4-Dibromobenzene	2, 5	84		
	138–140	1,2,4,5-Tetrachlorobenzene	3	99		
			3, 6	227		

a. For preparation, see page 589.
b. For preparation, see page 589.

TABLE 12   Hydrocarbons: Alkenes

Bp	Compound	Bp	Compound
34	Isoprene	149–150	1,5-Cyclooctadiene
83	Cyclohexene	152	DL-α-Pinene
116	5-Methyl-2-norbornene	160	Bicyclo[4.3.0]nona-3,7-diene
122–123	1-Octene	165–167	(−)-β-Pinene
126–127	4-Vinyl-1-cyclohexene	165–169	α-Methylstyrene
132–134	2,5-Dimethyl-2,4-hexadiene	181	1-Decene
141	5-Vinyl-2-norbornene	181	Indene
143	1,3-Cyclooctadiene	251	1-Tetradecene
145	4-Butylstyrene	274	1-Hexadecene
145–146	Cycloctene	349	1-Octadecene
145–146	Styrene		

TABLE 13   Hydrocarbons: Aromatic

			Melting Point of Derivatives		
			Nitro[a]		Picrate[b]
Bp	Mp	Compound	Position	Mp	Mp
80		Benzene	1, 3	89	84
111		Toluene	2, 4	70	88
136		Ethylbenzene	2, 4, 6	37	96
138		p-Xylene	2, 3, 5	139	90
138–139		m-Xylene	2, 4, 6	183	91
143–145		o-Xylene	4, 5	118	88
145		4-t-Butylstyrene	2, 4	62	
145–146		Styrene			
152–154		Cumene	2, 4, 6	109	
163–166		Mesitylene	2, 4	86	97
			2, 4, 6	235	
165–169		α-Methylstyrene			
168		1,2,4-Trimethylbenzene	3, 5, 6	185	97
176–178		p-Cymene	2, 6	54	
189–192		4-t-Butyltoluene			
197–199		1,2,3,5-Tetramethylbenzene	4, 6	181(157)	
203		p-Diisopropylbenzene			
204–205		1,2,3,4-Tetramethylbenzene	5, 6	176	92
207		1,2,3,4-Tetrahydronaphthalene	5, 7	95	
240–243		1-Methylnaphthalene	4	71	142
	34–36	2-Methylnaphthalene	1	81	116
	50–51	Pentamethylbenzene	6	154	131
	69–72	Biphenyl	4, 4′	237(229)	
	80–82	1,2,4,5-Tetramethylbenzene	3, 6	205	
	80–82	Naphthalene	1	61(57)	149
	90–95	Acenaphthene	5	101	161
	99–101	Phenanthrene			144(133)
	112–115	Fluorene	2	156	87(77)
			2, 7	199	
	214–217	Anthracene			138

a. For preparation, see page 589.
b. For preparation, see page 589.

TABLE 14   Ketones

Bp	Mp	Compound	Semicarbazone[a] Mp	2,4-Dinitrophenylhydrazone[b] Mp
56		Acetone	187	126
80		2-Butanone	136, 186	117
88		2,3-Butanedione	278	315
100–101		2-Pentanone	112	143
102		3-Pentanone	138	156
106		Pinacolone	157	125
114–116		4-Methyl-2-pentanone	132	95
124		2,4-Dimethyl-3-pentanone	160	88, 95
128–129		5-Hexen-2-one	102	108
129		4-Methyl-3-penten-2-one	164	205
130–131		Cyclopentanone	210	146
133–135		2,3-Pentanedione	122 (mono) 209 (di)	209
145		4-Heptanone	132	75
145		5-Methyl-2-hexanone	147	95
145–147		2-Heptanone	123	89
146–149		3-Heptanone	101	81
156		Cyclohexanone	166	162
162–163		2-Methylcyclohexanone	195	137
169		2,6-Dimethyl-4-heptanone	122	66, 92
169–170		3-Methylcyclohexanone	180	155
173		2-Octanone	122	58
191		Acetonylacetone	185 (mono) 224 (di)	257 (di)
202		Acetophenone	198	238
216		Phenylacetone	198	156
217		Isobutyrophenone	181	163
218		Propiophenone	182	191
226		4-Methylacetophenone	205	258
231–232		2-Undecanone	122	63
232		*n*-Butyrophenone	188	191
232		4-Chloroacetophenone	204	236
235		Benzylacetone	142	127
	35–37	4-Chloropropiophenone	176	223
	35–39	4-Phenyl-3-buten-2-one	187	227
	36–38	4-Methoxyacetophenone	198	228
	48–49	Benzophenone	167	238
	53–55	2-Acetonaphthone	235	262
	60	Desoxybenzoin	148	204
	76–78	3-Nitroacetophenone	257	228
	78–80	4-Nitroacetophenone		257
	82–85	9-Fluorenone	234	283
	134–136	Benzoin	206	245
	147–148	4-Hydroxypropiophenone		240

a. For preparation, see page 589.   b. For preparation, see page 589.

TABLE 15 Nitriles

Bp	Mp	Compound	Bp	Mp	Compound
77		Acrylonitrile	212		3-Tolunitrile
83–84		Trichloroacetonitrile	217		4-Tolunitrile
97		Propionitrile	233–234		Benzyl cyanide
107–108		Isobutyronitrile	295		Adiponitrile
115–117		*n*-Butyronitrile		30.5	4-Chlorobenzyl cyanide
174–176		3-Chloropropionitrile		32–34	Malononitrile
191		Benzonitrile		38–40	Stearonitrile
205		2-Tolunitrile		46–48	Succinonitrile
				71–73	Diphenylacetonitrile

TABLE 16 Nitro Compounds

Bp	Mp	Compound	Bp	Mp	Acetamide[a] Mp	Benzamide[b] Mp
					Amine Obtained by Reduction of Nitro Groups	
210–211		Nitrobenzene	184		114	160
225		2-Nitrotoluene	200		110	146
225		2-Nitro-*m*-xylene	215		177	168
230–231		3-Nitrotoluene	203		65	125
245		3-Nitro-*o*-xylene	221		135	189
245–246		4-Ethylnitrobenzene	216		94	151
	34–36	2-Chloro-6-nitrotoluene	245		157(136)	173
	36–38	4-Chloro-2-nitrotoluene		21	139(131)	
	40–42	3,4-Dichloronitrobenzene		72	121	
	43–50	1-Chloro-2,4-dinitrobenzene		91	242 (di)	178 (di)
	52–54	4-Nitrotoluene		45	147	158
	55–56	1-Nitronaphthalene		50	159	160
	83–84	1-Chloro-4-nitrobenzene		72	179	192
	88–90	*m*-Dinitrobenzene		63	87 (mono)	125 (mono)
					191 (di)	240 (di)

a. For preparation, see page 589.
b. For preparation, see page 589.

TABLE 17   Phenols

Bp	Mp	Compound	Derivatives	
			α-Naphthylurethane[a] Mp	Bromo[b] Mp
175–176		2-Chlorophenol	120	48 (mono) 76 (di)
181	42	Phenol	133	95 (tri)
202	32–34	p-Cresol	146	49 (di) 108 (tetra)
203		m-Cresol	128	84 (tri)
228–229		3,4-Dimethylphenol	141	171 (tri)
	32–33	o-Cresol	142	56 (di)
	42–43	2,4-Dichlorophenol		68
	42–45	4-Ethylphenol	128	
	43–45	4-Chlorophenol	166	90 (di)
	44–46	2,6-Dimethylphenol	176	79
	44–46	2-Nitrophenol	113	117 (di)
	49–51	Thymol	160	55
	62–64	3,5-Dimethylphenol		166 (tri)
	64–68	4-Bromophenol	169	95 (tri)
	74	2,5-Dimethylphenol	173	178 (tri)
	92–95	2,3,5-Trimethylphenol	174	
	95–96	1-Naphthol	152	105 (di)
	98–101	4-t-Butylphenol	110	50 (mono) 67 (di)
	104–105	Catechol	175	192 (tetra)
	109–110	Resorcinol	275	112 (tri)
	112–114	4-Nitrophenol	150	142 (di)
	121–124	2-Naphthol	157	84
	133–134	Pyrogallol	173	158 (di)

a. For preparation, see page 590.
b. For preparation, see page 590.

Isolation of Lycopene and β-Carotene

Prelab Exercise: The primary reason for low yields in this experiment is oxidation of the product during isolation. Once crystalline it is reasonably stable. Speculate on the primary products formed by photochemical air oxidation of lycopene.

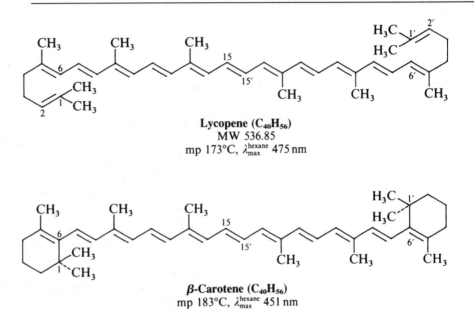

Lycopene (C₄₀H₅₆)
MW 536.85
mp 173°C, λ_{max}^{hexane} 475 nm

β-Carotene (C₄₀H₅₆)
mp 183°C, λ_{max}^{hexane} 451 nm

Lycopene, the red pigment of the tomato, is a C_{40}-carotenoid made up of eight isoprene units. β-Carotene, the yellow pigment of the carrot, is an isomer of lycopene in which the double bonds at C_1—C_2 and C_1'—C_2' are replaced by bonds extending from C_1 to C_6 and from C_1' to C_6' to form rings. The chromophore in each case is a system of eleven all-*trans* conjugated double bonds; the closing of the two rings renders β-carotene less highly pigmented than lycopene.

The colored hydrocarbons are materials for an interesting experiment in thin-layer chromatography (Chapter 9), but commercial samples are extremely expensive and subject to deterioration on storage. The isolation procedure in this chapter affords amounts of pigments that are unweighable, except on a microbalance, but more than adequate for the thin-layer chromatography experiment. It is suggested that half the students isolate lycopene and the other half β-carotene.

Lycopene and β-carotene from tomato paste and strained carrots

As interesting variations, try extraction of lycopene from commercial catsup or red bell peppers

Fresh tomato fruit contains about 96% water, and R. Willstätter and H. R. Escher isolated from this source 20 mg of lycopene per kg of fruit. They then found a more convenient source in commercial tomato paste, from which seeds and skin have been eliminated and the water content reduced by evaporation in vacuum to a content of 26% solids, and isolated 150 mg of lycopene per kg of paste. The expected yield in the present experiment is 0.75 mg.

A jar of strained carrots sold as baby food serves as a convenient source of β-carotene. The German investigators isolated 1 g of β-carotene per kg of "dried" shredded carrots of unstated water content.

The following procedure calls for dehydration of tomato or carrot paste with ethanol and extraction of the residue with dichloromethane, an efficient solvent for lipids.

Experiment

Chromatography of Lycopene and Carotene

A 5-g sample of tomato or carrot paste is transferred to the bottom of a 25 × 150-mm test tube with the aid of a paste dispenser,[1] made by connecting an 18-cm section of 11-mm glass tubing by means of a No. 00 rubber stopper and a short section of 5 mm glass tubing to a 50-cm length of 3-mm rubber tubing. The open end of the glass tube is calibrated to deliver 5 g of paste and marked with a file scratch.

These carotenoids are very sensitive to photochemical air oxidation. Protect solutions and solid from undue exposure to light.

Thrust the dispenser into a can of fruit paste and "suck up" paste to the scratch mark on the glass tube. Then wipe the glass tube with a facial tissue, insert it into a 25 × 150-mm test tube, and blow out the paste onto the bottom of the receiver. Wash the dispenser tube clean with water and leave it upright to drain for the next user.

Dehydration of fruit paste

CH$_2$Cl$_2$, bp 41°C, den 1.336

Add 10 mL of 95% ethanol and heat almost to boiling for 5 min on a steam bath with stirring. Then filter the hot mixture on a small Hirsch funnel. Scrape out the test tube with a spatula, let the tube drain thoroughly, and squeeze liquid out of the solid residue in the funnel with a spatula. Pour the yellow filtrate into a 125-mL Erlenmeyer flask. Then return the solid residue, with or without adhering filter paper, to the original test tube, add 10 mL of dichloromethane to effect an extraction, insert the condenser, and reflux the mixture for 3–4 min. Filter the yellow fitrate and add the filtrate to the storage flask. Repeat the extraction with two or three further 10-mL portions of dichloromethane, pour the combined extracts into a separatory funnel, add water and sodium chloride solution (to aid in layer separation),

---

1. To be supplied to each section of the class, along with a can opener.

and shake. Dry the colored lower layer over anhydrous sodium sulfate, filter the solution into a dry flask, and evaporate to dryness (aspirator tube or rotary evaporator).

The crude carotenoid is to be chromatographed on a 12-cm column of acid-washed alumina (Merck), prepared with petroleum ether (37–53°C) as solvent. Run out excess solvent, or remove it from the top of the chromatography column with a suction tube, dissolve the crude carotenoid in a few milliliters of toluene, and then transfer the solution onto the chromatographic column with a Pasteur pipette. Elute the column with petroleum ether, discard the initial colorless eluate, and collect all yellow or orange eluates together in a 50-mL Erlenmeyer flask. Place a drop of solution on a microscope slide and evaporate the rest to dryness (rotary evaporator or aspirator tube). Examination of the material spotted on the slide may reveal crystallinity. Then put a drop of concentrated sulfuric acid beside the spot and mix with a stirring rod. Compare the color of your test with that of a test on the other carotenoid.

Color test

Finally, dissolve the sample obtained by evaporating the petroleum ether in the least amount of dichloromethane, hold the Erlenmeyer in a slanting position for drainage into a small area, and transfer the solution with a Pasteur pipette to a 10 × 75-mm test tube or a 3-mL centrifuge tube. Add a boiling stone, hold the tube over a hot plate in a slanting position, and evaporate to dryness. Evacuate the tube at the aspirator, shake out the boiling stone, cork and label the test tube, and keep it in a dark place.

Look up the current prices of commercial lycopene[2] and β-carotene.[3] Note that β-carotene is in demand as a source of vitamin A and is manufactured by an efficient synthesis. No uses for lycopene have been found.

Cleaning Up   The ethanol used for dehydration of the plant material can be flushed down the drain along with the saturated sodium chloride solution. Recovered and unused dichloromethane should be placed in the halogenated organic waste container, the solvents used for TLC should be placed in the organic solvents container, and the drying agents, once free of solvents, can be placed in the nonhazardous solid waste container along with the used plant material and TLC plates.

2. Pfaltz and Bauer, Flushing, NY 11368.

3. Aldrich Chemical Co., Milwaukee, WI 53233.

72

Oleic Acid from Olive Oil

Prelab Exercise: Write a flow sheet indicating all operations needed to isolate pure oleic acid from olive oil.

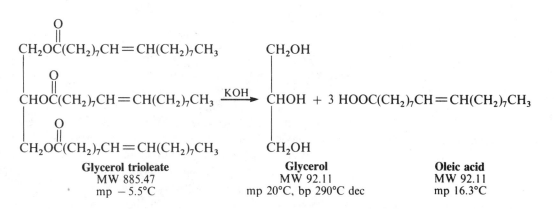

Glycerol trioleate	**Glycerol**	**Oleic acid**
MW 885.47	MW 92.11	MW 92.11
mp −5.5°C	mp 20°C, bp 290°C dec	mp 16.3°C

Saponification refers to alkaline hydrolysis of fats and oils to give glycerol and the alkali metal salt of a long chain fatty acid (a soap). In this experiment saponification of olive oil is accomplished in a few minutes by use of a solvent permitting operation at 160°C.[1] Of the five acids found in olive oil, listed in Table 1, three are unsaturated and two are saturated. Oleic acid and linoleic acid are considerably lower melting and more soluble in organic solvents than the saturated components, and when a solution of the

TABLE 1 Acids of Split Olive Oil (Typical Composition)

Acid	Formula	%	MW	mp (°C)
Oleic	$C_{18}H_{34}O_2$	64	282.45	13, 16
Linoleic	$C_{18}H_{32}O_2$	16	280.44	−5
Linolenic	$C_{18}H_{30}O_2$	2	278.42	liquid
Stearic	$C_{18}H_{36}O_2$	4	284.07	69.9
Palmitic	$C_{16}H_{32}O_2$	14	256.42	62.9

Note for the instructor

1. The experiment gains in interest if students are given the choice of olive oil from two sources (for example, Italy and Israel) and instructed to compare notes.

FIG. 1 Model of *n*-nonane molecule in a plastic cylinder, 14.3 cm in diameter.

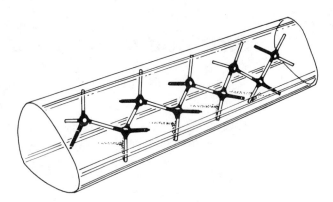

Urea inclusion complexes

hydrolyzate in acetone is cooled to −15°C, about half of the material separates as a crystallizate containing the two saturated compounds, stearic acid and palmitic acid.

The unsaturated acid fraction is then recovered from the filtrate and treated with urea in methanol to form the urea inclusion complex. Normal alkanes having seven or more carbon atoms form complexes with urea in which hydrogen-bonded urea molecules are oriented in a helical crystal lattice in such a way as to leave a cylindrical channel in which a straight-chain hydrocarbon fits. The guest molecule (hydrocarbon) is not bonded to the host (urea) but merely trapped in the channel. The cylindrical channel is of such a diameter (5.3 Å) as to accommodate a normal alkane, but not a thick, branched-chain hydrocarbon such as 2,2,4-trimethylpentane. In Fig. 1 a model of *n*-nonane to the scale 0.2 cm = 1 Å constructed from plastic-metal atoms[2] is inserted into a tightly fitting cellulose acetate cylinder, 14.3 cm in diameter, which defines the space occupied by the resting guest molecule. Table 2 shows that as the hydrocarbon chain is lengthened, more urea molecules are required to extend the channel, but the host-guest relationship is not stoichiometric. The higher saturated fatty acids form

TABLE 2 Molecules of Urea per Molecule of *n*-Alkane

Alkane	Ratio	Alkane	Ratio
C_6	No complex	C_{11}	8.7
C_7	6.1	C_{12}	9.7
C_8	7.0	C_{16}	12.3
C_9	7.3	C_{24}	18.0
C_{10}	8.3	C_{28}	21.2

2. L. F. Fieser, *J. Chem. Ed.*, **40**, 457 (1963); *ibid.*, **42**, 408 (1965).

urea-inclusion complexes in which the ratio of host to guest molecules is about the same as for the corresponding alkanes:

$$
\left.
\begin{array}{l}
\text{Myristic acid (C}_{14}) \\
\text{Palmitic acid (C}_{16}) \\
\text{Stearic acid (C}_{18})
\end{array}
\right\} \text{ requires }
\left\{
\begin{array}{l}
11.3 \\
12.8 \\
14.2
\end{array}
\right\} \text{ moles of urea}
$$

With the elimination of saturated acids from the olive oil hydrolyzate by crystallization from acetone, the problem remaining in isolation of oleic acid is to remove the doubly unsaturated linoleic acid. Models and cylinders show that the introduction of just one *cis* double bond is enough to widen the molecule to the extent that it can no longer be inserted into the 14.3-cm wide channel which accommodates *n*-alkanes (Fig. 1). However, a model of 3-nonyne likewise fails to fit into the 14.3-cm channel, and the fact that this acetylenic hydrocarbon nevertheless forms a urea complex[3] indicates that the channel is subject to some stretching, namely to a diameter of 16.2 cm, as in Fig. 2.

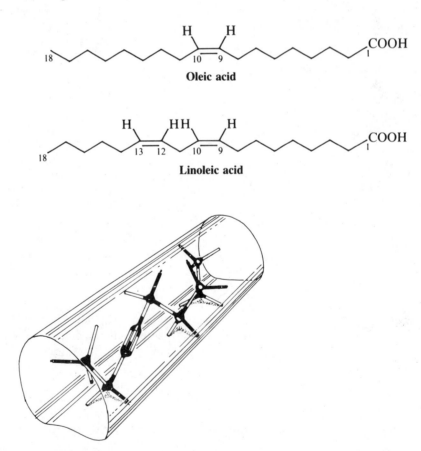

FIG. 2   3-Nonyne molecular model in a plastic cylinder, 16.2 cm in diameter.

3. J. Radell, J. W. Connolly, and L. D. Yuhas, *J. Org. Chem.*, **26,** 2022 (1961).

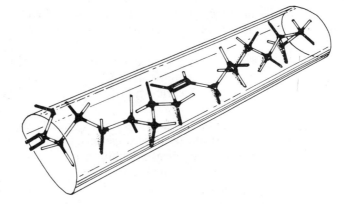

FIG. 3 Oleic acid model in a plastic cylinder, 16.2 cm in diameter.

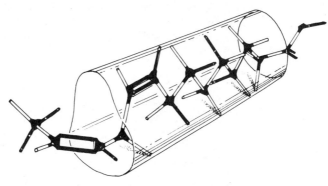

FIG. 4 Model of linoleic acid molecule does not have adequate space in the 16.2-cm cylinder.

From examination of the oleic acid model together with the 16.2-cm cylinder, it is evident that when the carbon atoms are arranged in the particular manner shown in Fig. 3 the oleic acid molecule can be accommodated in this channel. Careful study of the drawing in Fig. 4 will show that an attempt to insert the linoleic acid model into the same 16.2-cm cylinder meets with failure; the carboxy half of the molecule is accommodated, including the 9,10-double bond, but the 12,13-double bond imposes a stoppage and leaves a five-carbon tail projecting. Thus, when a solution of the two acids in hot methanol is treated with urea and let cool, crystals of the oleic acid complex separate and the linoleic acid is retained in the mother liquor.

Predict the host/guest ratio

The experiment has two objectives. One is to isolate pure oleic acid; the other is to determine the number of molecules of urea in the inclusion complex per molecule of the fatty acid. Make a prediction in advance.

Experiment

Isolation of Oleic Acid from Olive Oil

Pour about 10 g of olive oil into a tared 125-mL Erlenmeyer flask and adjust the weight to exactly 10.0 g using a Pasteur pipette. Add 2.3 g of potassium

*Handle hot triethylene glycol
with care*

Rapid saponification

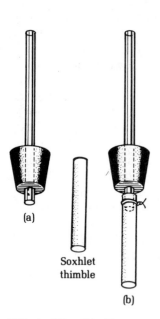

(a)

Soxhlet
thimble

(b)

FIG. 5   Filter thimble.

hydroxide pellets and 20 mL of triethylene glycol, insert a thermometer, and bring the temperature to 160°C by heating over an electrically heated sand bath or electric hot plate; the two layers initially observed soon merge. Then, by removing the flask from the hot plate and replacing it as required, keep the temperature at 160°C for 5 min to ensure complete hydrolysis, and cool the thick yellow syrup to room temperature. Add 50 mL of water, using it to rinse the thermometer, and acidify the soapy solution with 10 mL of concentrated hydrochloric acid. Cool to room temperature, extract the oil with ether, wash the ether with saturated sodium chloride solution, and dry the mixture over anhydrous sodium sulfate. Filter, to remove the drying agent, into a tared 125-mL Erlenmeyer filter flask. Evaporate the solution on the steam bath and evacuate at the aspirator until the weight is constant (9–10 g of acid mixture).

Add 75 mL of acetone to dissolve the oil in the filter flask and place the flask in an ice bath to cool while making preparations for quick filtration. An internal filter for the filter flask is made by moistening the hole of a No. 4 one-hole rubber stopper with glycerol and thrusting a 10-cm section of 9-mm glass tubing through it until the tube projects about 8 mm from the smaller end of the stopper (Fig. 5(a)). A 10 × 50-mm Soxhlet extraction thimble is then wired onto the projecting end (b).

When a mixture of the oleic acid crystals and mother liquor has been prepared as described below, the internal filtering operation is performed as follows: The side arm of a filter flask is connected with 5-mm rubber tubing to a rubber pressure bulb with valve, the stopper carrying the filter is inserted tightly into the flask, and a receiving flask (125 mL) is inverted and rested on the rubber stopper (Fig. 6(a)). The two flasks are grasped in the left hand, with the thumb and forefinger pressing down firmly on the rim of the empty flask to keep the stopper of the other flask in place. While operating the pressure bulb constantly with the right hand, turn the assembly slowly to the left until the side arm of the filter flask is slanting up and the delivery tube is slanting down (Fig. 6(b)). Squeeze the bulb constantly until the bulk of the liquid has been filtered and then stop. Remove the filtrate and place the filter flask in the normal position.

FIG. 6   Apparatus for filtering crystals from mother liquor. (a) Flasks with pressure bulb and filter thimble assembly; (b) position of flasks when filtering.

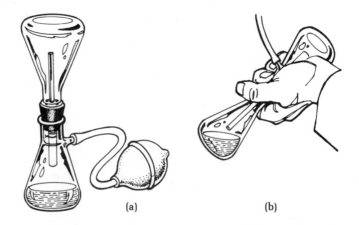

(a)   (b)

Avoid touching dry ice

Quick filtration at − 15°C

Crystallization of the saturated acids requires cooling in a dry ice-isopropyl alcohol bath at − 15°C for about 15 min, and this is done most conveniently in a beaker half-filled with isopropyl alcohol, mounted on a magnetic stirrer and provided with a toluene low temperature thermometer. Add crushed dry ice a little at a time to bring the bath temperature to − 15°C. The filter flask containing the acetone solution of acids is fitted with the rubber pressure bulb, but the stopper carrying the filter is not put in place until later. Place the filter flask in the cooling bath and swirl the mixture occasionally. Add more dry ice as required to maintain a bath temperature of − 15°C. After the first crystals appear, as a white powder, swirl and cool the solution an additional 10 min to let the whole mixture acquire the temperature of the bath. Then introduce the filter and inverted receiver and filter quickly by the technique described previously. Stop the process as soon as the bulk of liquid has been collected, for the solid will melt rapidly.[4]

Evaporate the solvent from the filtrate on the steam bath and evacuate the residual fraction of unsaturated acids at the aspirator until the weight is constant. The yield of a mixture of stearic and palmitic acids is 5–7 g. In a second 125-mL Erlenmeyer flask dissolve 11 g of urea in 50 mL of methanol and pour the solution onto the unsaturated acid fraction. Reheat to dissolve any material that separates and let the solution stand until a large crop of needles has separated. Then cool thoroughly in an ice bath with swirling, collect the product, and rinse the flask with filtrate. Press down the crystals and drain to promote rapid drying and spread out the needles on a large filter paper. After a few minutes, transfer the crystals to a fresh paper. The yield of colorless oleic acid-urea inclusion complex is 10–12 g. The complex does not have a characteristic melting point.

When the complex is fully dry, bottle a small sample, note carefully the weight of the remainder, and place it in a separatory funnel. Add 25 mL of water, swirl, and note the result. Then extract with ether for recovery of oleic acid. Wash the extract with saturated sodium chloride solution, filter the solution through sodium sulfate into a tared 125-mL Erlenmeyer flask, evaporate the solution on the steam bath, and pump out to constant weight on the steam bath. The yield of pure oleic acid is 2–3 g. From the weights of the complex and of the acid, calculate the number of moles of urea per mole of acid in the complex.

Determine the host/guest ratio

4. Alternative technique: Do the crystallization in an Erlenmeyer flask cooled in a salt-ice bath to − 15°C and filter by suction on a small Büchner funnel that has been chilled outdoors in winter weather or in a refrigerator freezing compartment.

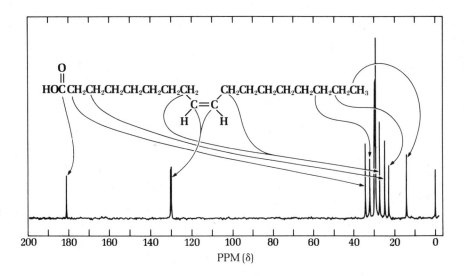

FIG. 7 ^{13}C nmr spectrum of oleic acid (22.6 MHz).

Questions

1. What is the effect of vigorous shaking when making the first ether extraction after the hydrolysis of olive oil?

2. How would you estimate the number of double bonds in a fat?

3. Why are unsaturated fats deposited in the extremities of animals in contrast to the torso, where saturated fats are deposited?

Reaction Kinetics: Williamson Ether Synthesis

Prelab Exercise: In the reaction $A + B \rightarrow C$ the concentration of B seems to have little effect on the rate of the reaction. Analysis shows the concentration of A to be 1×10^{-3} M at the beginning of the reaction. One hundred minutes later the concentration of A is 0.21×10^{-3} M. What is the order of the reaction? What is its rate?

$$CH_3I + CH_3-\underset{\underset{CH_3}{|}}{\overset{\overset{CH_3}{|}}{C}}-O^-K^+ \xrightarrow{(CH_3)_3COH} CH_3-\underset{\underset{CH_3}{|}}{\overset{\overset{CH_3}{|}}{C}}-O-CH_3 + K^+I^-$$

The formation of *t*-butyl methyl ether from methyl iodide and potassium *t*-butoxide is an example of the Williamson ether synthesis. By analyzing this reaction as it proceeds you can determine the rate of the reaction as well as the order. In this way you also may be able to infer a mechanism for the reaction.

Known quantities of reactants are allowed to react for known lengths of time; from time to time, samples of the reaction mixture are analyzed quantitatively. You can follow either the disappearance of the reactants or the appearance of the products. Analysis of the reaction might include gas chromatography, infrared or nmr spectroscopy (methyl iodide decrease or ether increase), or reaction with aqueous silver nitrate to precipitate insoluble silver iodide. We shall use still another method—titration of the unreacted base, potassium *t*-butoxide, with an acid. When using this method aliquots (known fractions) of the reaction mixture are removed and immediately diluted with ice water to stop the reaction. Unreacted *t*-butoxide ion will react with the water to form hydroxide ion and *t*-butyl alcohol; the base is titrated to a phenolphthalein end point with standard perchloric acid.

The rate of the reaction depends on the temperature. Ideally the reaction should be run in a constant temperature bath; otherwise wrap the reaction flask with a towel to shield it from drafts, keep it stoppered, and ensure that all reactants and apparatus are at room temperature (which should be above 25.5°C, the freezing point of *t*-butyl alcohol).

Experiments

Procedure

Avoid skin contact with methyl iodide and butoxide

Pipet 200 mL of a 0.110 M solution of methyl iodide in *t*-butyl alcohol using a pipetter, into a 300-mL Erlenmeyer flask, quickly followed by exactly 20.0 mL of ca. 0.5 M potassium *t*-butoxide in *t*-butyl alcohol (note the exact concentration of the potassium *t*-butoxide on the label). Swirl to ensure thorough mixing, and place the flask in the constant temperature bath (note exact temperature). Immediately withdraw a 10-mL aliquot ($\frac{1}{22}$ of the reaction mixture) with a pipette and run it into 20 mL of an ice and water mixture in a 250-mL Erlenmeyer flask. Between aliquots rinse the pipette with water and a few drops of acetone and dry by sucking air through it with an aspirator. It is not necessary to carry out the titration immediately, so label the flask. During the first hour withdraw 10-mL aliquots at about 10-min intervals (note the exact time in seconds when the pipette is half empty), then at 20-min intervals during the second hour. By the end of the second hour well over half of the reactants will have reacted to give the product. Add a drop of phenolphthalein indicator to each quenched aliquot and titrate with standardized perchloric acid to a very faint pink color when seen against a white background. Note the exact concentration of the perchloric acid; it should be near 0.025 M.

Cleaning Up   Combine all solutions, neutralize with dilute hydrochloric acid, and place in the halogenated organic waste container.

Mathematics of Rate Studies

In studying a reaction you must first determine what the products are. It has previously been found that the reaction of methyl iodide and potassium *t*-butoxide gives only the ether, but the equation tells you nothing about how the rate of the reaction is affected by the relative concentrations of the reactants, nor does it tell you anything about the mechanism of the reaction.

(a) First-Order Reaction.   The rate of a first-order reaction depends only on the concentration of one reactant A. For example, if

$$2A + 3B \longrightarrow \text{products}$$

and the reaction is first-order in A, despite what the equation seems to say the rate equation is

$$\text{Rate} = \frac{-d[A]}{dt} = k[A]$$

where [A] is the concentration of A, $-d[A]/dt$ is the rate of disappearance of A with time, and k is a proportionality constant known as "the rate constant." Experimentally we determine

$$a = \text{concentration of A at time zero}$$
$$x = \text{concentration of A which has reacted by time } t$$
$$a - x = \text{concentration of A remaining at time } t$$

The rate expression is then

$$\frac{dx}{dt} = k(a - x)$$

which upon integration gives

$$-\ln(a - x) = kt - \ln(a)$$

Changing signs and changing from base e to base 10 logarithms gives

$$\log(a - x) = \frac{-k}{2.303} t + \log(a)$$

This equation has the form of the equation for a straight line,

$$y = mx + b$$

If the reaction under study is first-order, a straight line will be obtained when the term $\log(a - x)$ at various times is plotted against time. The slope of the line is $-k/2.303$, from which the value of the rate constant can be found. It will have units of $(\text{time})^{-1}$.

(b) Second-Order Reaction. The rate of a second-order reaction depends on the concentration of two reactants. For example, if

$$2A + 3B \longrightarrow \text{products}$$

the rate equation would be

$$\text{Rate} = -\frac{d[A]}{dt} = -\frac{d[B]}{dt} = k_2[A][B]$$

Experimentally we determine

$$a = \text{concentration of A at time zero}$$
$$b = \text{concentration of B at time zero}$$
$$x = \text{concentration of A (and concentration of B),}$$
$$\text{which has reacted by time } t$$
$$a - x = \text{concentration of A remaining at time } t$$
$$b - x = \text{concentration of B remaining at time } t$$

The rate equation now becomes

$$\frac{dx}{dt} = k_2(a - x)(b - x)$$

which on integration gives

$$k_2 t = \frac{1}{a - b} \ln \frac{b(a - x)}{a(b - x)}$$

which upon rearrangement and conversion to base ten logarithms gives

$$k_2 t = \frac{2.303}{a - b} \log \frac{a - x}{b - x} + \frac{2.303}{a - b} \log \frac{b}{a}$$

The final term is a constant. Rearrangement into the form of an equation for a straight line gives

$$\log \frac{a - x}{b - x} = k_2 \left(\frac{a - b}{2.303} \right) t + \text{constant}$$

If the reaction being studied is second-order, a plot of $\log (a - x)/(b - x)$ against time will give a straight line with slope of $k_2(a - b)/2.303$. The value of the rate constant is given by

$$k_2 = \text{slope} \left(\frac{2.303}{a - b} \right)$$

Treatment of Data

The fact that some reaction occurred before the first aliquot was withdrawn is immaterial. It is only necessary to know the concentrations of both reactants at the time of withdrawal of the first aliquot (which we shall call time-zero). The amount of base present at time-zero is determined directly by titration. The amount of methyl iodide present is equal to the concentration originally present after mixing minus that which reacted before the first aliquot was withdrawn. The amount that reacted before the first aliquot was withdrawn is equal to the amount of butoxide that reacted between the time of mixing and the time the first aliquot was withdrawn.

Make plots of the log of the concentration of methyl iodide and the log of the concentration of potassium t-butoxide vs. time. If either gives a straight-line relationship, the reaction is first-order. If not, calculate the log $(a - x)/(b - x)$ for each aliquot and plot against time in seconds. If this plot gives a straight line (within your experimental error) for this reaction, you now have strong evidence that the reaction is second-order, i.e., first-order in each reactant. Obtain the rate constant, k, for this reaction at this temperature, from the slope of your plot. The units of k_2 are moles $s^{-1}L^{-1}$.

Computer Analysis of Data

The repeated calculation of log $(a - x)/(b - x)$ for each aliquot from titration data is, to say the least, tedious. A computer program can do this simple job. After a little experience you will be able to write a program that would take as input the concentrations of all reactants, milliliters of acid, the time, and then print out k. There are standard computer programs ("library programs") available that will accept your various values of ln $(a - x)/(b - x)$ and the corresponding times, calculate the mathematically "best" straight line through the points (the "least squares" line), and print out the slope and intercept of the line.

A program written in the BASIC computer language to calculate the value of ln $(a - x)/(b - x)$ for various values of x looks like this:

```
10   READ A, B
20   DATA 0.098, 0.479
30   READ X
40   DATA 0.007, 0,019, 0.025, 0.037, 0.049, 0.058, 0.062
50   LET Y = LOG((A - X)/(B - X))
60   PRINT A,B,X,Y
70   GO TO 30
80   END
```

In this example, $a = 0.098$, $b = 0.479$, and $x = 0.007, 0.019$, etc. The computer will READ for "a" the value of 0.098, because that is the first number it finds in the data printed in line 20. It reads 0.479 for "b" because that is the second number in line 20. The next command (line 30) says READ X and the computer reads the number 0.007 (line 40). We are trying to calculate ln $(a - x)/(b - x)$ so we set this equal to Y. The LET statement tells the computer to carry out the calculation substituting 0.098 for A, 0.479 for B, and 0.007 for X. LOG means natural logarithms to the computer so it carries out the computation and prints out the values of A, B, and X, which we supplied, and Y, which it calculates. The next command (line 70) says go back to line 30 and pick up new values for X, leaving the values for A and B the same. This time the computer automatically substitutes 0.019 in the equation, ignoring 0.007, which it has already used. Again it carries out the computation, prints out A, B, X, and Y, and keeps going through the cycle until it runs out of data in line 40.

Searching the Chemical Literature

In planning a synthesis or investigating the properties of compounds, the organic chemist is faced with the problem of locating information on the chemical and physical properties of substances that have been previously prepared as well as the best methods and reagents for carrying out a synthesis. With over 7,000,000 organic compounds known and new ones being reported at the rate of more than a third of a million per year, the task might seem formidable. However, the field of chemistry has developed one of the best information retrieval systems of all the sciences. In a university chemistry library it is possible to locate information on almost any known compound in a few minutes. Even a modest library will have information on several million different substances. The ultimate source of information is the primary literature: articles written by individual chemists and published in journals. The secondary literature consists of compilations of information taken from these primary sources. For example, to find the melting point of benzoic acid, one would naturally turn to a secondary reference containing a table of physical constants of organic compounds. If, on the other hand, one wished to have information about the phase changes of benzoic acid under high pressure, one would need to consult the primary literature. If the piece of information, such as a melting point, is crucial to an investigation, the primary literature should be consulted. Transcription errors occur in the preparation of secondary references.

Handbooks

For rapid access to information such as mp, bp, density, solubility, optical rotation, λ max, and crystal form, one turns first to the *Handbook of Chemistry and Physics* where information is found on some 15,000 organic compounds, including the Beilstein reference to each compound. These compounds are well known and completely characterized. The majority are commercially available. The *Merck Index* contains information on nearly 10,000 compounds, especially those of pharmaceutical interest. In addition to the usual physical properties, information and literature references to synthesis, isolation, and medicinal properties, such as toxicity data, are found. The last third of the book is devoted to such items as a long cross index of names (which is very useful for looking up drugs), a table of organic name reactions, an excellent section on first aid for poisons, a list of chemical poisons, and a listing of the locations of many poison control centers.

   A useful reference is the *Aldrich Catalog Handbook of Fine Chemicals,* available without charge from the Aldrich Company. Not only does this catalog list the prices of the more than 12,000 chemicals (primarily organic) as well as the molecular weights, melting points, boiling points, and optical rotations, but it also gives reference to ir and nmr spectral data and to Beilstein, *Merck Index,* and Fiesers' *Reagents.* In addition, for each chemical, the catalog provides the reference to the *Registry of Toxic Effects of Chemical Substances* (RTECS No.) and, if appropriate, the reference to Sax, *Dangerous Properties of Industrial Materials.* Special hazards are noted ("severe poison," "lachrymator," "corrosive"), and a reference is made to one of thirty different disposal methods for each chemical. The catalog also gives pertinent information (uses, physiological effects, etc.) with literature references for many compounds. A complete list of all chemicals sold commercially and the addresses of the companies making them is found in *Chem Sources,* published yearly.

After consulting these three single-volume references, one would turn to more comprehensive multivolume sources such as the *Dictionary of Organic Compounds,* Fifth Edition. This dictionary, still known as "Heilbron," the name of its former editor, now comprises seven volumes of specific information, with primary literature references, on the synthesis, reactions, and derivatives of more than 50,000 compounds. Rodd's *Chemistry of Carbon Compounds,* another valuable multivolume work with primary literature references, is organized by functional group rather than in dictionary form. Elsevier's *Encyclopedia of Organic Compounds* in about twenty volumes is an incomplete reference work on the chemical and physical properties of compounds. It is useful for those areas it covers. References to Elsevier are found in the *Handbook of Chemistry and Physics.*

Beilstein's *Handbuch der organischen Chemie* is certainly not a "handbook" in the American sense—it can occupy an entire alcove of a chemistry library! And the currently produced volumes must be among the most expensive contemporary works purchased by a library, for each book now costs more than $700. However, with characteristic German thoroughness, this reference covers every well-characterized organic compound that has been reported in the literature up to 1949. Along with a main reference set come three supplements covering the periods 1910–1919, 1920–1929, and 1930–1949. The fourth supplement covering 1950–1959 is not yet complete. Although written in German, this reference can provide much information even to those with no knowledge of the language. Physical constants and primary literature references are easy to pick out. Beilstein is organized around a complex classification scheme that is explained in *The Beilstein Guide* by O. Weissback, but the casual user can gain access through the *Handbook of Chemistry and Physics* and the *Aldrich Catalog.* In the *Handbook* the reference for 2-iodobenzoic acid is listed as $B9^2$, 239, which means the reference will be found on p. 239 of Vol. (Band) 9 of the second supplement (Zweites Ergänzungwerk, EII). In the *Aldrich Catalog* the reference is given as Beil **9**, 363, which indicates that information can be found in the main series on p. 363, Vol. 9. The system number assigned to each compound can be traced through the supplements. The easiest access to Beilstein itself is through the formula index (General Formelregister) of the second supplement. A rudimentary knowledge of German will enable one to pick out iodo (Jod) benzoic acid (Saure), for example.

Chemical Abstracts

Secondary references are incomplete, some suffer from transcription errors, and the best—Beilstein—is twenty years behind in its survey of the organic literature, so one must often turn to the primary chemical literature. However, there are more than 14,000 periodicals where chemical information might appear. Chemists are fortunate in having an index, *Chemical Abstracts,* that covers this huge volume of literature very promptly and publishes biweekly abstracts of each article. The information in each abstract is then compiled into author, chemical substance, general subject, formula, and patent indexes. Since publication began in 1907, nomenclature has changed. Now no trivial names are used, so acetone appears as 2-propanone, and *o*-cresol as benzene, 1-methyl, 2-hydroxy. It takes some experience to adapt to these changing names and even now *Chemical Abstracts* does not follow the IUPAC rules exactly. The formula index is useful because it provides not only reference to specific abstracts but also a correct name that can be found in the chemical substances index. Currently indexes are published twice each year. These are then grouped into five-year collective indexes. The index for the period 1987–1991 occupies more than 80 volumes and costs $22,000. The yearly subscription price is now $7800.

A complete search of the chemical literature would entail use of *Chemischer Zentralblatt,* the oldest abstract journal, which first appeared in 1830. Collective and formula indexes started in 1930 go up to 1969. These collective indexes are more reliable than those of *Chemical Abstracts* for the period 1930–1939.

Title Indexes

As long as a year can elapse between the time a paper appears and the time an abstract is published. Three publications help fill the gap by providing much the same information: a list of the titles, authors, and keywords that appear in titles of papers recently published—or, in some cases, papers about to be published. These publications are *Chemical Titles, Current Chemical Papers,* and *Current Contents.*

Science Citation Index

Science Citation Index is unique in that it allows for a type of search not possible with any other index—a search forward in time. For example, if one wanted to learn what recent applications have been made of the cuprous chloride catalyzed coupling of terminal acetylenes (Chapter 50) first reported by Stansbury and Proops, one would look up their paper [*J. Org.* **27,** 320 (1962)] in *Science Citation Index* and find there a list of those articles subsequent to 1962 in which an author cited the work of Stansbury and Proops in a footnote.

Planning a Synthesis

In planning a synthesis the organic chemist is faced with at least three overlapping considerations: the chemical reactions, the reagents, and the experimental procedure to be employed. For students an advanced text-book such as Carey and Sundberg, *Advanced Organic Chemistry,* Vols. 1 and 2, or March's *Advanced Organic Chemistry,* 3rd Ed., 1985, which has literature references, might be a place to start. It is also instructive for the beginning synthetic chemist to read about some elegant and classical syntheses that have been carried out in the past. For this purpose see Anand, Bindra, and Ranganathan, *Art in Organic Synthesis,* for some of the most elegant syntheses carried out through 1970. Similarly, Bindra and Bindra, *Creativity in Organic Synthesis,* Vol. 1, and Fleming, *Selected Organic Syntheses,* are compendia of elegant syntheses. For natural product synthesis see the excellent series by Ap Simon, *The Total Synthesis of Natural Products,* Vols. 1–6. House's *Modern Synthetic Reactions,* 2nd ed., is a more comprehensive source of information with several thousand references to the original literature.

If it appears that a particular reaction will be employed in the synthesis, then *Organic Reactions* should be consulted. Over 100 preparative reactions, with examples of experimental details, are covered in great detail in some thirty volumes. Theilheimer's 40 volumes of *Synthetic Methods of Organic Chemistry* is organized according to the types of bonds being made or broken and, because it is published annually, serves as a means for continually updating all synthetic procedures. Buehler and Pearson, *Survey of Organic Synthesis,* in two volumes, is a good summary of synthetic methods, classified by functional group. Also the annual *Newer Methods of Preparative Organic Chemistry* and *Annual Reports in Organic Synthesis* should be consulted. The latter is an excellent review of new reactions in a given year, organized by reaction type. The Fiesers' unique *Reagents for Organic Synthesis,* a biannual series since 1967, critically surveys the reagents employed to carry out organic synthesis. Included are references to the original literature, *Organic Syntheses* (see below), the critical reviews that have been written about various reagents, and suppliers of reagents. Smith's comprehensive indices (1987) make the series even more valuable. The index of reagents according to type of reaction is very useful when planning a synthesis. Other good references are *Compendium of Organic Synthetic Methods,* Vols. 1–6, and Nakanishi et al., *Natural Products Chemistry,* Vols. 1 and 2.

In order to operate on one functional group without affecting another functional group it is often necessary to put on a protective group. Two books deal with this rather specialized subject: McOmie's *Protective Groups in Organic Chemistry* and Greene's *Protective Groups in Organic Synthesis.*

Before carrying out the synthesis itself one should consult *Organic Syntheses,* an annual series since 1921, which is grouped in six collective volumes with reaction and reagent indexes. *Organic Syntheses* gives detailed procedures for carrying out more than 1000 different reactions. Each synthesis is submitted for review and then the procedure is sent to an independent laboratory to be checked. The reactions, unlike many that are reported in the primary literature, are carried out a number of times and therefore can be relied upon to work. Laboratory techniques are covered in Bates and Schaefer's *Research Techniques in Organic Chemistry* and in the fourteen volumes of Weissberger et al., *Technique of Organic Chemistry,* an uneven, multiauthor compendium. An excellent advanced text on the detailed mechanisms of many representative reactions is Lowry and Richardson's *Mechanism and Theory in Organic Chemistry,* 2nd ed.

Modern Chemistry Library

Although the chemistry library as we now know it probably will not change markedly in the next decade, in one aspect it has already changed. It is now possible to search a large part of the chemical literature using a personal computer connected via telephone lines to a commercial data base held in a large computer. By rapidly and efficiently calling into a data base containing *Chemical Abstracts, Science Citation Index,* and several other indexes, a librarian trained to use the system can acquire information for the chemist much more rapidly than he or she could do by a manual search. In some cases, the information obtained would be impossible to get by conventional means. For example, to locate all references to the nmr spectra of insulin during the period 1976–1981 using the Lockheed DIALOG® system, one first asks for the number of references to nmr found in *Chemical Abstracts* during that period. The answer is printed out within a few seconds: 12,000. Next one asks how many references to insulin. The answer: 20,000. The next query asks the computer to cross the two lists for references common to both. The answer: 5. One can then ask that the references be printed out. At some expense, the entire abstract for each reference can be printed out. It is less expensive to have the abstracts typed off-line at the computer center and sent by mail.

Wiswesser Line Notation

Using a computer it is also possible to search for partial structures, an option not available with a manual search. For example, one could ask for a list of all compounds having a four-membered ring with a carbonyl group in the ring. These compounds, when fused to another ring system, will not appear in an index under the name cyclobutanone. Wiswesser line notation, a linear computer-intelligible way of encoding organic structures, can be searched by computer for partial structures. The notation for cyclobutenone is L4V BHJ. The V is the symbol for a carbonyl group; the 4 indicates a four-membered ring. These elements will be found in the line notation of bigger and more complex molecules that have four-membered rings containing a carbonyl group.

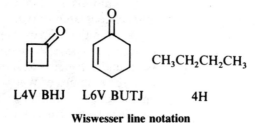

$$CH_3CH_2CH_2CH_3$$

L4V BHJ   L6V BUTJ   4H

Wiswesser line notation

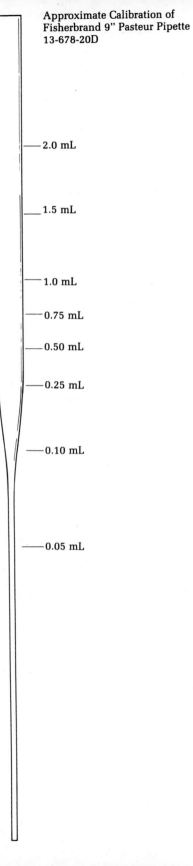

Approximate Calibration of
Fisherbrand 9" Pasteur Pipette
13-678-20D

—— 2.0 mL

—— 1.5 mL

—— 1.0 mL

—— 0.75 mL

—— 0.50 mL

—— 0.25 mL

—— 0.10 mL

—— 0.05 mL

Index